Advanced Principles of Avionics

Advanced Principles of Avionics

Edited by
Sam Dillon

STATES
ACADEMIC PRESS
www.statesacademicpress.com

Published by States Academic Press,
109 South 5th Street,
Brooklyn, NY 11249, USA

ISBN: 978-1-63989-019-4

Cataloging-in-Publication Data

Advanced principles of avionics / edited by Sam Dillon.
p. cm.
Includes bibliographical references and index.
ISBN 978-1-63989-019-4
1. Avionics. 2. Electronics in transportation. 3. Aeronautics. I. Dillon, Sam.
TL695 .A38 2022
629.135--dc23

For information on all States Academic Press publications
visit our website at www.statesacademicpress.com

STATES
ACADEMIC PRESS

Contents

Preface

This book has been a concerted effort by a group of academicians, researchers and scientists, who have contributed their research works for the realization of the book. This book has materialized in the wake of emerging advancements and innovations in this field. Therefore, the need of the hour was to compile all the required researches and disseminate the knowledge to a broad spectrum of people comprising of students, researchers and specialists of the field.

Avionics is the category of electronic systems that are used in aircraft, artificial satellites and spacecrafts. Avionic systems involve communication, navigation, and collision-avoidance system. It also includes the display and management of multiple systems that are fitted to aircraft to perform individual and specific functions. These functions can be simple as well as complex in nature. Communication system helps in connecting the flight deck to the passengers and the flight deck to the ground. Navigation systems help in calculating the position automatically and displaying it to the flight crew on moving map displays. Collision-avoidance system deals with air traffic control. It can detect the location of the nearby aircraft and provide instructions for avoiding a mid-air collision. This book aims to shed light on some of the advanced principles in the field of avionics. It strives to provide a fair idea about this discipline and to help develop a better understanding of the latest advances within this field. It is appropriate for students seeking detailed information in this area as well as for experts.

At the end of the preface, I would like to thank the authors for their brilliant chapters and the publisher for guiding us all-through the making of the book till its final stage. Also, I would like to thank my family for providing the support and encouragement throughout my academic career and research projects.

Editor

1

Air Data Sensor Fault Detection with an Augmented Floating Limiter

Fabio Balzano,[1] Mario L. Fravolini,[1] Marcello R. Napolitano,[2] Stéphane d'Urso,[2] Michele Crispoltoni,[1] and Giuseppe del Core[3]

[1]*Department of Engineering, University of Perugia, Perugia 06125, Italy*
[2]*Department of Mechanical and Aerospace Engineering, West Virginia University, Morgantown, WV 26506, USA*
[3]*Department of Science and Technology, Parthenope University of Naples, Naples 80143, Italy*

Correspondence should be addressed to Mario L. Fravolini; mario.fravolini@unipg.it

Academic Editor: Kenneth M. Sobel

Although very uncommon, the sequential failures of all aircraft Pitot tubes, with the consequent loss of signals for all the dynamic parameters from the Air Data System, have been found to be the cause of a number of catastrophic accidents in aviation history. This paper proposes a robust data-driven method to detect faulty measurements of aircraft airspeed, angle of attack, and angle of sideslip. This approach first consists in the appropriate selection of suitable sets of model regressors to be used as inputs of neural network-based estimators to be used online for failure detection. The setup of the proposed fault detection method is based on the statistical analysis of the residual signals in fault-free conditions, which, in turn, allows the tuning of a pair of floating limiter detectors that act as time-varying fault detection thresholds with the objective of reducing both the false alarm rate and the detection delay. The proposed approach has been validated using real flight data by injecting artificial ramp and hard failures on the above sensors. The results confirm the capabilities of the proposed scheme showing accurate detection with a desirable low level of false alarm when compared with an equivalent scheme with conventional "a priori set" fixed detection thresholds. The achieved performance improvement consists mainly in a substantial reduction of the detection time while keeping desirable low false alarm rates.

1. Introduction

The Air Data System (ADS) [1] is a critical component of the conventional suite of sensors for both manned and unmanned aircraft. The ADS provides direct measurements of the aircraft true airspeed; typically connected with the ADS are the vanes for measuring the important aerodynamic angles known as angle of attack and angle of sideslip. Since the Pitot probes and α-β vanes are installed on the fuselage and/or wings, they are exposed to external atmospheric agents; in general, these devices are robust to even extreme weather conditions. However, there are very specific weather conditions which can lead to peculiar ice crystals obstructing the tiny conducts of the ADS, thus inducing faulty sensor measurements. Incorrect measurements of airspeed could lead to unrecoverable flight conditions, such as in the cases of Air France Flight 447 [2] and the NASA X-31 experimental aircraft [3]. Additional causes of crashes included erroneous covering of the Pitot tubes (AeroPerù 757) [4] and formations of insect nests inside the static taps (Birgenair Flight 310) [5] as well as icing of the angle-of-attack vane (XL Airways Germany Flight 888T) [6]. At present, the conventional approach to provide fault tolerance for flight sensors is based on Hardware Redundancy (HR) [7] which basically consists in the installation of multiple sensors (initially 4, later reduced to 3) measuring the same parameter. These sensors are continuously monitored to check their status, and their values are continuously compared. The use of Built-In-Testing (BIT) voting schemes allows the detection and the isolation of a faulty sensor. An alternative approach is given by Analytical Redundancy (AnR) [8]. AnR basically involves the use of virtual sensors, such as conventional state estimators, neural networks, Kalman Filters, special cases of Kalman Filters (such as Extended Kalman Filters and Unscented Kalman Filters), and/or other predictive models, to provide alternative estimates of the parameter from the faulty sensor. Recently, an AnR scheme based on machine learning techniques has been proposed in [9] for the failure detection and correction of the airspeed sensor of an aircraft.

Through data from other functional sensors (measuring parameter highly correlated to the parameter associated with the faulty sensor) used as inputs, the virtual sensor can estimate the target measurement which when compared with data provided by the sensor under scrutiny produces the residual signal. The residual signal is clearly very important for fault detection purposes; in fact, an alarm status can be declared when the residual signal exceeds some detection thresholds. It should be emphasized that the approach proposed in this paper is completely data-driven [10, 11] and does not require any specific aircraft-dependent dynamic mathematical model. In summary, the innovative aspects of this research effort are the complete independence from any model-based requirement along with the introduction and the validation of "ad hoc" designed adaptive thresholds (floating limiters) that are applied to the whitened and filtered residuals with the goal of improving the performance in terms of fault detectability, fault detection delay, and low false alarm rate when compared with conventional fixed-threshold schemes [12–15].

Recently, the authors of this paper have also proposed in [16] another data-driven approach for the design of robust fault detection filters for aircraft air data sensors. In [16], the robustness of the scheme is achieved through the adoption of parametric interval prediction models that are derived by the solution of a convex optimization problem. Instead, in the present study, the robustness of the failure detection scheme was achieved through time-varying fault detection thresholds that are derived online by a simple floating limiter filter that is based on the windowed mean and variance of the whitened residual.

The paper is organized as follows. Section 2 describes the identification model for the neural networks used in this study. Section 3 briefly introduces the P92 Tecnam aircraft from which the flight data were recorded. Section 4 outlines the data-driven modelling of the true airspeed, angle-of-attack α, and angle-of-sideslip β sensors. Sections 5 and 6 are dedicated to the analysis of the residuals and to threshold computation, respectively. Finally, Section 7 provides the results along with an assessment of the performance of the proposed scheme in terms of false alarm rate and detection delay compared with a conventional fixed-threshold fault detection system.

2. Prediction Model: Feedforward Neural Network

This section describes the neural network-based model used for the estimation of airspeed $\mathrm{TAS}(k)$, $\alpha(k)$, and $\beta(k)$ as functions of other available measurements. Analytical Redundancy approaches are based on the use of online estimators that are supposed to provide an estimation of the sensor measurement. Several types of estimators can be found in the literature, such as Kalman Filters [17], Unscented Kalman Filters [18], Extended Kalman Filters [19], FIR filter estimator [20], and neural networks adaptive and nonadaptive [21, 22]. Specifically, in this effort, a nonautoregressive model [23–25] has been used. The model is essentially a nonlinear time series model, also known as nonlinear regression [26] modelled as follows:

$$\tilde{y}(k) = f_{\mathrm{NR}}(X(k)) + \varepsilon(k), \tag{1}$$

where $\tilde{y}(k)$ is the $y(k)$ (target) estimation, $X(k)$ is the final set of regressors (see Sections 4 and 5), and $\varepsilon(k)$ is a modelling error. Note that the NR model does not depend on previous values of$y(k)$; this guarantees that the estimation is not affected by the occurrence of a fault on $y(k)$. Thus, the NR model is able to provide a reliable long-term (multistep ahead) estimation independently of the occurrence of the fault; on the contrary, an adaptive model (i.e., NARX model) estimation is influenced by the previous occurrence of a fault on (k).

FIGURE 1: Semiautonomous P92 Tecnam.

3. P92 Aircraft and Flight Data

The proposed scheme has been validated using flight data from the semiautonomous P92 Tecnam aircraft shown in Figure 1.

The P92 is a general aviation single-engine light aircraft with a classic high-wing configuration [27]. During the test flights, the aircraft was remotely controlled through a data link from a ground station. The on-board pilot was operative only during the take-off and landing phases. The nominal aircraft mass is 600 kg (1324 lb); the propulsion is provided by a Rotax 912 ULS, 74 kW (68 hp) with a two-blade fix pitch propeller, capable of providing a max cruise speed of 219 km/h (118 kts) and an operational ceiling of 4267 (14,000 ft). A total of 8 flights were recorded for approx. 3.5 hours of flight time. The available flight data were divided into two groups, that is, training data (used for the goal of training a pseudo-online learning neural network used as AnR-based virtual sensors) and validation data (used for the assessment of the performance of the estimation scheme). Since the recorded flight data were not completely homogenous, a different number of flights were used for the robust estimation of the TAS and of the aerodynamic angles α and β.

For the TAS and β analysis, two flights were used for training purposes and five flights for validation purposes. Instead, for the α analysis, two flights were used for training

TABLE 1

(a) TAS training and validation flight statistics

	Max $V(k)$ (m/s)	Mean $V(k)$ (m/s)	Std $V(k)$ (m/s)	Flight duration (min)
TR-FLIGHT-1	37.1	31.3	2.6	34.8
TR-FLIGHT-2	35.5	30.5	1.1	26.8
VAL-FLIGHT-1	37.3	33.4	2.8	16.5
VAL-FLIGHT-2	37.1	32.1	2	17.3
VAL-FLIGHT-3	33.8	31.1	0.7	18
VAL-FLIGHT-4	34.4	31	1.02	28.3
VAL-FLIGHT-5	38	32	1.5	30.8

(b) α training and validation flight statistics

	Max $\alpha(k)$ (deg)	Mean $\alpha(k)$ (deg)	Std $\alpha(k)$ (deg)	Flight duration (min)
TR-FLIGHT-1	14.1	6.4	1.9	34.8
TR-FLIGHT-2	12.1	6.3	0.9	26.8
VAL-FLIGHT-1	11.4	4.9	1.7	16.5
VAL-FLIGHT-2	14	5.6	1.4	17.3
VAL-FLIGHT-3	9.3	6.4	0.5	18
VAL-FLIGHT-4	10.3	4.6	1.5	20.3
VAL-FLIGHT-5	9.1	6.2	0.7	28.3
VAL-FLIGHT-6	10.5	5.8	0.9	30.8

(c) β training and validation flight statistics

	Max $\beta(k)$ (deg)	Mean $\beta(k)$ (deg)	Std $\beta(k)$ (deg)	Flight duration (min)
TR-FLIGHT-1	2.9	−0.1	1.5	34.8
TR-FLIGHT-2	8.3	0.3	1.4	26.8
VAL-FLIGHT-1	1.6	−0.7	1.1	17.3
VAL-FLIGHT-2	9.8	0.6	1.1	18
VAL-FLIGHT-3	7.1	1.3	1.5	20.3
VAL-FLIGHT-4	6.2	1.5	0.7	19.9
VAL-FLIGHT-5	6.5	0.9	1.2	30.8

purposes and six flights for validation purposes. Data statistics for all the flights are shown in Tables 1(a), 1(b), and 1(c).

For all flights (both training and validation), only the portion of the flight at "cruise" altitude was considered. In fact, the take-off and the initial climbing, as well as the descent and the landing, were not considered since they are associated with specific aerodynamic configurations (including, for example, the use of flaps and/or high angles-of-attack conditions). This decision was made because it was clear that not enough data were available for training the neural estimators at these conditions since only a few minutes for each flight are available. Therefore, only flight data with altitude higher than $h(k) > 250$ m were considered for the purpose of this study. The flight parameters were recorded with a sampling time of $T_C = 0.1$ s.

4. Data-Driven Modeling of the Aircraft Airspeed Velocity, Angle of Attack, and Angle of Sideslip

The true airspeed (TAS), the angle of attack (α), and the angle of sideslip (β) are all estimated as functions of a combination of measurements from all other sensors (excluding, of course, TAS, α, and β). Furthermore, the scheme does not require any information, data, or functional relationships such as the aircraft equations of motion. In other words, the approach can be applied without any loss of generality to any type of aircraft. Thus, the proposed approach can be classified as "data-driven" instead of "model-driven." It should be clarified that in all cases, the GPS velocity was not considered. In fact, the GPS velocity and the TAS are strongly related through the relationship $\text{gps}(k) + V\text{wind}(k) = V(k)$. Therefore, a wind gust could result as an additive fault $f(t)$ of the same amplitude.

Thus, TAS, α, and β are estimated as function of other measurements using the model structure described in Section 2. The initial vector of regressors $X(k)$ for TAS, α, and β has been set simply selecting the signals that characterize the aircraft kinematic and dynamic response as well as considering the correlation coefficient with the target signal. The resulting regressors are

$$
\begin{aligned}
X_{\mathrm{TAS0}}(k) &= [\varphi(k), \vartheta(k), \psi(k), \alpha(k), \beta(k), P(k), Q(k), \\
&\quad R(k), \mathrm{Nx}(k), \mathrm{Ny}(k), \mathrm{Nz}(k), \delta_E(K), \\
&\quad \delta_A(K), \delta_R(K), T_{\mathrm{TH}}(k), h(k), T_{\mathrm{AIR}}(K)], \\
X_{\alpha 0}(k) &= [\vartheta(k), h(k), \mathrm{Nx}(k), \mathrm{Ny}(k), \mathrm{Nz}(k), \\
&\quad P(k), Q(k), R(k), \mathrm{TAS}(k), \beta(k), \psi(k), \\
&\quad \delta_E(K), \delta_A(K), \delta_R(K), T_{\mathrm{TH}}(k), \mathrm{CR}(k)], \\
X_{\beta 0}(k) &= [h(k), \mathrm{Nx}(k), \mathrm{Ny}(k), \mathrm{Nz}(k), P(k), \\
&\quad Q(k), R(k), \mathrm{TAS}(k), \alpha(k), \varphi(k), \\
&\quad \vartheta(k), \psi(k), \delta_E(K), \delta_A(K), \delta_R(K), \\
&\quad \mathrm{Tr}(k), W(k), \mathrm{CR}(k)],
\end{aligned}
\tag{2}
$$

where ϕ, ϑ, and ψ are the aircraft roll, pitch, and yaw angles, respectively; α and β are the angles of attack and sideslip, respectively; P, Q, and R are the aircraft angular rates; Nx, Ny, and Nz are the axial load factors; δ_E, δ_A, and δ_R are the elevator, aileron and rudder deflections, respectively; T_{TH} is the throttle; h is the altitude; T_{AIR} is the air temperature; T AS is the true aircraft airspeed; CR is the aircraft climbing rate; Tr is the track angle; and W is the wind direction. Note that in more general cases, additional control surface deflections could be introduced.

To cope with possible nonlinearities and dynamic effects, each vector $X(k)$ of potential regressors was augmented with the quadratic terms (only pure quadratic terms were added and not mixed terms), resulting in

$$X_{\mathrm{TAS0}}(k)|_{\mathrm{AUG}} = \left[X_{\mathrm{TAS0}}(k), X^2_{\mathrm{TAS0}}(k)\right], \quad (3)$$

$$X_{\alpha 0}(k)|_{\mathrm{AUG}} = \left[X_{\alpha 0}(k), X^2_{\alpha 0}(k)\right], \quad (4)$$

$$X_{\beta 0}(k)|_{\mathrm{AUG}} = \left[X_{\beta 0}(k), X^2_{\beta 0}(k)\right]. \quad (5)$$

5. Regressor Selection

A Matlab-based automated procedure known as Stepwise Selection [28] has been successfully used in this effort for the selection of the regressors. This scheme iteratively adds and removes variables relying on the *t-statistics* of the estimated regression coefficients. Starting from the initial set of regressors in (3), the final set of selected regressors provided by the Stepwise Selection procedure is given by

$$\begin{aligned} X_{\mathrm{TAS}}(k) &= \left[\alpha(k), Q(k), R(k), \mathrm{Nx}(k), \mathrm{Nz}(k), T_{\mathrm{TH}}(k), h(k), \psi^2(k), \alpha^2(k), \delta_E{}^2(K)\right], \\ X_\alpha(k) &= \left[\vartheta(k), h(k), \mathrm{Nx}(k), \mathrm{Nz}(k), P(k), Q(k), \mathrm{TAS}(k), \delta_R(K), T_{\mathrm{TH}}(k), \mathrm{CR}(k), \mathrm{Nz}^2, \mathrm{TAS}^2(k), T_{\mathrm{TH}}{}^2(k)\right], \\ X_\beta(k) &= \left[\mathrm{Nx}(k), \mathrm{Ny}(k), P(k), Q(k), R(k), \mathrm{TAS}(k), \varphi(k), \delta_A(K), \delta_R(K), \mathrm{Tr}(k), W(k), \mathrm{Ny}^2(k), \mathrm{TAS}^2(k), \delta_R{}^2(K), \delta_E{}^2(K)\right]. \end{aligned} \quad (6)$$

Specifically, Figures 2(a)–2(c) show the variable selection process by the stepwise selection method for the three sensors. The index of the selected variables is represented by the red dots in the upper section of each figure while in the lower section the RMSE prediction error (evaluated on the training data) as function of the algorithm iteration is reported.

The correlations between each regressor selected by the stepwise process and the target signal are reported, for each of the sensors, in Tables 2(a), 2(b), and 2(c):

6. Multilayer Perceptron Models

The estimation model introduced in Section 2 is given by

$$\tilde{y}(k) = f_{\mathrm{NR}}(X(k)). \quad (7)$$

For each sensor TAS, α, and β, the $f_{\mathrm{NR}}(k)$ function was mapped through a neural network architecture. After considering different options, the neural topology selected for this study is the well-known multilayer perceptron with one hidden layer. The specific architecture for each neural approximator is shown in Table 3.

Table 4 summarizes the statistical results achieved for the training flight data of the residual signals which are defined as the difference between the real values of the measured signal and its estimation $(r(k) = y(k) - \tilde{y}(k))$.

In a recent study [16], the authors compared the performance of neural network models with those of simple linear regression models using the same set of regressors. In that study, it was verified that the mapping performances of the linear regression models are worse than those provided by the neural network models. This was the main motivation of the use of the neural network models in the present study. A possible motivation of this fact could be that the unknown identifying functions are not linear in the selected regressors; therefore, neural estimators work better than linear regression models.

7. Failure Modeling and Residual Whitening

Failure modeling is a very important task in the overall design of fault tolerant schemes [29–31]. In the literature, several fault modeling approaches can be found, such as bias, freezing, drift, loss of accuracy, and calibration error [32]. In this effort, without any loss of generality, only two specific classes of failures have been considered, that is, step failure and ramp failures to simulate the injection of hard or soft failures. For each of these cases, the magnitude of the failure was also varied to test the fault sensitivity. Clearly, hard and soft failures provide different challenges in terms of detectability. The residual signal is defined as the difference between the sensor output and the estimate of the same signal, that is,

$$r(k) = y(k) - \tilde{y}(k). \quad (8)$$

At nominal conditions (no fault), the $r(k)$ signal should theoretically approach the statistical zero. Unfortunately, this is not true in practice due to unavoidable system and measurement noise as well as estimation errors. Thus, the goals of the fault detection scheme are to detect the occurrence of the failure, as soon as possible after its occurrence, along with the minimization of false alarm rates. From a detailed statistical analysis, it became evident that the residual signal is characterized by a high level of autocorrelation caused by uncertainties at low frequencies and noise. Therefore, in its raw form, the residual signal is not a suitable tool since it contains high levels of colored noise. In fact, the residual signal should be uncorrelated for optimal FD. Therefore, a whitening filter has been designed to remove as much correlation as possible from the residual. For our purposes, it was assumed, as shown in (9), that the correlated sequence can be modeled as an autoregressive (AR) process of order N [33]:

$$\sum_{i=1}^{N} d_i \cdot r(k-i) + \varepsilon_r(k), \quad (9)$$

where d_i are AR parameters and $\varepsilon_r(k)$ is white noise. The term $\sum_{i=1}^{N} d_i \cdot r(k-i)$ can be considered as a one-step-ahead prediction of $r(k)$ and $\varepsilon_r(k) = r(k) - \tilde{r}(k)$ as the prediction error. The d_i parameters were estimated through the Least Squares method to the mean squared error $E^2 = L^{-1}\sum_L \varepsilon(k)^2$ of the training data. Then, after the computing of the d_i

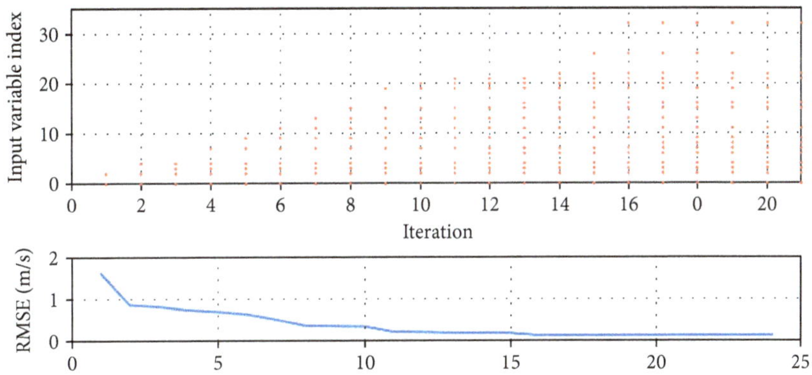

(a) TAS variable selection process

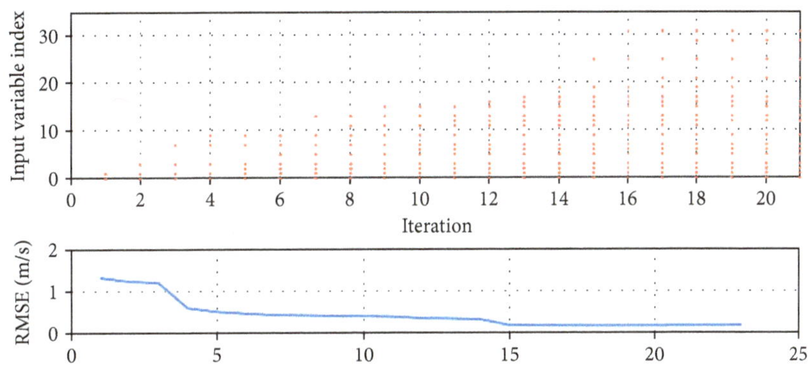

(b) Alpha variable selection process

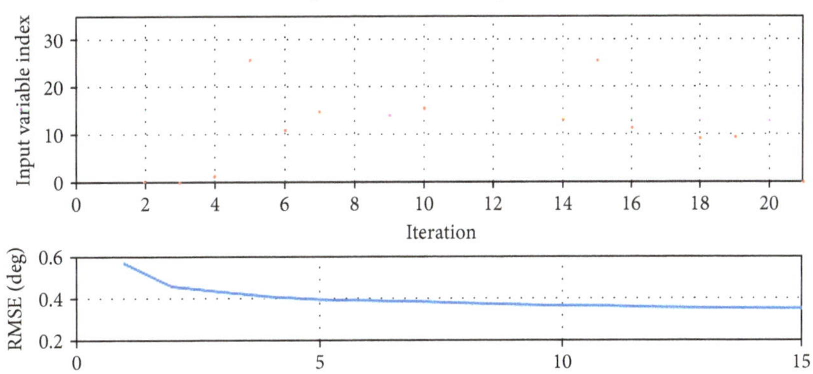

(c) Beta variable selection process

Figure 2

parameters, the $\varepsilon_r(k)$ can be simply derived as output of the linear filter:

$$\varepsilon_r(z) = W_{\text{FIR}}(z) r(z), \quad (10)$$

where the discrete time transfer function can be easily obtained as

$$W_{\text{FIR}}(z) = 1 - d_1 z^{-1} - d_2 z^{-2} - d_3 z^{-3} - -d_N z^{-N}. \quad (11)$$

The $\varepsilon_r(z)$ signal is therefore labeled as "whitened residual" $r_W(k)$.

The correct length of the whitening filter was experimentally determined by evaluating $r_W(k)$ for increasing values of the filter length N, applying the Ljung-Box test [34] to verify whether the whitened residual exhibits a significant autocorrelation. Table 5 shows the final values of N selected by this design process.

A comparison between the residuals before and after the whitening filtering is shown in Figures 3(a), 3(b), 4(a), 4(b), 5(a), and 5(b).

8. Design of the EWMA Filter

Following the design of the whitening filter, the design of the fault detection filters was performed using tools from the statistical process control (SPC). In this context, the exponentially weighted moving average (EWMA) filter is a well-known chart used to detect the presence of small shifts

Table 2

(a) Correlation between selected regressors and TAS

α	Q	R	Nx	Nz	T_{TH}	h	ψ^2	α^2	${\delta_E}^2$
−0.9	0.05	0.13	0.73	0.2	0.35	0.24	−0.83	−0.15	0.36

(b) Correlation between selected regressors and α

ϑ	h	Nx	Nz	P	Q	TAS	δ_R	T_{TH}	CR	Nz^2	TAS^2	${T_{TH}}^2$
0.58	−0.005	−0.64	0.05	0.09	0.15	−0.9	−0.01	−0.4	−0.16	0.06	−0.88	−0.32

(c) Correlation between selected regressors and β

Nx	Ny	P	Q	R	TAS	φ	δ_A	δ_R	Tr	W	Ny^2	TAS^2	${\delta_R}^2$	${\delta_E}^2$
0.1	0.9	0.1	−0.1	−0.02	0.14	0.1	0.5	0.07	0.007	0.1	−0.4	0.1	0.04	0.12

Table 3: Neural network architectures.

	Number of neurons in the hidden layer	Number and type of output neurons
TAS(K)	7 sigmoidal neurons	1 linear neuron
$\alpha(K)$	7 sigmoidal neurons	1 linear neuron
$\beta(k)$	8 sigmoidal neurons	1 linear neuron

Table 4: Training performance results. Statistics of residual signals based on TR-FLIGHT-1 and TR-FLIGHT-2.

	Mean $(r(k))$	Std $(r(k))$	Max $(r(k))$
TAS(K)	$3.6 * 10^{-3}$	0.1129	1.9240
$\alpha(K)$	$2.1 * 10^{-4}$	0.1030	1.2362
$\beta(k)$	$-7.5 * 10^{-6}$	0.1992	1.8095

Table 5: Filter length and whitening filter design.

	N length
TAS(k)	25
$\alpha(k)$	55
$\beta(k)$	30

in process variables. The EWMA chart tracks the exponentially weighted moving average mean of all the previous samples, so that the most recent samples are weighted more heavily than the previous ones. Even though normality of the distribution is usually the basic assumption in many EWMA charts, this chart has exhibited enough robustness even with non-normally distributed data [35]. The output of the filter $z(k)$ is the output of a first-order linear digital filter taking as input the whitened residual, that is,

$$z(k) = (1-\lambda)\cdot z(k-1) + \lambda \cdot r_W(k)\; 0 < \lambda < 1. \quad (12)$$

Clearly, the selection of the appropriate value of the smoothing parameter λ is critical. Larger values of λ are associated with a larger weight of the current input on the current output; conversely, smaller values of λ are associated with larger weights of all the previous input samples on the current output. The output mean and standard deviation can be computed, at stationary condition, using (14).

$$\mu_z = \mu_r, \quad (13)$$

$$\sigma_z^2 = \sigma_r^2\left(\frac{\lambda}{2-\lambda}\right), \quad (14)$$

where μ is the mean and σ is the standard deviation.

A standard commonly adopted procedure for the selection of λ was not found in the literature. In this effort, the Least Squares method [36] was used, with the goal of finding an optimized value of λ so that the sum of the sample squared errors (SSE) of the one-step prediction error $e_z(k) = r_w(k) - z(k-1)$ is minimized.

Figure 5 shows sample squared errors as function of λ for the $z(k)$ EWMA signals produced by the $r_w(k)$ signals that originate from the three AD models evaluated on the training data (SSE are all shifted so that SSE(1) = 0 for all the three filters).

Based on Figure 6, the tuning of the EWMA filter was performed selecting the value of λ such that the SSE is minimized. Thus, the minimum values of SSE were obtained with the values of λ in Table 6:

9. Data-Driven Threshold Setting

For FD purposes, conventional approaches include the use of fixed thresholds or varying (floating) thresholds. Several studies have been conducted using both the approaches [12, 13, 37, 38], typically resulting in a trade-off between them. Fixed thresholds are easier to set since they depend on a lower number of parameters. On the other hand, fixed thresholds may not allow detection following the occurrence of small amplitude failures. The setup of floating thresholds

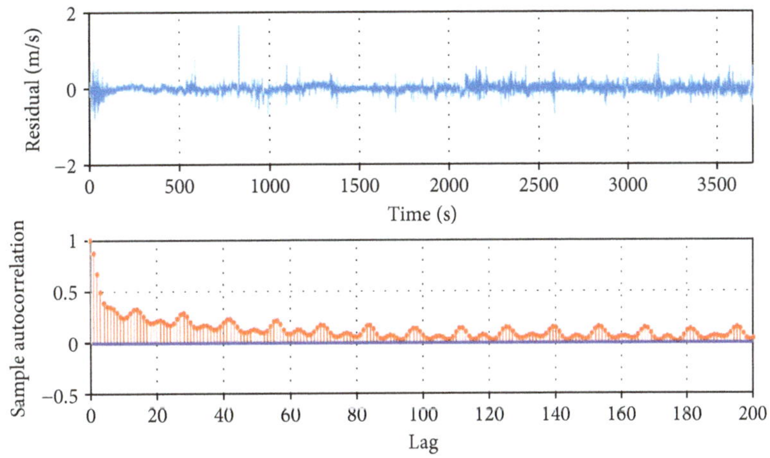

(a) TAS residual and its autocorrelation

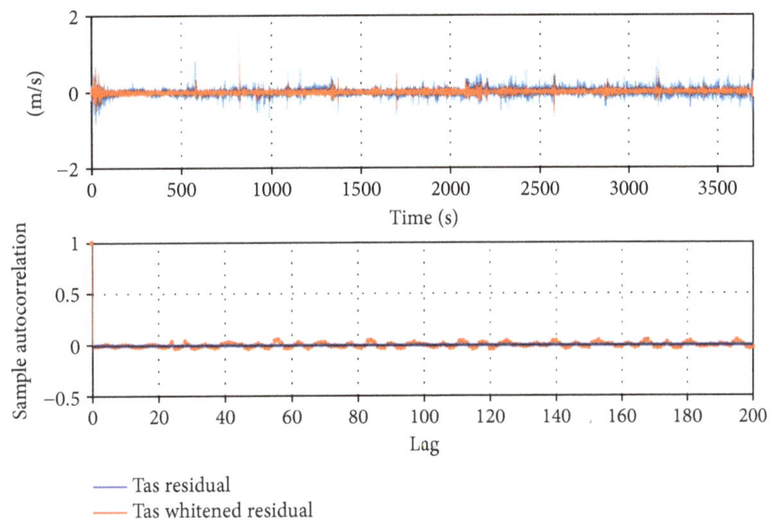

(b) TAS whitened residual and its autocorrelation

FIGURE 3

is, of course, more complex; however, they provide in general more robustness and better performance in terms of detectability, detection delay, and false alarm rate. Clearly, their success requires a careful design. A typical problem of floating (adaptive) thresholds can be given by an excessively fast adaptation to the failure that could result in a missed fault detection. In this effort, floating (adaptive) thresholds [39] are defined, as follows:

$$\begin{aligned} \mathrm{FT_{UP}} &= \bar{z} + \eta \cdot \sigma(z) + b, \\ \mathrm{FT_{LOW}} &= \bar{z} - \eta \cdot \sigma(z) - b, \end{aligned} \tag{15}$$

where $\bar{z}$ and $\sigma(z)$ are the mean value and the standard deviation, respectively, of the EWMA-filtered residual, both computed through a sliding window of length 1800 (180 sec), while η and b are design parameters. Once again, a standard procedure for setting η and b could not be found in the literature; therefore, an "ad hoc" sensitivity study was conducted using the training data to find the optimal values with the goal of minimizing false alarms. Specifically, a grid search was performed over the two parameters, with η in the range 0 to 10 (with a 0.5 step increment) and b in the range 0 to 3 (with a 0.05 step increment). For each grid point, the experimental P_{FA} was computed on the training data retaining only the couples of values (η and b) leading to an experimental $P_{FA} < 0.0001$. Next, between these couples, the best was selected as the one minimizing the following index:

$$j = \frac{\eta}{\sigma(z)} + b. \tag{16}$$

Clearly a small value for J implies a small band for the floating limiter. This is positive from a fault detection point of view. Table 7 shows the optimized values of η and b achieved applying the proposed procedure:

The results of this design are summarized in Figures 7(a)–7(c), where the experimental probability plots derived from the training data are provided.

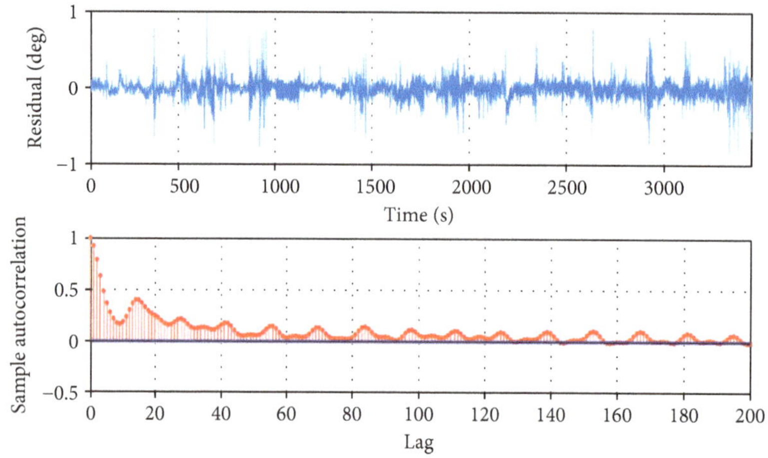

(a) Alpha residual and its autocorrelation

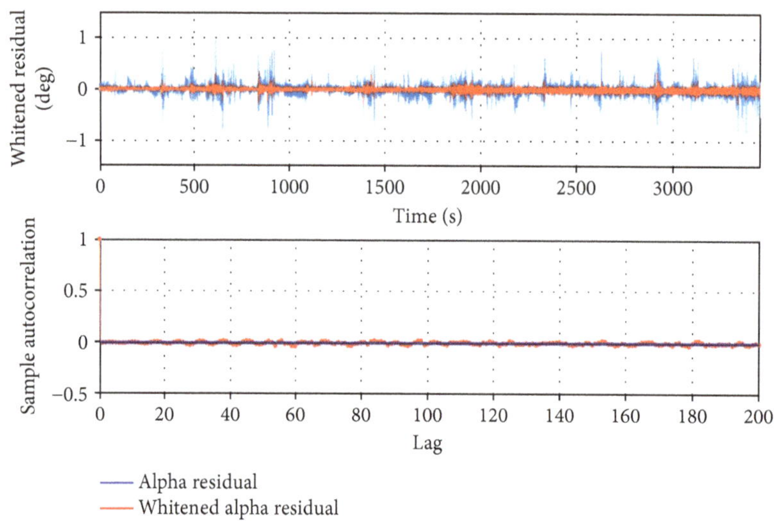

(b) Alpha whitened residual and its autocorrelation

FIGURE 4

In each of the Figures 7(a)–7(c), there are three lobes. The central lobe is the probability distributions of EWMA outputs, while the blue lobes are lower threshold probability distributions and the red lobes are upper threshold probability distributions. Reducing false alarms requires an intersection between the probability distribution of the EWMA filter and the probability distributions of the thresholds as small as possible. If the EWMA output does not exceed significantly the thresholds, this entails that those high false alarm rates are avoided.

10. Results

Fault detection during validation tests was performed using 5 flights for the true airspeed (TAS), 6 flights for the angle of attack, and 5 flights for the angle of sideslip. The testing was conducted in simulated real time using a Simulink model. For each flight, an artificial additive fault was injected at $t = 700$ s with the detectors set using the values of λ, η, and b reported in Tables 6 and 7. Without any loss of generality, two failures were considered, that is, a "fast" step and a "slow" ramp (the duration of the ramp is 10 sec, then for $t > 10$, the amplitude remains constant at the value for $t = 10$; in this case, for magnitude (or "size") of the fault, we conventionally mean the steady state value reached by the fault signal for $t = 10$ s) with modeling parameters shown in Table 8.

A critical parameter for a FD scheme is the number of false alarms. Clearly, a lower false alarm level implies higher robustness against system and measurement noise. The evaluation of the FD scheme was conducted in two steps; in the first step, the scheme was tested on every validation flight in a fault-free condition (without injecting any fault). This is due to the fact that false alarms are independent from the fault injection; therefore, false alarm rate was evaluated on the whole length of the flight assuming no fault. During the second step, a fault was injected to evaluate the detection delay. Tables 9(a), 9(b), and 9(c) show, for the three sensors, the number of false alarms in terms of number of samples along with the "longest continuous stay out of thresholds" time in seconds.

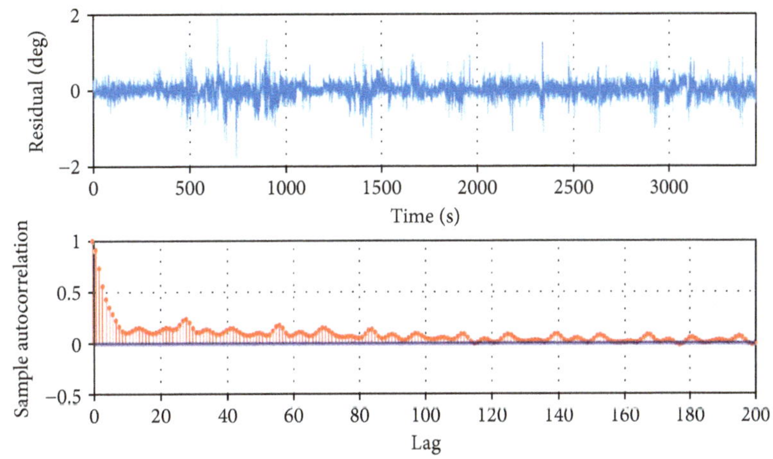

(a) Beta residual and its autocorrelation

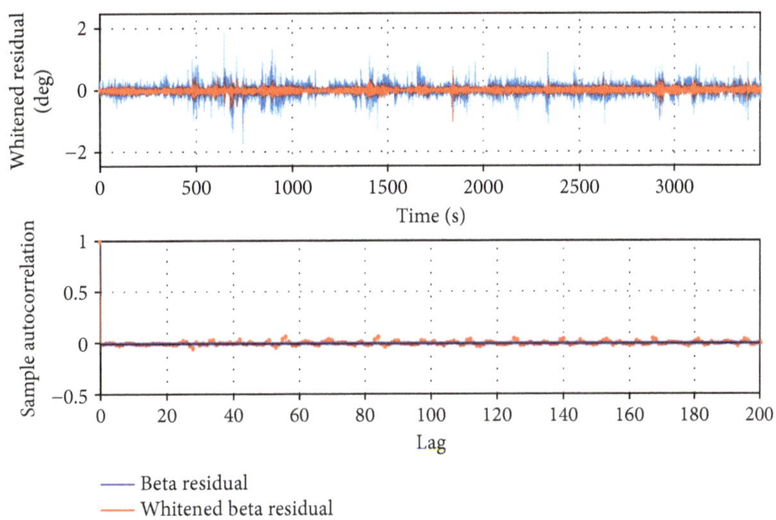

(b) Beta whitened residual and its autocorrelation

FIGURE 5

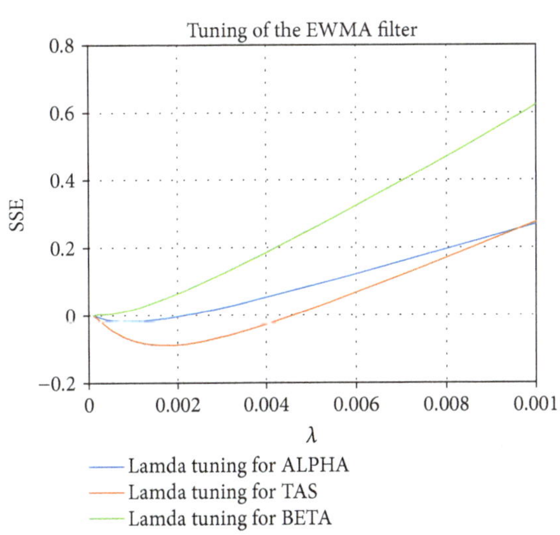

FIGURE 6: EWMA tuning.

TABLE 6: Values of the λ parameter for an optimized design of the EWMA filter.

	λ
TAS	0.002
α	0.001
β	0.0009

TABLE 7: Optimal design of the floating thresholds.

	η	b	P_{FA}
TAS	6	0.005	0.0001
A	7	0.003	0.0001
B	7	0.0075	0.0001

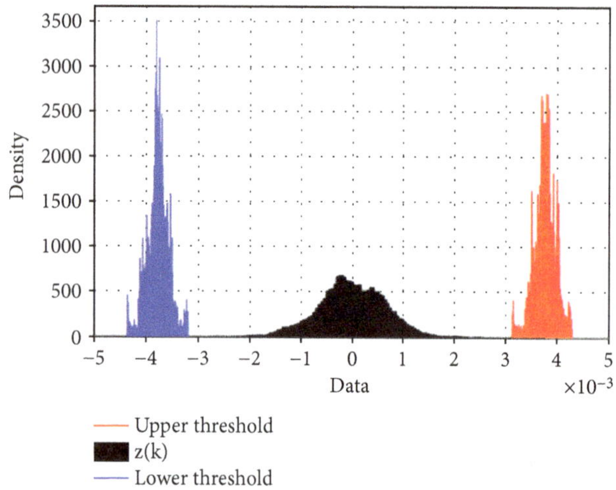

(a) Threshold design for TAS: probability plot

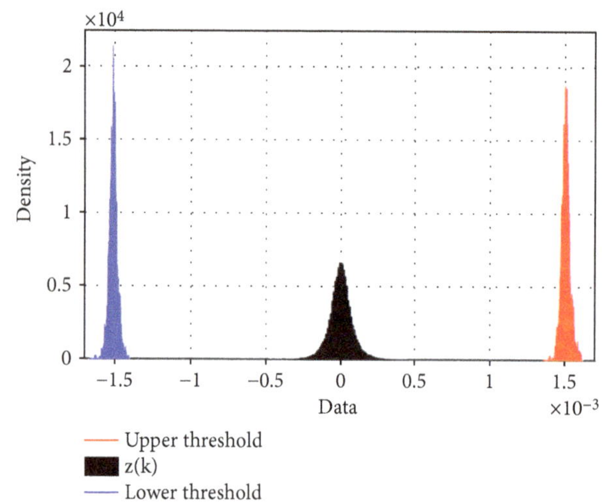

(b) Threshold design for α: probability density

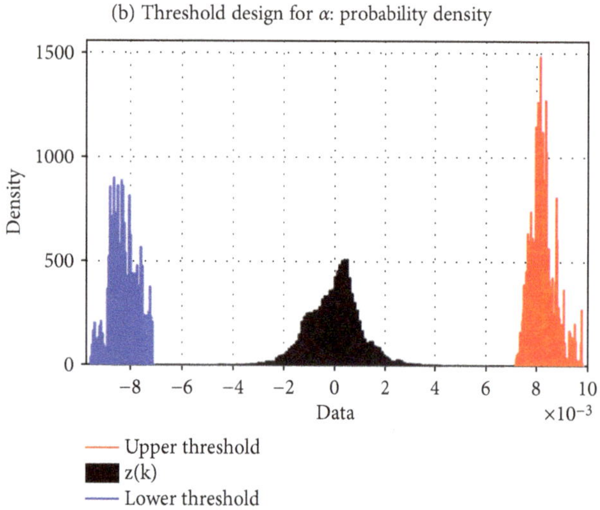

(c) Threshold design for beta: probability density

FIGURE 7

Other critical indexes for assessing the performance of a FD scheme are the fault detection delay and the minimum amplitude detectable fault. These indexes are defined as follows:

TABLE 8: Step magnitude and ramp slope for fault modeling.

	Step (magnitude)	10 s ramp (magnitude)
TAS (m/s)	2	0.2
α (deg)	3	0.3
β (deg)	3	0.3

TABLE 9

(a) Performance for TAS: robustness against false alarms

Airspeed	Flight duration (min)	False alarms (number of samples)	Maximum stay out of thresholds (sec)
Flight-1	17	7	0.4
Flight-2	17.3	0	0
Flight-3	18	0	0
Flight-4	28.3	0	0
Flight-5	30.8	22	0.6

(b) Performance for α: robustness against false alarms

Angle of attack	Flight duration (min)	False alarms (number of samples)	Maximum stay out of thresholds (sec)
Flight-1	17	35	0.4
Flight-2	17.3	8	0.3
Flight-3	18	0	0
Flight-4	20.3	34	0.6
Flight-5	28.3	0	0
Flight-6	30.8	56	0.7

(c) Performance for β: robustness against false alarms

Angle of sideslip	Flight duration (min)	False alarms (number of samples)	Maximum stay out of thresholds (sec)
Flight-1	17.3	0	0
Flight-2	18	0	0
Flight-3	20.3	40	1.5
Flight-4	19.9	42	1.6
Flight-5	30.8	120	2.2

(i) Fault Detection Delay. This is the time elapsed from the injection of the fault to the moment when the detector detects a "true fault."

(ii) Minimum Amplitude Detectable Fault. This is the minimum amplitude of the fault that produces the detection of a "true fault."

In this study, a fault detection is considered a "true fault" if, considering an observation period of 60 s following the failure, the fault detection thresholds are crossed for a time

TABLE 10

(a) Detection delay for step and ramp TAS fault: $\Delta V = 2$ m/s

Airspeed	Step fault detection delay (sec)	Ramp fault detection delay (sec)
Flight-1	0.4	3.8
Flight-2	0.4	4.2
Flight-3	0.5	4.7
Flight-4	0.1	4
Flight-5	0.4	3.5

(b) Detection delay for step and ramp α fault: $\Delta\alpha = 3$ deg

Angle of attack	Step fault detection delay (sec)	Ramp fault detection delay (sec)
Flight-1	0.3	4
Flight-2	0.1	7.1
Flight-3	0.1	8
Flight-4	0.1	8.3
Flight-5	0.1	6.9
Flight-6	0.1	7.2

(c) Detection delay for step and ramp β fault $\Delta\beta = 3$ deg

Angle of sideslip	Step fault detection delay (sec)	Ramp fault detection delay (sec)
Flight-1	2.5	7.5
Flight-2	2.4	7.6
Flight-3	0.8	6.5
Flight-4	1.7	6.9
Flight-5	1.6	7.7

duration longer than 4 s during the observation period; otherwise, the detection is considered not reliable and the true fault is not declared. This logic implies that the minimum decision time for a fault declaration is 60 seconds.

Tables 10(a), 10(b), and 10(c) show the performance for TAS, α, and β in terms of detection delay, while Figures 8(a), 8(b), 9(a), 9(b), 10(a), and 10(b) show the threshold evolution during the injection of the fault.

The minimum amplitude detectable fault was computed by incrementally decreasing the amplitude of the fault injected until it was no longer possible to declare a "true fault" within the predefined observation period.

The results for the minimum amplitude detectable faults are shown in Table 11:

Tables 12(a), 12(b), and 12(c) summarize the performance of the FD scheme with the minimum amplitude detectable along with the associated fault detection delay.

11. Comparison between Adaptive- and Fixed-Threshold Approaches

As discussed in the literature, a very interesting area of research is the performance comparison in terms of specific FD parameters between "fixed" and "adaptive" thresholds. Considering the fixed-threshold computation approach discussed in [37] as baseline, in this study, the comparison was performance using the same training and validation data for both approaches, reporting the performance in terms of false alarms and detection delay. For the purpose of being completely objective and conducting a fair comparison, fixed thresholds have been computed starting from the residual signals using the same neural estimators and the same whitening filters to obtain a whitened residual. The difference between the two approaches starts after the EWMA filtering (designed with the same λ parameters). In fact, for the fixed-threshold approach, the EWMA output was processed to find optimized values for upper and lower limits. In fact, during the training, the UCL and LCL were derived from data depending on the desired probability of false alarm P_{FA}, where the P_{FA} is defined as the probability that $z(k)$ exceeds the detection thresholds in a fault-free condition. Defining $F(z)$ as the experimental cumulative PDF of $z(k)$, the fault detection thresholds are defined as follows:

$$\begin{aligned} \text{UCL} &= \min\ (\text{z}) | F(z) \geq (1 - P_{\text{FA}}), \\ \text{LCL} &= \max\ (\text{z}) | F(z) \leq P_{\text{FA}}. \end{aligned} \tag{17}$$

$P_{FA} = 0.0001$ was set the same way in both approaches. Also, the validation tests were conducted on the same flights used in Section 10, using the same fault amplitude and the same injection time for the failure. Tables 13, 14, and 15 show validation results in terms of false alarms and detection delay:

As far as soft failures are concerned, the detection delay analysis was performed using the minimum amplitude detectable fault used in Section 10 and reported in Table 11. The achieved results are shown in Tables 15(a), 15(b), and 15(c):

Analyzing the results, it is evident that detecting fault featuring adaptive limiters could save from 0% to 50% of the time, with the same false alarm rate in the case of hard failures. Substantially different results were achieved when comparing the two approaches in the case of soft failures. In fact, adaptive thresholds allow a much lower detection delay versus fixed thresholds (up to a ratio of 1:30). In some cases, the occurrence of a soft failure was not even detected using fixed thresholds.

Note: the study presented in this paper is specifically focused on the design of experimental FD filters. Therefore, the problem of fault isolation is not considered at all in the study. Indeed, if the fault signature matrix for the residuals for TAS, α, and β are computed, it can be immediately seen that the three faults are not strongly isolable since each residual depends on TAS, α, and β. Different approaches can be used to deal with the fault isolation problem, such as

(1) to perform a fault sensitivity analysis to investigate whether the fault has a different effect on the amplitudes of three residuals and then exploit this information for discriminating the faulty signal.

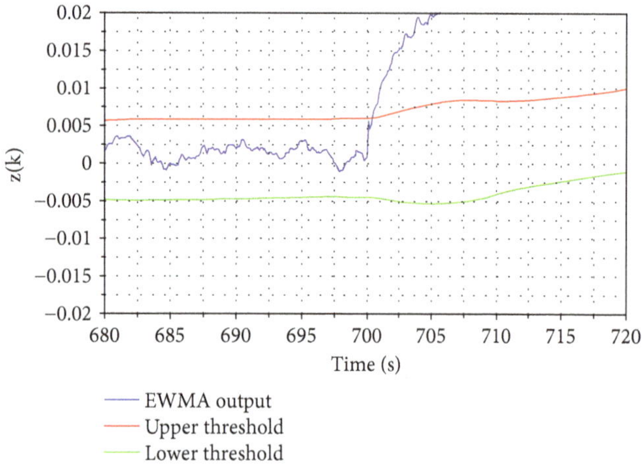

(a) TAS, zoom on fault detection, step fault at time 700 s injected in Flight-1

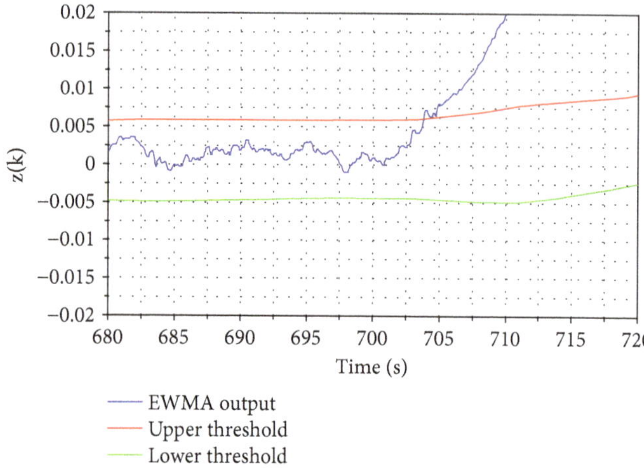

(b) TAS, zoom on fault detection, ramp fault at time 700 s injected in Flight-1

Figure 8

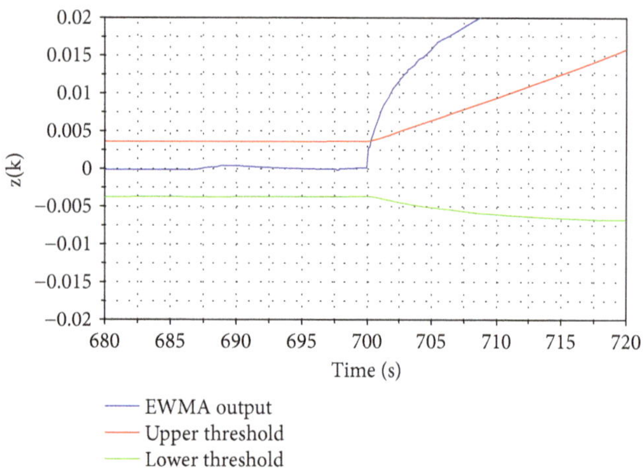

(a) α, zoom on fault detection, step fault at time 700 s injected in Flight-1

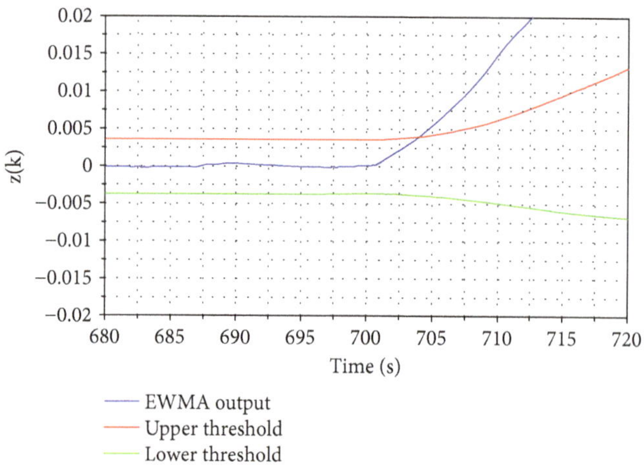

(b) α, zoom on fault detection, ramp fault at time 700 s injected in Flight-1

Figure 9

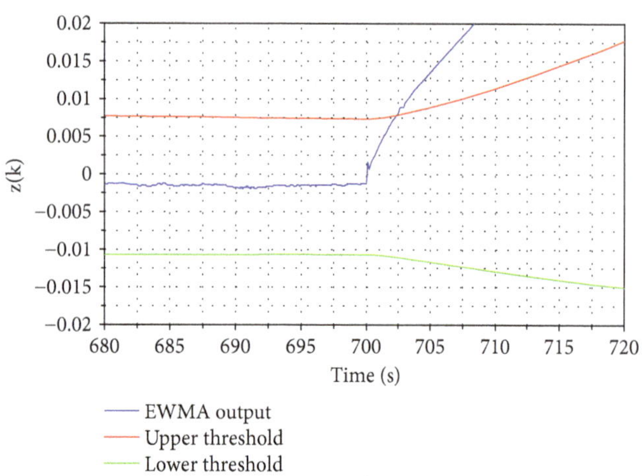

(a) β, zoom on fault detection, step fault at time 700 s injected in Flight-1

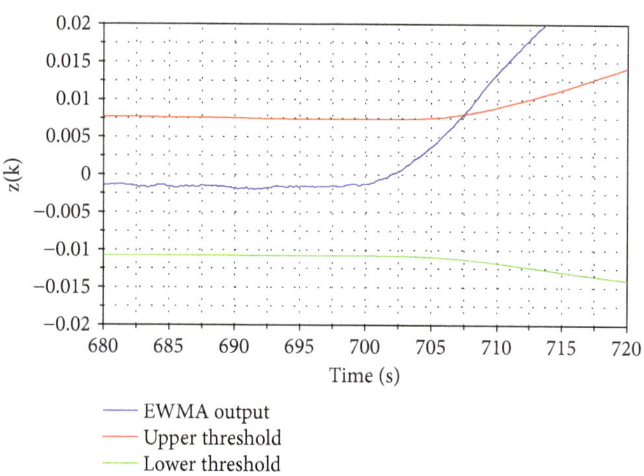

(b) β, zoom on fault detection, step fault at time 700 s injected in Flight-1

Figure 10

TABLE 11: Minimum amplitude detectable faults.

	Step (magnitude)	10 s ramp (magnitude)
TAS (m/s)	0.7	0.07
α (deg)	0.9	0.1
β (deg)	1.3	0.13

TABLE 12

(a) Detection delay for step and ramp TAS fault

Airspeed	Step fault detection delay (sec)	Ramp fault detection delay (sec)
Flight-1	3.7	15.4
Flight-2	1.9	10.1
Flight-3	10.2	11.3
Flight-4	1.9	9.4
Flight-5	1.4	13.8

(b) Detection delay for step and ramp α fault

Angle of attack	Step fault detection delay (sec)	Ramp fault detection delay (sec)
Flight-1	15.5	15
Flight-2	2.8	8.8
Flight-3	10.4	10.8
Flight-4	6.2	10.6
Flight-5	5.5	9.6
Flight-6	5.7	18.2

(c) Detection delay for step and ramp β fault

Angle of sideslip	Step fault detection delay (sec)	Ramp fault detection delay (sec)
Flight-1	17.3	18.3
Flight-2	13.1	18.6
Flight-3	15	25.5
Flight-4	7.6	13.3
Flight-5	9.8	16.4

TABLE 13

(a) Comparison between fixed-threshold and adaptive-threshold false alarms for TAS

TAS	Flight duration (min)	False alarms (num. of samples) (adaptive)	Max stay out of thresholds (sec) (adaptive)	False alarms (num. of samples) (fixed)	Max stay out of thresholds (sec) (fixed)
FL-1	17	7	0.4	5	0.3
FL-2	17.3	0	0	2	0.2
FL-3	18	0	0	0	0
FL-4	28.3	0	0	3	0.2
FL-5	30.8	22	0.6	18	0.5

(b) Comparison between fixed-threshold and adaptive false alarms delay for α

α	Flight duration (min)	False alarms (num. of samples) (adapt)	Max stay out of thresholds (sec) (adapt)	False alarms (num. of samples) (fixed)	Max stay out of thresholds (sec) (fixed)
FL-1	17	35	0.4	40	0.7
FL-2	17.3	8	0.3	12	0.4
FL-3	18	0	0	2	0.2
FL-4	20.3	34	0.6	40	0.5
FL-5	28.3	0	0	5	0.5
FL-6	30.8	56	0.7	60	1.2

(c) Comparison between fixed-thresholds and adaptive false alarms delay for β

β	Flight duration (min)	False alarms (num. of samples) (adapt)	Max stay out of thresholds (sec) (adapt)	False alarms (num. of samples) (fixed)	Max stay out of thresholds (sec) (fixed)
FL-1	17.3	0	0	5	0.2
FL-2	18	0	0	4	.01
FL-3	20.3	40	1.5	45	1.3
FL-4	19.9	42	1.6	46	1.2
FL-5	30.8	120	2.2	80	0.5

(2) to exclude, a priori, from the set of regressors the reaming two signals (for instance the regressor set used for approximating the TAS signal does not include the α and β signals). In this case, we achieve, by construction, experimental models that produce a strong isolable failure signature matrix. Interested readers are referred to the recent paper by the authors in [16] where this last approach has been applied.

12. Conclusions

The goal of this study was the design of a robust adaptive threshold scheme able to deal with failures involving air data sensors. A completely data-driven approach for the design and the tuning of the FD schemes for true airspeed, angle-of-attack, and angle-of-sideslip sensors of an aircraft has been proposed. It has been shown that, through the use of this scheme, it is possible to improve significantly the robustness in terms of false alarms and failure detection delay. The data-driven design philosophy was used in many parts of the overall FD procedure that is in the regressor selection phase for the considered sensors, in the design of the withering and EWMA filters as long as for the computation of detection thresholds. Validation studies based on actual flight data

Table 14

(a) Comparison between fixed-threshold and adaptive-threshold detection delay for TAS

TAS	Step fault detection delay (adaptive)	Ramp fault detection delay (adaptive)	Step fault detection delay (fixed)	Ramp fault detection delay (fixed)
FL-1	0.4	3.8	0.7	4.8
FL-2	0.4	4.2	0.7	5.7
FL-3	0.5	4.7	0.8	6
FL-4	0.1	4	0.7	6
FL-5	0.4	3.5	0.8	5.6

(b) Comparison between fixed-threshold and adaptive-threshold detection delay for α

α	Step fault detection delay (adaptive)	Ramp fault detection delay (adaptive)	Step fault detection delay (fixed)	Ramp fault detection delay (fixed)
FL-1	0.3	4.0	3.6	9.3
FL-2	0.1	7.1	1.6	7.2
FL-3	0.1	8.0	0.6	5.5
FL-4	0.1	8.3	1.8	7.4
FL-5	0.1	6.9	1.2	6.8
FL-6	0.1	7.2	1.1	6

(c) Comparison between fixed-threshold and adaptive-threshold detection delay for β

β	Step fault detection delay (adaptive)	Ramp fault detection delay (adaptive)	Step fault detection delay (fixed)	Ramp fault detection delay (fixed)
FL-1	2.5	7.5	3	9
FL-2	2.4	7.6	3.6	9
FL-3	0.8	6.5	4.4	10
FL-4	1.7	6.9	6.5	6.8
FL-5	1.6	7.7	3.5	10

Table 15

(a) Comparison between fixed-threshold and adaptive-threshold detection delay for TAS: soft failures

TAS	Step fault detection delay (adaptive)	Ramp fault detection delay (adaptive)	Step fault detection delay (fixed)	Ramp fault detection delay (fixed)
FL-1	3.7	15.4	7.6	19.4
FL-2	1.9	10.1	15.3	19.2
FL-3	10.2	11.3	31.5	30
FL-4	1.9	9.4	16.3	17
FL-5	1.4	13.8	27.8	28

(b) Comparison between fixed-threshold and adaptive-threshold detection delay for α: soft failures

α	Step fault detection delay (adaptive)	Ramp fault detection delay (adaptive)	Step fault detection delay (fixed)	Ramp fault detection delay (fixed)
FL-1	15.5	15	Not detected	40.9
FL-2	2.8	8.8	86.2	85.6
FL-3	10.4	10.8	10.7	12.6
FL-4	6.2	10.6	41.2	40.8
FL-5	5.5	9.6	Not detected	51.4
FL-6	5.7	18.2	26.7	26

(c) Comparison between fixed-threshold and adaptive-threshold detection delay for β: soft failures

β	Step fault detection delay (adaptive)	Ramp fault detection delay (adaptive)	Step fault detection delay (fixed)	Ramp fault detection delay (fixed)
FL-1	17.3	18.3	19	20
FL-2	13.1	18.6	21	23
FL-3	15	25.5	43.8	48.1
FL-4	7.6	13.3	5.9	11.2
FL-5	9.8	16.4	46.7	42.8

have shown good overall performance in the presence of hard and soft failures and the possibility to detect very small faults. Additionally, a comparison with a fixed-threshold approach was performed. The results of the comparison revealed that for hard failures, there is a limited difference between the approaches. In contrast, a different trend has been observed in the case of soft failures. Specifically, a significant detection delay has been observed in the case of fixed thresholds, leading also to the "no detection at all" situation in some cases.

The capabilities of this data-driven FD approach coupled with the flexibility of the adaptive threshold provided by the floating limiters have proven to be substantially better than fixed-threshold methods in terms of FD performance. This last aspect makes the proposed scheme appealing for the application to a variety of sensor failures for systems where physical redundancy in the sensors is not available.

Conflicts of Interest

The authors Fabio Balzano, Mario L. Fravolini, Marcello R. Napolitano, Stéphane D'Urso, Michele Crispoltoni, and Giuseppe del Core declare that there is no conflict of interest regarding the publication of this paper.

References

[1] R. P. Collinson, *Introduction to Avionics Systems*, Springer Science and Business, 2013.

[2] Bureau d'Enquetes et d'Analyses pour la securite de l'aviation civile, "Re-port 3: on the accident on 1 June 2009 to the Airbus A330-203 registered F-GZCP operated by Air France flight AF 447 Rio de Janeiro–Paris, France," https://www.bea.aero/docspa/2009/f-cp090601.en/pdf/f-cp090601.en.pdf.

[3] J. Levine, *J.X-31's loss*, NASA X-Press, 2004, http://www.nasa.gov/centers/dryden/news/X-Press/stories/2004/013004/new_x31.html.

[4] J. T. McKenna, "Blocked static ports eyed in Aeroperu 757 crash," *Aviation Week & Space Technology*, vol. 145, no. 20, p. 76, 1996.

[5] Dirección General de Aeronáutica Civil (Dominican Republic), "Reporte final accidente aereo Birgenair, vuelo ALW-301," 1996, http://www.fss.aero/accident-reports/.

[6] Bureau d'Enquêtes et d'Analyses pour la Sécurité de l'Aviation Civile, "Report of the accident on 27 November 2008 off the coast of Canet-Plage (66) to the Airbus A320-232 registered D-AXLA operated by XL Airways Germany," https://www.bea.aero/.

[7] E. Dubrova, *Fault-Tolerant Design*, Springer, Berlin, 2013.

[8] P. M. Frank, "Fault diagnosis in dynamic systems using analytical and knowledge-based redundancy: a survey of some new results," *Automatica*, vol. 26, no. 3, pp. 459–474, 1990.

[9] R. C. Swischuk and D. L. Allaire, "A machine learning approach to aircraft sensor error detection and correction," in *Proceeding of the 2018 AIAA/ASCE/AHS/ASC Structures, Structural Dynamics, and Materials Conference*, p. 1164, Kissimmee, FL, USA, 2018.

[10] S. X. Ding, *Data-Driven Design of Fault Diagnosis and Fault-Tolerant Control Systems*, Springer, 2014.

[11] S. Simani, C. Fantuzzi, and R. J. Patton, *Model-Based Fault Diagnosis, in Dynamic Systems Using Identification Techniques*, Springer-Verlag, 2002.

[12] N. Talebi, M. A. Sadrnia, and A. Darabi, "Robust fault detection of wind energy conversion systems based on dynamic neural networks," *Computational intelligence and neuroscience*, vol. 2014, Article ID 580972, 13 pages, 2014.

[13] J. Qi, X. Zhao, Z. Jiang, and J. Han, "An adaptive threshold neural-network scheme for rotorcraft UAV sensor failure diagnosis," in *Advances in Neural Networks - ISNN 2007*, D. Liu, S. Fei, Z. Hou, H. Zhang, and C. Sun, Eds., vol. 4493 of Lecture Notes in Computer Science, Springer, Berlin, Heidelberg, 2007.

[14] J. M. Lucas and M. S. Saccucci, "Exponentially weighted moving average control schemes: properties and enhancements," *Technometrics*, vol. 32, no. 1, pp. 1–12, 1990.

[15] J. S. Hunter, "The exponentially weighted moving average," *Journal of Quality Technology*, vol. 18, no. 4, pp. 203–210, 2018.

[16] M. L. Fravolini, M. R. Napolitano, G. D. Core, and U. Papa, "Experimental interval models for the robust fault detection of aircraft air data sensors," *Control Engineering Practice*, vol. 78, pp. 196–212, 2018.

[17] Q. Li, R. Li, K. Ji, and W. Dai, "Kalman filter and its application," in *2015 8th International Conference on Intelligent Networks and Intelligent Systems (ICINIS)*, pp. 74–77, Tianjin, China, 2015.

[18] E. A. Wan and R. Van Der Merwe, "The unscented Kalman filter for nonlinear estimation," in *Proceedings of the IEEE 2000 Adaptive Systems for Signal Processing, Communications, and Control Symposium (Cat. No.00EX373)*, Lake Louise, Alberta, Canada, 2000.

[19] F. Keisuke, "Extended Kalman filter," *Reference Manual*, 2013, http://www-jlc.kek.jp/2003oct/subg/offl/kaltest/doc/ReferenceManual.pdf.

[20] S. Zhao, F. Liu, and Y. S. Shmaliy, "Optimal FIR estimator for discrete time-variant state-space model," in *2014 11th International Conference on Electrical Engineering, Computing Science and Automatic Control (CCE)*, pp. 1–5, Campeche, Mexico, 2014.

[21] K. Hornik, M. Stinchcombe, and H. White, "Multilayer feedforward networks are universal approximators," *Neural Networks*, vol. 2, no. 5, pp. 359–366, 1989.

[22] D. Svozil, V. Kvasnicka, and J. Pospichal, "Introduction to multi-layer feed-forward neural networks," *Chemometrics and Intelligent Laboratory Systems*, vol. 39, no. 1, pp. 43–62, 1997.

[23] N. Oliver, *Nonlinear System Identification: from Classical Approaches to Neural Networks and Fuzzy Models*, Springer-Verlag, 2001.

[24] I. Rivals and L. Personnaz, "MLPs (mono layer polynomials and multi-layer perceptrons) for nonlinear modeling," *Journal of Machine Learning Research*, vol. 3, pp. 1383–1398, 2003.

[25] M. W. Gardner and S. R. Dorling, "Artificial neural networks (the multilayer perceptron)—a review of applications in the atmospheric sciences," *Atmospheric Environment*, vol. 32, no. 14–15, pp. 2627–2636, 1998.

[26] J. A. Farrell and M. M. Polycarpou, *Adaptive Approximation Based Control: Unifying Neural, Fuzzy and Traditional Adaptive Approximation Approaches*, Wiley, 2006.

[27] https://www.tecnam.com/it/aircraft/p92-eaglet-adv-ul/.

[28] N. R. Draper and H. Smith, *Applied Regression Analysis*, Wiley-Interscience, Hoboken, NY, USA, 1998.

[29] N. Nyberg, "A general framework for fault diagnosis based on statistical hypothesis testing," in *Proceedings of the 12th International Workshop on Principles of Diagnosis*, pp. 135–142, Sansicario, Italy, 2001.

[30] J. Gertler, "Analytical redundancy methods in fault detection and isolation; survey and synthesis," in *Proceedings of the IFAC Fault Detection, Supervision and Safety for Technical Processes*, pp. 9–21, Baden, Germany, 1991.

[31] J. Chen and R. J. Patton, *Robust Model-Based Fault Diagnosis for Dynamic Systems*, Kluwer, 1999.

[32] H. Alwi, C. Edwards, and C. P. Tan, "Fault tolerant control and fault detection and isolation," in *Fault Detection and Fault-Tolerant Control Using Sliding Modes. Advances in Industrial Control*, Springer, 2011.

[33] D. C. Montgomery and C. M. Mastrangelo, "Some statistical process control methods for autocorrelated data," *Journal of Quality Technology*, vol. 23, no. 3, pp. 179–193, 2018.

[34] G. M. Ljung and G. E. P. Box, "On a measure of lack of fit in time series models," *Biometrika*, vol. 65, no. 2, pp. 297–303, 1978.

[35] N. Ye, S. Vilbert, and Q. Chen, "Computer intrusion detection through EWMA for autocorrelated and uncorrelated data," *IEEE Transactions on Reliability*, vol. 52, no. 1, pp. 75–82, 2003.

[36] J. S. Oakland, *Statistical Process Control*, Butterworth-Heinemann, Burlington, MA, USA, 2007.

[37] M. L. Fravolini, G. del Core, U. Papa, P. Valigi, and M. R. Napolitano, "Data-driven schemes for robust fault detection of air data system sensors," *IEEE Transactions on Control Systems Technology*, no. 99, pp. 1–15, 2017.

[38] S. A. Raka and C. Combastel, "Fault detection based on robust adaptive thresholds: a dynamic interval approach," *Annual Reviews in Control*, vol. 37, no. 1, pp. 119–128, 2013.

[39] M. G. Perhinschi, M. R. Napolitano, G. Campa, B. Seanor, J. Burken, and R. Larson, "An adaptive threshold approach for the design of an actuator failure detection and identification scheme," *IEEE Transactions on Control Systems Technology*, vol. 14, no. 3, pp. 519–525, 2006.

2

An Improved Nondominated Sorting Genetic Algorithm III Method for Solving Multiobjective Weapon-Target Assignment

You Li,[1] Yingxin Kou,[1] and Zhanwu Li[1,2]

[1]*Aeronautics and Astronautics Engineering College, Air Force Engineering University, Xi'an, Shaanxi 710038, China*
[2]*College of Electronic Communication, Northwestern Polytechnical University, Xi'an 710072, China*

Correspondence should be addressed to Zhanwu Li; afeulzw@189.cn

Academic Editor: Mahmut Reyhanoglu

Multiobjective weapon-target assignment is a type of NP-complete problem, and the reasonable assignment of weapons is beneficial to attack and defense. In order to simulate a real battlefield environment, we introduce a new objective—the value of fighter combat on the basis of the original two-objective model. The new three-objective model includes maximizing the expected damage of the enemy, minimizing the cost of missiles, and maximizing the value of fighter combat. To solve the problem with complex constraints, an improved nondominated sorting algorithm III is proposed in this paper. In the proposed algorithm, a series of reference points with good performances in convergence and distribution are continuously generated according to the current population to guide the evolution; otherwise, useless reference points are eliminated. Moreover, an online operator selection mechanism is incorporated into the NSGA-III framework to autonomously select the most suitable operator while solving the problem. Finally, the proposed algorithm is applied to a typical instance and compared with other algorithms to verify its feasibility and effectiveness. Simulation results show that the proposed algorithm is successfully applied to the multiobjective weapon-target assignment problem, which effectively improves the performance of the traditional NSGA-III and can produce better solutions than the two multiobjective optimization algorithms NSGA-II and MPACO.

1. Introduction

With the rapid development of military air combat, the weapon-target assignment (WTA) problem has attracted worldwide attention [1]. The WTA problem is a classic scheduling problem that aims to assign weapons to maximize military effectiveness and meet all constraints. So, it is important to find a proper assignment of weapons to targets.

The study of the WTA problem can be traced back to the 1950s and 1960s when Manne [2] and Day [3] built the model of the WTA problem. From the perspective of the quantity of objective functions, Hosein and Athans [4] classify the WTA problem into two classes: the single-objective weapon-target assignment problem and the multiple-objective weapon-target assignment (MWTA) problem. When taking the time factor into account, Galati and Simaan [5] divide the WTA problem into two categories: dynamic weapon-target assignment problem and static weapon-target assignment problem. The current research status of various WTA problems are summarized in Table 1.

In contrast to the single-objective weapon-target assignment problem, MWTA can take different criterions into consideration that are more in line with real combat decision making. In this paper, we mainly focus on the static multiobjective weapon-target assignment (SMWTA) problem, which aims at finding proper static assignments.

The combination of simulation and optimization algorithms to solve the SMWTA problem is not new. At present, a number of studies address this problem. In Liu et al. [16], an improved multiobjective particle swarm optimization (MOPSO) was used to solve the SMWTA problem with two objective functions: maximum enemy damage probability and minimum total firepower unit. The specific example they used contains only 7 platforms and 10 targets.

Table 1: Summary of variant metaheuristic algorithms and implementation of various WTA [6].

Researchers	Year	Metaheuristic algorithm	Implementation (WTA)
Lee et al. [6]	2002	IS + ACO	Static single objective
Lee et al. [7]	2002	GA	Static single objective
Lee et al. [8]	2003	GA + ACO	Static single objective
Galati and Simaan [5]	2007	Tabu	Dynamic single objective
Lee [9]	2010	VLSN	Static single objective
Xin et al. [10]	2010	VP + tabu	Dynamic single objective
Li and Dong [11]	2010	DPSO + SA	Dynamic single objective
Chen et al. [12]	2010	SA	Static single objective
Xin et al. [13]	2011	Rule-based heuristic	Dynamic multiobjective
Fei et al. [14]	2012	Auction algorithm	Static single objective
Bogdanowicz et al. [15]	2013	GA	Static single objective
Liu et al. [16]	2013	MOPSO	Static multiobjective
Zhang et al. [17]	2014	MOEA/D	Static multiobjective
Ahner and Parson [18]	2015	Dynamic programming	Dynamic multiobjective
Li et al. [19]	2015	NSGA-II, MOEA/D	Static multiobjective
Dirik et al. [20]	2015	MILP	Dynamic multiobjective
Hongtao and Fengju [21]	2016	CSA	Static single objective
Li et al. [22]	2016	MDE	Dynamic multiobjective
Li et al. [23]	2017	MPACO	Static multiobjective

Zhang et al. [17] proposed a decomposition-based evolutionary multiobjective optimization method based on the MOEA/D algorithm. Considering the constraints of attack resource and damage probability, a mathematic model on weapon-target assignment was formulated. Both the proposed repair method and appropriate decomposition approaches can effectively improve the performance of the algorithm. But the algorithm has not been tested on a large-scale WTA problem, and it has a low convergence speed.

In the work of Li et al. [19], a new optimization approach for the MWTA problem was developed based on the combination of two types of multiobjective optimizers: NSGA-II (domination-based) and MOEA/D (decomposition-based) enhanced with an adaptive mechanism. Then, a comparison study among the proposed algorithms, NSGA-II and MOEA/D, on solving instances of a three-scale MWTA problem was performed, and four performance metrics were used to evaluate each algorithm. They only applied the proposed adaptive mechanism to the MWTA problem, but they did not verify the behavior of the proposed adaptive mechanism on standard problems. In addition, they also considered the next step to solve the MWTA problem with an improved version of NSGA-II (called NSGA-III [24]).

In our previous work [23], we proposed a modified Pareto ant colony optimization (MPACO) algorithm to solve the bi-objective weapon-target assignment (BOWTA) problem and introduce the pilot operation factor into a WTA mathematic model. The proposed algorithm and two multi-objective optimization algorithms NSGA-II and SPEA-II were applied to solve different scales of instances. Simulation results show that the MPACO algorithm is successfully applied in the field of WTA, which improves the performance of the traditional Pareto ant colony optimization (P-ACO) algorithm effectively and produces better solutions than the other two algorithms.

Although the above methods have remarkable effects on solving the SMWTA problem, all of them considered two objectives, maximizing the expected damage of the enemy and minimizing the cost of missiles, without considering the attack power. Due to the fact that fighters cannot destroy the targets at once, we put forward the value of fighter combat to evaluate the ability of sustained operational capability. On the basis of the original double-objective model, we propose the three-objective model, which is closer to real air combat. The new three-objective model includes maximizing the expected damage of the enemy, minimizing the cost of missiles, and maximizing the value of fighter combat. As the number of objectives is increased from two to three, the performance of evolutionary multiobjective algorithms (EMOAs) may deteriorate. They face some difficulties as follows: (i) A large fraction of the population is nondominated. (ii) The evaluation of diversity is computationally complex. (iii) The recombination operation may be inefficient [25]. Recently, EMOAs like the nondominated sorting genetic algorithm III (NSGA-III) [26] have been proposed to deal with these difficulties and scale with the number of objectives.

In 2014, Deb and Jain [24, 26] proposed a reference point-based many-objective evolutionary algorithm following the NSGA-II framework (namely, NSGA-III). The basic framework of the NSGA-III remains similar to the NSGA-II algorithm, but the NSGA-III improves the ability to solve the multiobjective optimization problem (MOP) by changing the selection mechanism of its predecessor. Namely, the main difference is the substitution of crowding distance for a selection based on well-distributed and adaptive reference points. These reference points help maintain the diversity of population members and also allow the NSGA-III to perform well on MOP with differently scaled objective values. This is an advantage of the NSGA-III and another reason why we choose the NSGA-III algorithm to solve the SMWTA problem.

The NSGA-III has been successfully applied to real-world engineering problems [27, 28] and has several proposed variants, such as combining different variation operators [29], solving monoobjective problems [30], and integrating alternative domination schemes [31]. As far as we know, none of the previous related work has studied the MWTA problem of three objective functions and applied the NSGA-III algorithm to solve the MTWA problem.

In this paper, we have proposed an improved NSGA-III (I-NSGA-III) for solving the SMWTA problem. The proposed algorithm is used to seek better Pareto-optimal solutions between maximizing the expected damage, minimizing the cost, and maximizing the value of fighter combat. Based on the framework of the original NSGA-III, the proposed algorithm is devised with several attractive features to enhance the optimization performance, including an improvement strategy of reference points and an online operator selection mechanism.

Improvement Strategy of Reference Points. We can see from studies [24, 26] that reference points of the original NSGA-III are uniformly distributed on a hyperplane to guide solutions to converge. The locations of these reference points are predefined, but the true Pareto front of the SMWTA problem is unknown beforehand. So the mismatches between the reference points and the true Pareto front may degrade the search ability of the algorithms. If appropriate reference points can be continuously generated during the evolution according to information provided by the current population, it will be possible to achieve a solution set with good performances. Therefore, we add the improvement strategy of reference points to the original NSGA-III algorithm, that is, continuously generating good reference points and eliminating useless reference points.

Online Operator Selection Mechanism. Crossover and mutation operators used in the evolutionary process of optimization with NSGA-III can generate offspring solutions to update the population and seriously affect search capability. The task of choosing the right operators depends on experience and knowledge about the problem. The online operator selection mechanism proposed in this paper aims to automatically select operators from a pool with simulated binary crossover, DE/rand/1, and nonlinear differential evolution crossover. Different crossover operators can be selected online according to the information of generations. Another benefit of this mechanism is that the operator choice can adapt to the search landscape and improve the quality of Pareto-optimal solutions.

The rest of this paper is organized as follows. Section 2 reviews the related work. In Section 3, a new mathematical model of the SMWTA problem and assumption descriptions are presented. Section 4 provides the introduction of the NSGA-III algorithm and presents the proposed I-NSGA-III for solving the WTA problem. Detailed improvements of the proposed algorithm are also introduced in Section 4. Section 5 is divided into two subsections as follows: (i) In order to verify the proposed algorithm, five state-of-the-art algorithms are considered for comparison studies. (ii) The proposed algorithm and others, like NSGA-III [9], MPACO [7], and NSGA-II [16], are tested on the SMWTA problem. Section 6 concludes the paper and presents a direction for future work.

2. Related Work

Many realistic problems contain several (two or more) conflicting objectives that are to be minimized or maximized simultaneously [32]. Most single-objective optimization problems can find only one solution (others may lack the appropriate conditions), but multiobjective optimization problems (MOPs) can find a set of Pareto solutions that consider all the objective functions and constraints. Generally, multiobjective optimization can be presented as follows [33]:

$$\begin{aligned} y &= f(x) = [f_1(x), f_2(x), \dots, f_m(x)], \\ e(x) &= [e_1(x), e_2(x), \dots, e_k(x)] \geq 0, \\ x &= (x_1, x_2, \dots, x_n) \in X, \\ y &= (y_1, y_2, \dots, y_m) \in Y, \end{aligned} \quad (1)$$

where X denotes the space of decision variables and Y denotes the space of objective functions.

There have been various studies on multiobjective evolutionary optimization besides NSGA-III, such as a set-based genetic algorithm for interval many-objective optimization problems, set-based many-objective optimization guided by a preferred region, a many-objective evolutionary algorithm using a one-by-one selection strategy, and many-objective evolutionary optimization based on reference points. The review on the related work will be further enriched if these studies are included.

2.1. Set-Based Evolutionary Optimization. The goal of a multiobjective evolutionary algorithm (MOEA) is to seek a Pareto solution set which is well converged, evenly distributed, and well extended. If a set of solutions and its performance indicators are taken as the decision variable and objectives of a new optimization problem, respectively, it is more likely that a Pareto-optimal set that satisfies the performance indicators will be obtained. Based on this idea, a many-objective optimization (MaOP) can be transformed into an MOP with two or three objectives, and then a series of set-based evolutionary operators are employed to solve the transformed MOP [34]. Compared with the traditional MOEAs, set-based MOEAs have two advantages: (i) the new objectives are used to measure a solution set and (ii) each individual of the set-based evolutionary optimization is a solution set consisting of several solutions of the original problem.

Researchers have carried out studies on set-based MOEAs, including the frameworks, the methods of transforming objectives, the approaches for comparing set-based individuals, and so on The first set-based MOEA was proposed by Bader et al. [35]. In their work, solutions in a population are firstly divided into a number of solution sets of the same size, and then the hypervolume indicator is adopted to assess the performance of those sets. In the method proposed by Zitzler et al. [36], not only is the preference relation between a pair of set-based individuals defined, but the representation of preferences, the design of the algorithm, and the performance evaluation are also incorporated into a framework. Bader et al. [35] presented a set-based Pareto dominance relation and designed a fitness function reflecting the decision maker's preference to effectively solve MaOPs. A comparison of results with traditional MOEAs shows that the proposed method is effective. Besides, Gong et al. [37] also presented a set-based genetic algorithm for interval MaOPs based on hypervolume and imprecision.

2.2. Local Search-Based Evolutionary Optimization. Ishibuchi and Murata [38] firstly proposed the study combining MOEA and local search method, IM-MOGLS for short. In their work, a weighted sum approach is used to combine all objectives into one objective. After generating offspring by genetic operators, a local search is conducted starting from each new individual, optimizing the combined objectives. Based on IM-MOGLS, Ishibuchi et al. [39] add more selectivity to its starting points. Ishibuchi and Narukawa [40] proposed a hybrid of NSGA-II and local search. Knowles and Corne [41] combined local search with PAES. The use of gradients appeared in the Pareto descent method (PDM) proposed by Harada et al. in [42] and in the work of Bosman [43]. One of the few applications of achievement scalarization functions (ASFs) in MOEA area was done by Sindhya et al. [44]. MOEA was combined with a rough set-based local search in the work of Santana-Quintero et al. [45]. A genetic local search algorithm for multiobjective combinatorial optimization (MOCO) was proposed by Jaszkiewicz [46]. Firstly, Pareto ranking and a utility function are applied to obtain the best solutions. Secondly, pairs of solutions are selected randomly to undergo recombination. Finally, local search is applied to offspring pairs. In their study, MOCO is used to successfully solve the traveling salesperson problem (TSP).

The above studies combine an MOEA with classical local search methods; however, none of them applies the local search method to theoretically identify poor solutions in a population. Abouhawwash et al. [47] proposed a new hybrid MOEA procedure that first identified poorly converged nondominated solutions and then improved it by using an ASF-based local search operator. They encouraged researchers to pay more attention to the Karush Kuhn Tucker proximity metric (KKTPM) and other theoretical optimality properties of solutions in arriving at better multiobjective optimization algorithms.

2.3. Reference Point-Based Evolutionary Optimization. Existing reference point-based approaches usually adopt only one reference point to represent the decision maker's ideal solution. Wierzbicki [48] firstly proposed a reference point approach in which the goal is to achieve a Pareto solution closest to a supplied reference point of aspiration level based on solving an achievement scalarization problem. Deb and Sundar [32] introduced the decision maker's preference to find a preferred set of solutions near the reference point. Decomposition strategies have also been incorporated into reference point approaches to find preferred regions in the method proposed by Mohammadi et al. [49].

Up to date, there is only a few researches on achieving the whole Pareto-optimal solution set by employing multiple reference points. Figueira et al. [50] proposed a parallel multiple reference point approach for MOPs. In their work, the reference points are generated by estimating the bounds of the Pareto front, and solutions near each reference point can be obtained in parallel. This priori method is very convenient; however, the later evolution process increases the computational complexity. Wang et al. [51] proposed a preference-inspired coevolutionary approach. Although solutions and reference points are optimized simultaneously during the evolution process, the fitness value of an individual is calculated by the traditional Pareto dominance. In the work done by Deb and Jain [24], a hyperplane covering the whole objective space is obtained according to the current population, then a set of well-distributed reference points are generated on the hyperplane. However, the Pareto fronts of most practical problems are not uniformly distributed in the whole objective space, and it is necessary to adopt reference points which are adaptive to various problems. Liu et al. [52] proposed a reference point-based evolutionary algorithm (RPEA) method. However, the value of δ which seriously affects the performance of the algorithm is a constant during evolution. In addition, the Tchebychev approach is only adopted in this study, and it may not be appropriate to all kinds of problems.

2.4. Indicator-Based Evolutionary Optimization. Compared with the above approaches, indicator-based evolutionary algorithms (IBEAs) [53] adopt a single indicator which accounts for both convergence and distribution performances of a solution set. Because solutions can be selected one by one based on the performance indicator, the algorithm is also called one-by-one selection evolutionary optimization. The hypervolume is usually adopted as the indicator in IBEAs. However, the computational complexity for calculating hypervolume increases exponentially as the number of objectives increases. So it is hard to be used to solve MaOPs. To address this issue, Bader and Zitzler [54] proposed an improved hypervolume-based algorithm—HypE. In their work, Monte Carlo simulations are applied to estimate the hypervolume. This method can save computational resources while ensuring the accuracy of the estimated hypervolume. In recent years, some algorithms [55, 56] are also proposed to enhance the computational efficiency of IBEAs for solving MaOPs.

Motivated by simultaneously measuring the distance of the solutions to the Pareto-optimal front, and maintaining a sufficient distance between each other, Liu et al. [57] proposed a many-objective evolutionary algorithm using a one-by-one selection strategy, 1by1EA for short. However, there are two issues in this algorithm. (i) In 1by1EA, the contour lines formed by the convergence indicator have a similar shape to that of the Pareto-optimal front of an optimization problem. However, the shape of the Pareto-optimal front of a practical optimization problem is frequently unknown beforehand. (ii) The algorithm does not include a mechanism that can adaptively choose an appropriate convergence indicator or use an ensemble of multiple convergence indicators during the evolution. Based on the above two issues, we have improved the original NSGA-III.

3. Problem Formulation

The WTA formation can be described as finding a proper assignment of weapon units to target units as illustrated in Figure 1. Some formulation of the problem, including the assumptions and the new three-objective mathematical model, are introduced in this section.

3.1. Assumption Description. In this research, to establish a reasonable WTA mathematical model, the following assumptions can be defined:

Assumption 1. We assume that the mathematical model is composed of F fighters, M missiles, and T targets and the opposing groups are not necessarily equal in quantity. (Each fighter is equivalent to one platform, which possess different kinds and quantities of missiles).

Assumption 2. Each fighter can use different missiles to attack one target. (Each missile can only attack one target).

Assumption 3. The distributed unit total of each type of missile cannot exceed the number of assigned missile unit resources in a military air operation.

Assumption 4. We assume that the probability of a kill, which is labeled as q_{ij}, between the missile (ith unit of M) and the unit being attacked (jth unit of T) is provided.

Assumption 5. If the target is within the work area, a missile can be assigned effectively. If not, the missile is not.

3.2. Mathematical Model. Multiobjective WTA optimization is used to seek a balance among the maximum expected damage, minimum missile consumption, and maximum combat value. Thus, definitions and constraints related to the optimization model are shown as follows.

Definition 1. One introduces a new objective—the value fighters engage in combat on the basis of the bi-objective WTA model that one established in the literature [23]. The multiobjective model of WTA is to maximize the total

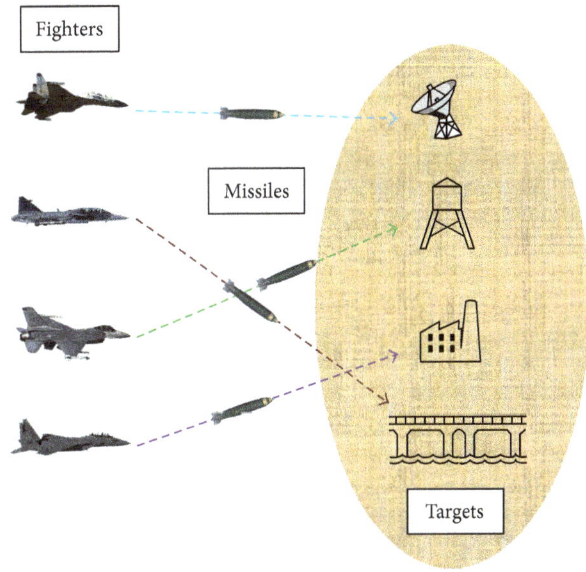

FIGURE 1: Illustration of the WTA problem.

effectiveness of attack, minimize the cost of missiles, and maximize the value of fighter combat. The mathematical functions of the model are shown as

$$
\begin{aligned}
f_1 &= \max \sum_{j=1}^{n}\left[1-\prod_{i=1}^{m}\left(1-\rho_k q_{ij}\right)^{x_{ij}}\right],\\
f_2 &= \min \sum_{i=1}^{m}\sum_{j=1}^{n} c_i x_{ij},\\
f_3 &= \max \sum_{k=1}^{s}\left(1-\frac{1}{A_k}w_k\right),\\
f &= \min\left(\frac{1}{f_1}, f_2, \frac{1}{f_3}\right).
\end{aligned}
\quad (2)
$$

Definition 2. A_k represents the number of missiles carried by the kth fighter, and w_k represents the number of missiles that have been launched on the kth fighter.

Definition 3. $q_{ij} \in (0,1)$ represents the kill probability of each $M(i=1,2,\ldots,m)$ missile attacking different $T(j=1,2,\ldots,n)$ targets.

Definition 4. The decision table $X=[x_{ij}]_{M\times T}$ can be described as Table 2, where x_{ij} is a Boolean value and represents whether i missile is assigned to j target. The relationship between the Boolean value and missile allocation is shown in Table 3.

Definition 5. The constant c_i represents the cost of ith missile in this paper.

Definition 6. Based on the real battlefield, we assume that the number of missiles per fighter is not more than 4.

TABLE 2: The decision table of WTA.

	T_1	T_2	T_3	$\cdots$	T_n
M_1	x_{11}	x_{12}	x_{13}	$\cdots$	x_{1n}
M_2	x_{21}	x_{22}	x_{23}	$\cdots$	x_{2n}
M_3	x_{31}	x_{32}	x_{33}	$\cdots$	x_{3n}
M_4	x_{41}	x_{42}	x_{43}	$\cdots$	x_{4n}
M_5	x_{51}	x_{52}	x_{53}	$\cdots$	x_{5n}
M_6	x_{61}	x_{62}	x_{63}	$\cdots$	x_{6n}
$\vdots$	$\vdots$	$\vdots$	$\vdots$	x_{ij}	$\vdots$
M_m	x_{l1}	x_{l2}	x_{l3}	$\cdots$	x_{mn}

TABLE 3: The relationship between Boolean value and missile allocation.

Missile allocation	Boolean value
i missile of k fighter unit is assigned to target j	$x_{ij}=1$
i missile of k fighter unit is not assigned to target j	$x_{ij}=0$

Definition 7. ρ_k is the pilot operation factor proposed by our previous article [23] ($\rho_k \in (0,1)$).

Constraint. Three constraints that the above function variables must satisfy are shown in Table 4.

4. Nondominated Sorting Genetic Algorithm III

4.1. Introduction of NSGA-III. The basic framework of the NSGA-III algorithm remains similar to the NSGA-II algorithm with significant changes in its mechanism [24]. But unlike in NSGA-II, a series of reference points are introduced to improve the diversity and convergence of NSGA-III. The proposed improved NSGA-III is based on the structure of NSGA-III; hence, we give a description of NSGA-III here.

The algorithm starts with an initial population (feasible solutions) of size N and a series of widely distributed G-dimensional reference points. p represents the division and is given by the user. Das and Dennis's systematic approach [58] is used to place reference points on the normalized hyperplane having an intercept of one on each axis. The total number of reference points (H) in an M-objective problem is given by

$$H=\binom{G+p-1}{p}. \tag{3}$$

The NSGA-III uses a set of reference directions to maintain diversity among solutions. A reference direction is a ray starting at the origin point and passing through a reference point, as illustrated in Figure 2. The population size N is chosen to be the smallest multiple of four greater than H, with the idea that for every reference direction, one population member is expected to be found [30].

Let us suppose the parent population at the generation t is P_t (of size N). The offspring population Q_t having N members is obtained by recombination and mutation of P_t. C_t is the combination of parent and offspring population (of size $2N$). To preserve elite members, C_t is sorted to different nondomination levels ($L_1, L_2, \ldots, L_l$). Thereafter, individuals of each nondomination level are selected to construct a new population S_t, starting from the first nondomination level L_1, until the size of $S_t \geq N$ (the first time larger than N). Suppose that the last level included is the lth level. In most situations, individuals of the lth level are only sorted partially by the diversity maintenance operator. This is achieved by computing crowding distance for the lth level in NSGA-II. But the NSGA-III replaces it with reference direction-based niching. Before the above operation, objective values are normalized by formulas (4), (5), and (6).

$$f'_g(x)=f_g(x)-z_g^{\min}, \tag{4}$$

$$\mathrm{ASF}(x,w)=\frac{\max_{g=1}^{G} f'_g(x)}{w_g}, \tag{5}$$

$$f_g^n(x)=\frac{f'_g(x)}{a_g-z_g^{\min}}, \tag{6}$$

where the ideal point of the population S_t is determined by identifying the minimum value ($z_g^{\min}$), for each objective function $g=1,2,\ldots,G$ and by constructing the ideal point $\bar{z}=(z_1^{\min}, z_2^{\min}, \ldots, z_G^{\min})$. $f'_g(x)$ denotes the translated objective functions ($x \in S_t$). $\mathrm{ASF}(x,w)$ denotes the extreme point value in each objective axis, and w_g is the weight vector of each objective (when $w_g=0$, it will be replaced by a small number 10^{-6}). $f_g^n(x)$ denotes the normalized objective functions, and a_g represents the intercept of the gth objective axis.

After the normalizing operation, the original reference points calculated by formulas (4) and (6) lie on this normalized hyperplane. In order to associate each population member with a reference point, the shortest perpendicular distance between each individual of S_t and each reference line is calculated.

Finally, a niche-preservation strategy is employed to select individuals from lth level that are associated with each reference point. α_h represents the number of population members from $P_{t+1}=S_t/L_l$ connected to the hth reference point. The specific niche-preservation strategy is shown as follows:

(1) The reference point set $J_{\min}=\{h : \arg\min_h \alpha_h\}$ with the min α_h is identified. When $|J_{\min}|>1$, one ($\bar{h} \in J_{\min}$) is selected at random.

(2) According to the value of $\alpha_{\bar{h}}$, two cases are discussed.

   (i) $\alpha_{\bar{h}}=0$

      (a) If one or more members in front the lth level are associated with the $\bar{h}$th reference point,

Table 4: Detailed information about the constraints.

Constraint	Explanation	Formula
Missile assignment constraint	At most, one missile is assigned to one target	$\sum_{i=1}^{m} x_{ij} \geq 1 \quad \forall j = 1, 2, \ldots, n.$
Quantity assignment constraint	One missile can attack only one target	$\sum_{j=1}^{n} x_{ij} \leq 1 \quad \forall i = 1, 2, \ldots, m.$
Parameter constraint	The total allocation does not exceed the total number of missiles	$\sum_{i=1}^{m} \sum_{j=1}^{n} x_{ij} \leq m.$

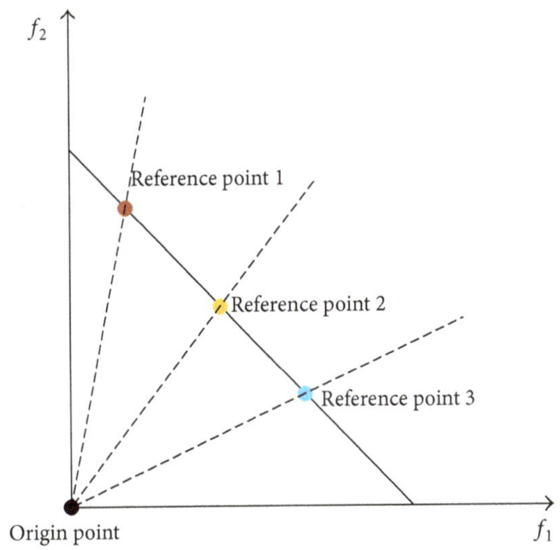

Figure 2: Three reference points are shown on a normalized reference line for a two-objective problem.

the one having the shortest perpendicular distance from the hth reference line is added to P_{t+1}. The count $\alpha_{\bar{h}}$ will also add one.

(b) If one member in the front lth level is associated with the hth reference point, the hth reference point will be ignored for the current generation.

(ii) $\alpha_{\bar{h}} \geq 1$

A randomly chosen member from the front lth level that is associated with the hth reference point is added to P_{t+1}, and the count $\alpha_{\bar{h}}$ is then incremented by one.

After counts are updated, the above procedure will be repeated for individuals to fill the entire P_{t+1}.

The flow chart of the NSGA-III algorithm is shown in Figure 3.

4.2. Improvement Strategy of Reference Points. In this section, an improvement strategy of reference points is proposed. The strategy mainly includes two parts: (i) generation of new reference points and (ii) elimination of useless reference points.

NSGA-III requires a set of reference points to be supplied before the algorithm can be applied [26]. If the user does not put forward specific requirements for the Pareto-optimal front, a structured set of points created by Das and Dennis's approach [58] is located on a normalized hyperplane. NSGA-III was originally designed to find Pareto-optimal points that are closer to each of these reference points, and the positions of the structured reference points are predefined at the beginning. However, the true Pareto front of a practical optimization problem (like the MWTA problem) is usually unknown beforehand, so the preset reference points may not reflect the development trend of the true Pareto-optimal front. In this study, a set of reference points with good performances in convergence and distribution are created by making full use of information provided by the current population. Due to the increase of the new reference points, the total number of reference points increases, and the computational complexity of the algorithm increases simultaneously. In order to keep the convergence speed of the algorithm, we propose the elimination mechanism of the reference point.

4.2.1. Generation of New Reference Points. We can learn from the above NSGA-III algorithm that the P_{t+1} population is created and the niche count α for different supplied reference points is updated after the niche operation. All reference points are expected to be useful in finding nondominated fronts ($\alpha = 1$ for every reference point). If the hth reference point is not associated with any population member, the niche count of the hth reference point (α_h) is zero. If the NSGA-III will never find an associated population member for the hth reference point, the hth reference point is considered useless. It is then better to replace the hth reference point with a new reference point that correctly reflects the direction of the Pareto-optimal front. However, we do not know if the hth reference point is eventually useful in advance. Under this circumstance, we simply add a set of reference points by adopting the formulas (7) and (8). The scale of new reference points equals the number of $\alpha = 0$ reference points.

$$h_{\text{new}}^{g} = \min_{x \in P_t} f_g(x) + \varepsilon_g, \tag{7}$$

$$\varepsilon_g - \delta\left(f_g^{\max} - f_g^{\min}\right), \tag{8}$$

where δ is a random number that belongs to $(0, 1)$ and $h_{\text{new}}^{g} = (h_{\text{new}}^{1}, h_{\text{new}}^{2}, \ldots, h_{\text{new}}^{G})$ denotes a new reference point for the objective g. $f_g^{\max}(x)$ and $f_g^{\min}(x)$ are the maximal and minimal values, respectively, of the gth objective.

The pseudo code can be demonstrated as follows:

In many cases, the total number of reference points will be greatly increased by the above operations, and many of the new reference points eventually become

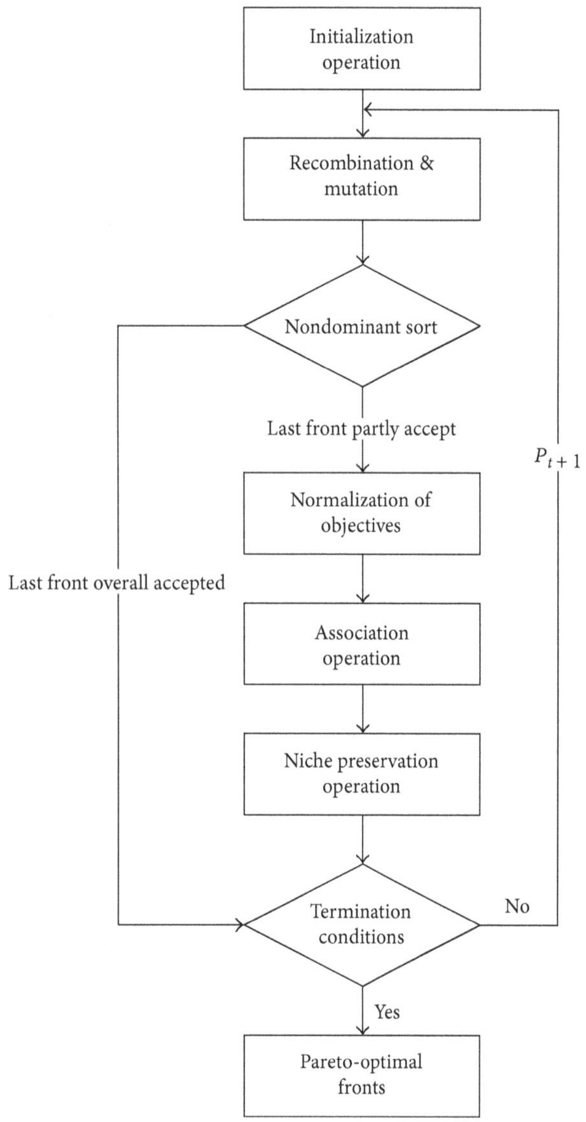

Figure 3: A flow chart of the NSGA-III algorithm.

useless. With the large increase of reference points, the algorithm will also be slowed down. Thus, we consider keeping the convergence speed of the algorithm by eliminating useless reference points, as described in the following subsection.

4.2.2. Elimination of Useless Reference Points. After new reference points are generated by the above operation, the niche count α of all reference points will be recalculated and recorded. Note that the total value of niche count α is equal to population size N (namely, $\sum_{h=1}^{H} \alpha_h = N$). Ideally, each reference point is exactly associated with one solution from the P_{t+1} population; that is, solutions are well-distributed among the reference points. Then, all reference points for $\alpha = 0$ are removed from the total reference point set H. However, in order to maintain uniform distribution of the reference points, the reference points obtained by Das and Dennis's systematic approach [58] will not be deleted. Thus, the existing reference points consist of two parts: (i) the original reference points (even if their niche count is zero) and (ii) all $\alpha = 1$ reference points.

Based on the niche count α of the respective reference points and information provided by the current population, the reference point set is adaptively redefined by generation and elimination operations. The improvement strategy for the reference points is intended to maintain diversity and guide the solutions closer to the Pareto-optimal front.

4.3. Online Operator Selection Mechanism. In multiobjective evolutionary algorithms (MOEAs), crossover and mutation operators are used to generate offspring solutions to update the population, and it can seriously affect search capability. In the original NSGA-III algorithm, only a simulated binary crossover (SBX) operator is adopted, and different crossover operators cannot be selected online according to the information of generations. Therefore, we propose a strategy based on the performance of previous generations to select operators adaptively from a pool, that is, an online operator selection (OOS) mechanism. According to a study in the literature [59], adaptive operator selection can select an appropriate strategy to adapt to the search landscape and produce better results.

Based on the information collected by the credit assignment methods, the OOS is applied to select operators for generating new solutions. In this paper, we use a probability method that uses a roulette wheel-like process for selecting an operator to solve the dilemma of exploration and exploitation.

Probability matching (PM) is one of the famous probability operator selection methods. The formula for calculating the probability of operator op being selected at next generation $t + 1$ is shown below [60]:

$$d_{\mathrm{op}}(t+1) = d_{\min} + (1 - K \cdot d_{\min}) \cdot \frac{\lambda_{\mathrm{op}}(t+1)}{\sum_{i=1}^{K} \lambda_i(t+1)}, \quad (9)$$

where K is the number of operators and $d_{\min}$ is the minimal probability of any operator. λ_{op} is the quality associated with operator op. Clearly, the sum of probabilities for all operators is 1 ($\sum_{i=1}^{K} \lambda_i(t+1) = 1$). If one operator gets rewards during many generations and the others' rewards are almost 0, its maximum selection probability $d_{\max}$ is equal to $1 + (1 - K)d_{\min}$.

The pool ζ in the proposed algorithm is composed of three well-known strategies: simulated binary crossover (SBX), DE/rand/1, and nonlinear differential evolution crossover (NDE). The parent individuals x_1, x_2, and x_3 are randomly selected from the population P_t in the original NSGA-III algorithm. The small difference in our algorithm is that the first parent individual x_1 of all strategies is equal to the current solution. Three operators are shown as below.

4.3.1. Simulated Binary Crossover. The simulated binary crossover (SBX) operator is proposed by Deb and Agrawal and is found to be particularly useful in problems where the upper and lower bounds of the multiobjective global optimum are not known a priori [61]. Two offspring solutions, y_1 and y_2, are created from two parent solutions, x_1 and x_2, by formulas (10) and (11) as follows:

$$y_1 = \frac{1}{2}[(1-\beta_1)x_1 + (1+\beta_1)x_2], \tag{10}$$

$$y_2 = \frac{1}{2}[(1+\beta_2)x_1 + (1-\beta_2)x_2], \tag{11}$$

where the spread factor β_i is defined as the ratio of the absolute difference in offspring values to that of the parents and is a random number; β_i can be calculated by the formula (12) as follows:

$$\beta_i = \begin{cases} (2\varphi_i)^{1/(\eta_c+1)}, & \varphi_i \leq \frac{1}{2}, \\ [2(1-\varphi_i)]^{-(1/(\eta_c+1))}, & \text{otherwise}, \end{cases} \tag{12}$$

where φ_i is a random number between [0, 1], and η_c is the crossover parameter given by the user.

4.3.2. DE/rand/1. The DE/rand/1 operator is one of the most commonly used DE variants [62], and all different solutions are randomly chosen from the population. So, this strategy does not generate biased or special search directions, and then a new direction is selected at random each time. The DE/rand/1 strategy can be defined by formula (13).

$$y = x_1 + U \cdot (x_2 - x_3), \tag{13}$$

where U represents the mutation scaling factor.

4.3.3. Nonlinear Differential Evolution Crossover. The nonlinear differential evolution crossover (NDE) strategy was presented in the literature [63] for the MOEA/D framework and was a hybrid crossover operator based on polynomials. The advantage of this strategy is that it ignores the values of the crossover rate and the mutation scaling factor. The offspring can be generated by formula (14).

$$y = \psi^2 \mu_1 + \psi \mu_2 + \mu_3, \tag{14}$$

where ψ is generated according to an interpolation probability (V_{itr}) [63] and can be defined by formula (15).

$$\psi = \begin{cases} \text{rand}[0,2], & V_{\text{itr}} \geq 1, \\ \text{rand}[2,3], & \text{otherwise}. \end{cases} \tag{15}$$

The parameters μ_1, μ_2, and μ_3 are given by formulas (16), (17), and (18) accordingly.

$$\mu_1 = \frac{x_3 - 2x_2 + x_1}{2}, \tag{16}$$

$$\mu_2 = \frac{4x_2 - 3x_3 - x_1}{2}, \tag{17}$$

$$\mu_3 = x_3. \tag{18}$$

4.4. The Proposed NSGA-III Algorithm. In this paper, we add the improvement strategy of reference points and online operator selection mechanism to the NSGA-III framework. The pseudo code of the proposed NSGA-III is shown in Algorithm 2, where differences with the original NSGA-III are set in italic and bold-italic.

Compared with the original NSGA-III, the proposed algorithm has two main parts. Bold-italic marks are the improvement strategy of the reference points, and italic marks are the online operator selection mechanism.

(1) In the bold-italic parts, the reference points that satisfy the condition (Section 4.2.2) are first removed according to the niche count α of all reference points. Second, new reference points are generated by referring the process of Algorithm 1. Based on the niche count α of each reference point and information provided by the current population, the reference point set is adaptively redefined by generation and elimination operations.

(2) There are two differences between our proposed algorithm and the original NSGA-III in the italic parts.

   (i) The first difference is the selection of the operator to be used (Step 4), which occurs based on the probabilities associated with each operator. With the success probability of an operator increasing, the most successful operator will be selected more often, and the quality of the solutions will also be improved theoretically. Because the same operator is selected by a deterministic approach during all recombination (Step 5) of the same generation, an undesirable bias should be introduced to select the best performance operator in the initial generations. So, stochastic selection mechanisms are applied in this paper.

   (ii) Other differences are the calculation of the rewards (Step 22) and the update of information with each operator (Step 23).

      (a) In this paper, the op in the pool ζ has associated a probability $d_{\text{op}}(t)$ of being selected at generation t by formula (9). The adapted process is based on its quality $\lambda_{\text{op}}(t)$, which is updated according to a reward $e_{\text{op}}(t)$.

      The reward $e_{\text{op}}(t)$ adopted in this paper is a Boolean value ($e_{\text{op}}(t) = \{0, 1\}$) in Step 22. If x is generated by the operator op, x does not belong to P_t but belongs to P_{t+1}; then, $e_{\text{op}}(t) = 1$; otherwise, $e_{\text{op}}(t)$ is equal to 0.

```
Input: P_{t+1}, α, δ
Output: h^g_new
After niche operation, P_{t+1} is created and α is updated;
for h^g_new = h^1_new → h^G_new do
    for g = 1 → G do
        According to formula (7) and (8), then generate the reference points h^g_new based on these selected individuals;
        check position h^g_new;
        if h^g_new lie on the first quadrant then
            check existence h^g_new;
        if h^g_new doesn't exist in then
            store h^g_new;
        end if
    end for
end for
Output h^g_new.
```

Algorithm 1: Generation of New Reference Points.

```
Input: P_0, I_max, N,H
Output: Pareto solutions
Initialize uniform distribution reference points H;
t = 0;
while termination conditions are not satisfied (t < I_max) do
        op←Select operator();
        C_t = Q_t ∪ Recombination & Mutation (P_t);
        (L_1, L_2, ..., L_l) = Non-dominated-sort (C_t);
        i = 1;
        repeat
            P_{t+1} = P_{t+1} ∪ L_i;
            i = i + 1;
        until |P_{t+1}| ≥ N;
        if |P_{t+1}| = N then
            break;
        else
            Normalize the objectives;
            Delete the useless reference points;
            Associate each solution in P_{t+1} with a reference point;
            Compute niche count α of reference points;
            Fill P_{t+1} with N − |P_{t+1}| solutions from L_i using niching information;
            Generate new reference points;
        end if
        Calculate operator rewards (P_t, P_{t+1});
        Update operator information ();
        t = t + 1;
end while
return Pareto-optimal front.
```

Algorithm 2: The Proposed NSGA-III Algorithm.

(b) After calculating the rewards, the quality $\lambda_{op}(t)$ of the operator op available in the pool can be updated by formula (19):

$$\lambda_{op}(t+1) = (1-\theta)\cdot\lambda_{op}(t) + \theta\cdot e_{op}(t), \quad (19)$$

where θ is the adaption rate ($\theta \in (0,1)$).

Finally, the operator selection probabilities are updated by formula (9).

5. Experiment Results and Analysis

5.1. Test Problem. In order to verify the proposed algorithm, five state-of-the-art algorithms are considered for comparison studies. They are NSGA-III-OSD [64], NSGA-III [24], NSGA-

II [65], MPACO [23], and MOEA/D [66]. NSGA-III and NSGA-II are the traditional NSGA algorithms. NSGA-III-OSD is an improved version of the NSGA-III based on objective space decomposition. In MOEA/D algorithm, a predefined set of weight vectors is applied to maintain the solution diversity. MPACO is the algorithm we proposed before. All algorithms are tested with 4 different benchmark MaOP problems known as DTLZ1 to DTLZ4 [67].

For DTLZ, 4 instances (DTLZ1–4) with 3, 5, 8, 10, and 15 objectives are used. Based on the work of Deb et al. [67], the number of decision variables is set as $DV = G + r - 1$, where $r = 5$ for DTLZ1 and $r = 10$ for DTLZ2–4. According to the work [68], the number of decision variables is set as $DV = p\ v + dv$. The parameters the position-related variable pv and the distance-related value dv are set to $2 \times (G - 1)$ and 20, respectively. The main characteristics of all benchmark problems are shown in Table 5.

In this subsection, inverted generational distance (IGD) indicator [69] is used to evaluate the quality of a set of obtained nondominated solutions. This indicator can measure the convergence and diversity of solutions simultaneously. Smaller IGD values indicate that the last nondominated population has better convergence and coverage of the Pareto front. Each algorithm is conducted 30 runs independently on each test problem. Aiming to be as fair as possible, in each run, all the comparison algorithms perform the same maximum iteration I_{max} as shown in Table 6. The population size used in this study for different numbers of objectives is shown in Table 7.

The six algorithms considered in this study need to set some parameters, and five of them are shown in Table 8. The parameters of the MPACO algorithm can be found in the literature [23].

Comparison results of I-NSGA-III with five other MOEAs in terms of IGD values on different objectives of DTLZ1–4 test problems are presented in Tables 9–12. It shows both the median and standard deviation of the IGD values on 30 independent runs for the six compared MOEAs, where the best median and standard deviation are highlighted in italic.

Based on the statistical results of DTLZ1, we can see that I-NSGA-III shows better performance than the other five MOEAs on three-, eight- and ten-objective test problems. For five- and fifteen-objective problems, it achieves the second smallest IGD value. Furthermore, NSGA-III can obtain the best IGD value on the fifteen-objective test problem and MPACO can obtain the best IGD value on the five-objective test problem. NSGA-II can deal with three-objective instances but works worse on more than three objectives.

Based on the statistical results of DTLZ2, we can see that the performance of I-NSGA-III, NSGA-III-OSD, and MOEA/D is comparable in this problem. The NSGA-III is worse than two improved algorithms (I-NSGA-III and MOEA/D) and MOEA/D, however, is better than MPACO. Based on the IGD values obtained by NSGA-II on three- and five-objective problems, we find that this algorithm can perform well. But when the number of objectives increases, NSGA-II still works worst among the six algorithms.

Based on the statistical results of DTLZ3, we find that I-NSGA-III performs significantly best among six algorithms on different objectives of the DTLZ3 problem. NSGA-III-OSD can achieve a similar IGD value as NSGA-III. Although MOEA/D can achieve the close to smallest IGD value on the three- and five-objective test problems, it significantly worsens on more than five-objective test problems. MPACO can defeat MOEA/D on eight-, ten- and fifteen-objective test problems. In addition, NSGA-II is still helpless for more than three-objective problems.

Based on the statistical results of DTLZ4, we find that I-NSGA-III performs significantly better than the other five MOEAs on almost all DTLZ4 test problems, except the eight-objective instance. Furthermore, the standard deviation of IGD obtained by I-NSGA-III shows that the proposed

Table 5: Characteristics of test problems.

Test problem	Characteristics
DTLZ1	Linear, multimodal
DTLZ2	Concave
DTLZ3	Concave, multimodal
DTLZ4	Concave, biased

Table 6: Maximum iteration for each test problem.

	The number of objectives				
	3	5	8	10	15
DTLZ1	400	600	750	1000	1500
DTLZ2	250	350	500	750	1000
DTLZ3	1000	1000	1000	1500	2000
DTLZ4	600	1000	1250	2000	3000

Table 7: Number of population size.

The number of objectives	The size of populations	Divisions
3	92	{12}
5	212	{6}
8	156	{3,2}
10	276	{3,2}
15	136	{2,1}

Table 8: Parameter setting of each algorithm.

	η_c	η_m	p_c	p_m	Other parameters
I-NSGA-III			—		$d_{min} = 0.1$ $U = 0.5$
	30				$\mu = 0.2$
NSGA-III-OSD [64]		20		$1/DV$	$\mu = 0.7$
			1.0		$\theta = 5$
NSGA-III [24]					—
NSGA-II [65]					—
MOEA/D [66]	20				$T = 20$ $\delta = 0.9$ $\theta = 5$

Table 9: Median and standard deviation of the IGD values achieved by each algorithm on DTLZ1 (the best medians are in italic font).

	G	I-NSGA-III	NSGA-III-OSD	NSGA-III	MPACO	MOEA/D	NSGA-II
DTLZ1	3	*3.687e − 04*	4.087e − 04	4.056e − 04	7.063e − 04	5.709e − 04	1.861e − 03
		7.68e − 05	2.14e − 04	1.76e − 03	1.25e − 04	5.75e − 04	1.16e − 02
	5	3.492e − 04	2.863e − 03	3.119e − 03	*3.428e − 04*	3.781e − 04	2.321e − 02
		1.10e − 04	2.45e − 03	2.62e − 03	*4.06e − 03*	7.05e − 05	3.22e − 03
	8	*2.559e − 03*	3.125e − 03	3.356e − 03	3.863e − 03	3.414e − 03	4.023e − 02
		2.280e − 04	2.32e − 03	4.31e − 03	3.29e − 03	2.50e − 02	1.26e − 01
	10	*2.942e − 03*	3.243e − 03	3.397e − 03	4.163e − 03	4.596e − 03	5.873e − 02
		3.62e − 04	1.28e − 02	2.87e − 03	6.44e − 03	6.54e − 03	4.13e − 02
	15	2.212e − 02	2.315e − 02	*2.080e − 02*	2.273e − 02	2.271e − 02	4.873e + 01
		2.15e − 03	2.76e − 02	*2.87e − 04*	2.15e − 02	6.56e − 03	3.32e + 01

Table 10: Median and standard deviation of the IGD values achieved by each algorithm on DTLZ2 (the best medians are in italic font).

	G	I-NSGA-III	NSGA-III-OSD	NSGA-III	MPACO	MOEA/D	NSGA-II
DTLZ2	3	*4.921e − 03*	4.944e − 03	5.212e − 03	6.898e − 03	5.021e − 03	7.915e − 03
		6.09e − 05	1.45e − 04	1.08e − 04	1.94e − 03	1.01e − 04	1.865e − 03
	5	*2.540e − 02*	2.549e − 02	2.593e − 02	2.821e − 02	2.548e − 02	3.815e − 02
		3.17e − 04	4.24e − 05	3.37e − 04	2.25e − 03	1.58e − 03	9.29e − 03
	8	*3.438e − 02*	3.459e − 02	3.461e − 02	4.045e − 02	3.451e − 02	6.992e − 02
		2.01e − 03	1.46e − 03	1.23e − 03	1.18e − 02	3.43e − 03	4.97e − 02
	10	3.563e − 02	3.573e − 02	3.581e − 02	4.092e − 02	*3.552e − 02*	7.865e − 02
		1.56e − 03	1.322e − 03	2.19e − 03	8.63e − 03	*9.83e − 04*	7.65e − 03
	15	4.638e − 02	*4.618e − 02*	4.656e − 02	5.252e − 02	4.622e − 02	3.561e + 01
		2.22e − 03	*2.23e − 03*	1.89e − 03	8.50e − 03	9.99e − 04	1.42e + 01

Table 11: Median and standard deviation of the IGD values achieved by each algorithm on DTLZ3 (the best medians are in italic font).

	G	I-NSGA-III	NSGA-III-OSD	NSGA-III	MPACO	MOEA/D	NSGA-II
DTLZ3	3	*6.164e − 03*	6.456e − 03	6.353e − 03	8.129e − 03	6.192e − 03	8.092e − 03
		1.10e − 04	6.54e − 03	4.14e − 03	1.96e − 03	5.33e − 04	2.74e − 02
	5	*2.639e − 02*	2.783e − 02	2.679e − 02	2.896e − 02	2.658e − 02	6.013e − 02
		7.93e − 05	6.12e − 03	3.77e − 03	3.83e − 03	5.75e − 04	5.18e − 01
	8	*3.582e − 02*	3.632e − 02	3.812e − 02	4.040e − 02	4.378e − 02	6.816e − 02
		1.65e − 03	7.12e − 03	8.49e − 03	6.91e − 03	1.69e − 01	1.63e − 01
	10	*3.666e − 02*	3.706e − 02	3.797e − 02	4.260e − 02	5.684e − 02	8.562e − 02
		1.11e − 03	3.25e − 03	5.74e − 03	1.12e − 02	6.79e − 02	2.11e − 01
	15	*4.636e − 02*	6.509e − 02	6.488e − 02	5.403e − 02	7.613e − 02	9.866e + 01
		2.13e − 03	8.32e − 02	3.42e − 02	6.25e − 02	2.04e − 01	7.11e + 01

Table 12: Median and standard deviation of the IGD values achieved by each algorithm on DTLZ4 (the best medians are in italic font).

	G	I-NSGA-III	NSGA-III-OSD	NSGA-III	MPACO	MOEA/D	NSGA-II
DTLZ4	3	*6.165e − 03*	7.171e − 03	1.436e − 02	1.851e − 02	5.514e − 02	6.411e − 02
		2.14e − 05	3.45e − 04	3.73e − 03	2.13e − 01	2.64e − 03	3.58e − 01
	5	*2.696e − 02*	2.919e − 02	2.733e − 02	3.057e − 02	4.622e − 02	5.536e − 02
		6.38e − 04	6.51e − 04	6.57e − 04	1.58e − 03	1.53e − 01	1.99e − 01
	8	4.747e − 02	*4.74e − 02*	4.753e − 02	5.222e − 02	7.999e − 02	9.623e − 02
		9.73e − 04	*8.56e − 04*	9.83e − 04	6.75e − 02	2.46e − 02	1.25e − 01
	10	*4.535e − 02*	4.621e − 02	5.09e − 02	5.899e − 02	7.92e − 02	8.478e − 02
		1.53e − 04	6.19e − 05	1.87e − 02	6.02e − 03	1.87e − 02	9.56e − 02
	15	*5.831e − 02*	5.893e − 02	6.777e − 02	6.581e − 02	9.01e − 02	9.969e − 02
		1.21e − 02	1.37e − 02	3.91e − 02	1.46e − 02	3.52e − 01	7.48e − 02

algorithm is rather robust. NSGA-III-OSD shows the closest overall performance to I-NSGA-III. MOEA/D is significantly worse than MPACO. NSGA-II still has no advantage on the DTLZ4 test problem.

The above results show that the proposed I-NSGA-III could perform well on almost all the instances in DTLZ1–4, and the obtained solution sets have good convergence and diversity. In the next subsection, the proposed algorithm will be applied to solve an SMWTA problem.

5.2. SMWTA Problem

5.2.1. Parameter Setting. For the MWTA problem, the population size N of the I-NSGA-III algorithm is 150, and the maximum number of iterations $I_{\max}$ is 200. According to the approach in the literature [58], the total number of reference points H is set to 120. The DE scaling factor U, polynomial mutation probability p_m, and PM minimum probability $d_{\min}$ are used in this work, as they are frequently used in the literature [68]. Some parameters for the I-NSGA-III are shown in Table 13.

Table 14 shows the number of reference points H, population size N, and maximum iteration $I_{\max}$ for different algorithms. Other parameters in the NSGA-III algorithm are the same as those in the literature [24]. The parameters of the MPACO algorithm and the NSGA-II algorithm can be found in the literature [23].

Table 13: Parameters of I-NSGA-III algorithm (Part).

Parameter	Value
η_c	30
η_m	20
$d_{\min}$	0.1
p_m	0.07
U	0.5
H	120
N	150
$I_{\max}$	200

Table 14: Number of reference points, population size, and maximum number of iterations for all algorithms.

	I-NSGA-III	NSGA-III	MPACO	NSGA-II
Reference points (H)	120	120		
Population size (N)	150	120	150	150
Maximum iteration($I_{\max}$)	200			

5.2.2. Simulation Environment. We can see from Table 15 that the proposed algorithm has been implemented in C++ on a CPU Intel(R) Core(TM) i5-4460T with 1.90 GHz and 8 GB of RAM. The operating system is Windows 7 64-bit.

5.2.3. Numerical Experiments and Analysis. We use the same specific instance as in our previous work [23] to verify the performance of the algorithm. The instance includes 4 fighters that carry different numbers of missiles (12 missiles in total) and 10 targets. Appendix A shows missile damage probability $p_{ij}(i = 1, \ldots, 12.j = 1, \ldots, 10)$, pilot operation factor $\rho_k(k = 1, \ldots, 4)$, and cost of missiles c_i.

First, an enumeration approach [19] is employed to get a set of evenly distributed true optimal solutions and thus obtain evenly distributed true Pareto solutions (PSs) for the specific instance. Second, in order to verify the applicability and feasibility of the proposed algorithm, we apply I-NSGA-III, NSGA-III, MPACO, and NSGA-II to find PSs in the instance. The statistical results are shown in Figures 4–7.

Figures 1–4 show the distribution of the true PSs solved by the enumeration method and final PSs obtained by I-NSGA-III, NSGA-III, MPACO, and NSGA-II. We can see that the optimization results of the I-NSGA-III algorithm are obviously better than those of the other algorithms because the I-NSGA-III can find close and even PSs in the objective space. It is evident that the algorithm can guarantee the quality of solutions. So the I-NSGA-III for optimizing the SMWTA is well verified to be feasible.

In Figure 1, 150 evenly distributed solutions in the real Pareto front are found, while only 120 reference points are preset. Since the reference point and the solution are one-to-one correspondence, the effectiveness of the improvement strategy of reference points—generating new reference points and eliminating useless reference points operations—is demonstrated. This strategy increases the number of reference points from 120 to 150, improves the efficiency of the algorithm, and finds more Pareto-optimal solutions that meet the requirements. Meanwhile, due to the adoption of the online operator selection mechanism in the I-NSGA-III algorithm, the search landscape is reduced and the quality of the solutions is improved. However, in Figure 2, the original NSGA-III algorithm can only find 95 solutions on the premise of the initial 120 reference points, and only a small number of individuals are located at the Pareto front. Therefore, the original NSGA-III is less efficient than the algorithm proposed in this paper for solving the SMWTA problem. As we can see from Figures 3 and 4, although MPACO is superior to NSGA-II to some extent, the two algorithms are obviously inferior to the I-NSGA-III algorithm.

We analyze results from Appendix B and Appendix C as follows:

When funds for national defense are sufficient and detailed enemy information are available, we can choose Scheme 1, which costs the most money and obtains the greatest expected damage to the enemy to complete a fatal attack. As we are in a repressive state of military power, we can accomplish the task with only one attack in Scheme 1. However, this situation is rare in real combat. When we have only a small amount of information about the enemy or it is difficult to launch a large-scale attack, we choose one scheme among Schemes 147, 149, and 150 that can achieve maximum fighter combat value to launch a probing attack. In these three schemes, taking into account the least cost in terms of money and the greatest expected damage value, we should choose Scheme 150.

Table 15: Parameters of simulation environment.

Computer specifications	Intel(R) Core(TM) i5-4460T CPU at 1.90 GHz with 8.00 GB RAM
Operating system	Windows 7 (x64) operating system
Language	C++
Software	Microsoft Visual C++ 6.0

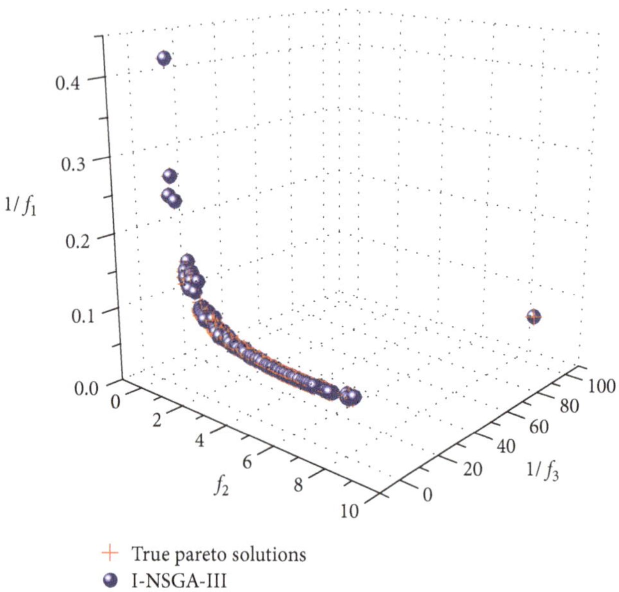

Figure 4: Plots of the true PSs and the final PSs found by I-NSGA-III on the specific instance.

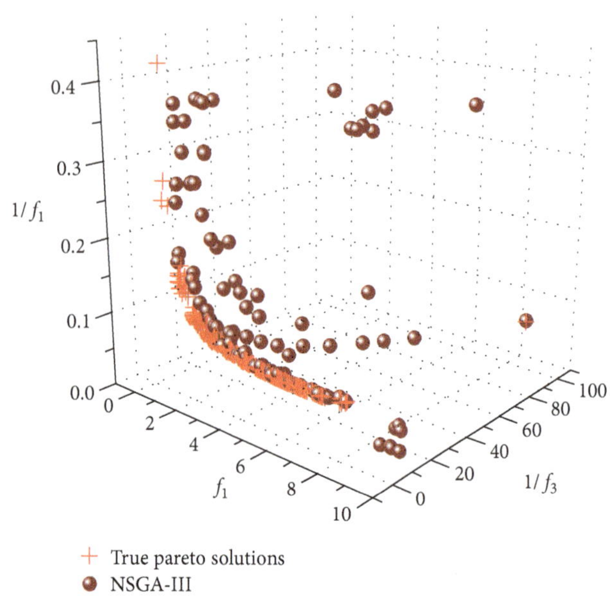

Figure 5: Plots of the true PSs and the final PSs found by NSGA-III on the specific instance.

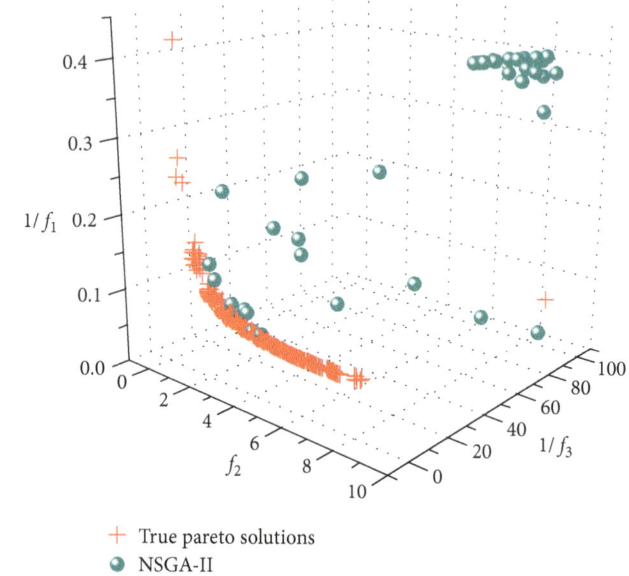

Figure 6: Plots of the true PSs and the final PSs found by NSGA-II on the specific instance.

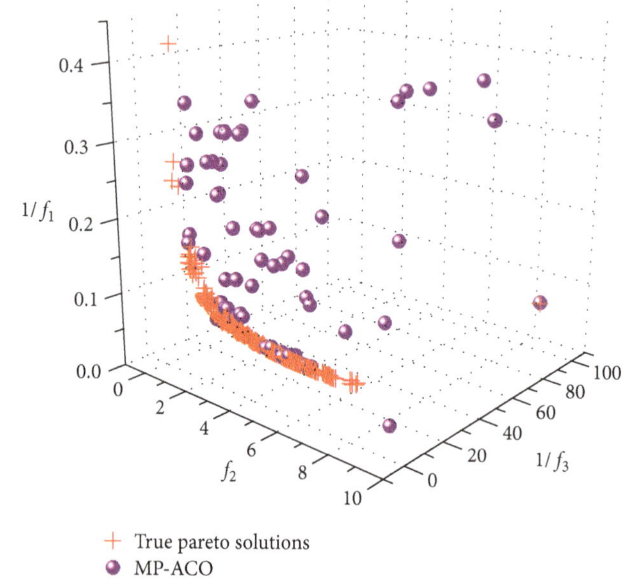

Figure 7: Plots of the true PSs and the final PSs found by MPACO on the specific instance.

Considering that funds are insufficient, we can only choose Scheme 148. Considering that all targets must be allocated and that the cost of missiles should be minimized, we can only choose Scheme 4.

Solving the SMWTA problem is the foundation of the dynamic multiobjective weapon-target assignment problem (DMWTA). The goal of the DMWTA is to provide a set of near-optimal or acceptable real-time decisions in real air combat. So, the time performance of algorithms is also an important index. In the end part, we test four algorithms on the specific instance in 30 runs and record the iteration

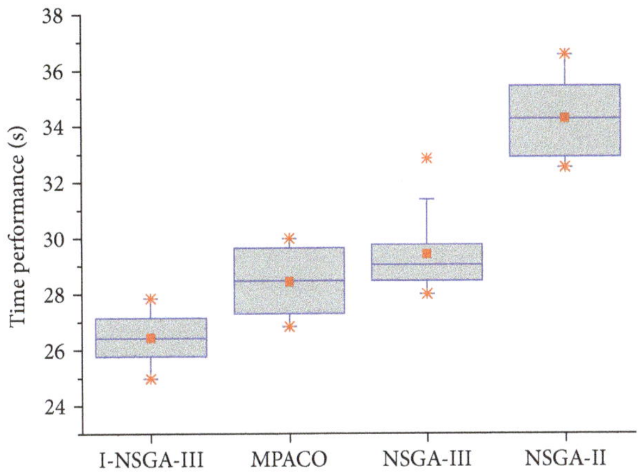

FIGURE 8: Time performance of four different algorithms. * represent extreme outliers.

time of each algorithm. The statistical results of time performances are shown in Figure 8.

In real air combat situations, pilots often make deadly decisions within seconds or even within milliseconds. We can see from Figure 5 that I-NSGA-III has a time advantage in solving the special instance compared with other algorithms, and the time performance of the NSGA-II is worst among the four algorithms. Although improvement of the strategy of reference points affects the iteration rate, an appropriate strategy can be selected by an online operator selection mechanism to improve the mutation efficiency and the quality of solutions. Compared with the original NSGA-III algorithm, the improvement strategy of reference points plays a more important role than the online operator selection mechanism in the time performance field.

In this section, we do some work as follows: Firstly, we use four classic test problems (DTLZ1–4) to evaluate the proposed algorithm and compare it with the other five state-of-the-art algorithms. Secondly, we test the four different algorithms on a specific example, verify the applicability and feasibility of the proposed algorithm, give a comparison study among the four algorithms, and show the corresponding distribution results in Appendix B and Appendix C. Thirdly, we show the time performance of four algorithms in 30 runs. To summarize, I-NSGA-III has been proved to be an effective technique for the SMWTA optimization problem and is obviously the best among the four algorithms.

6. Conclusion

We apply NSGA-III to the WTA problem and propose the I-NSGA-III to solve SMWTA in this paper. The main contributions of the thesis are summarized as follows: on the one hand, the expected damage to the enemy and the cost of missiles are taken into account from a practical viewpoint; in terms of the other objective—the value of fighter combat was introduced to make the model in line with real air combat. In this paper, an improvement strategy of reference points and an online operator selection mechanism are proposed and embedded into the original NSGA-III algorithm to improve the performances of the I-NSGA-III algorithm. The experiments have shown that I-NSGA-III can find better Pareto solutions than the other three algorithms for the SMWTA problem. More importantly, I-NSGA-III is more suitable for solving the problem from the time performance viewpoint.

However, we have mainly studied the SMWTA problem; few studies have focused on dynamic problems, which are more instructive to real air combat. In recent years, more and more studies have begun to pay attention to DWTA problems. A further study on this topic is one of our future tasks.

Appendix

Appendix A contains three data tables. These three tables come from our previous work [23] and are used to express the data used in the instance to verify the performance of the proposed algorithm. In Table 16, each data represents the damage probability of different missiles attacking different targets. In Table 17, ρ_k is the pilot operation factor and represents the talent, training time, and operation stability of the pilot which may affect the attack performance. In Table 18, c_i represents the cost of ith missile (c_i). The larger the value is, the more the missile costs.

The data in Appendix B represent the results and the corresponding distribution by I-NSGA-III. In each scheme, the first three columns represent the values of the three objective functions, and the latter twelve columns represent the corresponding results of the WTA distribution. (As an example, in Scheme 33, $1/f_1 = 0.07168$, $f_2 = 5.72$, and $1/f_3 = 0.70588$. Missile 2 is assigned to Target 3, Missiles 3 and 4 are assigned to Target 7, Missiles 6 and 10 are assigned to Target 9, Missiles 8 and 12 are assigned to Target 5, and Missile 9 is assigned to Target 1).

In order to compare the statistical results of all algorithms, the statistical results obtained by NSGA-III, MPACO, and NSGA-II are given in Appendix C and shown in Figures 6–8.

A. The Value of the Specific Example Used in This Paper

TABLE 16: Missile damage probability.

Missile unit	Target 1	2	3	4	5	6	7	8	9	10
1	0.82	0.85	0.78	0.75	0.52	0.88	0.44	0.76	0.72	0.56
2	0.56	0.72	0.88	0.46	0.64	0.47	0.68	0.45	0.48	0.75
3	0.45	0.62	0.54	0.73	0.84	0.76	0.78	0.42	0.53	0.65
4	0.56	0.42	0.76	0.84	0.73	0.83	0.86	0.62	0.78	0.82
5	0.45	0.58	0.81	0.44	0.63	0.59	0.78	0.77	0.65	0.70
6	0.46	0.61	0.55	0.68	0.75	0.83	0.73	0.66	0.82	0.48
7	0.66	0.71	0.65	0.44	0.86	0.79	0.59	0.85	0.53	0.56
8	0.56	0.42	0.76	0.84	0.73	0.72	0.44	0.75	0.48	0.47
9	0.88	0.78	0.44	0.67	0.56	0.86	0.58	0.65	0.73	0.42
10	0.56	0.88	0.68	0.45	0.75	0.73	0.61	0.76	0.84	0.78
11	0.84	0.54	0.44	0.42	0.65	0.56	0.71	0.55	0.45	0.88
12	0.83	0.76	0.84	0.62	0.82	0.75	0.42	0.68	0.57	0.54

TABLE 17: Pilot operation factor table.

Fighter		F_1			F_2			F_3			F_4	
Missile unit	M_1	M_2	M_3	M_4	M_5	M_6	M_7	M_8	M_9	M_{10}	M_{11}	M_{12}
p_k		0.96			0.95			0.98			0.93	

TABLE 18: The cost of each missile.

Missile unit	M_1	M_2	M_3	M_4	M_5	M_6	M_7	M_8	M_9	M_{10}	M_{11}	M_{12}
c_i	0.62	0.63	0.69	0.80	0.72	0.90	0.96	0.68	0.72	0.65	0.66	0.65

B. The Value of PSs and Results of WTA Distribution

TABLE 19

	$1/f_1$	f_2	$1/f_3$	M_1	M_2	M_3	M_4	M_5	M_6	M_7	M_8	M_9	M_{10}	M_{11}	M_{12}
1	0.06822	8.68000	100.00000	9	3	5	7	7	9	5	5	1	9	7	3
2	0.06860	8.02000	3.00000	9	3	7	7	7	9	5	3	1	9	0	5
3	0.06853	8.00000	4.00000	9	3	5	7	7	9	5	0	1	9	7	3
4	0.06867	7.96000	2.00000	9	3	7	7	0	9	5	3	1	9	7	5
5	0.06879	7.88000	2.00000	9	3	7	0	7	9	5	3	1	9	7	5
6	0.06880	7.72000	4.00000	9	3	5	7	7	9	0	3	1	9	7	5
7	0.06912	7.34000	1.71429	9	3	5	7	7	9	5	0	1	9	0	3
8	0.06917	7.37000	1.50000	9	3	5	7	7	9	5	3	1	9	0	0
9	0.06924	7.30000	1.20000	9	3	7	7	0	9	5	3	1	9	0	5
10	0.06936	7.12000	1.71429	9	3	5	7	7	0	5	3	1	9	0	9
11	0.06956	7.27000	1.20000	9	3	0	7	0	9	5	3	1	9	7	5
12	0.06956	7.06000	1.33333	9	3	7	7	0	0	5	3	1	9	7	5
13	0.06957	7.23000	1.20000	9	3	7	0	7	9	5	3	1	0	7	5
14	0.06961	7.22000	1.20000	9	3	7	0	7	9	5	3	1	9	0	5
15	0.06973	7.16000	1.00000	9	3	7	0	0	7	5	3	1	9	7	5
16	0.06987	6.92000	1.33333	9	3	5	0	7	7	0	3	1	9	7	5
17	0.06991	6.69000	1.09091	9	3	5	7	7	9	5	0	1	0	0	3
18	0.06996	6.72000	1.00000	9	3	5	7	7	9	5	3	1	0	0	0
19	0.06997	6.82000	2.00000	9	3	5	7	9	0	0	5	1	9	7	3
20	0.07002	6.44000	1.20000	9	3	5	7	7	0	5	0	1	9	0	3
21	0.07004	6.65000	0.85714	9	3	7	7	0	9	5	3	1	0	0	5
22	0.07015	6.40000	0.92308	9	3	7	7	0	0	5	3	1	9	0	5
23	0.07041	6.57000	0.85714	9	3	7	0	7	9	5	3	1	0	0	5
24	0.07053	6.32000	0.92308	9	3	7	0	7	0	5	3	1	9	0	5
25	0.07056	6.35000	0.92308	9	3	5	7	0	9	0	5	1	0	7	3
26	0.07060	6.10000	1.00000	9	3	5	7	0	0	0	3	1	9	7	5
27	0.07087	6.51000	0.75000	9	3	7	0	0	9	5	5	1	0	7	3
28	0.07094	6.26000	0.80000	9	3	7	0	0	0	5	3	1	9	7	5
29	0.07130	6.00000	0.66667	9	3	7	7	0	9	5	5	1	0	0	0
30	0.07142	5.75000	0.70588	9	3	7	7	0	0	5	5	1	9	0	0
31	0.07152	5.94000	0.70588	9	3	0	7	0	9	5	0	1	0	7	5
32	0.07165	5.67000	0.75000	9	3	5	7	0	9	0	0	1	0	7	5
33	0.07168	5.72000	0.70588	0	3	7	7	0	9	0	5	1	9	0	5
34	0.07169	5.92000	0.66667	9	3	7	0	7	9	5	5	1	0	0	0

TABLE 19: Continued.

	$1/f_1$	f_2	$1/f_3$	M_1	M_2	M_3	M_4	M_5	M_6	M_7	M_8	M_9	M_{10}	M_{11}	M_{12}
35	0.07192	5.45000	0.75000	9	3	5	7	0	0	0	5	1	9	7	0
36	0.07200	5.83000	0.63158	9	3	7	0	0	9	5	0	1	0	7	5
37	0.07211	5.41000	0.75000	9	3	0	7	0	0	0	5	1	9	7	5
38	0.07212	5.58000	0.66667	9	3	7	0	0	0	5	0	1	9	7	5
39	0.07224	5.61000	0.63158	9	3	7	0	0	0	5	5	1	9	7	0
40	0.07247	5.55000	0.63158	9	3	7	0	0	9	0	5	1	0	7	5
41	0.07278	5.88000	0.60000	9	0	7	0	0	9	5	5	1	0	7	3
42	0.07306	5.32000	0.57143	9	3	7	7	0	9	5	0	1	0	0	0
43	0.07319	5.07000	0.60000	9	3	7	7	0	0	5	0	1	9	0	0
44	0.07341	5.29000	0.57143	9	3	0	7	0	9	5	0	1	0	7	0
45	0.07347	5.24000	0.57143	9	3	7	0	7	9	5	0	1	0	0	0
46	0.07374	5.17000	0.52174	9	3	7	0	0	7	5	0	1	9	0	0
47	0.07400	5.02000	0.60000	9	3	5	7	0	9	0	0	1	0	7	0
48	0.07404	4.93000	0.54545	9	3	7	0	0	0	5	0	1	9	7	0
49	0.07413	4.77000	0.63158	9	3	5	7	0	0	0	0	1	9	7	0
50	0.07451	4.79000	0.60000	9	3	5	7	0	0	0	5	1	9	0	0
51	0.07490	4.87000	0.54545	9	3	5	0	0	7	0	0	1	9	7	0
52	0.07502	5.23000	0.50000	9	0	7	0	0	9	5	3	1	0	7	0
53	0.07520	4.95000	0.52174	9	7	7	0	0	0	5	3	1	9	0	0
54	0.07522	4.81000	0.60000	9	0	5	7	0	0	0	5	1	9	0	3
55	0.07538	4.62000	0.57143	9	3	7	0	0	0	0	0	1	9	7	5
56	0.07611	4.63000	0.48000	9	3	0	7	0	9	5	0	1	0	0	0
57	0.07625	4.38000	0.50000	9	3	0	7	0	0	5	0	1	9	0	0
58	0.07674	4.36000	0.50000	9	3	5	7	0	9	0	0	1	0	0	0
59	0.07689	4.11000	0.52174	9	3	5	7	0	0	0	0	1	9	0	0
60	0.07728	4.68000	0.46154	9	0	0	7	0	9	5	3	1	0	0	0
61	0.07742	4.04000	0.54545	9	3	5	7	0	0	0	0	0	9	0	1
62	0.07772	4.52000	0.44444	9	3	7	0	0	9	5	0	1	0	0	0
63	0.07786	4.27000	0.46154	9	3	7	0	0	0	5	0	1	9	0	0
64	0.07809	4.16000	0.50000	9	0	5	7	0	0	0	3	1	9	0	0
65	0.07895	4.57000	0.42857	9	0	7	0	0	9	5	3	1	0	0	0
66	0.07910	4.32000	0.44444	9	0	7	0	0	0	5	3	1	9	0	0
67	0.07920	4.21000	0.46154	9	3	7	0	0	9	0	0	1	0	0	5
68	0.07987	3.90000	0.50000	9	3	7	0	0	0	0	0	0	9	1	5
69	0.08025	3.99000	0.46154	9	3	7	0	0	0	0	5	1	9	0	0
70	0.08042	3.97000	0.48000	9	3	7	0	0	0	0	0	1	0	7	5
71	0.08140	3.73000	0.42857	9	3	0	7	0	0	5	0	1	0	0	0
72	0.08235	3.39000	0.46154	9	3	5	7	0	0	0	0	0	9	0	0
73	0.08269	3.90000	0.38710	0	3	7	0	0	9	5	0	1	0	0	0
74	0.08275	3.78000	0.41379	9	0	0	7	0	0	5	3	1	0	0	0
75	0.08308	3.60000	0.44444	9	3	0	7	0	9	0	0	0	0	0	5
76	0.08325	3.62000	0.40000	9	3	7	0	0	0	5	0	1	0	0	0
77	0.08347	3.55000	0.41379	9	3	7	0	0	0	5	0	0	9	0	0
78	0.08351	3.51000	0.42857	9	0	5	7	0	0	0	3	1	0	0	0
79	0.08372	3.44000	0.44444	9	0	5	7	0	0	0	3	0	9	0	0
80	0.08408	3.95000	0.37500	0	0	7	0	0	9	5	3	1	0	0	0
81	0.08424	3.85000	0.40000	0	0	7	0	0	9	5	0	0	9	0	3
82	0.08436	3.67000	0.40000	0	0	7	0	0	0	5	0	1	9	0	3
83	0.08472	3.85000	0.38710	9	0	7	0	0	9	5	3	0	0	0	0
84	0.08491	3.49000	0.41379	1	3	7	0	0	9	0	0	0	0	0	5
85	0.08518	3.24000	0.42857	9	3	7	0	0	0	0	0	0	9	0	5

TABLE 19: Continued.

	$1/f_1$	f_2	$1/f_3$	M_1	M_2	M_3	M_4	M_5	M_6	M_7	M_8	M_9	M_{10}	M_{11}	M_{12}
86	0.08599	3.34000	0.40000	9	3	7	0	0	0	0	5	1	0	0	0
87	0.08693	3.29000	0.37500	0	3	0	7	0	9	5	0	0	0	0	0
88	0.08755	3.01000	0.38710	9	3	0	7	0	0	5	0	0	0	0	0
89	0.08840	2.74000	0.40000	9	3	5	7	0	0	0	0	0	0	0	0
90	0.08904	3.18000	0.35294	0	3	7	0	0	9	5	0	0	0	0	0
91	0.08911	3.06000	0.37500	9	0	0	7	0	0	5	3	0	0	0	0
92	0.08964	2.79000	0.40000	0	0	5	7	0	0	0	0	0	9	0	3
93	0.08999	2.79000	0.38710	9	0	5	7	0	0	0	3	0	0	0	0
94	0.09133	2.95000	0.35294	9	0	7	0	0	0	5	3	0	0	0	0
95	0.09158	2.98000	0.35294	0	0	7	0	0	0	5	3	0	9	0	0
96	0.09169	2.84000	0.36364	9	3	5	0	0	7	0	0	0	0	0	0
97	0.09267	2.92000	0.35294	0	0	7	0	0	9	0	3	0	0	0	5
98	0.09288	2.62000	0.36364	9	3	7	0	0	0	0	5	0	0	0	0
99	0.09365	2.67000	0.36364	0	0	7	0	0	0	0	3	0	9	0	5
100	0.09366	3.20000	0.34286	0	0	0	0	0	9	5	3	0	0	7	0
101	0.09528	2.87000	0.35294	0	3	0	0	0	9	0	5	0	0	7	0
102	0.09603	2.59000	0.36364	9	3	0	0	0	0	0	5	0	0	7	0
103	0.09658	2.61000	0.36364	9	0	0	0	0	0	0	3	0	0	7	5
104	0.09662	2.74000	0.35294	0	0	7	0	0	0	0	3	9	0	0	5
105	0.09706	2.55000	0.37500	3	7	0	0	0	0	0	0	0	9	0	5
106	0.09733	2.93000	0.34286	0	3	0	0	0	7	0	5	9	0	0	0
107	0.09777	2.98000	0.34286	9	0	0	0	0	0	5	3	7	0	0	0
108	0.09789	2.71000	0.35294	3	0	7	0	0	0	0	5	9	0	0	0
109	0.09955	2.65000	0.35294	9	3	0	0	0	0	0	5	7	0	0	0
110	0.10275	2.38000	0.34286	9	0	0	7	0	0	5	0	0	0	0	0
111	0.10392	2.11000	0.35294	9	0	5	7	0	0	0	0	0	0	0	0
112	0.10424	2.14000	0.35294	0	0	5	7	0	0	0	0	0	9	0	0
113	0.10445	2.35000	0.34286	0	0	0	7	0	9	0	0	0	0	0	5
114	0.10571	2.27000	0.32432	9	0	7	0	0	0	5	0	0	0	0	0
115	0.10695	2.10000	0.34286	9	0	0	7	0	0	0	5	0	0	0	0
116	0.10779	2.05000	0.35294	9	5	0	7	0	0	0	0	0	0	0	0
117	0.10850	2.21000	0.32432	9	0	5	0	0	7	0	0	0	0	0	0
118	0.10883	1.99000	0.33333	0	0	7	0	0	0	0	0	0	9	0	5
119	0.10884	2.52000	0.31579	0	0	0	0	0	9	5	0	0	0	7	0
120	0.11016	1.99000	0.32432	9	0	7	0	0	0	0	5	0	0	0	0
121	0.11105	1.94000	0.33333	9	5	7	0	0	0	0	0	0	0	0	0
122	0.11152	2.58000	0.30769	0	0	0	0	0	7	5	0	9	0	0	0
123	0.11280	1.93000	0.33333	9	0	0	0	0	0	0	0	0	0	7	5
124	0.11319	1.90000	0.33333	9	7	0	0	0	0	0	0	0	0	0	5
125	0.11357	2.24000	0.31579	0	0	0	0	0	9	0	5	0	0	7	0
126	0.11463	1.96000	0.32432	9	0	0	0	0	0	0	5	0	0	7	0
127	0.11503	1.93000	0.32432	9	7	0	0	0	0	0	5	0	0	0	0
128	0.11560	1.91000	0.33333	9	5	0	0	0	0	0	0	0	0	7	0
129	0.11648	2.30000	0.30769	0	0	0	0	0	7	0	5	9	0	0	0
130	0.11774	1.92000	0.33333	9	0	0	0	0	0	0	0	0	7	0	5
131	0.11968	2.02000	0.31579	9	0	0	0	0	0	0	5	7	0	0	0
132	0.13725	1.70000	0.30769	0	0	0	7	0	9	0	0	0	0	0	0
133	0.13881	1.42000	0.31579	9	0	0	7	0	0	0	0	0	0	0	0
134	0.14259	1.59000	0.29268	0	0	7	0	0	9	0	0	0	0	0	0
135	0.14490	1.34000	0.30000	0	0	7	0	0	0	0	0	0	9	0	0
136	0.15017	1.56000	0.29268	0	0	0	0	0	9	0	0	0	0	7	0

Table 19: Continued.

	$1/f_1$	f_2	$1/f_3$	M_1	M_2	M_3	M_4	M_5	M_6	M_7	M_8	M_9	M_{10}	M_{11}	M_{12}
137	0.15204	1.28000	0.30000	9	0	0	0	0	0	0	0	0	0	7	0
138	0.15274	1.25000	0.30000	9	7	0	0	0	0	0	0	0	0	0	0
139	0.15276	1.86000	0.28571	0	0	0	0	0	9	5	0	0	0	0	0
140	0.15531	1.62000	0.28571	0	0	0	0	0	7	0	0	9	0	0	0
141	0.15858	1.55000	0.29268	0	0	0	0	0	9	0	0	0	0	0	5
142	0.16067	1.27000	0.30000	9	0	0	0	0	0	0	0	0	0	0	5
143	0.16082	1.38000	0.29268	0	0	0	0	0	0	0	0	9	0	7	0
144	0.16223	1.58000	0.28571	0	0	0	0	0	9	0	5	0	0	0	0
145	0.16441	1.30000	0.29268	9	0	0	0	0	0	0	5	0	0	0	0
146	0.17473	1.40000	0.28571	0	0	0	0	0	0	0	5	9	0	0	0
147	0.24888	0.90000	0.26667	0	0	0	0	0	9	0	0	0	0	0	0
148	0.25407	0.62000	0.27273	9	0	0	0	0	0	0	0	0	0	0	0
149	0.27956	0.72000	0.26667	0	0	0	0	0	0	0	0	9	0	0	0
150	0.42517	0.68000	0.26667	0	0	0	0	0	0	0	9	0	0	0	0

C. The Statistical Results of NSGA-III, MPACO, and NSGA-II

Table 20

NSGA-III	$1/f_1$	f_2	$1/f_3$
1	0.06952	8.68	100
2	0.07019	7.72	4
3	0.07036	8.02	3
4	0.07093	6.82	2
5	0.07122	7.37	1.5
6	0.07145	6.47	1.09091
7	0.0719	7.16	1
8	0.07208	6.02	1
9	0.07256	6.2	0.8
10	0.07341	5.3	0.66667
11	0.0741	6.48	0.63158
12	0.07627	4.62	0.57143
13	0.07792	5.3	0.54545
14	0.07802	5.6	0.52174
15	0.07985	4.87	0.54545
16	0.08105	5.85	0.5
17	0.0817	4.95	0.44444
18	0.08333	3.97	0.48
19	0.08402	4.27	0.4
20	0.08459	3.89	0.48
21	0.0884	3.21	0.42857
22	0.09031	4.21	0.4
23	0.09554	3.56	0.35294
24	0.09752	3.36	0.4
25	0.10173	3.36	0.35294
26	0.10595	3.67	0.34286
27	0.10874	3.55	0.34286
28	0.10988	2.99	0.31579

Table 20: Continued.

NSGA-III	$1/f_1$	f_2	$1/f_3$
29	0.11019	2.89	0.34286
30	0.1147	2.71	0.31579
31	0.11756	2.65	0.32432
32	0.12685	2.55	0.33333
33	0.13384	2.24	0.31579
34	0.15216	2.03	0.29268
35	0.16145	1.92	0.3
36	0.17051	1.99	0.29268
37	0.17871	1.27	0.3
38	0.18988	1.34	0.29268
39	0.25602	1.27	0.27273
40	0.27956	1.34	0.26667
41	0.1	7.8	42.5
42	0.1	7.2	33.5
43	0.1	6.8	27.5
44	0.1	6.2	18.5
45	0.1	5.6	12.5
46	0.1	5	7.1
47	0.16	4.4	4.675
48	0.16	3.8	2.975
49	0.1	4.4	5.9
50	0.16	3.2	1.775
51	0.1	3.8	5.6625
52	0.1	3.2	5.6125
53	0.22	3.4	2.6125
54	0.22	2.8	0.625
55	0.37577	0.996	3.34975
56	0.06623	7.3388	3.34975

TABLE 20: Continued.

NSGA-III	$1/f_1$	f_2	$1/f_3$
57	0.24409	2.1668	3.34975
58	0.20791	2.7876	3.34975
59	0.17189	3.5728	3.34975
60	0.13616	4.561	3.34975
61	0.10086	5.7982	3.34975
62	0.20791	2.7876	3.34975
63	0.20791	8.7876	3.34975
64	0.38368	2.3574	3.34975
65	0.35316	0.996	3.34975
66	0.28033	1.678	3.34975
67	0.0352	9.9994	3.34975
68	0.40638	8.8566	3.34975
69	0.38431	8.337	3.34975
70	0.38979	8.547	3.34975
71	0.32185	2.349	3.34975
72	0.03548	9.6992	3.34975
73	0.13622	4.5612	3.34988
74	0.13634	4.5616	3.35023
75	0.03543	9.6266	3.56358
76	0.3168	1.295	3.57137
77	0.14243	4.0068	3.60527
78	0.13617	4.561	3.61382
79	0.28181	1.8256	3.70112
80	0.06495	9.9334	3.74135
81	0.05946	9.998	3.77945
82	0.03234	9.2458	3.78025
83	0.28132	1.678	3.7807
84	0.35609	1.433	3.8949
85	0.38276	2.0834	4.91208
86	0.05692	9.7142	5.08312
87	0.14241	6.1792	5.13159
88	0.3838	8.7272	5.54011
89	0.31968	1.9758	6.90731
90	0.37994	7.7678	8.09498
91	0.38328	2.2392	8.89547
92	0.38012	1.3922	9.5236
93	0.40638	8.8566	9.88592
94	0.38335	8.6474	61.00714
95	0.35273	0.996	86.38796

TABLE 21

MPACO	$1/f_1$	f_2	$1/f_3$
1	0.06952	8.68	100
2	0.07256	6.2	0.8
3	0.07341	5.3	0.66667
4	0.0741	6.48	0.63158
5	0.07627	4.62	0.57143
6	0.07792	5.3	0.54545

TABLE 21: Continued.

MPACO	$1/f_1$	f_2	$1/f_3$
7	0.07802	5.6	0.52174
8	0.07985	4.87	0.54545
9	0.08105	5.85	0.5
10	0.09752	3.36	0.4
11	0.10173	3.36	0.35294
12	0.10595	3.67	0.34286
13	0.10874	3.55	0.34286
14	0.10988	2.99	0.31579
15	0.11019	2.89	0.34286
16	0.1147	2.71	0.31579
17	0.17051	1.99	0.29268
18	0.17871	1.27	0.3
19	0.18988	1.34	0.29268
20	0.25602	1.27	0.27273
21	0.27956	1.34	0.26667
22	0.03346	9.2458	3.34975
23	0.35273	0.9962	3.35025
24	0.15292	9.0056	3.56627
25	0.2462	2.3204	3.64173
26	0.20956	3.0006	3.67611
27	0.31718	1.437	3.77513
28	0.24848	2.4014	3.79922
29	0.28457	1.876	3.82353
30	0.17966	4.1672	4.10582
31	0.25343	6.60826	4.11647
32	0.13914	2.5606	4.14042
33	0.28639	2.0884	4.35678
34	0.14175	3.698	4.46223
35	0.14277	2.987	4.51727
36	0.08449	2.43	4.55926
37	0.17589	4.6238	4.7343
38	0.08002	2.0406	4.7849
39	0.14984	5.929	4.83138
40	0.12355	7.461	4.8318
41	0.14186	6.0588	4.98583
42	0.28437	2.3816	5.13553
43	0.18122	4.935	5.1871
44	0.21446	3.8735	5.25744
45	0.21328	3.9618	5.30667
46	0.24787	9.3158	5.31994
47	0.19132	5.1524	5.51549
48	0.32606	3.1232	5.76914
49	0.21922	4.3486	6.03888
50	0.17991	5.6686	6.33581
51	0.1799	5.6758	6.34759
52	0.32229	2.2544	6.54919
53	0.29146	5.6102	6.90432
54	0.32349	2.3852	7.04616
55	0.32211	2.4028	7.11377
56	0.32594	2.9264	9.26279

Table 21: Continued.

MPACO	$1/f_1$	f_2	$1/f_3$
57	0.36071	2.9998	13.2611
58	0.37503	6.6086	45.73848
59	0.37731	7.084	51.68306
60	0.34764	9.1058	56.91861
61	0.34774	9.1278	57.17808
62	0.34857	5.3214	59.41081
63	0.38565	8.1212	66.20195

Table 22

NSGA-II	$1/f_1$	f_2	$1/f_3$
1	0.1909	5.455	5.98965
2	0.145	6.875	6.01718
3	0.2138	4.338	6.01943
4	0.2458	2.358	3.71603
5	0.2866	5.658	4.24055
6	0.2113	5.455	4.83263
7	0.0840	4.27	0.4
8	0.0845	3.89	0.48
9	0.1338	2.24	0.31579
10	0.1521	2.03	0.29268
11	0.1009	5.7982	61.439
12	0.4026	7.8266	61.853
13	0.4026	7.8266	61.853
14	0.4030	7.931	63.3753
15	0.4033	8.013	64.5881
16	0.3863	9.349	64.9901
17	0.404	8.212	67.5876
18	0.3515	9.987	67.8457
19	0.4041	8.236	67.9489
20	0.4034	8.04	69.4217
21	0.0686	7.9656	69.50514
22	0.39702	9.0474	72.81714
23	0.40559	8.6368	74.26286
24	0.4045	8.3316	74.43411
25	0.39136	9.1984	75.73334
26	0.40568	8.661	77.74505
27	0.3862	9.349	77.8771
28	0.0569	9.64	77.8771
29	0.4068	8.971	79.7996
30	0.2171	2.787	81.1023
31	0.4075	9.169	83.2120
32	0.38	7.797	83.7143
33	0.3862	9.349	86.3879
34	0.3999	8.848	86.3879
35	0.38637	9.349	86.3879
36	0.10173	3.36	0.35294
37	0.10595	3.67	0.34286

Table 22: Continued.

NSGA-II	$1/f_1$	f_2	$1/f_3$
38	0.10874	3.55	0.34286
39	0.10988	2.99	0.31579

Conflicts of Interest

The authors declare that there is no conflict of interest regarding the publication of this paper.

References

[1] R. N. Rai and N. Bolia, "Optimal decision support for air power potential," *IEEE Transactions on Engineering Management*, vol. 61, no. 2, pp. 310–322, 2014.

[2] A. S. Manne, "A target-assignment problem," *Operations Research*, vol. 6, no. 3, pp. 346–351, 1958.

[3] R. H. Day, "Allocating weapons to target complexes by means of nonlinear programming," *Operations Research*, vol. 14, no. 6, pp. 992–1013, 1966.

[4] P. A. Hosein and M. Athans, *Preferential Defense Strategies. Part I: The Static Case*, MIT Laboratory for Information and Decision Systems, Cambridge, MA, USA, 1990.

[5] D. Galati and M. Simaan, "Effectiveness of the Nash strategies in competitive multi-team target assignment problems," *IEEE Transactions on Aerospace and Electronic Systems*, vol. 43, no. 1, pp. 126–134, 2007.

[6] Z. J. Lee, C. Y. Lee, and S. F. Su, "An immunity-based ant colony optimization algorithm for solving weapon–target assignment problem," *Applied Soft Computing*, vol. 2, no. 1, pp. 39–47, 2002.

[7] Z. J. Lee, S. F. Su, and C. Y. Lee, "A genetic algorithm with domain knowledge for weapon-target assignment problems," *Journal of the Chinese Institute of Engineers*, vol. 25, no. 3, pp. 287–295, 2002.

[8] Z.-J. Lee and W.-L. Lee, "A hybrid search algorithm of ant colony optimization and genetic algorithm applied to weapon-target assignment problems," in *Intelligent Data Engineering and Automated Learning*, pp. 278–285, Springer, Berlin, Heidelberg, 2003.

[9] M.-Z. Lee, "Constrained weapon–target assignment: enhanced very large scale neighborhood search algorithm," *IEEE Transactions on Systems, Man, and Cybernetics - Part A: Systems and Humans*, vol. 40, no. 1, pp. 198–204, 2010.

[10] B. Xin, J. Chen, J. Zhang, L. Dou, and Z. Peng, "Efficient decision makings for dynamic weapon-target assignment by virtual permutation and tabu search heuristics," *IEEE Transactions on Systems, Man, and Cybernetics, Part C (Applications and Reviews)*, vol. 40, no. 6, pp. 649–662, 2010.

[11] Y. Li and Y. Dong, "Weapon-target assignment based on simulated annealing and discrete particle swarm optimization in cooperative air combat," *Acta Aeronautica Et Astronautica Sinica*, vol. 31, no. 3, pp. 626–631, 2010.

[12] P. Chen, B.-j. Shen, L.-s. Zhou, and Y.-w. Chen, "Optimized simulated annealing algorithm for thinning and weighting large planar arrays," *Journal of Zhejiang University SCIENCE C*, vol. 11, no. 4, pp. 261–269, 2010.

[13] B. Xin, J. Chen, Z. Peng, L. Dou, and J. Zhang, "An efficient rule-based constructive heuristic to solve dynamic weapon-target assignment problem," *IEEE Transactions on Systems, Man, and Cybernetics - Part A: Systems and Humans*, vol. 41, no. 3, pp. 598–606, 2011.

[14] A. G. Fei, L. Y. Zhang, and Q. J. Ding, "Multi-aircraft cooperative fire assignment based on auction algorithm," *Systems Engineering and Electronics*, vol. 34, no. 9, pp. 1829–1833, 2012.

[15] Z. R. Bogdanowicz, A. Tolano, K. Patel, and N. P. Coleman, "Optimization of weapon–target pairings based on kill probabilities," *IEEE Transactions on Cybernetics*, vol. 43, no. 6, pp. 1835–1844, 2013.

[16] X. Liu, Z. Liu, W. S. Hou, and J. H. Xu, "Improved MOPSO algorithm for multi-objective programming model of weapon-target assignment," *Systems Engineering and Electronics*, vol. 35, no. 2, pp. 326–330, 2013.

[17] Y. Zhang, R. N. Yang, J. L. Zuo, and X. Jing, "Weapon-target assignment based on decomposition-based evolutionary multi-objective optimization algorithms," *Systems Engineering and Electronics*, vol. 36, no. 12, pp. 2435–2441, 2014.

[18] D. K. Ahner and C. R. Parson, "Optimal multi-stage allocation of weapons to targets using adaptive dynamic programming," *Optimization Letters*, vol. 9, no. 8, pp. 1689–1701, 2015.

[19] J. Li, J. Chen, B. Xin, and L. H. Dou, "Solving multi-objective multi-stage weapon target assignment problem via adaptive NSGA-II and adaptive MOEA/D: A comparison study," in *2015 IEEE Congress on Evolutionary Computation (CEC)*, pp. 3132–3139, Sendai, Japan, May 2015.

[20] N. Dirik, S. N. Hall, and J. T. Moore, "Maximizing strike aircraft planning efficiency for a given class of ground targets," *Optimization Letters*, vol. 9, no. 8, pp. 1729–1748, 2015.

[21] L. Hongtao and K. Fengju, "Adaptive chaos parallel clonal selection algorithm for objective optimization in WTA application," *Optik - International Journal for Light and Electron Optics*, vol. 127, no. 6, pp. 3459–3465, 2016.

[22] N. Li, W. Huai, and S. Wang, "The solution of target assignment problem in command and control decision-making behaviour simulation," *Enterprise Information Systems*, vol. 11, pp. 1–19, 2017.

[23] Y. Li, Y. Kou, Z. Li, A. Xu, and Y. Chang, "A modified Pareto ant colony optimization approach to solve biobjective weapon-target assignment problem," *International Journal of Aerospace Engineering*, vol. 2017, Article ID 1746124, 14 pages, 2017.

[24] K. Deb and H. Jain, "An evolutionary many-objective optimization algorithm using reference-point-based nondominated sorting approach, part I: solving problems with box constraints," *IEEE Transactions on Evolutionary Computation*, vol. 18, no. 4, pp. 577–601, 2014.

[25] H. Ishibuchi, N. Tsukamoto, and Y. Nojima, "Behavior of evolutionary many-objective optimization," in *Tenth International Conference on Computer Modeling and Simulation (uksim 2008)*, pp. 266–271, Cambridge, UK, April 2008.

[26] H. Jain and K. Deb, "An evolutionary many-objective optimization algorithm using reference-point based nondominated sorting approach, part II: handling constraints and extending to an adaptive approach," *IEEE Transactions on Evolutionary Computation*, vol. 18, no. 4, pp. 602–622, 2014.

[27] W. Mkaouer, M. Kessentini, A. Shaout et al., "Many-objective software remodularization using NSGA-III," *ACM Transactions on Software Engineering and Methodology*, vol. 24, no. 3, pp. 1–45, 2015.

[28] F. Chen, J. Zhou, C. Wang, C. Li, and P. Lu, "A modified gravitational search algorithm based on a non-dominated sorting genetic approach for hydro-thermal-wind economic emission dispatching," *Energy*, vol. 121, pp. 276–291, 2017.

[29] Y. Zhu, J. Liang, J. Chen, and Z. Ming, "An improved NSGA-III algorithm for feature selection used in intrusion detection," *Knowledge-Based Systems*, vol. 116, pp. 74–85, 2017.

[30] H. Seada and K. Deb, "U-NSGA-III: a unified evolutionary optimization procedure for single, multiple, and many objectives: proof-of-principle results," in *International Conference on Evolutionary Multi-Criterion Optimization*, pp. 34–49, Guimarães, Portugal, 2015.

[31] Y. Yuan, H. Xu, and B. Wang, "An improved NSGA-III procedure for evolutionary many-objective optimization," in *Proceedings of the 2014 conference on Genetic and evolutionary computation - GECCO '14*, pp. 661–668, New York, NY USA, July 2014.

[32] K. Deb and J. Sundar, "Reference point based multi-objective optimization using evolutionary algorithms," in *Proceedings of the 8th annual conference on Genetic and evolutionary computation - GECCO '06*, pp. 635–642, Seattle, WA, USA, July 2006.

[33] R. M. F. Alves and C. R. Lopes, "Using genetic algorithms to minimize the distance and balance the routes for the multiple traveling salesman problem," *2015 IEEE Congress on Evolutionary Computation (CEC)*, 2015, pp. 3171–3178, Sendai, Japan, May 2015.

[34] D. W. Gong, X. F. Ji, and X. Y. Sun, "Solving many-objective optimization problems using set-based evolutionary algorithms," *Acta Electronica Sinica*, vol. 42, no. 1, pp. 77–83, 2014.

[35] J. Bader, D. Brockhoff, S. Welten, and E. Zitzler, "On using populations of sets in multiobjective optimization," in *International Conference on Evolutionary Multi-Criterion Optimization*, pp. 140–154, Nantes, France, April 2009.

[36] E. Zitzler, L. Thiele, and J. Bader, "On set-based multiobjective optimization," *IEEE Transactions on Evolutionary Computation*, vol. 14, no. 1, pp. 58–79, 2010.

[37] D. Gong, J. Sun, and Z. Miao, "A set-based genetic algorithm for interval many-objective optimization problems," *IEEE Transactions on Evolutionary Computation*, vol. 22, no. 1, pp. 47–60, 2018.

[38] H. Ishibuchi and T. Murata, "A multi-objective genetic local search algorithm and its application to flowshop scheduling," *IEEE Transactions on Systems, Man and Cybernetics, Part C (Applications and Reviews)*, vol. 28, no. 3, pp. 392–403, 1998.

[39] H. Ishibuchi, T. Yoshida, and T. Murata, "Balance between genetic search and local search in memetic algorithms for multiobjective permutation flowshop scheduling," *IEEE Transactions on Evolutionary Computation*, vol. 7, no. 2, pp. 204–223, 2003.

[40] H. Ishibuchi and K. Narukawa, "Performance evaluation of simple multiobjective genetic local search algorithms on multiobjective 0/1 knapsack problems," in *Proceedings of the 2004 Congress on Evolutionary Computation (IEEE Cat. No.04TH8753)*, pp. 441–448, Portland, OR, USA, June 2004.

[41] J. D. Knowles and D. W. Corne, "M-PAES: a memetic algorithm for multiobjective optimization," in *Proceedings of the 2000 Congress on Evolutionary Computation. CEC00 (Cat. No.00TH8512)*, pp. 325–332, La Jolla, CA, USA, July 2000.

[42] K. Harada, J. Sakuma, K. Ikeda, I. Ono, and S. Kobayashi, "Local search for multiobjective function optimization: Pareto descent method," *Transactions of the Japanese Society for Artificial Intelligence*, vol. 21, no. 4, pp. 350–360, 2006.

[43] P. A. N. Bosman, "On gradients and hybrid evolutionary algorithms for real-valued multiobjective optimization," *IEEE Transactions on Evolutionary Computation*, vol. 16, no. 1, pp. 51–69, 2012.

[44] K. Sindhya, K. Deb, and K. Miettinen, "A local search based evolutionary multi-objective optimization approach for fast and accurate convergence," in *International Conference on Parallel Problem Solving from Nature*, pp. 815–824, Dortmund, Germany, September 2008.

[45] L. V. Santana-Quintero, A. G. Hernández-Díaz, J. Molina, C. A. Coello Coello, and R. Caballero, "DEMORS: a hybrid multi-objective optimization algorithm using differential evolution and rough set theory for constrained problems," *Computers & Operations Research*, vol. 37, no. 3, pp. 470–480, 2010.

[46] A. Jaszkiewicz, "Genetic local search for multi-objective combinatorial optimization," *European Journal of Operational Research*, vol. 137, no. 1, pp. 50–71, 2002.

[47] M. Abouhawwash, H. Seada, and K. Deb, "Towards faster convergence of evolutionary multi-criterion optimization algorithms using Karush Kuhn Tucker optimality based local search," *Computers & Operations Research*, vol. 79, pp. 331–346, 2017.

[48] A. P. Wierzbicki, "The use of reference objectives in multiobjective optimization," in *Multiple criteria decision making theory and application*, pp. 468–486, Springer, Berlin Heidelberg, 1980.

[49] A. Mohammadi, M. N. Omidvar, and X. Li, "Reference point based multi-objective optimization through decomposition," in *2012 IEEE Congress on Evolutionary Computation*, pp. 1–8, Brisbane, QLD, Australia, June 2012.

[50] J. R. Figueira, A. Liefooghe, E. G. Talbi, and A. P. Wierzbicki, "A parallel multiple reference point approach for multi-objective optimization," *European Journal of Operational Research*, vol. 205, no. 2, pp. 390–400, 2010.

[51] R. Wang, R. C. Purshouse, and P. J. Fleming, "Preference-inspired coevolutionary algorithms for many-objective optimization," *IEEE Transactions on Evolutionary Computation*, vol. 17, no. 4, pp. 474–494, 2013.

[52] Y. Liu, D. Gong, X. Sun, and Y. Zhang, "Many-objective evolutionary optimization based on reference points," *Applied Soft Computing*, vol. 50, pp. 344–355, 2017.

[53] E. Zitzler and S. Künzli, "Indicator-based selection in multiobjective search," in *International Conference on Parallel Problem Solving from Nature*, pp. 832–842, Birmingham, UK, September 2004.

[54] J. Bader and E. Zitzler, "HypE: an algorithm for fast hypervolume-based many-objective optimization," *Evolutionary Computation*, vol. 19, no. 1, pp. 45–76, 2011.

[55] M. Wagner and F. Neumann, "A fast approximation-guided evolutionary multi-objective algorithm," in *Proceeding of the fifteenth annual conference on Genetic and evolutionary computation conference - GECCO '13*, pp. 687–694, Amsterdam, Netherlands, July 2013.

[56] R. H. Gómez and C. A. Coello Coello, "Improved metaheuristic based on the R2 indicator for many-objective optimization," in *Proceedings of the 2015 on Genetic and Evolutionary Computation Conference - GECCO '15*, pp. 679–686, Madrid, Spain, July 2015.

[57] Y. Liu, D. Gong, J. Sun, and Y. Jin, "A many-objective evolutionary algorithm using a one-by-one selection strategy," *IEEE Transactions on Cybernetics*, vol. 47, no. 9, pp. 2689–2702, 2017.

[58] I. Das and J. E. Dennis, "Normal-boundary intersection: a new method for generating the Pareto surface in nonlinear multicriteria optimization problems," *SIAM Journal on Optimization*, vol. 8, no. 3, pp. 631–657, 1998.

[59] K. Li, Á. Fialho, S. Kwong, and Q. Zhang, "Adaptive operator selection with bandits for a multiobjective evolutionary algorithm based on decomposition," *IEEE Transactions on Evolutionary Computation*, vol. 18, no. 1, pp. 114–130, 2014.

[60] W. Gong, Á. Fialho, Z. Cai, and H. Li, "Adaptive strategy selection in differential evolution for numerical optimization: an empirical study," *Information Sciences*, vol. 181, no. 24, pp. 5364–5386, 2011.

[61] K. Deb and R. B. Agrawal, "Simulated binary crossover for continuous search space," *Complex Systems*, vol. 9, no. 3, pp. 115–148, 1995.

[62] S. Biswas, S. Kundu, and S. Das, "Inducing niching behavior in differential evolution through local information sharing," *IEEE Transactions on Evolutionary Computation*, vol. 19, no. 2, pp. 246–263, 2015.

[63] K. Sindhya, S. Ruuska, T. Haanpää, and K. Miettinen, "A new hybrid mutation operator for multiobjective optimization with differential evolution," *Soft Computing*, vol. 15, no. 10, pp. 2041–2055, 2011.

[64] X. Bi and C. Wang, "An improved NSGA-III algorithm based on objective space decomposition for many-objective optimization," *Soft Computing*, vol. 21, no. 15, pp. 4269–4296, 2017.

[65] K. Deb, A. Pratap, S. Agarwal, and T. Meyarivan, "A fast and elitist multiobjective genetic algorithm: NSGA-II," *IEEE Transactions on Evolutionary Computation*, vol. 6, no. 2, pp. 182–197, 2002.

[66] Q. Zhang and H. Li, "MOEA/D: a multiobjective evolutionary algorithm based on decomposition," *IEEE Transactions on Evolutionary Computation*, vol. 11, no. 6, pp. 712–731, 2007.

[67] K. Deb, L. Thiele, M. Laumanns, and E. Zitzler, "Scalable test problems for evolutionary multiobjective optimization," in *Evolutionary Multiobjective Optimization*, Advanced Information and Knowledge Processing, A. Abraham, L. Jain, and R. Goldberg, Eds., pp. 105–145, Springer, London, 2005.

[68] Y. Yuan, H. Xu, and B. Wang, "An experimental investigation of variation operators in reference-point based many-objective optimization," in *Proceedings of the 2015 on Genetic and Evolutionary Computation Conference - GECCO '15*, pp. 775–782, New York, NY, USA, July 2015.

[69] E. Zitzler, L. Thiele, M. Laumanns, C. M. Fonseca, and V. G. da Fonseca, "Performance assessment of multiobjective optimizers: an analysis and review," *IEEE Transactions on Evolutionary Computation*, vol. 7, no. 2, pp. 117–132, 2003.

Unscented Particle Smoother and its Application to Transfer Alignment of Airborne Distributed POS

Xiaolin Gong,[1,2] Xiaorui Zheng,[1] Xing-Gang Yan,[3] and Zhaoxing Lu[1,4]

[1]*School of Instrumentation Science and Optoelectronics Engineering, Beihang University, Beijing 100191, China*
[2]*Science and Technology on Inertial Laboratory, Beijing 100191, China*
[3]*School of Engineering and Digital Arts, University of Kent, Canterbury CT2 7NT, UK*
[4]*Xi'an Institute of Hi-tech, Xi'an 710025, China*

Correspondence should be addressed to Xiaolin Gong; gongxiaolin@buaa.edu.cn

Academic Editor: Hikmat Asadov

This paper deals with the problem of state estimation for the transfer alignment of airborne distributed position and orientation system (distributed POS). For a nonlinear system, especially with large initial attitude errors, the performance of linear estimation methods will degrade. In this paper, a nonlinear smoothing algorithm called the unscented particle smoother (UPS) is proposed and utilized in the off-line transfer alignment of airborne distributed POS. In this algorithm, the measurements are first processed by the forward unscented particle filter (UPF) and then a backward smoother is used to achieve the improved solution. The performance of this algorithm is compared with that of a similar smoother known as the unscented Rauch-Tung-Striebel smoother. The simulation results show that the UPS effectively improves the estimation accuracy and this work offers a new off-line transfer alignment approach of distributed POS for multiantenna synthetic aperture radar and other airborne earth observation tasks.

1. Introduction

Airborne synthetic aperture radar (SAR) is an important tool for earth observation. Multiantenna SAR can achieve higher accuracy of interferometric processing and ground moving target indication through multibaseline interferometry technology (i.e., interferometric SAR), and it has become an important research direction of radar remote sensing [1]. Usually, these antennae are installed on both sides of the wing of the aircraft, and the motion parameters of each antenna are needed to complete the motion compensation [2]. As a result, airborne distributed position and orientation system (distributed POS) is developed, which is used to provide the time and space reference information for all points of multiantenna SAR or other requirements [3].

Airborne distributed POS is generally composed of four parts: a high-precision POS (main system), a few low-precision inertial measurement units (IMU, subsystem), a POS computer system (PCS), and postprocessing software [4, 5]. The high precision main system is usually installed in the cabin or underpart of the aircraft and performs the inertial/satellite integrated estimation with the position and velocity of global navigation satellite system (GNSS) as the measurement information. The subsystem, which is placed as close as possible to the measurement center of the load, is used to measure the acceleration and angular velocity which are sent to the PCS. In the PCS, the high precision motion information of main system is used as the reference signal to achieve the motion information of subsystem by transfer alignment technology. The nature of transfer alignment is to estimate and compensate the calculation error of motion parameters of the subsystem by using the high precision information of the main system [6]. It shows that the estimation method is crucial for distributed POS.

It should be noted that distributed POS is a nonlinear system. In some emergency situations, the distributed POS is expected to try to shorten the preparation time on the ground and even start up in flight, which will bring the uncertain errors and the misalignment of initial attitude is not small. These make the nonlinearity of the system to be further increased. Thus, the performance of transfer alignment based on the linear model and linear estimation method will degrade [7].

The unscented Kalman filter (UKF) is a typical nonlinear filtering method. It uses a deterministic sampling approach named as sigma points to propagate nonlinear systems and has been discussed in many literatures for the inertial/satellite integration system [8–10]. UKF is a representative filter to overcome the flaw of truncated error in extended Kalman filter (EKF) [11]. However, the UKF method is an approximation of linear minimum variance estimation and does not apply to general non-Gaussian distributions [12, 13]. Another popular solution strategy for general filtering problems is particle filter (PF) which is based on the recursive Bayesian estimation and suitable for the nonlinear and non-Gaussian distributed system [14]. The estimated accuracy of PF is higher than the accuracy of using EKF or UKF alone [7]. PF uses a set of sampling points extracted from the posterior probability to express its distribution. Choosing a reasonably recommended probability density is the core of PF. The closer the recommended density selected is to the true density, the better the filter effect is. Otherwise, it is worse or even divergent [7, 15]. The UKF is able to more accurately propagate the mean and covariance of the Gaussian approximation to the state distribution than the EKF. So PF is combined with UKF and formed a new PF called unscented particle filter (UPF) [15], in which the UKF is used for proposal distribution generation and the performance of the filtering algorithm is improved by taking into account the influence of the measured values on the state estimation.

The above UKF and PF are based on the idea of forward filtering, where only the observation information of the current moment and the previous time are used. The other methods, such as smoothing estimation, can make full use of all the observation information to estimate the state of each moment, and is particularly suitable for off-line data processing. Since the smoother uses more observations than the filter, although it cannot be used in real time, the accuracy of the optimal smoothing algorithm is theoretically higher than that of the KF [16]. The unscented Rauch-Tung-Striebel smoother (URTSS) was first proposed by Särkkä in 2008 [17]. This smoother takes the benefit over unscented transformation as well as the smoothing iterative characteristics, which makes the estimation accuracy much better than that of the UKF. In ref. [18], the URTSS is applied to the off-line integrated estimation of the inertial/satellite integration system and shows an obvious accuracy advantage over extended R-T-S smoother in attitude estimation. In particle smoothing, ref. [19] proposed a particle smoothing algorithm which is a direct extension of the PF. This method has the advantage of fixed interval smoothing, but it cannot express the state variables in the past very well. Then, the forward-backward smoother and two-filter smoother based on PF are proposed [20]. This kind of smoothing algorithm needs to carry on the particle filter first and store the particles with weight selected in the approximate process and then perform the smoothing process. However, there is still the problem of particle degeneracy.

For the airborne earth observation system, there are two modes of image processing: real time and off-line. In the off-line cases, there is no high requirement for fast data processing in real-time imaging mode. The smoother can be employed to provide a better solution for off-line image processing. Motivated by these, a smoother called unscented particle smoother (UPS) is proposed in this paper, which is then applied to the off-line transfer alignment of airborne distributed POS. The UPS is made by taking the advantages of the UKF, PF, and smoothing. In the UPS, UKF is used to generate the proposed distributions to overcome the shortage of particle degradation in PF. Then, the forward-backward smoother using all observations is further combined together to obtain higher estimation accuracy. The UPS can be applied to nonlinear and non-Gaussian noise systems as well. There is no need to keep still to obtain the initial motion information of the main system and subsystem before flying off.

The rest of this paper is organized as follows. In Section 2, the UPS are proposed. In Section 3, the design of UPS for airborne distributed POS is given in detail. In Section 4, the performance of the UPS method is demonstrated and compared with that of the URTSS in estimation accuracy by a simulation test. Finally, conclusions are given in Section 5.

2. Unscented Particle Smoothing Algorithm

The proposed UPS has two structures: a forward filter and a backward smoother. The forward filter is UPF used to obtain the filtering state estimation. In UPF, the proposed density is determined by UKF, which not only solves the problem of the degradation of particles but also enables particles to get the latest a posteriori information of the measurement vector when they are updated, which is helpful for particles to move toward the area with high likelihood. Then, the backward smoother is conducted after the forward filter to modify the importance weights to achieve the smoothed state estimation.

Suppose the discrete form of the n_x-state system equation and the observation equation is as follows:

$$\begin{aligned} \mathbf{x}_k &= f(\mathbf{x}_{k-1}, k-1) + \boldsymbol{\Gamma}_{k-1}\mathbf{w}_{k-1}, \\ \mathbf{y}_k &= h(\mathbf{x}_k, k) + \mathbf{v}_k, \end{aligned} \tag{1}$$

where $\mathbf{x}_k \in R^{n_x}$ represents the state vector of the system at time k and $\mathbf{y}_k \in R^{n_y}$ denotes the measurement vector at time t_k. The functions f and h describe state and measurement models, respectively [8, 21]; $\boldsymbol{\Gamma}_{k-1}$ is the state noise distribution matrix, and w_{k-1} and v_k represent the state and measurement noises, respectively, which can be described by the corresponding probability density functions $p(w_{k-1})$

and $p(v_k)$. In this work, w_{k-1} and v_k are assumed to the uncorrelated white Gaussian noise. The mean value of w_{k-1} and v_k is zero and their variance matrices are $\mathbf{Q}_{k-1}$ and $\mathbf{R}_k$, respectively [8, 22].

Based on (1), the process of UPS is described as follows.

Step 1. Initialization ($k = 0$) [7, 15]. Suppose the initial state variable $x_0 \sim p(x_0)$, and the covariance matrix is $\mathbf{P}_0$. For $i = 1, 2, \ldots, N$, draw the states (particles) $x_0^{(i)}$ from the prior probability density $p(x_0)$ and set $x_0^{(i)} \sim N(\widehat{x}_0^{(i)}, \mathbf{P}_0^{(i)})$, where $\widehat{x}_0^{(i)} = x_0^{(i)}$ and $\mathbf{P}_0^{(i)} = \mathbf{P}_0$. N is the number of particles.

Step 2. Importance sampling step (for $i = 1, 2, \ldots, N$, $k = 1, 2, 3, \ldots, T$).

(1) Update the particles with the UKF:

(i) Calculate sigma points

$$\chi_{k-1(0)}^{(i)} = \widehat{x}_{k-1}^{(i)},$$

$$\chi_{k-1(j)}^{(i)} = \widehat{x}_{k-1}^{(i)} + \left(\sqrt{(n_x + \lambda)\mathbf{P}_{k-1}^{(i)}}\right)_{(j)} \quad j = 1, 2, \ldots, n_x,$$

$$\chi_{k-1(j)}^{(i)} = \widehat{x}_{k-1}^{(i)} - \left(\sqrt{(n_x + \lambda)\mathbf{P}_{k-1}^{(i)}}\right)_{(j)} \quad j = n_x + 1, n_x + 2, \ldots, 2n_x, \tag{2}$$

where the sigma point set $S = \{\chi_{k-1(j)}^{(i)}, W_j^{(l')};\ j = 0, 1, \ldots, 2n_x, l' \in (m, c)\}$ is composed of the sigma points $\chi_{k-1(j)}^{(i)}$ and their respective mean (m) and covariance (c) weights $W_j^{(l')}$. $\lambda = \alpha^2(n_x + \kappa) - n_x$ denotes a scaling parameter. The parameter α determines the spread of samples that is usually set between 10^{-4} and 1. The constant κ is typically set to 0 or $3 - n_x$ where n_x is the state dimension. The corresponding weights are given by:

$$W_0^{(m)} = \frac{\lambda}{(n_x + \lambda)},$$

$$W_0^{(c)} = \frac{\lambda}{(n_x + \lambda)} + \left(1 - \alpha^2 + \beta\right),$$

$$W_j^{(m)} = W_j^{(c)} = \frac{0.5}{(n_x + \lambda)} \quad j = 1, 2, \ldots, 2n_x, \tag{3}$$

where the parameter β is used to incorporate prior knowledge of the distribution (for Gaussian distribution, $\beta = 2$ is optimal).

(ii) Time update (propagate particle into future)

$$\chi_{k|k-1(j)}^{(i)} = f\left(\chi_{k-1(j)}^{(i)}, k - 1\right),$$

$$\widehat{x}_{k|k-1}^{(i)} = \sum_{j=0}^{2n_x} W_j^{(m)} \chi_{k|k-1(j)}^{(i)},$$

$$\mathbf{P}_{k|k-1}^{(i)} = \sum_{j=0}^{2n_x} W_j^{(c)} \left[\chi_{k|k-1(j)}^{(i)} - \widehat{x}_{k|k-1}^{(i)}\right] \cdot \left[\chi_{k|k-1(j)}^{(i)} - \widehat{x}_{k|k-1}^{(i)}\right]^{\mathrm{T}} + \mathbf{\Gamma}_{k-1}\mathbf{Q}_k\mathbf{\Gamma}_{k-1}^{\mathrm{T}},$$

$$\gamma_{k|k-1(j)}^{(i)} = h\left(\chi_{k|k-1(j)}^{(i)}, k\right),$$

$$\widehat{y}_{k|k-1}^{(i)} = \sum_{j=0}^{2n_x} W_j^{(m)} \gamma_{k|k-1(j)}^{(i)}, \tag{4}$$

where $\mathbf{P}_{k|k-1}^{(i)}$ denotes the prior error covariance.

(iii) Measurement update (incorporate new observation)

$$\mathbf{P}_{\widehat{y}_k\widehat{y}_k}^{(i)} = \sum_{j=0}^{2n_x} W_j^{(c)} \left[\gamma_{k|k-1(j)}^{(i)} - \widehat{y}_{k|k-1}^{(i)}\right] \cdot \left[\gamma_{k|k-1(j)}^{(i)} - \widehat{y}_{k|k-1}^{(i)}\right]^{\mathrm{T}} + \mathbf{R}_k,$$

$$\mathbf{P}_{\widehat{x}_k\widehat{y}_k}^{(i)} = \sum_{j=0}^{2n_x} W_j^{(c)} \left[\chi_{k|k-1(j)}^{(i)} - \widehat{x}_{k|k-1}^{(i)}\right] \cdot \left[\gamma_{k|k-1(j)}^{(i)} - \widehat{y}_{k|k-1}^{(i)}\right]^{\mathrm{T}},$$

$$K_k^{(i)} = \mathbf{P}_{\widehat{x}_k\widehat{y}_k}^{(i)} \mathbf{P}_{\widehat{y}_k\widehat{y}_k}^{(i)} - 1,$$

$$\widehat{x}_k^{(i)} = \widehat{x}_{k|k-1}^{(i)} + K_k^{(i)}\left(y_k - \widehat{y}_{k|k-1}^{(i)}\right),$$

$$\mathbf{P}_k^{(i)} = \mathbf{P}_{k|k-1}^{(i)} - K_k^{(i)}\mathbf{P}_{\widehat{y}_k\widehat{y}_k}\left(K_k^{(i)}\right)^{\mathrm{T}}, \tag{5}$$

where $\widehat{y}_{k|k-1}^{(i)}$ denotes the predicted measurement, $\mathbf{P}_{\widehat{y}_k\widehat{y}_k}^{(i)}$ denotes the covariance of predicted measurement error, and $\mathbf{P}_{\widehat{x}_k\widehat{y}_k}^{(i)}$ denotes the cross-covariance matrix between $\widehat{x}_{k|k-1}^{(i)}$ and $\widehat{y}_{k|k-1}^{(i)}$. $K_k^{(i)}$ denotes the filter gain.

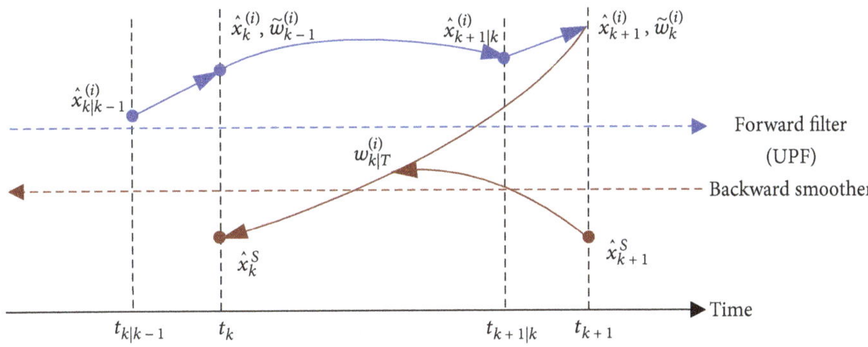

FIGURE 1: The principle diagram of UPS.

(2) Importance sampling step.

For $i = 1, 2, \ldots, N$, sample $\tilde{x}_k^{(i)} \sim q(x_k^{(i)} \mid x_{0:k-1}^{(i)}, y_{1:k}) = N(\hat{x}_k^{(i)}, \mathbf{P}_k^{(i)})$ and set $\tilde{x}_k^{(i)} \triangleq (x_{0:k-1}^{(i)}, \tilde{x}_k^{(i)})$ and $\tilde{\mathbf{P}}_{0:k}^{(i)} \triangleq (\mathbf{P}_{0:k-1}^{(i)}, \tilde{\mathbf{P}}_k^{(i)})$.

(3) Update and normalize the importance weights.

For $i = 1, 2, \ldots, N$, evaluate the importance weights up to a normalizing constant:

$$\tilde{w}_k^{(i)} = w_{k-1}^{(i)} \frac{p\left(y_k \mid \tilde{x}_k^{(i)}\right) p\left(\tilde{x}_k^{(i)} \mid x_{k-1}^{(i)}\right)}{q\left(\tilde{x}_k^{(i)} \mid x_{0:k-1}^{(i)}, y_{1:k}\right)}. \quad (6)$$

For $i = 1, 2, \ldots, N$, normalize and store the importance weights:

$$\tilde{w}_k^{(i)} = w_k^{(i)} \left[\sum_{i=1}^{N} w_k^{(i)}\right]^{-1}. \quad (7)$$

Compute the filtering state estimation $\hat{x}_k^F$:

$$\hat{x}_k^F = \sum_{i=1}^{N} \tilde{w}_k^{(i)} \tilde{x}_k^{(i)}. \quad (8)$$

Step 3. Using the importance weights stored in the forward filtering to conduct the backward smoothing recursion. Set $w_{T|T}^{(i)} = \tilde{w}_{T|T}^{(i)}$. For $k = T-1, \ldots, 1$ and $i = 1, 2, \ldots, N$, evaluate

$$w_{k|T}^{(i)} = w_{k+1|T}^{(i)} \left[\sum_{j=1}^{N} w_{k+1|T}^{(j)} \frac{p\left(\tilde{x}_{k+1}^{(j)} \mid \tilde{x}_k^{(i)}\right)}{\sum_{m=1}^{N} \tilde{w}_k^{(m)} p\left(\tilde{x}_{k+1}^{(j)} \mid \tilde{x}_k^{(m)}\right)}\right]. \quad (9)$$

(1) Resampling

For $\tilde{x}_k^{(i)}$ $(i = 1, 2, \ldots, N)$, Using sequential importance resampling (SIR) [15] to complete resampling to obtain N random samples $\mathbf{x}'^{(i)}_{0:k}$, and set $\mathbf{w}_k^{(i)} = \tilde{\mathbf{w}}_k^{(i)} = 1/N$.

(2) Compute the smoothed state estimation $\hat{x}_k^S$

$$\hat{x}_k^S = \frac{1}{N} \sum_{i=1}^{N} x'^{(i)}_k. \quad (10)$$

The principle diagram of UPS is shown in Figure 1. In summary, there are two processes: a forward filtering process and a backward smoothing process. The measurements are first processed by the forward filter and the importance weight $\tilde{w}_k^{(i)}$ should be stored in the forward filter stage. And then, a separate backward smoothing pass is used to modify the importance weights for obtaining the smoothing solution.

3. UPS Design for Airborne Distributed POS

The design of UPS for airborne distributed POS is given in this section. Firstly, the definition of coordinate frames is introduced. i and e denote the earth-center inertial frame and the earth-centered earth-fixed frame, respectively. The navigation frames of the main system and subsystem are both topocentric frames (the z-axis parallel to the upward vertical, the x-axis pointing eastward, and the y-axis pointing northward), represented by n and n_s, respectively. The body frames of the main system and subsystem are denoted by b and b_s. A detailed description of these coordinate frames is available in [18, 23].

Figure 2 show the block diagram of transfer alignment in distributed POS based on UPS.

In Figure 2, the strapdown navigator calculates the attitude (heading, pitch, and roll), velocity, and position expressed in n_s-frame by using the angular rates and

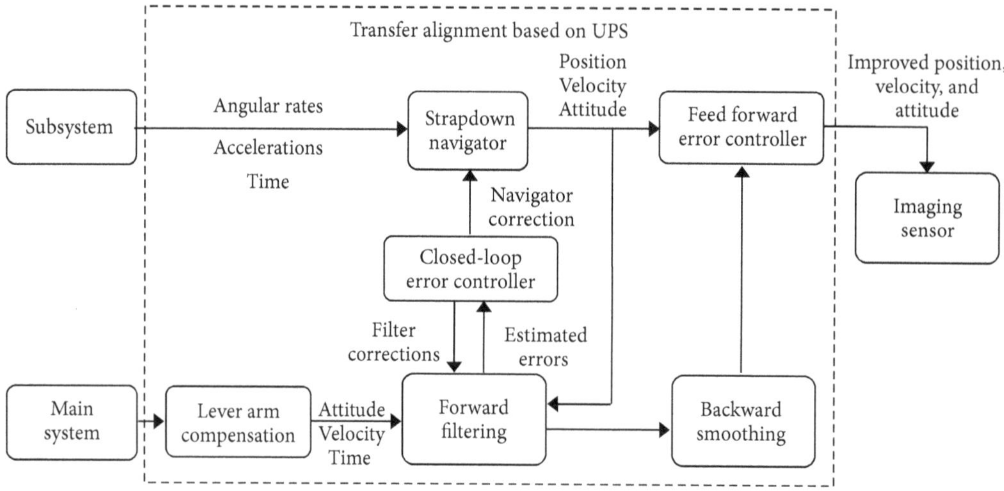

FIGURE 2: The block diagram of transfer alignment in distributed POS based on UPS.

accelerations of subsystem. Since the strapdown navigation is an integration process, any errors of the accelerometers and gyroscopes will integrate into slowly growing errors on velocity, position, and attitude. The differences of velocity and attitude between the main system and the strapdown navigator are used in the forward filtering to estimate the growing velocity, position, and attitude errors in the strapdown navigator. The closed-loop error controller is used to correct the result of strapdown navigator by using the estimation of forward filtering. After the end of the forward filter, the backward smoother is performed to compensate the estimation of forward filtering, and the improved estimation results are used to correct the result of strapdown navigator in the feed forward error controller. The improved attitude, velocity, and position are outputted to the imaging sensor at last.

3.1. Nonlinear Inertial Navigation Error Model. The inertial navigation error equations are the basis of the mathematical model for transfer alignment. According to the definitions above, the nonlinear SINS error model of the subsystem based on angle error, which includes attitude error equation, velocity error equation, position error equation, and inertial senor constant error equation, is given by [24]:

$$\begin{aligned}
\dot{\boldsymbol{\phi}}^{n_s} &= \left(\mathbf{I} - \mathbf{C}_{n_s}^{p}\right)\boldsymbol{\omega}_{in_s}^{n_s} + \delta\boldsymbol{\omega}_{in_s}^{n_s} - \mathbf{C}_{b_s}^{n_s}\boldsymbol{\varepsilon}^{b_s}, \\
\delta\dot{\mathbf{V}}^{n_s} &= \left(\mathbf{I} - \mathbf{C}_{p}^{n_s}\right) f^{n_s} - \left(2\delta\omega_{ie}^{n_s} + \delta\omega_{en_s}^{n_s}\right) \\
&\quad \times \mathbf{V}^{n_s} - \left(2\delta\omega_{ie}^{n_s} + \delta\omega_{en_s}^{n_s}\right) \times \delta\mathbf{V}^{n_s} + \widehat{\mathbf{C}}_{b}^{n_s}\boldsymbol{\nabla}^{b_s}, \\
\delta\dot{L} &= \frac{\delta V_{\mathrm{N}}}{R_{\mathrm{M}} + H} - \frac{V_{\mathrm{N}}}{(R_{\mathrm{M}} + H)^2}\delta H, \\
\delta\dot{\lambda} &= \frac{\sec L}{R_{\mathrm{N}} + H}\delta V_{\mathrm{E}} + \frac{V_{\mathrm{E}} \sec L \tan L}{R_{\mathrm{N}} + H}\delta L - \frac{V_{\mathrm{E}} \sec L}{(R_{\mathrm{N}} + H)^2}\delta H, \\
\delta\dot{H} &= \delta V_{\mathrm{U}}, \\
\dot{\varepsilon}_{c}^{b_s} &= 0, \\
\dot{\boldsymbol{\nabla}}_{c}^{b_s} &= 0,
\end{aligned} \tag{11}$$

where $\boldsymbol{\phi}^{n_s} = [\phi_{\mathrm{E}} \ \ \phi_{\mathrm{N}} \ \ \phi_{\mathrm{U}}]^{\mathrm{T}}$ denotes the attitude error vector in n_s-frame, the subscripts E, N, and U denote the east, north, and up components, respectively; $\omega_{in_s}^{n_s}$ denotes the rotation velocity of the n_s-frame relative to the i-frame expressed in n_s-frame with error $\delta\omega_{in_s}^{n_s}$; $\mathbf{C}_{b_s}^{n_s}$ denotes the coordinate transformation matrix from the b_s-frame to the n_s-frame; ε^{b_s} denotes the gyro drift of subsystem in b_s-frame, which consists of random constant drift $\varepsilon_c^{b_s}$ and the Gaussian white noise $\omega_\varepsilon^{b_s}$ with $\varepsilon_c^{b_s} = [\varepsilon_x^{b_s} \, \varepsilon_y^{b_s} \, \varepsilon_z^{b_s}]^{\mathrm{T}}$ and $\omega_\varepsilon^{b_s} = [\omega_{\varepsilon_x}^{b_s} \, \omega_{\varepsilon_y}^{b_s} \, \omega_{\varepsilon_z}^{b_s}]^{\mathrm{T}}$; $\mathbf{V}^{n_s} = [V_{\mathrm{E}} \ \ V_{\mathrm{N}} \ \ V_{\mathrm{U}}]^{\mathrm{T}}$ denotes the velocity vector in n_s-frame with error $\delta\mathbf{V}^{n_s} = [\delta V_{\mathrm{E}} \ \ \delta V_{\mathrm{N}} \ \ \delta V_{\mathrm{U}}]^{\mathrm{T}}$; $f^{n_s} = [f_{\mathrm{E}} \ \ f_{\mathrm{N}} \ \ f_{\mathrm{U}}]^{\mathrm{T}}$ denotes the specific force measured by the accelerometers of subsystem expressed in n_s-frame; $\omega_{ie}^{n_s}$ denotes the rotation velocity of the e-frame relative to the i-frame expressed in n_s-frame with error $\delta\omega_{ie}^{n_s}$; $\omega_{en_s}^{n_s}$ denotes the rotation velocity of the n_s-frame relative to the e-frame expressed in n_s-frame with error $\delta\omega_{en_s}^{n_s}$; $\boldsymbol{\nabla}^{b_s}$ denotes the accelerometer bias of subsystem in b_s-frame, which consists of random constant bias $\boldsymbol{\nabla}_c^{b_s}$ and the Gaussian white noise $\omega_\nabla^{b_s}$ with $\boldsymbol{\nabla}_c^{b_s} = [\nabla_x^{b_s} \, \nabla_y^{b_s} \, \nabla_z^{b_s}]^{\mathrm{T}}$ and $\omega_\nabla^{b_s} = [\omega_{\nabla_x}^{b_s} \, \omega_{\nabla_y}^{b_s} \, \omega_{\nabla_z}^{b_s}]^{\mathrm{T}}$; $\sec L = 1/\cos L$. The symbols δL, $\delta\lambda$, and δH denote the latitude error, longitude error, and height error, respectively; and R_{M} and R_{N} denote the meridian and transverse radius of curvature, respectively.

The nonlinear terms of attitude and velocity error equations are $(\mathbf{I} - \mathbf{C}_{n_s}^{p})\omega_{in_s}^{n_s}$ and $(\mathbf{I} - \mathbf{C}_{p}^{n_s})f^{n_s}$ in (11), respectively, and the nonlinear matrix $\mathbf{C}_{n_s}^{p} = (\mathbf{C}_{p}^{n_s})^{\mathrm{T}}$ is expressed as follows:

$$\mathbf{C}_{n_s}^{p} = \begin{bmatrix} \cos\phi_N \cos\phi_U - \sin\phi_E \sin\phi_N \sin\phi_U & \cos\phi_N \sin\phi_U + \sin\phi_E \sin\phi_N \cos\phi_U & -\cos\phi_E \sin\phi_N \\ -\cos\phi_E \sin\phi_U & \cos\phi_E \cos\phi_U & \sin\phi_E \\ \sin\phi_N \cos\phi_U + \sin\phi_E \cos\phi_N \sin\phi_U & \sin\phi_N \sin\phi_U - \sin\phi_E \cos\phi_N \cos\phi_U & \cos\phi_E \cos\phi_N \end{bmatrix}. \quad (12)$$

3.2. Model of Rigid Misalignment Angle and Flexure Angle. During the transfer alignment, there are flexure angles and rigid misalignment angles between the main system and the subsystem. These angles cannot be measured accurately and the flexure angles vary with the time. Therefore, the flexure angles and rigid misalignment angles should be modeled and estimated.

The models of rigid misalignment angles and flexure angles are shown in the following equations:

$$\begin{aligned} &\dot{\rho}_j = 0, \\ &\ddot{\theta}_j + 2\beta_j\dot{\theta}_j + \beta_j^2\theta_j = \eta_j \quad j = x, y, z, \end{aligned} \quad (13)$$

where ρ_j denotes the rigid misalignment angle of subsystem; θ_j denotes the flexure angle and described by the second-order Markov process [25]; $\beta_j = 2.146/\tau_j$ and τ_j is the correlation time; x, y, and z are the axes of b_s-frame; η_j denotes the Gaussian white noise with covariance $Q_{\eta_j} = 4\beta_j^3\sigma_j^2$; and σ_j^2 denotes the covariance of θ_j; $Q_\eta = [Q_{\eta_x}\ Q_{\eta_y}\ Q_{\eta_z}]^{\mathrm{T}}$.

3.3. Design of the Smoother for Transfer Alignment of Distributed POS

3.3.1. State Equation. On the basis of models established in Subsections 3.1 and 3.2, let

$$\begin{aligned} x &= [x_1 \quad x_2]^{\mathrm{T}}, \\ x_1 &= \left[\phi_E \quad \phi_N \quad \phi_U \quad \delta V_E \quad \delta V_N \quad \delta V_U \quad \delta L \quad \delta\lambda \quad \delta H \quad \varepsilon_x^{b_s} \quad \varepsilon_y^{b_s} \quad \varepsilon_z^{b_s} \quad \nabla_x^{b_s} \quad \nabla_y^{b_s} \quad \nabla_z^{b_s}\right]^{\mathrm{T}}, \\ x_2 &= \left[\rho_x \quad \rho_y \quad \rho_z \quad \theta_x \quad \theta_y \quad \theta_z \quad \dot{\theta}_x \quad \dot{\theta}_y \quad \dot{\theta}_z\right]^{\mathrm{T}}. \end{aligned} \quad (14)$$

Then, the continuous-time system model can be given by

$$\dot{x} = f(x, \mathrm{t}) + G(\mathrm{t})\mathbf{W}_I(\mathrm{t}), \quad (15)$$

where the nonlinear function $f(x, \mathrm{t})$ is determined by error model of transfer alignment in Subsections 3.1 and 3.2, the vector $\mathbf{W}_I(\mathrm{t})$ represents the system noise. $\mathbf{W}_1 = \left[\omega_{\varepsilon_x}^{b_s} \quad \omega_{\varepsilon_y}^{b_s} \quad \omega_{\varepsilon_z}^{b_s} \quad \omega_{\nabla_x}^{b_s} \quad \omega_{\nabla_y}^{b_s} \quad \omega_{\nabla_z}^{b_s} \quad \eta_x \quad \eta_y \quad \eta_z\right]^{\mathrm{T}}$ is the zero-mean Gaussian white noise vector with covariance Q_I which consists of covariance Q_ε of gyro random drift, covariance Q_∇, of accelerometer random bias, and Q_η.

3.3.2. Measurement Equation. The differences of velocity and attitude between the main system and the subsystem solutions are considered as the measurement vector. The measurement vector can be given as

$$\mathbf{y}(t) = \left[\delta\psi \quad \delta\theta \quad \delta\gamma \quad \delta V_E' \quad \delta V_N' \quad \delta V_U'\right]^{\mathrm{T}}, \quad (16)$$

where $\delta\psi$, $\delta\theta$, and $\delta\gamma$ denote the differences of heading angle, pitch angle, and roll angle between the main system and the subsystem; $\delta V_E'$, $\delta V_N'$, and $\delta V_U'$ denote the differences of the east, north, and up velocities between the main system and the subsystem accounting for the lever arm between the main system and subsystem [25].

The measurement vector can be expressed linearly by the state vector shown in (14). Hence, the linear measurement model can be given by

$$\mathbf{y}(t) = \mathbf{H}\mathbf{x}(t) + \mathbf{v}(t), \quad (17)$$

where the measurement noise $\mathbf{v}(t) = \left[v_{\delta\psi} \quad v_{\delta\theta} \quad v_{\delta\gamma} \quad v_{\delta V_E'} \quad v_{\delta V_N'} \quad v_{\delta V_U'}\right]^{\mathrm{T}}$ is the white noise vector of the solution of the main system with zero mean and covariance matrix $\mathbf{R}$. The measurement matrix $\mathbf{H}(t)$ can be given as

$$\mathbf{H}(t) = \begin{bmatrix} \mathbf{H}_1 & 0_{3\times3} & 0_{3\times9} & \mathbf{H}_2 & \mathbf{H}_3 & 0_{3\times3} \\ 0_{3\times3} & \mathbf{I}_{3\times3} & 0_{3\times9} & 0_{3\times3} & 0_{3\times3} & 0_{3\times3} \end{bmatrix}_{6\times24}. \quad (18)$$

Here, the attitude matrix of main system $\mathbf{C}_b^n$ is denoted as $\mathbf{T}_a$. $\mathbf{T}_a$, $\mathbf{H}_1$, $\mathbf{H}_2$, and $\mathbf{H}_3$ are as follows:

$$\mathbf{C}_b^n = \begin{bmatrix} \cos\gamma_m \cos\varphi_m - \sin\gamma_m \sin\theta_m \sin\varphi_m & -\cos\theta_m \sin\varphi_m & \sin\gamma_m \cos\varphi_m + \cos\gamma_m \sin\theta_m \sin\varphi_m \\ -\cos\gamma_m \sin\varphi_m + \sin\gamma_m \sin\theta_m \cos\varphi_m & \cos\theta_m \cos\varphi_m & \sin\gamma_m \sin\varphi_m - \cos\gamma_m \sin\theta_m \cos\varphi_m \\ -\sin\gamma_m \cos\theta_m & \sin\theta_m & \cos\gamma_m \cos\theta_m \end{bmatrix}, \tag{19}$$

where φ_m, θ_m, and γ_m denote the heading, pitch, and roll of the main system, respectively.

$$\mathbf{H}_1 = \begin{bmatrix} \dfrac{T_a^{(12)}T_a^{(32)}}{\left(T_a^{(12)}\right)^2 + \left(T_a^{(22)}\right)^2} & \dfrac{T_a^{(22)}T_a^{(32)}}{\left(T_a^{(12)}\right)^2 + \left(T_a^{(22)}\right)^2} & -1 \\ -\dfrac{T_a^{(22)}}{\sqrt{1-(T_a^{(32)})^2}} & \dfrac{T_a^{(12)}}{\sqrt{1-(T_a^{(32)})^2}} & 0 \\ \dfrac{T_a^{(21)}T_a^{(33)} - T_a^{(31)}T_a^{(23)}}{\left(T_a^{(33)}\right)^2 + \left(T_a^{(31)}\right)^2} & \dfrac{T_a^{(31)}T_a^{(13)} - T_a^{(11)}T_a^{(33)}}{\left(T_a^{(33)}\right)^2 + \left(T_a^{(31)}\right)^2} & 0 \end{bmatrix},$$
$$\mathbf{H}_2 = \mathbf{H}_3 = \begin{bmatrix} \dfrac{T_a^{(12)}T_a^{(23)} - T_a^{(13)}T_a^{(22)}}{\left(T_a^{(12)}\right)^2 + \left(T_a^{(22)}\right)^2} & 0 & \dfrac{T_a^{(11)}T_a^{(22)} - T_a^{(12)}T_a^{(21)}}{\left(T_a^{(12)}\right)^2 + \left(T_a^{(22)}\right)^2} \\ \dfrac{T_a^{(33)}}{\sqrt{1-(T_a^{(32)})^2}} & 0 & -\dfrac{T_a^{(31)}}{\sqrt{1-(T_a^{(32)})^2}} \\ -\dfrac{T_a^{(31)}T_a^{(32)}}{\left(T_a^{(33)}\right)^2 + \left(T_a^{(31)}\right)^2} & 1 & -\dfrac{T_a^{(32)}T_a^{(33)}}{\left(T_a^{(33)}\right)^2 + \left(T_a^{(31)}\right)^2} \end{bmatrix}. \tag{20}$$

The UPS for transfer alignment of airborne distributed POS postprocessing can be summarized into two parts: the forward filtering solution and the backward smoothing solution.

Figure 3 shows the data flaw of UPS used in airborne distributed POS and the relation between the state vector estimation of forward filtering $\hat{x}_k^F$ and backward recursion $\hat{x}_k^S$ at time k and $k+1$. In forward filtering solution, the filter results are applied to the nominal trajectory during each filtering step and will be reset to zero after the correction of strapdown navigation (velocity, position, and attitude). The function of the backward smoothing solution is to compensate for the estimation of the forward filter stored during the forward filtering process. The filtering and smoothing are calculated at the main system rate. And after each estimation points, a strapdown inertial navigation algorithm is performed to provide the navigation information.

4. Simulation and Analysis

In order to analyze the performance of the proposed UPS, numerical simulation is provided and the comparison with the URTSS using the simulated data of a distributed POS based on one flight trajectory is presented as well.

4.1. Design of Simulation. In this subsection, a typical "S + U" shaped flight trajectory of airborne earth observation is designed. Figure 4 shows the plane trajectory and trajectory parameters, respectively. In Figure 4, the AB and CD segment can be considered as the imaging segment.

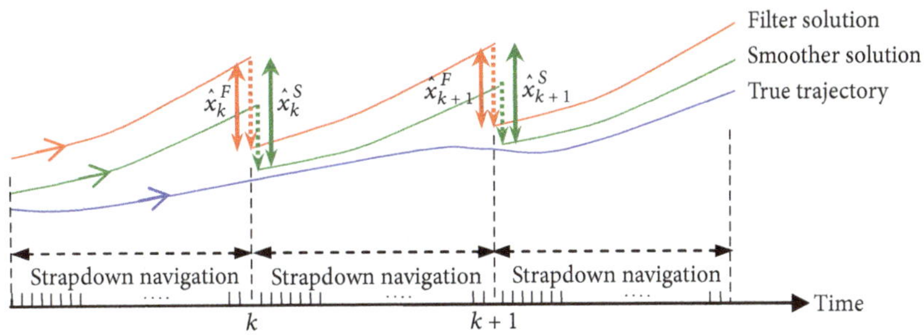

FIGURE 3: The data flaw of UPS used in airborne distributed POS.

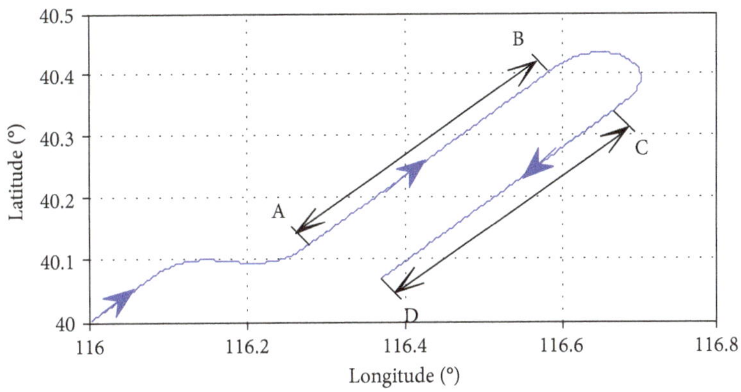

FIGURE 4: Plane trajectory with S-shaped maneuver plus U-shaped flight.

Parameters of flight trajectory are listed as follows: initial latitude is 40°, longitude is 116°, and height is 500 m; initial flight velocity is 100 m/s; and initial heading angle, pitch angle, and roll angle are 40°, 0°, and 0°, respectively. Firstly, the aircraft flies 100 seconds at constant velocity; secondly, the aircraft turns 60° clockwise (100 seconds) and then turns 60° anticlockwise (100 seconds); and thirdly, the aircraft flies 400 seconds at constant velocity, finally turns 180° clockwise (100 seconds), and continually flies 400 seconds at constant velocity. Total flight time is 1300 s.

4.1.1. Data Generation. A trajectory generator is used to generate the theoretical data of the scheduled flight trajectory, which includes position, velocity, attitude, and the output data of gyros and accelerometers. The real outputs of the main system are obtained by adding the corresponding measurement noise to the theoretical position, velocity, and attitude. Then, the theoretical outputs of gyros and accelerometers are converted by rigid misalignment angles and flexure angles, and the constant noise and random noise are added to be the inertial sensor outputs of the subsystem. Meanwhile, the motion parameter benchmarks of subsystem can be obtained through transforming theoretical position, velocity, and attitude by flexure angle.

4.1.2. Parameters of the Main System and Subsystem. The measurement noises of the main system at heading, pitch, roll, and velocity are 0.005°(1σ), 0.0025°(1σ), 0.0025° (1σ), and 0.005m/s (1σ), respectively. Both gyro constant drift and random drift of subsystem are 0.01°/h. Both accelerometer constant bias and random bias of the subsystem are 50μg.

4.1.3. Other Parameters. The rigid misalignment angle of the subsystem relative to the main system is $\rho = [0.5° \quad 0.5° \quad 0.5°]^{\mathrm{T}}$, and the lever arm between the main system and subsystem is $r^{b_m} = [5m \quad 0.1m \quad 0.1m]^{\mathrm{T}}$. For multiantenna SAR, these antennae are fixed under both sides of the wing. So the flexure angle rotated around the vertical axis is large, while the flexure angles around the other two axes are small. Accordingly, the correlation time of the second-order Markov processes is selected as 2 s, 5 s, and 2 s, and the covariances of flexure angles are 0.01, 0.15, and 0.01, respectively. The curves of flexure angle and flexure angle rate selected in this simulation are shown in Figures 5 and 6. All the initial attitude errors are 40°. The particle number is 100. The data update rate of the subsystem is 100 Hz and the filter frequency is 1 Hz.

4.2. Simulation Results and Analysis. The differences between theoretical motion parameters of subsystem and their estimates obtained from estimation, called estimate errors, are used to assess and compare the performance of UPS and URTSS.

Figures 7–10 show the estimate error curves of UPS and URTSS, including the estimate errors of attitude, velocity, position, and baseline. The means of root mean square error (RMSE) and standard deviation (STD) values of motion information estimate errors in imaging segments AB and

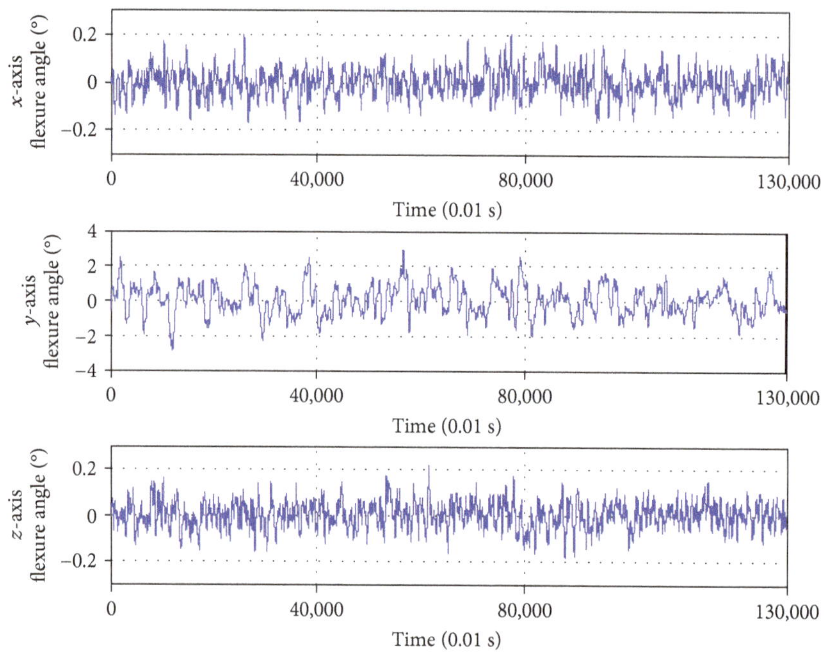

FIGURE 5: Flexure angle on x-axis, y-axis, and z-axis.

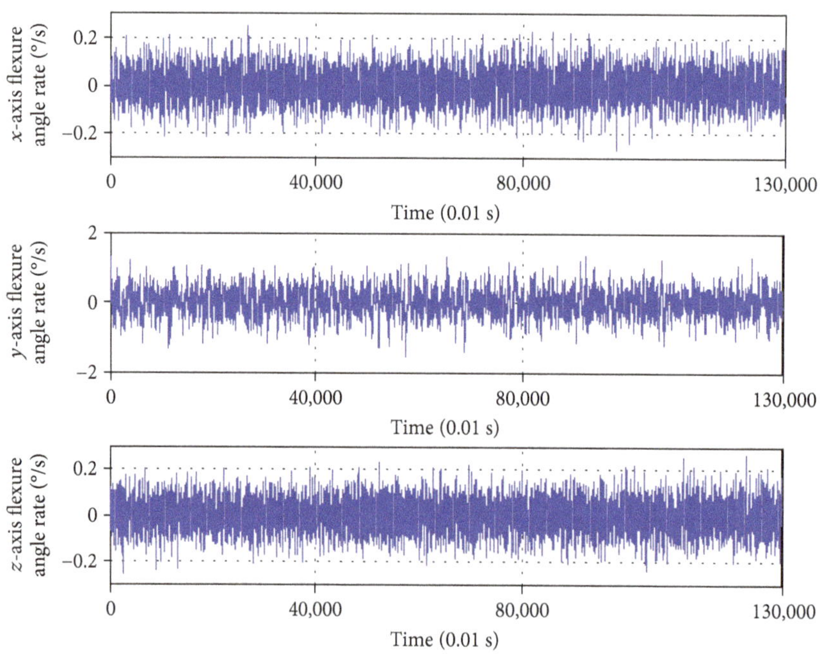

FIGURE 6: Flexure angle rate on x-axis, y-axis, and z-axis.

CD are calculated and shown in Table 1. The calculation formulas of RMSE and STD are given as follows:

$$\mathrm{RMSE} = \sqrt{\frac{\sum_{i=1}^{n'} (e_i)^2}{n'}}, \qquad \mathrm{STD} = \sqrt{\frac{\sum_{i=1}^{n'} (e_i - \bar{e})^2}{n'}}, \tag{21}$$

where e_i denotes the ith estimate error, n' denotes the number of estimate error, and $\bar{e}$ denotes the mean of all estimate error.

For attitude estimation, from Figure 7 and Table 1, we can see that there is a large and obvious trend item in the heading estimation error of URTSS, and the estimation error of URTSS on pitch and roll is prone to be affected by maneuver and their value are bigger than UPS even after 180° clockwise. Compared with URTSS, when the

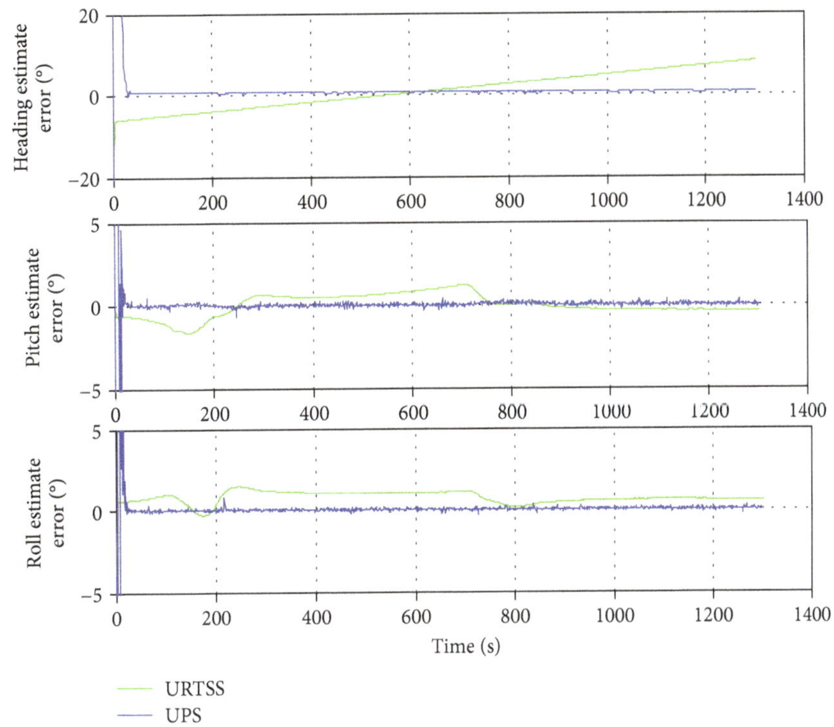

FIGURE 7: Attitude estimate errors of UPS and URTSS.

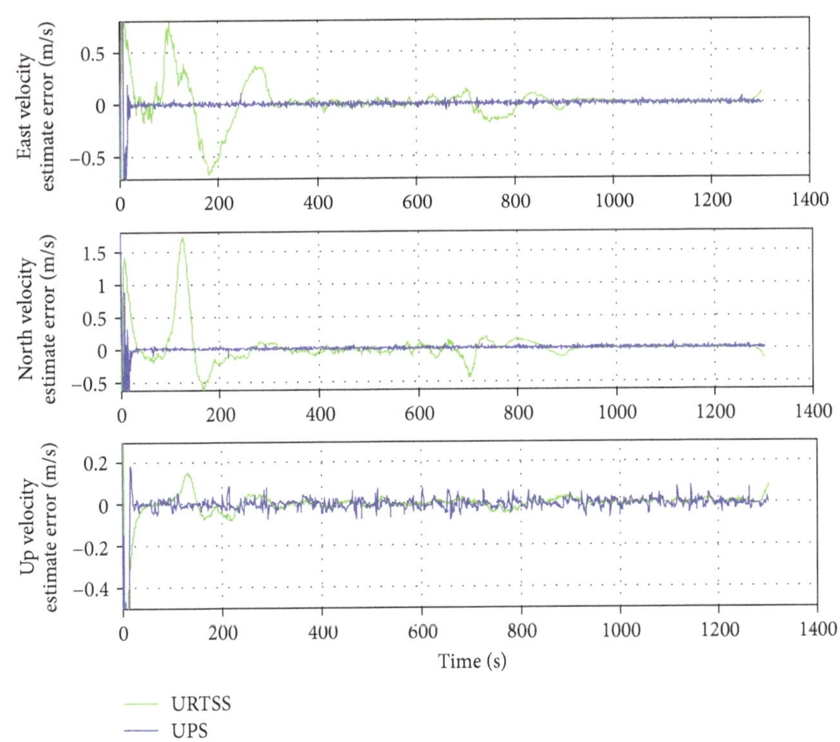

FIGURE 8: Velocity estimate errors of UPS and URTSS.

initial misalignment angle is 40°, the attitude estimation error of UPS is significantly reduced and the error curves are more stable both on AB and CD segments; the estimation accuracy on heading and pitch of UPS is obviously better than that of URTSS, and only the STD on roll of UPS is slightly lower than that of URTSS.

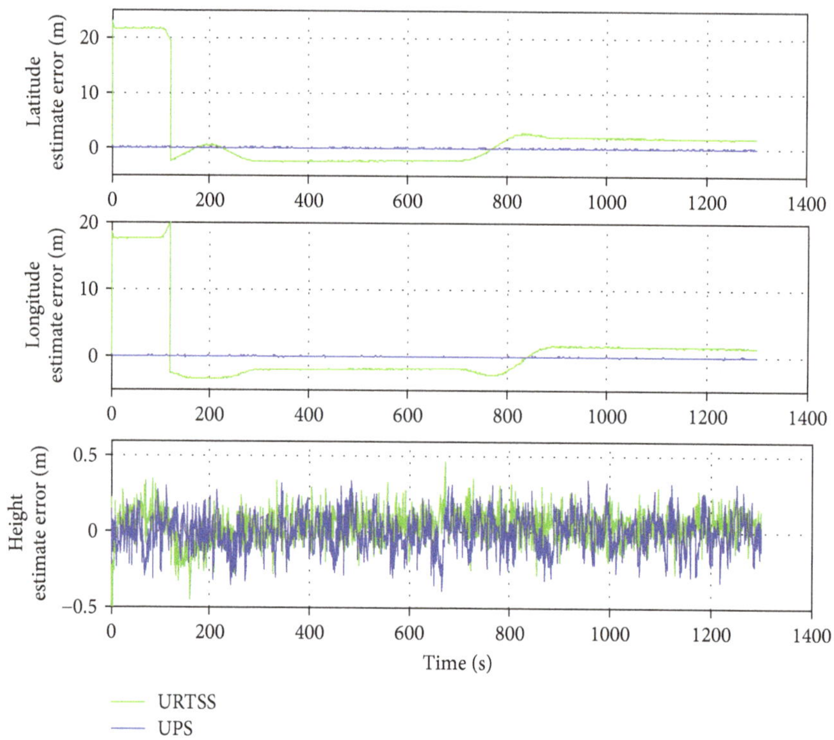

FIGURE 9: Position estimate errors of UPS and URTSS.

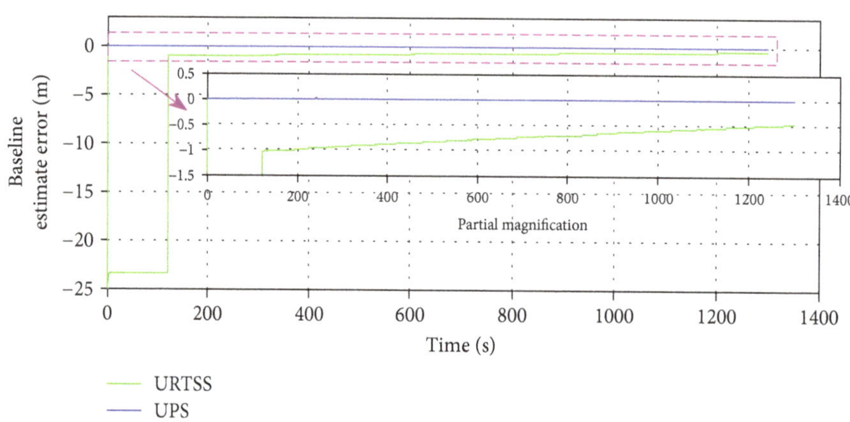

FIGURE 10: Baseline estimate errors of UPS and URTSS.

For velocity estimation, Figure 8 and Table 1 show that the estimation accuracy on east velocity and north velocity of UPS is better than that of URTSS and the error curves of URTSS are prone to be affected by maneuver, especially in the initial 'S' shape flight, the estimation error of velocity of URTSS is unstable and very large; the estimation accuracy on up velocity of UPS is lower than that of URTSS.

For position estimation, Figure 9 and Table 1 show that the estimation accuracy on the latitude and longitude of UPS is better than that of URTSS, and the estimation accuracy on the height of UPS is lower than that of URTSS, which corresponds to the up velocity error result. The estimation error of latitude and longitude of URTSS in the initial 'S' shape flight (from 0 s to 300 s) is very large due to the large velocity error in this stage.

Besides, the space distance between the main system and subsystem (i.e., the baseline between two antennas of SAR) is a very important parameter for interference precision of interferometric SAR and the higher the measurement accuracy of the baseline, the higher the object's digital elevation model accuracy of the interferometric SAR. Here, the baseline error is calculated by theoretical position and real position output of the main system and subsystem. From Figure 10 and Table 1, we can see that the baseline error curve of UPS is more stable and the baseline estimation accuracy of UPS is better than that of URTSS both on RMSE and STD.

Table 1: Means of RMSE and STD values of estimate errors.

Parameter		URTSS		UPS	
		RMSE	STD	RMSE	STD
Attitude error	Heading (°)	3.7874	1.2914	0.4959	0.0217
	Pitch (°)	0.5166	0.1218	0.0766	0.0781
	Roll (°)	0.8238	0.0371	0.0569	0.0568
Velocity error	East (m/s)	0.0264	0.0245	0.0133	0.0134
	North (m/s)	0.0403	0.0414	0.0225	0.0148
	Up (m/s)	0.0125	0.0124	0.0224	0.0211
Position error	Latitude (m)	2.1444	0.0990	1.0077	0.0529
	Longitude (m)	1.7280	0.0878	0.0510	0.0495
	Height (m)	0.1078	0.0987	0.1294	0.0126
Baseline error	(m)	0.7039	0.0532	0.0014	0.0008

From the simulation result, we can see that the estimation accuracy of motion parameters of the proposed method is higher than that of the URTSS as a whole.

5. Conclusion

In this paper, a nonlinear smoother called UPS has been proposed to deal with the off-line transfer alignment estimation of airborne distributed POS. In the UPS, UKF, PF, and smoothing are combined together to provide the respective advantages from each of them to improve estimation accuracy. The validity of the proposed algorithm is verified by simulation test and comparison with URTSS on the estimation accuracy of motion parameters and baseline. The simulation results show that UPS can improve estimation accuracy and has more adaptability to the maneuver than that of URTSS. This method is a better choice for off-line transfer alignment in distributed POS with large misalignment and will greatly improve the resolution of flexible baseline interferometric SAR imaging. The next work is that the practical flight experiment based on long flexible baseline will be implemented to validate the performance of this method.

Conflicts of Interest

The authors declare that they have no conflicts of interest.

Acknowledgments

This work was supported in part by the National Natural Science Foundation of China (Grant nos. 61473020 and 61721091), International (Regional) Cooperation and Communication Project (Grant no. 61661136007), and Fundamental Research Funds for the Central Universities.

References

[1] J. Yang, W. Su, and H. Gu, "High-resolution multiple-input-multiple-output-inverse synthetic aperture radar imaging based on sparse representation," *IET Radar, Sonar & Navigation*, vol. 10, no. 7, pp. 1277–1285, 2016.

[2] L. Gao, Y. Zeng, and L. Wang, "Design of Array for Airborne Forward Looking SAR Based on Compressed Sensing," in *2014 IEEE International Conference on Signal Processing, Communications and Computing (ICSPCC)*, pp. 473–476, Guilin, China, August 2014.

[3] L. T. Zeng, Y. Huai, Z. Y. Li, M. D. Xing, and Y. Liang, "Two-dimensional autofocus technique for high-resolution spotlight synthetic aperture radar," *IET Signal Processing*, vol. 10, no. 6, pp. 699–707, 2016.

[4] D. Li, M. Zhong, and D. Guo, "Error Detection and Compensation in Transfer Alignment for the Distributed POS," in *2013 25th Chinese Control and Decision Conference (CCDC)*, pp. 4194–4199, Guiyang, China, May 2013.

[5] X. Gong and J. Zhang, "An Innovative Transfer Alignment Method Based on Federated Filter for Airborne Distributed POS," *Measurement*, vol. 86, pp. 165–181, 2016.

[6] L. Song, Z. Duan, and J. Sun, "Application of H∞ filter on the angular rate matching in the transfer alignment," *Discrete Dynamics in Nature and Society*, vol. 2016, Article ID 4859475, 9 pages, 2016.

[7] J. Sun, X. S. Xu, Y. T. Liu, T. Zhang, and Y. Li, "Initial alignment of large azimuth misalignment angles in SINS based on adaptive UPF," *Sensors*, vol. 15, no. 9, pp. 21807–21823, 2015.

[8] D. Wang, H. Lv, and J. Wu, "In-flight Initial Alignment for Small UAV MEMS-based Navigation Via Adaptive Unscented Kalman Filtering Approach," *Aerospace Science and Technology*, vol. 61, pp. 73–84, 2017.

[9] J. L. Crassidis, "Sigma-point Kalman filtering for integrated GPS and inertial navigation," *IEEE Transactions on Aerospace and Electronic Systems*, vol. 42, no. 2, pp. 750–756, 2006.

[10] G. Chang and M. Liu, "Hybrid Kalman and Unscented Kalman filters for INS/GPS integrated system considering constant lever arm effect," *Journal of Central South University*, vol. 22, no. 2, pp. 575–583, 2015.

[11] A. Farina, B. Ristic, and D. Benvenuti, "Tracking a Ballistic Target: Comparison of Several Nonlinear Filters," *IEEE Transactions on Aerospace and Electronic Systems*, vol. 38, no. 3, pp. 854–867, 2002.

[12] H. Liu, L. Yu, C. Ruan, and Z. Zhou, "Tracking air-to-air Missile Using Proportional Navigation Model with Genetic Algorithm Particle Filter," *Mathematical Problems in Engineering*, vol. 2016, Article ID 3921608, 11 pages, 2016.

[13] I. Moir, A. G. Seabridge, and M. Jukes, *Military Avionics Systems*, Thomson Press, New Delhi, India, 2005.

[14] N. J. Gordon, D. J. Salmond, and A. F. M. Smith, "Novel approach to nonlinear/non-Gaussian Bayesian state estimation," *IEE Proceedings F Radar and Signal Processing*, vol. 140, no. 2, p. 107, 1993.

[15] R. V. D. Merwe, A. Doucet, N. D. Freitas, and E. Wan, "The Unscented Particle Filter," in *Proceedings of the 13th International Conference on Neural Information Processing Systems*, pp. 563–569, Hong Kong, China, 2006.

[16] J. S. Meditch, "A survey of data smoothing for linear and nonlinear dynamic systems," *Automatica*, vol. 9, no. 2, pp. 151–162, 1973.

[17] S. Särkkä, "Unscented Rauch–Tung–Striebel Smoother," *IEEE Transactions on Automatic Control*, vol. 53, no. 3, pp. 845–849, 2008.

[18] X. Gong, R. Zhang, and J. Fang, "Application of unscented R-T-S smoothing on INS/GPS integration system post processing for airborne earth observation," *Measurement*, vol. 46, no. 3, pp. 1074–1083, 2013.

[19] G. Kitagawa, "Monte Carlo filter and smoother for non-Gaussian nonlinear state space models," *Journal of Computational and Graphical Statistics*, vol. 5, no. 1, pp. 1–25, 1996.

[20] J. Frydendall, "Particle Filters with Applications," Technical Report, University of Denmark, DTU Informatics, No. 2009–02, 2009.

[21] D. Jeon and Y. Eun, "Distributed asynchronous multiple sensor fusion with nonlinear multiple models," *Aerospace Science and Technology*, vol. 39, pp. 692–704, 2014.

[22] Y. Meng, S. Gao, and W. Wang, "Unscented Particle Filter Based Gaussian Process Regression for IMU/BDS Train Integrated Positioning," in *2016 IEEE Information Technology, Networking, Electronic and Automation Control Conference*, pp. 1070–1073, Chongqing, China, May 2016.

[23] J. Cheng, T. Wang, D. Guan, and M. Li, "Polar Transfer Alignment of Shipborne SINS with a Large Misalignment Angle," *Measurement Science & Technology*, vol. 27, no. 3, article 035101, 2016.

[24] D. Goshen-Meskin and I. Y. Bar-itzhack, "Unified approach to inertial navigation system error modeling," *Journal of Guidance, Control and Dynamics*, vol. 15, no. 3, pp. 648–653, 1992.

[25] Y. Lu and X. Cheng, "Random misalignment and lever arm vector online estimation in shipborne aircraft transfer alignment," *Measurement*, vol. 47, no. 1, pp. 756–764, 2014.

4

Performance Trades for Multiantenna GNSS Multisensor Attitude Determination Systems

Nathan A. Tehrani and Jason N. Gross

Department of Mechanical and Aerospace Engineering, West Virginia University, Morgantown, WV, USA

Correspondence should be addressed to Jason N. Gross; jason.gross@mail.wvu.edu

Academic Editor: Linda L. Vahala

We present various performance trades for multiantenna global navigation satellite system (GNSS) multisensor attitude estimation systems. In particular, attitude estimation performance sensitivity to various error sources and system configurations is assessed. This study is motivated by the need for system designers, scientists, and engineers of airborne astronomical and remote sensing platforms to better determine which system configuration is most suitable for their specific application. In order to assess performance trade-offs, the attitude estimation performance of various approaches is tested using a simulation that is based on a stratospheric balloon platform. For GNSS errors, attention is focused on multipath, receiver measurement noise, and carrier-phase breaks. For the remaining attitude sensors, different performance grades of sensors are assessed. Through a Monte Carlo simulation, it is shown that, under typical conditions, sub-0.1-degree attitude accuracy is available when using multiple antenna GNSS data only, but that this accuracy can degrade to degree level in some environments warranting the inclusion of additional attitude sensors to maintain the desired level of accuracy. Further, we show that integrating inertial sensors is more valuable whenever accurate pitch and roll estimates are critical.

1. Introduction

For any airborne sensing platform, the pointing accuracy is dependent on and can be limited to the accuracy of the onboard attitude solution [1, 2]. As such, a key to high pointing accuracy is a robust attitude-determination system. This paper outlines the development, simulation, and testing of a multisensor attitude determination algorithm intended for airborne astronomical and remote sensing platforms.

Attitude determination using multiantenna global navigation satellite system (GNSS) observations is a well-established technology, first proposed by Cohen and Parkinson in 1991 for spacecraft applications [3]. It was also adapted for aircraft use [4] and tested by the same author [5]. Multiantenna GNSS attitude determination has been tested on ground, waterborne, and flight vehicles [6], and the technology has matured to the level that multiple commercially available products [7, 8] are available. The technology has been used for remote sensing platforms since shortly after its proposal [9], and it is in use on multiple stratospheric balloon platforms [10].

There has been considerable effort to simulate gyroscope-free attitude determination using 3-axis magnetometers, 2-axis sun sensors, or both, for spacecraft applications [11]. Highlights include the use of a magnetometer-only sun-pointing algorithm by Ahn and Lee in 2003 [12]. The magnetometer-derived attitude was within 3° of a gyroscope-derived "truth." Psiaki modeled an orbit- and attitude-determination algorithm [13]. Using a 10 nT 3-axis magnetometer and a 0.005°-σ sun sensor, this approach showed less than 0.1° error in all axes. Crassidis and Markley created a sun sensor and magnetometer Kalman filter and showed that a magnetometer-only attitude estimate is markedly improved (error reduced by approximately half) with the inclusion of sun sensor data [11]. Considering the increasing presence of small uninhabited aerial system (UAS) applications in remote sensing, sensor fusion promises high attitude performance with low-cost, lightweight, and compact hardware [10, 14].

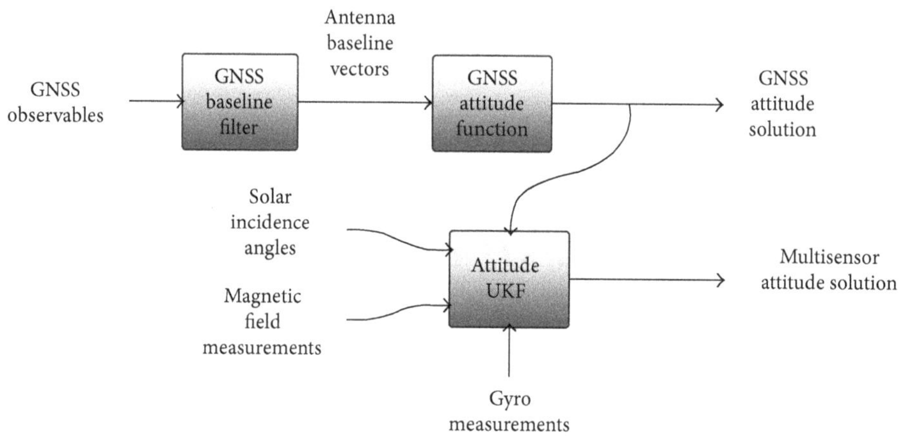

FIGURE 1: Block diagram showing the three main estimators: antenna baseline filter, GNSS-only attitude estimator, and multisensor attitude estimator. This figure is an updated version of what was presented in the conference paper version of this work [16].

This paper outlines the design of a GNSS-based attitude estimator that is then optionally augmented with various other attitude sensors for use in other airborne remote sensing or astronomy applications and offers the contribution of assessing performance sensitivity to various design configuration and common error sources. This paper makes up part of the first author's graduate thesis [15] and is a significant extension upon our previous conference paper [16] in which we improve upon the presented algorithm formulation and revamp our Monte Carlo simulation study design. Through a simulation that is built upon multiple sensor models, the GNSS-only-based attitude solution is shown to work well, but is significantly improved when additional sensors are optimally fused. Specifically, we show the benefits of including inertial, sun sensor, and magnetometer measurements. This paper is expected to aid system designers and scientific investigators as they propose or implement an attitude determination system that is required to meet a certain accuracy level.

The remainder of this paper is organized as follows. In Section 2, the design of the baseline estimation filter and attitude estimation filter is discussed. In Section 3, the simulation environment and the sensor data simulation are discussed. In Section 4, the performance of the GNSS-based and multisensor attitude estimators is presented and discussed. Section 5 summarizes this study's findings and discusses future work.

2. Attitude Estimation

2.1. Algorithm Overview. Figure 1 shows the overall algorithm considered in this study. First, a carrier-phase differential GNSS filter, as detailed in Section 2.2, estimates the baseline separations between antennas. Next, this information is used within a GNSS-only multiple antenna attitude estimator as described in Section 2.3. Finally, the resulting estimated attitude state is optionally fused with a multisensor estimator that also incorporates information from inertial sensors, magnetometers, and sun sensors, as discussed in Section 2.4.

2.2. Antenna Baseline Filter. This attitude estimation algorithm begins with antenna baseline determination using a Kalman filter. The Kalman filter estimator is a linear state estimator which was developed by Kalman, Swerling, and Bucy in a series of papers which detail its formulation and implementation [17–19]. A full derivation of the Kalman filter can be found in Crassidis and Markley and Groves [11, 20]. This section will focus on the state, covariance, and tuning of the Kalman filter applied in this study.

This implementation uses pseudorange and carrier-phase differential GNSS (CD-GNSS) measurements to estimate the relative position vectors between the antennas [21]. The state vector, $\mathbf{x}$, for this filter consists of the relative position vector components between antennas A and B, $\mathbf{x}_{A,B}$, $\mathbf{y}_{A,B}$, and $\mathbf{z}_{A,B}$, and a set of double-differenced pseudoranges and carrier-phase biases, $N_{A,B}$.

$$\mathbf{x}_{\text{baseline}} = \begin{bmatrix} \mathbf{x}_{A,B} \\ \mathbf{y}_{A,B} \\ \mathbf{z}_{A,B} \\ N_{A,B}^{1,k} \\ \vdots \\ N_{A,B}^{j,k} \end{bmatrix}. \tag{1}$$

The measurement models used to model the double-differenced carrier-phase observables follow the same approach outlined in [22], as is discussed next.

The model for an undifferenced GNSS pseudorange, ρ, and carrier-phase measurement, ϕ, (with units of carrier cycles) is given as [21]

$$\begin{aligned} \rho &= r + I_\rho + T_\rho + c(\delta t_u - \delta t_s) + \varepsilon_\rho, \\ \phi &= \lambda^{-1}\left[r + I_\phi + T_\phi\right] + \frac{c}{\lambda}(\delta t_u - \delta t_s) + N + \varepsilon_\phi, \end{aligned} \tag{2}$$

where λ is the wavelength corresponding to the frequencies $L1$ and $L2$ and expressed in meters. The geometric range r between the receiver and GNSS satellite is also expressed in meters, as are the ionospheric and tropospheric delays I and T. The speed of light c is expressed in meters per second. The clock biases of the receiver and satellite, δt_u and δt_s, respectively, are expressed in seconds. The unmodeled error sources, which include multipath and thermal noise, are included in ε in units of meters.

First, range and phase measurements for the master antenna A (antenna 1) and B (antennas 2, 3, or 4) are differenced to form single-differenced phase measurements:

$$\begin{aligned}\Delta\rho^j_{A,B} &= r^j_{A,B} + c\delta t_{A,B} + \varepsilon^j_{\rho,A,B},\\ \Delta\phi^j_{A,B} &= \lambda^{-1} r^j_{A,B} + \frac{c}{\lambda}\delta t_{A,B} + N^j_{A,B} + \varepsilon^j_{\phi,A,B}.\end{aligned} \tag{3}$$

Within (3), due to the very short baseline separation between the antennas, the atmospheric delays completely cancel along with any satellite clock bias and ephemeris errors. Next, the single differenced measurements are then differenced between satellites. For example, between satellite j and reference satellite k,

$$\begin{aligned}\nabla\Delta\rho^{j,k}_{f,A,B} &= r_{A/B_{k|k-1}} + \varepsilon^{j,k}_{\rho,A,B},\\ \nabla\Delta\phi^{j,k}_{f,A,B} &= -\lambda^{-1}\left(\underline{1}^j_A - \underline{1}^k_A\right)^T r_{A/B_{k|k-1}} + N^{j,k}_{A,B} + \varepsilon^{j,k}_{\phi,A,B},\end{aligned} \tag{4}$$

where the remaining receiver clock bias errors are eliminated, leaving only the unknown phase biases $N^{j,k}_{A,B}$, which are known to be integers.

Because the GPS and GLONASS satellite constellations operate at different frequencies, both a GPS and a GLONASS satellite are used as separate reference satellites [23]. GLONASS satellites operate using frequency division multiple access (FDMA), and the wavelength varies from satellite to satellite. The resulting interchannel bias is negligible when using like receivers, as in this model [24]. Because of this, double differences were only formed within each satellite constellation (i.e., GPS and GLONASS) and a different reference satellite was identified for each.

The observation matrix, $\mathbf{H}$, transforms the state $\mathbf{x}$ to predicted measurements. The first three rows of this filter's observation matrix consist of the difference of two three-component unit vectors, which point from the user to the satellite used in the double differencing and to a reference satellite, R.

$$\mathbf{H} = \begin{bmatrix} u^1_\mathrm{x} - u^R_\mathrm{x} & u^1_\mathrm{y} - u^R_\mathrm{y} & u^1_\mathrm{z} - u^R_\mathrm{z} & \lambda_{L1,1} & & & & \\ \vdots & \vdots & \vdots & & \ddots & & & \\ u^n_\mathrm{x} - u^R_\mathrm{x} & u^n_\mathrm{y} - u^R_\mathrm{y} & u^n_\mathrm{z} - u^R_\mathrm{z} & & & \lambda_{L1,n} & & \\ u^1_\mathrm{x} - u^R_\mathrm{x} & u^1_\mathrm{y} - u^R_\mathrm{y} & u^1_\mathrm{z} - u^R_\mathrm{z} & & & \lambda_{L2,1} & & \\ \vdots & \vdots & \vdots & & & & \ddots & \\ u^n_\mathrm{x} - u^R_\mathrm{x} & u^n_\mathrm{y} - u^R_\mathrm{y} & u^n_\mathrm{z} - u^R_\mathrm{z} & & & & & \lambda_{L2,n} \end{bmatrix}. \tag{5}$$

The measurement vector, $\mathbf{z}$, consists of double-differenced pseudorange and phase measurements for each satellite relative to the reference satellite, including measurements for each $L1$ and $L2$ frequencies:

$$\mathbf{z} = \left[\nabla\Delta\rho^{i,k}_{L1,A,B} \quad \nabla\Delta\rho^{i,k}_{L2,A,B} \quad \nabla\Delta\phi^{i,k}_{L1,A,B} \quad \nabla\Delta\phi^{i,k}_{L2,A,B}\right]. \tag{6}$$

Operating in parallel with the baseline estimation Kalman filter, the floating point estimated phase biases (for GPS satellites only); $N^{j,k}_{A,B}$ and their estimated error-covariance are fed into an integer ambiguity resolution algorithm. In particular, the least-squares ambiguity decorrelation adjustment (LAMBDA) method [25] is used to determine the integer biases and adjust the estimated relative positions.

The Kalman filter process noise, Q, and measurement noise, R, and initial error-covariance, P_0, assumptions selected for the differential GNSS baseline estimator are outlined in Table 1:

Table 1: Baseline filter assumed parameters, same as in [16].

Filter parameter	Assumed values
State error covariance P_0	Baseline states: 1 m Ambiguity states: 225 m
Measurement noise covariance R	$\sigma_\phi = 4 \cdot 10^{-4}$ m
Process noise covariance Q	Attitude states: in-run bias $\cdot\, 10^{-2}$ m/$\sqrt{s}$ Ambiguity states : 0 m/$\sqrt{s}$

2.3. Baseline to Attitude. Once the antenna relative baselines with respect to a master antenna are estimated using the baseline estimation filter, an earth-centered, earth-fixed (ECEF) antenna relative position matrix, $\mathbf{L}_\mathrm{e}$, is constructed at each epoch by vertically concatenating the estimate relative vectors of each of nonmaster antenna, as adopted from Cohen and Parkinson [3]:

$$\mathbf{L}_\mathrm{e} = \begin{bmatrix} \mathbf{x}_{2,e} & \mathbf{y}_{2,e} & \mathbf{z}_{2,e} \\ \mathbf{x}_{3,e} & \mathbf{y}_{3,e} & \mathbf{z}_{3,e} \\ \mathbf{x}_{4,e} & \mathbf{y}_{4,e} & \mathbf{z}_{4,e} \end{bmatrix}. \tag{7}$$

This matrix is used to estimate the platform attitude given by the antenna baseline vectors. The known body-centric antenna coordinate matrix, $\mathbf{L}_\mathrm{b}$, with its origin defined as the reference antenna's position, makes up the list of reference vectors.

Next, the singular value decomposition (SVD) method, as proposed by Markley and Mortari and described in (9), (10), and (11), is used to find the rotation matrix $\mathbf{C}^b_e$ between the ECEF, e, and body, b, frames [26]. This is then transformed into the local navigation frame using, n, the earth-to-nav rotation, which is dependent on the master antennas' latitude and longitude and is defined in many texts [20]:

$$\mathbf{C}^b_n = \mathbf{C}^e_n \mathbf{C}^b_e. \tag{8}$$

This solution requires the construction of a matrix $\mathbf{B}$ using the measured vectors $\mathbf{v}_i$ and reference vectors $\mathbf{w}_i$:

$$\mathbf{B} = \sum_{i=1}^{n} \mathbf{v}_i(\mathbf{w}_i)^T. \tag{9}$$

A SVD is performed on **B**, resulting in unitary matrices **U** and **V**. A diagonal, 3×3 matrix **M** is constructed using the determinants of **U** and **V**:

$$\mathbf{M} = \begin{bmatrix} 1 & & \\ & 1 & \\ & & |\mathbf{U}||\mathbf{V}| \end{bmatrix}. \tag{10}$$

By multiplying **U**, **M**, and **V**, one can find the rotation matrix **R**:

$$\mathbf{R} = \mathbf{UMV}^T. \tag{11}$$

The SVD rotation solution also provides a straightforward attitude error covariance matrix which was used as an error covariance matrix in the GNSS-only attitude filter and for the measurement covariance of the GNSS-attitude estimates when fusing the solution in a multisensor Kalman filter [27].

In order to obtain the attitude error covariance matrix, the matrix **B** is multiplied by the transpose of the nontransformed rotation matrix $\mathbf{C}_e^b$.

$$\mathbf{D} = \mathbf{B}\left(\mathbf{C}_e^b\right)^T. \tag{12}$$

D can then be used to find the inverse of the error covariance matrix:

$$P^{-1} = \mathrm{tr}(\mathbf{D}) * \mathbf{I}(3 \times 3) - \mathbf{D}. \tag{13}$$

This is then scaled by multiplying the nominal standard deviation of attitude error attributed to noise in the baseline measurements, in this case 0.01°, to obtain a measurement error covariance matrix to be used for GNSS measurements in the nonlinear filter.

The attitude dilution of precision, as proposed by Yoon and Lundberg, and will be used in the paper to assess trades with respect to multiple GNSS constellations, is a similar metric which assesses the ability to measure the Euler angles [28] depending on the GNSS satellite constellation geometry. It is defined as [28]

$$\mathrm{ADOP} = \sqrt{\mathrm{tr}\left[\left(n\mathbf{I} - \mathbf{SS}^T\right)^{-1}\right]}, \tag{14}$$

where n is the number of satellites in view, **I** is the 3×3 identity matrix, and **S** is a $3 \times N$ matrix comprising the unit vectors to each satellite, including the reference satellite [28].

2.4. Multisensor Unscented Kalman Filter. Finally, a nonlinear estimator is used for attitude determination using all sensor data, due to the nonlinear nature of attitude to sensor observation transformation. Several nonlinear attitude determination methods exist. The QUEST, or quaternion estimation method, seeks the unique quaternion solution for a set of vector measurements and reference vectors [29]. The RE-QUEST, or recursive QUEST, applies the same method, but rather than solving for a single epoch, it uses a filtering approach to find the time-varying attitude profile [30]. Other filtering techniques include the multiplicative extended Kalman filter (M-EKF) [31] and the quadratic extended Kalman filter (Q-EKF) [32].

In this paper, an unscented Kalman filter (UKF) was chosen for its nonlinear transformation ability. The UKF has been shown to perform nearly identically an EKF for GPS/INS attitude estimation applications [33, 34], but offers the implementation benefit of not requiring the analytic evaluation of linearized partial derivatives of the process and observation equations. The details of the UKF implementation followed in this study are offered in the tutorial paper by Rhudy et al. [34], and as such, these details are not discussed in detail herein. In this paper, an outline of the state vector, state prediction $f(x)$, and observation functions $h(x)$ for each measurement update is discussed.

This unscented Kalman filter utilizes inertial measurements for its state prediction, with bias estimation. The GNSS Euler angles, as well as magnetometer and sun sensor data, are used as measurement updates. Magnetometer measurement updates occur at each filter time step, which take place at 50 Hz intervals. GNSS and sun sensor updates occur at 10 Hz intervals.

The state vector, **x** estimated in the multisensor filter is given as

$$\mathbf{x}_U\mathrm{KF} = \begin{pmatrix} \phi \\ \theta \\ \psi \\ b_p \\ b_q \\ b_r \end{pmatrix}. \tag{15}$$

where ϕ, θ, and ϕ are the platform's roll, pitch, and yaw and $b_{p,q,r}$ is the time-varying biases of the IMU's roll rate, p, pitch rate, q, and yaw rate, r, gyroscopes.

Within the UKF framework, at each epoch, the state vector is expanded into a group of $2L+1$ sigma points, χ, where $L = 6$ is the length of the estimated state vector. For each group of sigma points l, the attitude states are predicted by integrating the IMU gyro data through the attitude kinematic equations [35]:

$$f(\phi,\theta,\psi)\colon \begin{bmatrix} \phi_i \\ \theta_i \\ \phi_i \end{bmatrix} = \begin{bmatrix} \phi_{i-1} \\ \theta_{i-1} \\ \phi_{i-1} \end{bmatrix} + \begin{bmatrix} 1 & t(\theta_{i-1})s(\phi_{i-1}) & t(\theta_{i-1})c(\phi_{i-1}) \\ 0 & c(\phi_{i-1}) & -s(\phi_{i-1})c(\theta_{i-1}) \\ 0 & \dfrac{s(\phi)}{c(\theta_{i-1})} & \dfrac{c(\phi_{i-1})}{c(\theta_{i-1})} \end{bmatrix} \cdot \left(\begin{bmatrix} p \\ q \\ r \end{bmatrix} - \begin{bmatrix} b_p \\ b_q \\ b_r \end{bmatrix} \right) \Delta t, \tag{16}$$

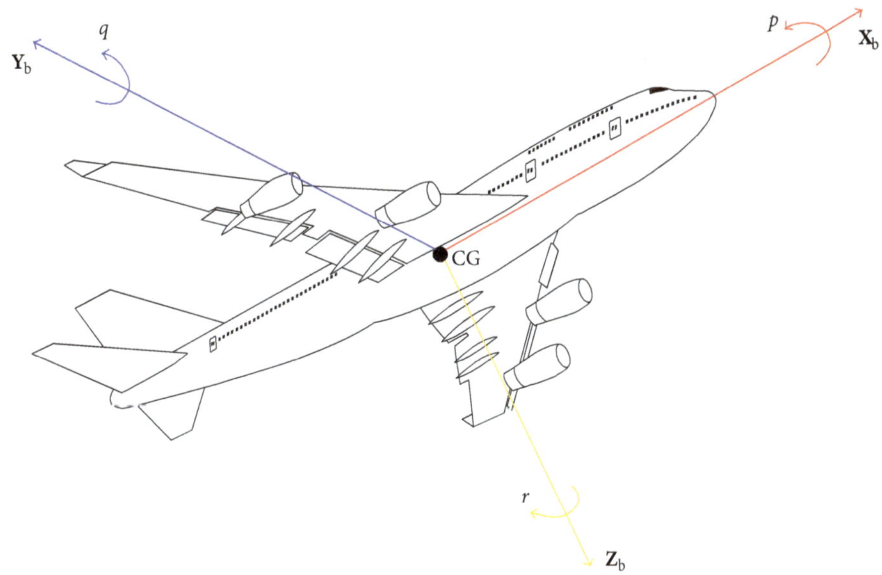

FIGURE 2: The roll, pitch, and yaw delta angles p, q, and r.

where $s(\cdot)$ represents sine, $c(\cdot)$ represents cosine, and $t(\cdot)$ represents tangent.

The delta angles p, q, and r, shown in Figure 2, are the measured delta angles from the gyroscope, corrected for the craft- and earth-rate rotations which were included in the IMU model.

$$\begin{bmatrix} p \\ q \\ r \end{bmatrix} = \begin{bmatrix} p' \\ q' \\ r' \end{bmatrix} - \left(\begin{bmatrix} p_c \\ q_c \\ r_c \end{bmatrix} + \begin{bmatrix} p_e \\ q_e \\ r_e \end{bmatrix} \right). \tag{17}$$

The earth-rate rotations p_e, q_e, and r_e can be found using the earth's rotation transformed to the body frame, then converting to Euler angles [20]

$$\Delta\theta_e = \Omega_E \begin{bmatrix} 0 & \sin(L) & 0 \\ -\sin(L) & 0 & -\cos(L) \\ 0 & \cos(L) & 0 \end{bmatrix} \mathbf{C}_n^b, \tag{18}$$

where Ω_E is earth's rotation rate and L is the craft's latitude. The craft-rate rotation component can be found in a similar way [20]

$$\Delta\theta_c = \begin{bmatrix} 0 & -\omega_{en,\mathbf{z}}^n & \omega_{en,\mathbf{y}}^n \\ \omega_{en,\mathbf{z}}^n & 0 & -\omega_{en,\mathbf{x}}^n \\ -\omega_{en,\mathbf{y}}^n & \omega_{en,\mathbf{x}}^n & 0 \end{bmatrix},$$

$$\omega_{en}^n = \begin{bmatrix} \dfrac{v_{eb,\mathrm{E}}^n}{(R_\mathrm{E}L + h)} \\ -\dfrac{v_{eb,\mathrm{N}}^n}{(R_\mathrm{N}L + h)} \\ -v_{eb,\mathrm{E}}^n \tan \dfrac{L}{(R_\mathrm{E}L + h)} \end{bmatrix}, \tag{19}$$

where $v_{eb,\mathrm{N}}^n$ and $v_{eb,\mathrm{E}}^n$ are the north and east components of the craft's velocity, h is the craft's altitude, and R_E is earth's radius.

Furthermore, ϕ_{i-1}, θ_{i-1}, and ψ_{i-1} are the previous epoch's roll, pitch, and yaw sigma points and are the first three elements of each column of χ, and $b_{p,q,r}$ is the sigma points corresponding to the estimated IMU bias, which is predicted as random walk parameters.

$$f(b_{p,q,r}) \colon \begin{bmatrix} b_{p,i} \\ b_{q,i} \\ b_{r,i} \end{bmatrix} = \begin{bmatrix} b_{p,i-1} \\ b_{q,i-1} \\ b_{r,i-1} \end{bmatrix} + \begin{bmatrix} w_{b,p} \\ w_{b,q} \\ w_{b,r} \end{bmatrix}. \tag{20}$$

The measurement-prediction matrix $\mathbf{\Psi}$ is populated by the predicted measurement vectors using each set of sigma-points in χ. Because measurements occur at different rates in this filter, it is necessary to have different measurement updates occur at different rates. For epochs coinciding with the sun sensor and GNSS attitude measurements, each column $\mathbf{\Psi}_i$ is as follows:

$$\mathbf{\Psi}_i = \begin{pmatrix} \mathbf{B}_{b,\mathbf{x}} \\ \mathbf{B}_{b,\mathbf{y}} \\ \mathbf{B}_{b,\mathbf{z}} \\ \angle_X \\ \angle_Y \\ \phi' \\ \theta' \\ \psi' \end{pmatrix}, \tag{21}$$

where B_b, $\angle_X$, and $\angle_Y$ are predicted magnetometer and sun sensor measurements based on the ith sigma point.

The observation equations use $\widehat{\mathbf{C}}_n^b$, the direction-cosine representation of the predicted attitude states $\widehat{\phi}$, $\widehat{\theta}$, and $\widehat{\psi}$. Modeling of the magnetometer measurements uses the attitude state multiplied by the earth's magnetic field vector:

$$h_{\mathbf{B}}(\phi,\theta,\psi):\ \overrightarrow{\mathbf{B}}_b = \widehat{\mathbf{C}}_n^b \overrightarrow{\mathbf{B}}_n. \tag{22}$$

Sun sensor measurements begin with the navigation-frame solar incidence vector $\mathbf{V}_{\text{sun},n}$, which are multiplied by the attitude state to find the body-frame incidence vector

$$\mathbf{V}_{\text{sun},b} = \widehat{\mathbf{C}}_n^b \mathbf{V}_{\text{sun},n}, \tag{23}$$

and the solar incidence angles $\angle_X$ and $\angle_Y$ are then calculated as follows:

$$\begin{aligned} h_{\angle_X}(\phi,\theta,\psi):\ \angle_X &= \frac{\pi}{2} + \operatorname{atan2}\frac{\mathbf{V}_{\text{sun},\mathbf{z},b}}{\mathbf{V}_{\text{sun},\mathbf{x},b}}, \\ h_{\angle_Y}(\phi,\theta,\psi):\ \angle_Y &= \frac{\pi}{2} + \operatorname{atan2}\frac{\text{sun}_{\mathbf{z},b}}{\text{sun}_{\mathbf{y},b}}, \end{aligned} \tag{24}$$

where atan2 is the four-quadrant tangent inverse.

The measurement update matrix $\mathbf{z}$ consists of the simulated sensor measurement at each filter epoch. These are similar in form to the columns of $\boldsymbol{\Psi}$:

$$\mathbf{z} = \begin{bmatrix} \mathbf{B}_{b,\mathbf{x}} \\ \mathbf{B}_{b,\mathbf{y}} \\ \mathbf{B}_{b,\mathbf{z}} \\ \angle_X \\ \angle_Y \\ \phi_{\text{GNSS}} \\ \theta_{\text{GNSS}} \\ \psi_{\text{GNSS}} \end{bmatrix}. \tag{25}$$

For epochs without GNSS and sun sensor measurements, $\boldsymbol{\Psi}$ and $\mathbf{z}$ consist only of magnetometer predictions and measurements.

3. Data Simulation

3.1. Flight Profile. The simulated flight data used in this study is based upon the recorded flight data of ANITA III balloon experiment. That is, to simulate a balloon flight, the onboard position and attitude solutions were accepted as truth for simulation purposes, and sensor readings with realistic measurement noise were simulated.

Figure 3 shows the Euler angle time histories during a two-hour segment of the ANITA III flight. As indicated in Figure 3, the platform had a small (<1°) oscillation in the roll and pitch axes and a constant rotation about the yaw axis. A variable starting location was used to investigate the effect of the lower GDOP and ADOP at high latitudes.

3.2. GNSS Observable Simulation. For each simulation run, four GNSS receivers were simulated with baseline separations of one-meter each, such that they are arranged in a square configuration. That is, the antennas were placed according to the following matrix:

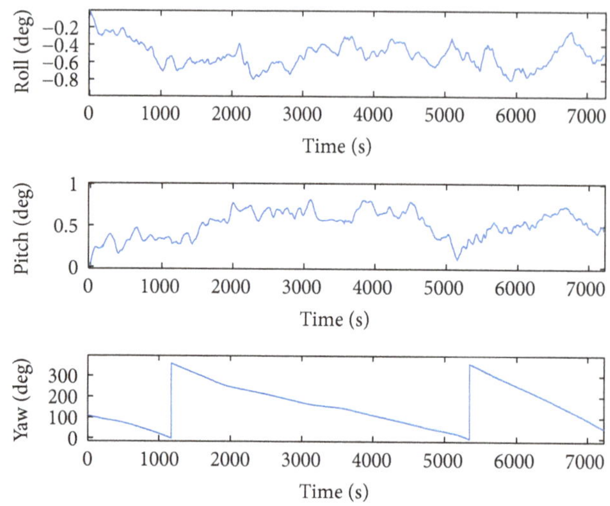

FIGURE 3: Attitude profile used in this work. This figure has been reproduced from the conference paper version of this work [16].

$$\mathbf{L}_b = \begin{bmatrix} \mathbf{x}_{2,\text{b}} & \mathbf{y}_{2,\text{b}} & \mathbf{z}_{2,\text{b}} \\ \mathbf{x}_{3,\text{b}} & \mathbf{y}_{3,\text{b}} & \mathbf{z}_{3,\text{b}} \\ \mathbf{x}_{4,\text{b}} & \mathbf{y}_{4,\text{b}} & \mathbf{z}_{4,\text{b}} \end{bmatrix}, \tag{26}$$

where $\mathbf{x}_{i,\text{b}}$, $\mathbf{y}_{i,\text{b}}$, and $\mathbf{z}_{i,\text{b}}$ are the body-centric coordinates of the ith antenna, $i = 1$, denoting the master antenna. GNSS carrier-phase data was simulated for each flight profile at a rate of 10 Hz using the MATLAB SatNav toolbox [36], which was modified by Watson et al. [37] to include additional GNSS error sources.

A number of deterministic and nondeterministic error sources are associated with GNSS measurements [21]. Fortunately, for attitude estimation applications, several of the primary GNSS error sources, including satellite and receiver clock biases and atmospheric delays, are canceled through the use of double-differenced GNSS observations [21]. However, two important error sources, namely, multipath reflections and carrier-phase breaks, or cycle slips, remain present. For example, when a metallic object reflects a GNSS signal onto the antenna, the multiple paths induce errors [21]. This could be a large problem on balloon-based scientific platforms, as the antennas are spaced closely and in close proximity to science payload. Thermal measurement noise in the receiver is another error source; it is actually amplified by double differencing GNSS data. As such, for this simulation study, multipath, carrier-phase breaks, and receiver thermal errors were assessed with respect to their effect on the attitude estimator's performance using the distributions indicated in Table 2.

3.3. Inertial Measurement Simulation. In addition to GNSS measurements, inertial measurement unit data was simulated for each flight profile and data at a sampling rate of 200 Hz. In particular, four grades of IMU triaxial rate gyroscope

Table 2: Sensor error source Monte Carlo simulation distribution parameters, same as in [16].

Error sources	Model parameters	Notes
Thermal noise	$\sigma_\rho = 0.32$ m, $\sigma_\phi = 0.16\lambda$	Linear scale factor randomly selected between 0 and 1
Multipath	1.0 intensity: $\sigma = 0.4$ m, $\tau = 15$ sec	Linear scale factor randomly selected between 0 and 2
Tropospheric delay	Percent of error assumed handled by broadcast correction	Modified Hopfield with linear scale factor randomly selected between 0.95 and 1.05
Ionospheric delay	First-order ionospheric effects mitigated with dual frequency	Linear scale factor randomly selected between 0.7 and 1
Carrier phase break	Likelihood set to 1 phase break per 24 minutes to 1 phase break per 240 minutes	
Gyroscope	In-run bias $\sigma = 9.6e^{-6}$(rad/sec), ARW = 0.2(deg/$\sqrt{\text{hr}}$)	Scaled Honeywell HG1700AG72 SF = (1/50, 1/200, 1/400)
Sun sensor	Zenith measurement noise $\sigma = 0.1$deg	Scaled solar MEMS ISSDX-60 SF = (1, 2, 3, 4)
Magnetometer	Measurement noise $\sigma = 2.67$ nT $\mathbf{A}_{\text{si}}$ terms scaled between 0.005 and 0.01 $\overrightarrow{\mathbf{B}}_{\text{hi}}$ terms scaled between 25 nT and 50 nT	ST LSM9DS0

and accelerometers were assessed in the simulation. To simulate the effect of IMU noise, these ideal measurements were then polluted with both a time-varying bias b_i with in-run stability $\sigma_{\text{in-run}}$ and measurement noise σ_{ARW}:

$$\begin{aligned} b_i &= b_{i-1} + X_1\sigma_{\text{in-run}}, \\ \Delta\theta_{i'} &= \Delta\theta_i + b_i + X_2\sigma_{\text{ARW}}, \end{aligned} \tag{27}$$

where X_1 and X_2 are normally distributed random numbers with a zero mean and unit standard deviation. The magnitude of these two noise terms was selected based on the grade of the inertial sensors assumed, which were varied as indicated in Table 2.

3.4. Magnetometer Simulation. Triaxial magnetometer data was also simulated for each flight based on the measurement models and uncertainties of available sensors [38]. In particular, the earth's magnetic field along the flight profile was calculated and sensor measurements were simulated by polluting these true values with random noise based on the measurement noises quoted by the manufacturers' specification sheets as indicated in Table 2 [38].

The magnetometer data consists of magnetic field intensity measurements ($\mathbf{B}_b$) in three orthogonal directions corresponding to the north, N, east, E, and down D axes in the body frame, b. This begins with $\mathbf{B}_\text{E}$, a vector constraining the simulated magnetic field intensities in the navigation frame, generated at each location along the flight path:

$$\overrightarrow{\mathbf{B}}_\text{E} = \begin{pmatrix} \mathbf{B}_{b,\text{N}} \\ \mathbf{B}_{b,\text{E}} \\ \mathbf{B}_{b,\text{D}} \end{pmatrix}. \tag{28}$$

The magnetic field was generated using the World Magnetic Model (WMM) [39] in an interface developed by Hardy et al. [40].

Body-frame magnetic field measurements are simulated by multiplying truth attitude (represented by the direction-cosine matrix $\mathbf{C}_n^b$) by the navigation-frame magnetic field:

$$\overrightarrow{\mathbf{B}}_b = \mathbf{C}_n^b \overrightarrow{\mathbf{B}}_E. \tag{29}$$

With three contributing error sources added: hard and soft iron errors and measurement noise, in a simplified method as described by Gebre-Egziabher et al. [41]:

$$\widehat{\mathbf{B}} = \mathbf{A}_{\text{si}} \overrightarrow{\mathbf{B}}_b + \overrightarrow{\mathbf{B}}_{\text{hi}}, \tag{30}$$

where $\mathbf{A}_{\text{si}}$ is a 3×3 matrix which describes the soft-iron error effect and $\overrightarrow{\mathbf{B}}_{\text{hi}}$ is a 3×1 vector containing the hard-iron offset, a magnetic field generated by ferromagnetic material on the platform. For this study, nominal values for $\mathbf{A}_{\text{si}}$ and $\overrightarrow{\mathbf{B}}_{\text{hi}}$ were used, based on the calibrations in the Gebre-Egziabher et al. paper [41]. Simulated measurement noise was then added to $\widehat{\mathbf{B}}$, corresponding to the precision level of the modeled magnetometer.

3.5. Sun Sensor Simulation. Simulated sun sensor data consists of solar incidence angles $\angle_X$ and $\angle_Y$ relative to the two horizontal body-frame axes X_b and Y_b. These were generated using the apparent solar azimuth θ_{sun} and elevation ϕ_{sun} calculated for each epoch of the flight duration. First, the solar azimuth and elevation values are transformed into a unit vector representing the sun's position in the sky with respect to the navigation frame, n:

$$\mathbf{V}_{\text{sun},n} = \left(\text{sun}_{\mathbf{x},n}\text{sun}_{\mathbf{y},n}\text{sun}_{\mathbf{z},n}\right). \tag{31}$$

This unit-vector is then transformed using the nav-to-body direction cosine matrix $\mathbf{C}_n^b$:

$$\mathbf{V}_{\text{sun},b} = \mathbf{C}_n^b \mathbf{V}_{\text{sun},n}. \tag{32}$$

Table 3: GNSS attitude performance—median RMS error.

Median errors	Roll (deg.)	Pitch (deg.)	Heading (deg.)
GPS only	0.44	0.42	0.16
GPS + GLONASS	0.28	0.26	0.12

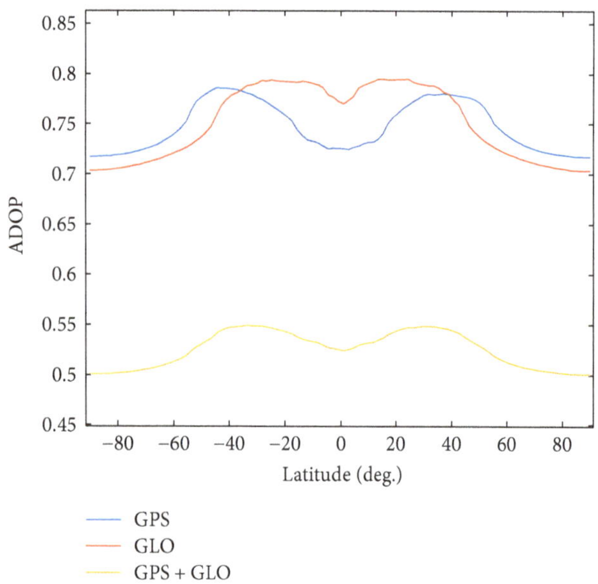

Figure 4: ADOP calculated along one meridian, averaged over a 24-hour period for the GPS constellation, for the GLONASS constellation, and for both constellations.

The incidence angles $\angle_X$ and $\angle_Y$ are calculated as follows:

$$\angle_X = \frac{\pi}{2} + \text{atan2}\frac{\text{sun}_{\mathbf{z},b}}{\text{sun}_{\mathbf{x},b}},$$
$$\angle_Y = \frac{\pi}{2} + \text{atan2}\frac{\text{sun}_{\mathbf{z},b}}{\text{sun}_{\mathbf{y},b}}, \quad (33)$$

as in the filter's measurement prediction step.

As with the magnetometer measurements, simulated measurement noise was added to the sun sensor measurements, corresponding to the performance of available sensors [42]. However, in the case of a sun sensor, as measurement noise increases at low solar elevations, the measurement noise was scaled according to solar elevation angle. Sun sensor measurements were simulated at 10 Hz intervals.

3.6. Error Sources and Monte Carlo. For this study, a total of 50 one-hour flight profiles were simulated in a Monte Carlo manner. In particular, the ECEF starting positions, magnitude of GNSS error sources, and quality of IMU, magnetometer, and sun sensor data were varied as indicated in Table 2. Note that by randomly varying the starting location, the GNSS constellation satellite geometry was randomized as well.

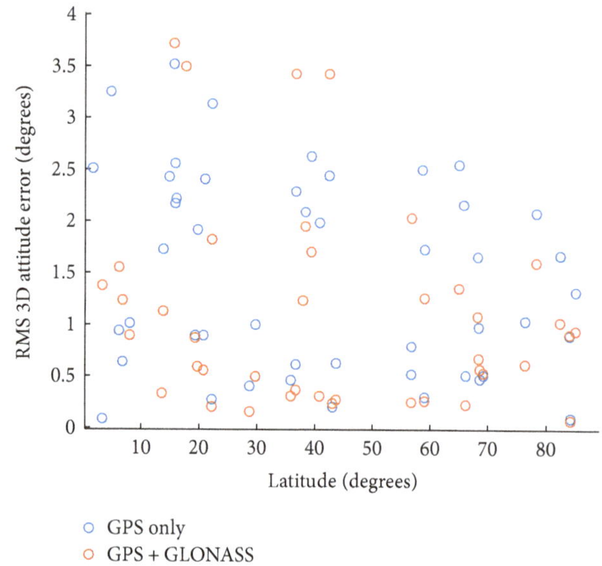

Figure 5: Comparison between GPS-only mode and GPS + GLONASS mode for all profiles, latitude shown.

Table 4: Unscented Kalman filter error statistics: median attitude error.

Median errors	Roll (deg.)	Pitch (deg.)	Heading (deg.)
GNSS + INS	0.59	0.38	0.64
INS + Mag + SS	0.0041	0.047	0.045
GNSS + INS + Mag + SS	0.048	0.053	0.047

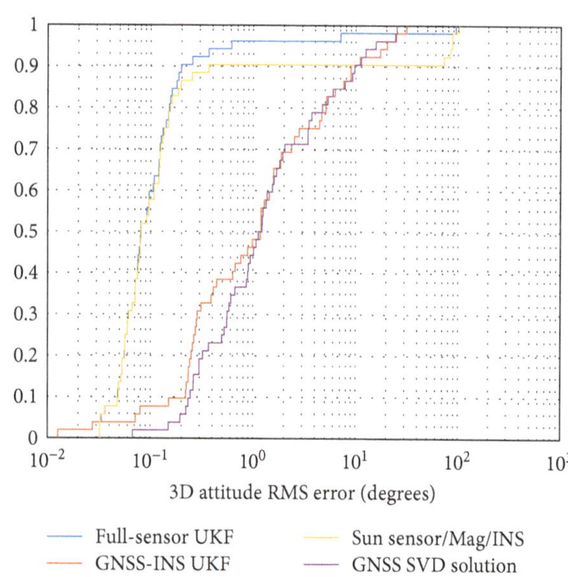

Figure 6: Comparison between GNSS-SVD solution and multisensor attitude filter in different modes—3-axis attitude error.

4. Results

4.1. GNSS Only. The GNSS-only attitude determination script was run in two modes, the first using GPS data only and the second adding GLONASS observables. The pitch,

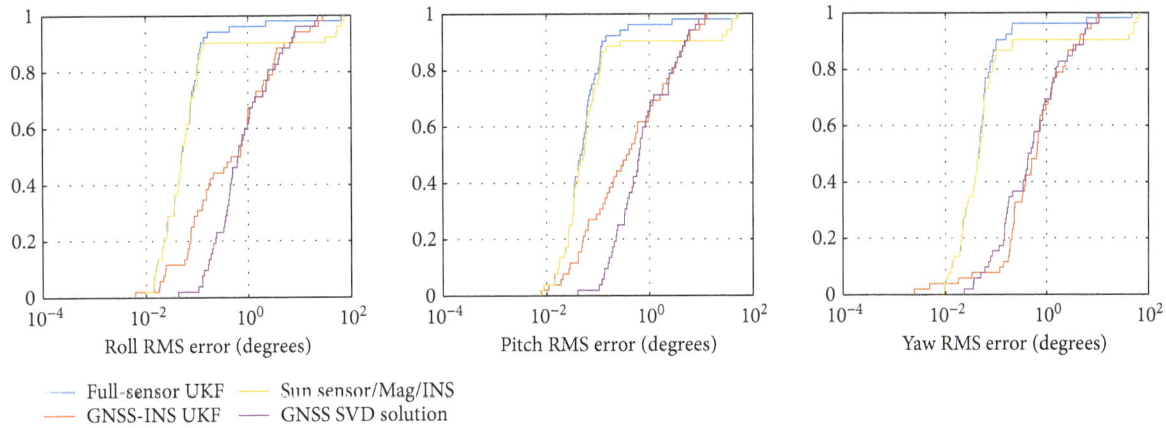

FIGURE 7: Comparison between GNSS-SVD solution and multisensor attitude filter in different modes—roll, pitch, and yaw error.

roll, and heading error statistics for both filter modes are presented in Table 3. These results include two simulations for which the baseline filter solution failed to converge, presumably due to carrier-phase break.

Using GLONASS as well as GPS satellites yielded a median performance improvement of 40 percent lower attitude error. In an Antarctic flight regime, fewer GNSS satellites are observable, and these are seen at lower elevations [43]. This can negatively impact the geometric dilution of precision (GDOP), a metric that describes the geometric diversity of satellite-receiver vectors [21] and also the attitude dilution of precision (ADOP), as defined in (14). Figure 4 shows the ADOP calculated by latitude for a single meridian, averaged over a full day. At high latitudes, the GLONASS constellation has a lower (and therefore more desirable) ADOP figure, and using both constellations yields a lower ADOP at all latitudes. Figure 5 shows the overall comparison between the GPS-only and GPS and GLONASS attitude solutions. For most profiles across the latitude range, the GPS + GLONASS solution outperformed the GPS-only solution.

4.2. GNSS and Multisensor Attitude Filter. Table 4 presents overall error statistics for the 50 trials for the GNSS + INS, GNSS + all sensors, and all sensors without GNSS, respectively.

Figure 6 shows the cumulative distribution of the 3D attitude error $= \sqrt{\phi^2 + \theta^2 + \psi^2}$ for the various filter configurations over the 50 simulated flights, and Figure 7 shows the corresponding roll, pitch, and yaw errors for the simulated flights.

In these tables, it is clear that using additional sensors in addition to GNSS can markedly improve performance. For example, Figure 8 shows the attitude estimation error for one example trial, in which the GNSS-only attitude is shown alongside the multisensor filters for comparison. When accurate pitch and roll estimates are critical, the inertial sensors offer more trials with significant accuracy increases, as evident when looking at the GNSS-INS-only solution. When the full suite of sensors is combined, 10% of the sun/INS/Mag solutions that drift considerably are

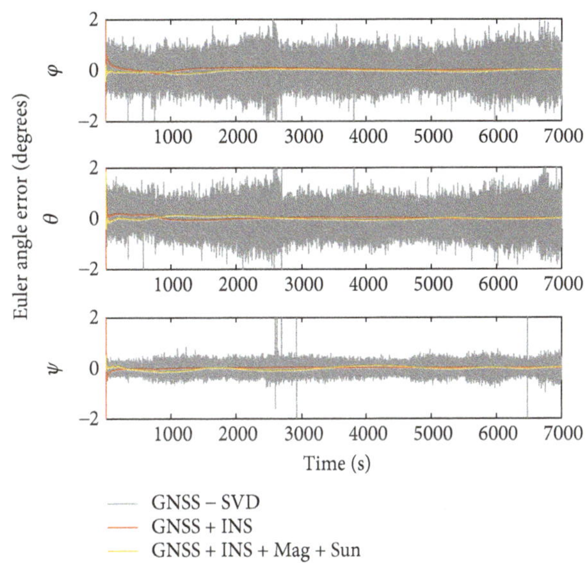

FIGURE 8: Roll, pitch, and heading errors for multisensor filter in GNSS + INS mode, GNSS + INS + Mag + SS mode, with GNSS-only result for comparison. This figure has been reproduced from the conference paper version of this work [16].

improved. As expected, the yaw estimate is greatly improved by the sun and magnetic measurements.

Of great interest is the algorithm's ability to handle carrier-phase breaks. For example, phase breaks could occur due to radiofrequency interference, such as during a data uplink/downlink transmission over the Iridium satellite constellation which operates very close to the GPS L1 frequency [44]. When a carrier-phase break occurs, it can fortunately be detected easily by a data editor [45]. As such, whenever this occurs, the baseline estimation filter resets the error covariance for the impacted carrier-phase ambiguities to a large value. The multisensor filter attitude determination performance was generally lower across the range of phase break likelihoods as shown in Figure 9. Notably, the multisensor UKF yielded a low error-level attitude solution for the few trials with considerably high GNSS-attitude errors.

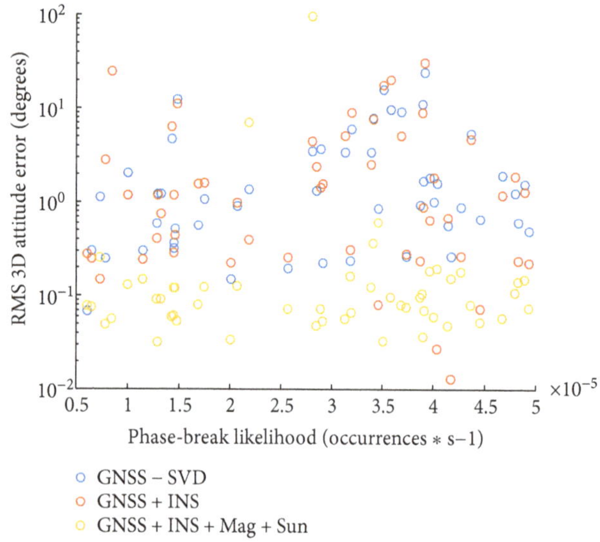

Figure 9: RMS attitude versus phase break likelihood for each trial.

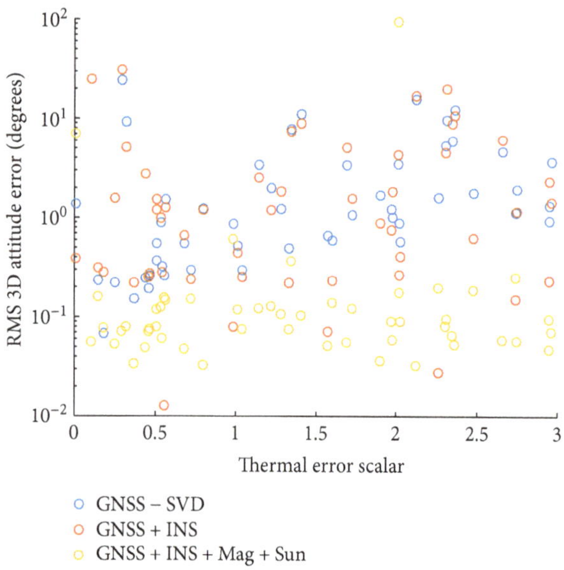

Figure 10: RMS attitude versus thermal error scale factor for each trial.

Also of interest is the estimator's performance with high receiver measurement thermal noise and GNSS multipath reflection errors. These could arise on a scientific platform depending on the GNSS antenna configuration relative to the other scientific instruments and antennas/Figures 10 and 11 show that the multisensor filter yields lower magnitude errors than the GNSS-only filter across both error scale ranges. Although an increasing level of multipath error did not noticeably affect the result of the GNSS-only filter performance, the multisensor filter performed better in nearly all trials.

Sensitivity to the ionospheric and tropospheric error contribution to the GNSS errors was not considered, as the short baseline between antennas led to cancellation of those error sources.

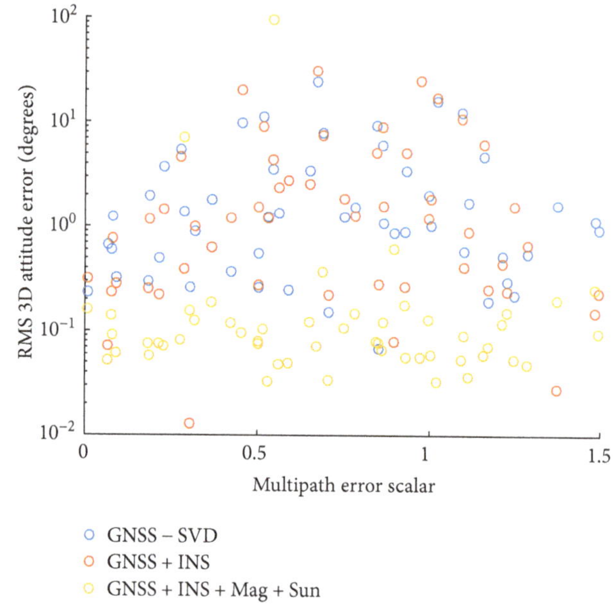

Figure 11: RMS attitude versus multipath error scale factor for each trial.

5. Conclusions

This study outlined the design of a GNSS-based attitude determination algorithm and its incorporation in a nonlinear attitude-determination filter which also utilized inertial measurements. Additional measurements from sun sensors and magnetometers were also proposed and added to the algorithm architecture. The algorithm was run in multiple modes with varying levels of measurement errors to assess how each measurement type contributed to the attitude determination performance. First, it was shown that the GNSS-only attitude solutions are consistently improved when GLONASS satellites are included in addition to GPS, owing to more observables and lower dilution of precision (especially in polar regions). Across all flight profiles, GLONASS observables yielded a median of 30 percent lower error in the pitch and roll axes, with closer performance in the yaw axis. The performance was somewhat mitigated when GNSS Euler angle measurements were combined with inertial measurements only. This was most evident in the yaw axis, further indicating tuning as a potential culprit. Incorporating sun sensor and magnetometer measurements yielded the best improvement to the nonlinear filter's performance, with median performance better than tenth-degree accuracy for a majority of the trials. The combined estimator showed nearly a uniform distribution and nearly consistent improvement when assessing sensitivity against two important GNSS error sources, phase breaks and multipath.

Conflicts of Interest

The authors declare that they have no conflicts of interest.

Acknowledgments

This work was supported in part by NASA West Virginia Space Grant Consortium Graduate Student Fellowship Program and the National Geospatial-Intelligence Agency Academic Research Program Grant no. HM0476-15-1-0004. This study is approved for NGA public release under case no. 17-839.

References

[1] U. Schonhoff, A. Klein, and R. Nordmann, "Attitude control of the airborne telescope SOFIA: μ-synthesis for a large scaled flexible structure," in *Proceedings of the 39th IEEE Conference on Decision and Control (Cat. No.00CH37187)*, pp. 3576–3581, Sydney, NSW, Australia, 2000.

[2] U. Schönhoff, P. Eisenträger, K. Wandner, H. J. Kaercher, and R. Nordmann, "End-to-end simulation of the image stability for the airborne telescope SOFIA," in *Astronomical Telescopes and Instrumentation*, pp. 380–389, Munich, Germany, 2000, SPIE - International Society for Optics and Photonics.

[3] C. E. Cohen and B. W. Parkinson, "Expanding the performance envelope of GPS-based attitude determination," in *ION GPS-91; Proceedings of the 4th International Technical Meeting of the Satellite Division of the Institute of Navigation*, pp. 1001–1011, Albuquerque, NM, USA, 1991.

[4] C. E. Cohen and B. W. Parkinson, "Aircraft applications of GPS-based attitude determination," in *Proceedings of the 5th International Technical Meeting of the Satellite Division of The Institute of Navigation (ION GPS 1992)*, pp. 775–782, Albuquerque, NM, USA, 1992.

[5] C. E. Cohen, B. W. Parkinson, and B. D. McNally, "Flight tests of attitude determination using GPS compared against an inertial navigation unit," *Navigation*, vol. 41, no. 1, pp. 83–97, 1994.

[6] P. J. G. Teunissen, G. Giorgi, and P. J. Buist, "Testing of a new single-frequency GNSS carrier phase attitude determination method: land, ship and aircraft experiments," *GPS Solutions*, vol. 15, no. 1, pp. 15–28, 2011.

[7] "ADU5," Magellan Professional, 6 2007, rev. C.

[8] NovAtel, "Attitude-pitch/roll/yaw," http://www.novatel.com/solutions/attitude/#whatWeOffer.

[9] C. K. Toth and D. A. Grejner-Brzezinska, "Performance analysis of the airborne integrated mapping system (aims)," *International Archives of Photogrammetry and Remote Sensing*, vol. 32, pp. 320–326, 1997.

[10] N. N. Gandilo, P. A. R. Ade, M. Amiri et al., "Attitude determination for balloon-borne experiments," in *SPIE Astronomical Telescopes + Instrumentation*, p. 91452U, Montréal, QC, Canada, 2014, SPIE - International Society for Optics and Photonics.

[11] J. L. Crassidis and F. L. Markley, "Predictive filtering for attitude estimation without rate sensors," *Advances in the Astronautical Sciences*, vol. 93, pp. 1021–1038, 1996.

[12] H.-S. Ahn and S.-H. Lee, "Gyroless attitude estimation of the sun-pointing mode satellite," in *International Conference on Control, Automation and Systems*, pp. 22–25, Gyeongju, Korea, October 2003.

[13] M. L. Psiaki, "Autonomous low-earth-orbit determination from magnetometer and sun sensor data," *Journal of Guidance, Control, and Dynamics*, vol. 22, no. 2, pp. 296–304, 1999.

[14] J. Everaerts, "The use of unmanned aerial vehicles (UAVS) for remote sensing and mapping," *The International Archives of the Photogrammetry, Remote Sensing and Spatial Information Sciences*, vol. 37, no. 2008, pp. 1187–1192, 2008.

[15] N. Tehrani, *Characterization and Flight Test of Multi-Antenna GNSS, Multi-Sensor Attitude Determination for Stratospheric Balloon Platforms, [M.S. Thesis]*, West Virginia University, Morgantown, WV, USA, 2017.

[16] N. Tehrani and J. Gross, "Characterization of multi-antenna GNSS, multi-sensor attitude determination for stratospheric balloon platforms," in *AIAA Modeling and Simulation Technologies Conference*, p. 1554, Grapevine, Texas, 2017.

[17] P. Swerling, *A Proposed Stagewise Differential Correction Procedure for Satellite Tracking and Prediction*, Rand Corporation, Santa Monica, CA, USA, 1958.

[18] R. E. Kalman, "A new approach to linear filtering and prediction problems," *Journal of Basic Engineering*, vol. 82, no. 1, pp. 35–45, 1960.

[19] R. E. Kalman and R. S. Bucy, "New results in linear filtering and prediction theory," *Journal of Basic Engineering*, vol. 83, no. 1, pp. 95–108, 1961.

[20] P. D. Groves, *Principles of GNSS, Inertial, and Multisensor Integrated Navigation Systems*, Artech House, Boston, MA, USA, 2013.

[21] P. Misra and P. Enge, *Global Positioning System: Signals, Measurements and Performance*, Ganga-Jamuna Press, Lincoln, MA, USA, 2nd edition, 2006.

[22] J. N. Gross, Y. Gu, and M. B. Rhudy, "Robust UAV relative navigation with DGPS, INS, and peer-to-peer radio ranging," *IEEE Transactions on Automation Science and Engineering*, vol. 12, no. 3, pp. 935–944, 2015.

[23] A. E. Zinoviev, "Using GLONASS in combined GNSS receivers: current status," in *Proceedings of ION GNSS*, pp. 1046–1057, Long Beach, CA, USA, 2005.

[24] A. Al-Shaery, S. Zhang, and C. Rizos, "An enhanced calibration method of GLONASS inter-channel bias for GNSS RTK," *GPS Solutions*, vol. 17, no. 2, pp. 165–173, 2013.

[25] P. Teunissen, P. De Jonge, and C. Tiberius, "The LAMBDA method for fast GPS surveying," in *International Symposium "GPS Technology Applications"*, pp. 1–8, Bucharest, Romania, 1995.

[26] F. L. Markley and D. Mortari, "How to Estimate Attitude from Vector Observations," in *AAS/AIAA Astrodynamics Specialist Conference*, pp. 16–19, Girdwood, Alaska, August 1999.

[27] M. D. Shuster, "The generalized WAHBA problem," *The Journal of the Astronautical Sciences*, vol. 54, no. 2, pp. 245–259, 2006.

[28] S. Yoon and J. B. Lundberg, "Euler angle dilution of precision in GPS attitude determination," *IEEE Transactions on Aerospace and Electronic Systems*, vol. 37, no. 3, pp. 1077–1083, 2001.

[29] M. D. Shuster and S. D. Oh, "Three-axis attitude determination from vector observations," *Journal of Guidance, Control, and Dynamics*, vol. 4, no. 1, pp. 70–77, 1981.

[30] I. Y. Bar-Itzhack, "Request - a recursive quest algorithm for sequential attitude determination," *Journal of Guidance, Control, and Dynamics*, vol. 19, no. 5, pp. 1034–1038, 1996.

[31] F. L. Markley, "Attitude error representations for Kalman filtering," *Journal of Guidance Control and Dynamics*, vol. 26, no. 2, pp. 311–317, 2003.

[32] J. Villares and G. Vázquez, "The quadratic extended Kalman filter," in *Processing Workshop Proceedings, 2004 Sensor Array and Multichannel Signal*, pp. 480–484, Barcelona, Spain, 2004, IEEE.

[33] J. N. Gross, Y. Gu, M. B. Rhudy, S. Gururajan, and M. R. Napolitano, "Flight-test evaluation of sensor fusion algorithms for attitude estimation," *IEEE Transactions on Aerospace and Electronic Systems*, vol. 48, no. 3, pp. 2128–2139, 2012.

[34] M. Rhudy, Y. Gu, J. Gross, S. Gururajan, and M. R. Napolitano, "Sensitivity analysis of extended and unscented Kalman filters for attitude estimation," *Journal of Aerospace Information Systems*, vol. 10, no. 3, pp. 131–143, 2013.

[35] B. L. Stevens, F. L. Lewis, and E. N. Johnson, *Aircraft Control and Simulation: Dynamics, Controls Design, and Autonomous Systems*, John Wiley & Sons, Hoboken, NJ, USA, 2015.

[36] GPSoft, *Satellite Navigation Toolbox 3.0 User's Guide*, 2003.

[37] R. M. Watson, V. Sivaneri, and J. N. Gross, "Performance characterization of tightly-coupled GNSS precise point positioning inertial navigation within a simulation environment," in *AIAA Guidance, Navigation, and Control Conference*, p. 1869, San Diego, CA, USA, 2016.

[38] STMicroelectronics, "LSM9DS0," 2013, http://www.st.com/resource/en/datasheet/lsm9ds0.pdf.

[39] S. Maus, S. Macmillan, S. McLean et al., "The US/UK world magnetic model for 2010-2015," NOAA Tech. Rep. NESDIS/NGDC, Boulder, CO, USA, 2010.

[40] J. Hardy, J. Strader, J. N. Gross et al., "Unmanned aerial vehicle relative navigation in GPS denied environments," in *2016 IEEE/ION Position, Location and Navigation Symposium (PLANS)*, pp. 344–352, Savannah, GA, USA, 2016, IEEE.

[41] D. Gebre-Egziabher, G. H. Elkaim, J. David Powell, and B. W. Parkinson, "Calibration of strapdown magnetometers in magnetic field domain," *Journal of Aerospace Engineering*, vol. 19, no. 2, pp. 87–102, 2006.

[42] SolarMEMS, "ISSDX," 2010, http://www.solar-mems.com/wp-content/uploads/2017/02/ISSDX_Technical_Specifications.pdf.

[43] S. Cojocaru, E. Birsan, G. Batrinca, and P. Arsenie, "GPS-GLONASS-GALILEO: a dynamical comparison," *Journal of Navigation*, vol. 62, no. 01, pp. 135–150, 2009.

[44] K. Maine, C. Devieux, and P. Swan, "Overview of Iridium satellite network," in *WESCON/'95. Conference record. 'Microelectronics Communications Technology Producing Quality Products Mobile and Portable Power Emerging Technologies*, p. 483, San Francisco, CA, USA, 1995, IEEE.

[45] G. Blewitt, "An automatic editing algorithm for GPS data," *Geophysical Research Letters*, vol. 17, no. 3, pp. 199–202, 1990.

5

Allocation Optimization Strategy for High-Precision Control of Picosatellites and Nanosatellites

Bo Yang,[1] Jun Miao,[1] Zichen Fan,[1] Jun Long,[2] and Xuhui Liu[2]

[1]*School of Astronautics, Beihang University, Beijing, China*
[2]*Beijing Institute of Control Engineering, Beijing, China*

Correspondence should be addressed to Jun Miao; miaojun@buaa.edu.cn

Academic Editor: Enrico Cestino

The high-precision control of picosatellites and nanosatellites has always plagued the astronautics field. Aiming to change the status quo of the actuators not being able to meet the high-precision attitude control of picosatellites and nanosatellites, this article formulates a control allocation strategy for picosatellites and nanosatellites using the solid propellant microthruster array (SPMA). To solve the problem of the diversity and complexity of ignition combinations brought about by the high integration of the SPMA, the energy consumption factor of the optimal allocation is established, and the relationships of the array's energy consumption factor, the control accuracy, the number, and the ignition combinations of the thruster array are deduced. The optimization objective is introduced by Sherman-Morrison formula and singular value decomposition. Thus, the energy consumption problem is transformed into an integer programming problem, acquiring the control allocation strategy and the optimal thruster energy. Simulation results show that the proposed algorithm can effectively reduce the thrust energy consumption and improve the precision control, demonstrating the feasibility and efficiency of the proposed algorithm for picosatellites and nanosatellites.

1. Introduction

In the past years, a tremendous amount of work has been done on research and development of picosatellites and nanosatellites (PNS) because PNS can offer a wide variety of applications in space. PNS have significant scientific and cost advantages over using large, heavy satellites [1, 2]. Measurements of the lower thermosphere [3], earth observation [4], and technology demonstration are only a few examples of the potential of PNS. Till 2020, more than 400 PNS launches per year are estimated [5]. All such missions need micropropulsion systems to maneuver the satellites. PNS have strict requirements for quality, size, power cost, and so on, and therefore, miniature and affordable propulsion systems with low power are urgently needed for PNS. Various systems have been researched to control the miniaturized satellites, such as cold gas microthrusters, electrospray microthrusters, field emission microthrusters, and mono/bipropellant thrusters. The most commonly used actuators for PNS are magnetic coils and wheels [6–9]. Wheels can provide continuous output without using fuel, whereas, when wheels are working at a low angular velocity, there are frictional torques that may have a strong influence on PNS. Wheels are excellent for using in conventional satellites because of its accurate output. For the PNS, miniaturized wheels would not only increase the cost but also reduce the accuracy.

This paper proposes solid propellant microthruster array (SPMA) as an interesting choice for pico- and nanosatellites. SPMA has several main advantages: (1) it has simple structure and high reliability, (2) it is free from the frictional forces inherent to moving parts, (3) there is no propellant leakage with high propellant stability, and (4) driving voltage is low [10–14]. In SPMA, multiple independent addressing thrusters are highly integrated on the same chip (10^6 thrusters can be integrated on a 1 cm chip), which can deliver low thrust values and impulses in a short time. Simultaneously, our team has produced 10×10 scale microthruster arrays and 100×100 scale microthruster arrays with conventional fabrication methods [15]. Nevertheless, the

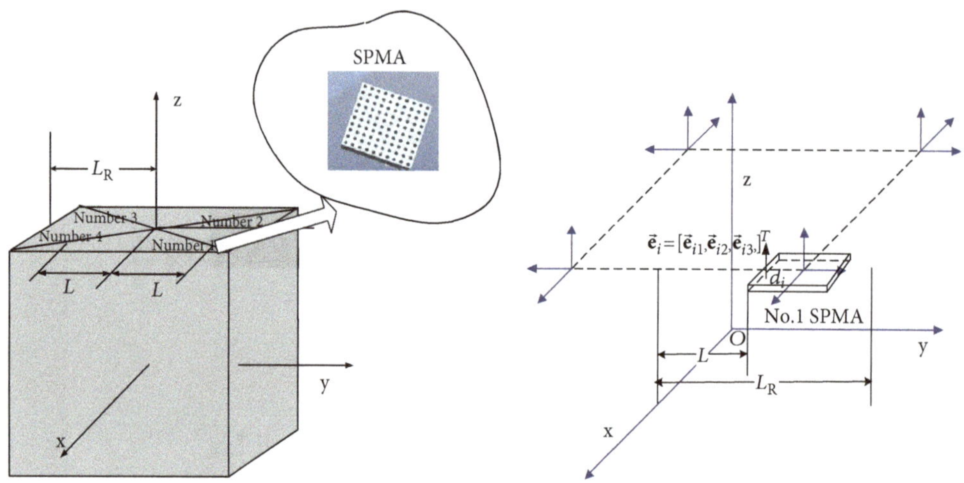

(a) Satellite structure diagram

(b) SPMA configuration diagram

Figure 1: SPMA in the satellite.

control allocation strategy for a large number of thrusters, misaligned and inappropriate positions for SPMA actuators, and irreproducible propellant and heat loss phenomena should be studied systematically before SPMA application.

Firstly, for a highly integrated actuator, a large number of thrusters have numerous combinations of ignition. Control torques produced by the thrusters in different locations are different and the thrusters in SPMA are not reproducible. Therefore, it is required that the actuators should not only satisfy the high-precision needs of satellite attitude control, but also have the smallest fuel consumption. For multiple and redundant thrusters in the SPMA, achieving the highest control accuracy with the least amount of fuel consumption is a dominant issue.

To solve the control allocation problems, there are several traditional methods: direct control allocation [16], daisy chaining [17], linear programming (LP) method [18], quadratic programming (QP) method [19], and pseudoinverse redistribution (PIR) method [20]. Numerous works also have been carried out on the improvement of allocation algorithms [21], the application of the technique to engineering [22], and the extension to theoretical approaches in the recent years [23]. Wang and Xie [24] designed an optimal thruster combinations table for the use of onboard real-time control allocation problems. In [25], a control allocation strategy is formulated as a min-max optimization problem, which ensures some robustness of system performances. Cristofaro and Johansen designed a fault-tolerant control allocation using unknown input observers [26]. Kirchengast et al. [27] investigated the impact of input matrix factorization on two algorithms (direct allocation and redistributed pseudoinverse) for constrained control allocation. Liu et al. [28] proposed an adaptive control allocation algorithm by updating the number of operating actuators. Although in the work of Liu et al., the number of operating actuators is changed, the total number of actuators is small, and the control effects of actuators in array are the same, so the simplified model reduces the analysis.

To summarize, previous scientific works are generally based on few thrusters with a fixed number of operating actuators. There is no existing method that can solve control allocation problems for a large and changed number of actuators. In this paper, we propose a control allocation optimization approach for PNS by dynamically analyzing and updating the number, ignition combinations, and impulse of actuators. The contributions of this research are presented as follows:

(1) For a large amount of thrusters like those in a highly integrated SPMA, the energy consumption of the thrusters is influenced by multiple factors such as the installation mode of SPMA, the ignition combinations, the unit impulse, and the precision control. It is difficult to quantitatively analyze the energy consumption, ignition combinations, and control accuracy. Therefore, this paper proposes an array energy consumption factor (AECF) to evaluate the performance of thruster combinations, establishing the relationship between the number of thrusters and the total energy consumption.

(2) An allocation optimization algorithm based on the AECF is formulated that can reduce the thrust energy consumption and improve control resolution.

2. Control Allocation Model of SPMA

For the cubic satellite shown in Figure 1, $O-xyz$ is the body coordinate system of satellite, the origin of the body coordinates locates at satellite's mass center, and principal axis coincides with inertial principal axes of satellite. The length of satellite is $2L_R$, and SPMA is installed in two symmetry planes perpendicular to the z-axis. As shown in Figure 1(a), the distance of each SPMA's edge from the satellite's symmetrical axis is L. The thruster arrays are installed on the diagonal line of satellite's surface, and the thruster serial number is shown in Figure 1(a).

Supposing each SPMA has $m \times m$, thrusters are represented as $\Gamma^{m\times m}$. The coordinates of the thruster in the SPMA can then be expressed as (i, j), $i \in [1, m]$, $j \in [1, m]$, $i, j \in Z$, and the position vector matrix of thruster array relative to the satellite centroid can be described as a matrix with the dimension of $1 \times N(N = m^2)$:

$[\mathbf{d}_{11}, \mathbf{d}_{12}, \ldots, \mathbf{d}_{1m}, \mathbf{d}_{21}, \mathbf{d}_{22}, \ldots, \mathbf{d}_{2m}, \mathbf{d}_{i1}, \mathbf{d}_{ij}, \ldots, \mathbf{d}_{im}, \ldots, \mathbf{d}_{mm}]$, where $\mathbf{d}_{ij}$ indicates the thruster position vector relative to the satellite centroid and can be expressed as $\mathbf{d}_{ij} = [x_{ij}, y_{ij}, z_{ij}]$. The corresponding thruster-generated triaxial component of the unit impulse is expressed as $\mathbf{e}_{ij} = [\mathbf{e}_{ijx}, \mathbf{e}_{ijy}, \mathbf{e}_{ijz}]^T$.

Δ_o indicates the unit distance of adjacent thrusters in the coordinate axis direction. The thruster control allocation model [29] is described as

$$\eta = \mathbf{AF}, \tag{1}$$

where $\eta \in R^n$ is the expected control command, and n is the degree of freedom (DOF) of the control requirement. $\mathbf{F}$ is the input control vector. $\mathbf{A}$ represents the thruster efficiency matrix.

The control torque of the thruster with the coordinate (i, j) for the centroid is

$$\eta_{ij} = (\mathbf{d}_{ij} \times \mathbf{e}_{ij})\mathbf{F}_{ij}. \tag{2}$$

The impulse matrix composed of all the thrusters can be represented as

$$\mathbf{F} = [\mathbf{F}_{11}, \mathbf{F}_{12}, \ldots, \mathbf{F}_{1m}, \mathbf{F}_{21}, \mathbf{F}_{22}, \ldots, \mathbf{F}_{2m}, \mathbf{F}_{i1}, \mathbf{F}_{ij}, \ldots, \mathbf{F}_{im}, \ldots, \mathbf{F}_{mm}]^T. \tag{3}$$

Equation (3) shows an impulse matrix composed of N working thrusters. Currently, each impulse from the thrusters in the same array is equal, specifically $\mathbf{F}_{11} = \mathbf{F}_{ij} = \cdots = \mathbf{F}_N$. The control torque at the satellite's centroid can be represented as

$$\eta = \sum_{i=1}^{m}\left(\sum_{j=1}^{m}\eta_{ij}\right) = \sum_{i=1}^{m}\sum_{j=1}^{m}((\mathbf{d}_{ij} \times \mathbf{e}_{ij})\mathbf{F}_{ij}) = \mathbf{AF}. \tag{4}$$

The efficiency matrix $\mathbf{A}$ can be written as follows:

$$\mathbf{A} = [\mathbf{h}_{11}, \mathbf{h}_{12}, \ldots, \mathbf{h}_{1m}, \mathbf{h}_{21}, \ldots, \mathbf{h}_{2m}, \mathbf{h}_{ij}, \ldots, \mathbf{h}_{mm}], \tag{5}$$

$$h_{ij} = [\mathbf{d}_{ij} \times \mathbf{e}_{ij}]^T = [y_{ij}\mathbf{e}_{ijz} - z_{ij}\mathbf{e}_{ijy}, z_{ij}\mathbf{e}_{ijx} - x_{ij}\mathbf{e}_{ijz}, x_{ij}\mathbf{e}_{ijy} - y_{ij}\mathbf{e}_{ijx}]^T. \tag{6}$$

For the satellite control system, the control dimension $n = 3$, $\mathbf{A} \in R^{3\times\lambda}$, and is the number of thrusters participating in control, $\lambda \in [0, N]$.

3. Energy Consumption Model of SPMA

In this section, the energy consumption model of the SPMA according to the control allocation model is established, and energy consumption factor of SMPA is deduced. Thus, the relationship among the quantity, the ignition combinations, and the energy consumption of the thrusters can be studied.

3.1. Deduce Energy Consumption Factor. Rewriting (1) as

$$\eta\eta^T = \mathbf{AFF}^T\mathbf{A}^T. \tag{7}$$

Supposing the control variance is σ^2 (representing the system error of actuator, namely, impulse error of thruster), $I_{n\times n}$ is an N-order unit matrix and thus

$$\text{cov}(\eta) = \sigma^2 I_{n\times n}, \tag{8}$$

writing (7) in the form of

$$\text{cov}(\eta) = \mathbf{A}(\text{cov}(\mathbf{F}))\mathbf{A}^T, \tag{9}$$

substituting (8) into (9) and systemize as follows:

$$\text{cov}(\mathbf{F}) = \sigma^2(\mathbf{A}^T\mathbf{A})^{-1}. \tag{10}$$

The trace on both sides of (10) and then SPMA energy consumption model can be obtained:

$$\text{tr}[\text{cov}(\mathbf{F})] = \sigma^2\text{tr}\left[(\mathbf{A}^T\mathbf{A})^{-1}\right]. \tag{11}$$

The left side of (11) represents the total required control impulse (thrust), namely, the energy consumption of the thruster in statistic. In this paper, $\text{tr}[(\mathbf{A}^T\mathbf{A})^{-1}] = \text{tr}[(\mathbf{AA}^T)^{-1}]$ is defined as the array energy consumption factor (AECF).

Setting matrix $\mathbf{K}$ as

$$\mathbf{K} = (\mathbf{A}^T\mathbf{A})^{-1} = \begin{bmatrix} \mathbf{K}_{11} & \mathbf{K}_{12} & \mathbf{K}_{13} \\ \mathbf{K}_{21} & \mathbf{K}_{22} & \mathbf{K}_{23} \\ \mathbf{K}_{31} & \mathbf{K}_{32} & \mathbf{K}_{33} \end{bmatrix}. \tag{12}$$

The AECF can be expressed as

$$\text{AECF} = \sqrt{\text{trace}(\mathbf{K})} = \sqrt{\mathbf{K}_{11} + \mathbf{K}_{22} + \mathbf{K}_{33}}. \tag{13}$$

Through the AECF's definition, it can be seen that the larger the AECF is, the larger the total energy consumption used for control will be. The smaller the AECF is, the smaller the total energy consumption used for control will be. Furthermore, according to (13) and the definition of matrix $\mathbf{A}$, it can be seen that the AECF is influenced by three indexes: the number of working thrusters, the ignition combinations, and the unit impulse of thrusters in SPMA.

3.2. Energy Optimization Allocation Strategy. When using the SPMA for high-precision control, because there are a large number of thrusters integrated in the array and under the condition of thruster redundancy configuration, allocation method that satisfies (1) is not unique. The total energy consumption will also be quite different when choosing a different number of thrusters to form ignition combinations. Therefore, to reduce the energy consumption of SPMA, the relationship between the number of working thrusters and value of AECF is studied; thus, an optimal energy allocation strategy based on AECF can be established.

Setting $\mathbf{A}_\lambda$ as the efficiency matrix of λ thrusters when being selected for ignition execution and randomly remove one thruster (supposing the coordinates of this thrust are (i, j)) from the λ thrusters, then the torque efficiency matrix of $\lambda - 1$ thrusters is $\mathbf{A}_{\lambda-1}$. The following relationship (14) can be obtained using (7):

$$\mathbf{A}_\lambda^T \mathbf{A}_\lambda = \mathbf{A}_{\lambda-1}^T \mathbf{A}_{\lambda-1} + h_{ij}^T h_{ij}, \tag{14}$$

variable $\mathbf{K}_\lambda$ is introduced as

$$\mathbf{K}_\lambda = \left(\mathbf{A}_\lambda^T \mathbf{A}_\lambda\right)^{-1} = \left(\mathbf{A}_{\lambda-1}^T \mathbf{A}_{\lambda-1} + h_{ij}^T h_{ij}\right)^{-1}. \tag{15}$$

Therefore, according to the Sherman-Morrison formula [30], (14) can be written as

$$\mathbf{K}_\lambda = \mathbf{K}_{\lambda-1} - \frac{\mathbf{K}_{\lambda-1} h_{ij}^T h_{ij} \mathbf{K}_{\lambda-1}}{1 + h_{ij} \mathbf{K}_{\lambda-1} h_{ij}^T}, \tag{16}$$

where $\mathbf{K}_{\lambda-1} = (\mathbf{A}_{\lambda-1}{}^T \mathbf{A}_{\lambda-1})^{-1}$, $(1 + h_{ij} \mathbf{K}_{\lambda-1} h^T{}_{ij})$ is the scalar quantity.

According to the definition of the AECF in Section 3.1, we can obtain

$$\begin{aligned} \text{AECF}_\lambda &= \text{trace}(\mathbf{K}_\lambda) = \text{trace}(\mathbf{K}_{\lambda-1}) - \text{trace}\left(\frac{\mathbf{K}_{\lambda-1} h_{ij}^T h_{ij} \mathbf{K}_{\lambda-1}}{1 + h_{ij} \mathbf{K}_{\lambda-1} h_{ij}^T}\right) \\ &= \text{AECF}_{l-1} - g_{ij}, \end{aligned} \tag{17}$$

where

$$g_{ij} = \text{trace}\left[\frac{\mathbf{K}_{\lambda-1} h_{ij}{}^T h_{ij} \mathbf{K}_{\lambda-1}}{1 + h_{ij} \mathbf{K}_{\lambda-1} h_{ij}{}^T}\right]. \tag{18}$$

From (17), it can be seen that AECF will add an item g_{ij} when reducing a thruster. Therefore, whether g_{ij} has a positive or negative value determines how the AECF is influenced by the number of working thrusters.

To determine the sign of g_{ij}, $\mathbf{A}_{\lambda-1}$ is performed on singular value decomposition (SVD):

$$\mathbf{A}_{\lambda-1} = USV^T, \tag{19}$$

where U and V^T, respectively, represent a $(\lambda - 1)$-order orthogonal matrix and a 3-order orthogonal matrix and S is a $(\lambda - 1) \times 3$-order diagonal matrix. Furthermore, $\mathbf{K}_{\lambda-1}$ can be written as

$$\mathbf{K}_{\lambda-1} = \left(\mathbf{A}_{\lambda-1}^T \mathbf{A}_{\lambda-1}\right)^{-1} = \left(VS^T U^T USV^T\right)^{-1} = V\left(S^T S\right)^{-1} V^T. \tag{20}$$

$S^T S$ is a 3-order positive diagonal matrix as follows:

$$S^T S = \begin{bmatrix} s_1^2 & & \\ & s_2^2 & \\ & & s_3^2 \end{bmatrix}, \tag{21}$$

where s_1, s_2, and s_3 are the eigenvalues of matrix $\mathbf{A}_{\lambda-1}$.

Since the orthogonal transformation has no influence on the matrix trace, (21) can be obtained:

$$\begin{aligned} \text{trace}\left(\mathbf{K}_{\lambda-1} h_{ij}^T h_{ij} \mathbf{K}_{\lambda-1}\right) &= \text{trace}\left[V\left(S^T S\right)^{-1} V^T h_{ij}^T h_{ij} V\left(S^T S\right)^{-1} V^T\right] \\ &= \text{trace}\left[\left(S^T S\right)^{-1} V^T h_{ij}^T h_{ij} V\left(S^T S\right)^{-1}\right], \end{aligned} \tag{22}$$

makes

$$h_{ij} V = [v_1, v_2, v_3]. \tag{23}$$

Substituting (22) into (21), we can obtain

$$\begin{aligned} \text{trace}\left(\mathbf{K}_{m-1} h_{ij}^T h_{ij} \mathbf{K}_{m-1}\right) &= \text{trace}\left[\left(S^T S\right)^{-1} V^T h_{ij}^T h_{ij} V\left(S^T S\right)^{-1}\right] \\ &= \text{trace}\left(\begin{bmatrix} \frac{1}{s_1^2} & & \\ & \frac{1}{s_2^2} & \\ & & \frac{1}{s_3^2} \end{bmatrix} \begin{bmatrix} v_1^2 & v_1 v_2 & v_1 v_3 \\ v_1 v_2 & v_2^2 & v_2 v_3 \\ v_1 v_3 & v_2 v_3 & v_3^2 \end{bmatrix} \begin{bmatrix} \frac{1}{s_1^2} & & \\ & \frac{1}{s_2^2} & \\ & & \frac{1}{s_3^2} \end{bmatrix}\right) \\ &= \frac{v_1^2}{s_1^4} + \frac{v_2^2}{s_2^4} + \frac{v_3^2}{s_3^4} > 0. \end{aligned} \tag{24}$$

We can also obtain

$$1 + h_{ij}\mathbf{K}_{\lambda-1}h_{ij}^T = 1 + h_{ij}V\left(S^TS\right)^{-1}V^Th_{ij}^T$$
$$= 1 + \left[v_1, v_2, v_3\right]\begin{bmatrix}\frac{1}{s_1^2} & & \\ & \frac{1}{s_2^2} & \\ & & \frac{1}{s_3^2}\end{bmatrix}\begin{bmatrix}v_1\\ v_2\\ v_3\end{bmatrix} \tag{25}$$
$$= 1 + \left(\frac{v_1^2}{s_1^2} + \frac{v_2^2}{s_2^2} + \frac{v_3^2}{s_3^2}\right) > 0.$$

Therefore, g_{ij} can be written as

$$g_{ij} = \frac{v_1^2/s_1^4 + v_2^2/s_2^4 + v_3^2/s_3^4}{1 + v_1^2/s_1^2 + v_2^2/s_2^2 + v_3^2/s_3^2} > 0. \tag{26}$$

Furthermore, we can determine that

$$\text{AECF}_\lambda = \text{AECF}_{\lambda-1} - g_{ij} < \text{AECF}_{\lambda-1}. \tag{27}$$

From (25) and (27), it can be seen that the AECF decreases with the increase of the number of working thrusters. Thus, when satisfying the same control variance σ^2, increase of the number of working thrusters can reduce the demand of unit thruster energy consumption (thrust/impulse), reduce the total energy consumption of SPMA, and improve the control resolution.

4. AECF Optimization Strategy

Based on the conclusion drawn from (27), the next step is optimizing the ignition combinations and achieving optimal energy consumption and precision control.

As deduced from Section 3, the smaller the AECF is, the lower the energy consumption is. From (25) and (27), we can see that the bigger the g_{ij} value brought about by the addition of ignition thruster is, the smaller the AECF and the energy consumption are; thus, the key to optimal energy consumption is finding the optimal g_{ij} value. To find this g_{ij} value, a simplified optimization criterion is established for AECF optimization algorithm. Firstly, basement of chain control allocation is introduced to calculate the initial value of the AECF.

4.1. Initial Value of Chain Basement. Chain basement algorithm considers each thruster in the thruster array as a basement and preferentially searches for the basement with a high control efficiency that meets the triaxial demand (force/torque). When one basement cannot complete the control force (torque), multiple sets of basements can be selected successively in a combined fashion to compensate.

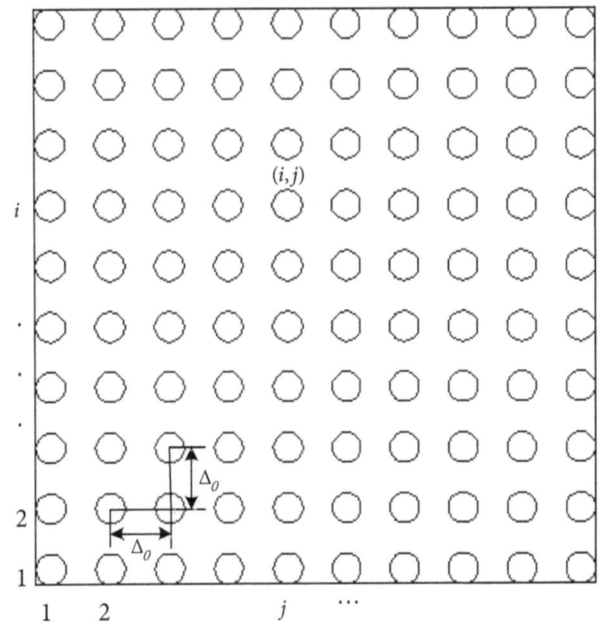

FIGURE 2: Front view of SPMA.

The chain basement control allocation method for λ thrusters is expressed as

$$C = \left[a_1 \quad a_2 \quad \cdots \quad a_\omega \quad \cdots a_\lambda\right] \cdot \left[u_1 \quad u_2 \quad \cdots u_\omega \quad \cdots u_\lambda\right], \quad \lambda \le N, \lambda, \omega \in Z, \omega \le \lambda, \tag{28}$$

where C is the expected control torque given by (1); u_i and a_i are, respectively, the control basement and the direction vector of the corresponding basements. When the first group of basements cannot satisfy the control requirements, the next group of basements is added according to the basement efficiency until the expected control range is satisfied.

The basement efficiency can be defined from the following equation:

$$\pi_k = \frac{C_\omega}{C_q}, \tag{29}$$

where C_ω is the control force (matrix) generated by the ωth thruster basement and C_q is the expected control force (matrix).

4.2. AECF Optimization Principle. Through the chain basement method, the initial χ thruster information $\mathbf{A}_\chi$ that satisfies the control requirements can be obtained. Equation (25) reflects the influence of interaction between the optimizing $(\chi+1)^{\text{th}}$ thruster and the initial value $\mathbf{A}_\chi$ on AECF. s_1, s_2, and s_3 can be determined by SVD from $\mathbf{A}_\chi$; thus, finding the g_{ij} value of the optimal $(\chi+1)^{\text{th}}$ thruster (also, supposing the

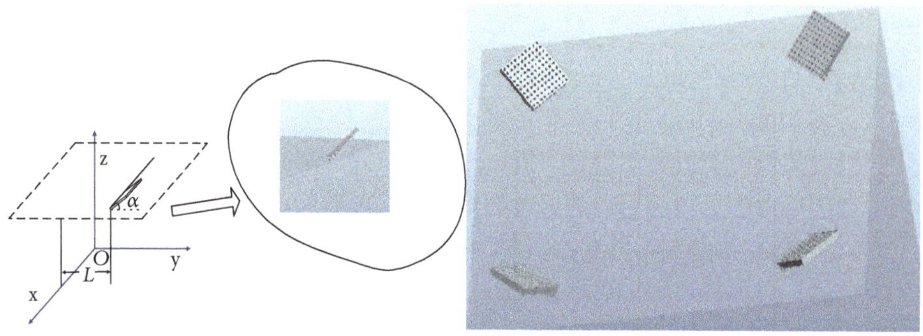

FIGURE 3: Array included angle.

coordinates for this thruster is (i, j)) can be changed to the following optimization problem.

$$J = \max \quad \frac{v_1^2/s_1^4 + v_2^2/s_2^4 + v_3^2/s_3^4}{1 + v_1^2/s_1^2 + v_2^2/s_2^2 + v_3^2/s_3^2}$$
$$\text{s.t.} \quad h_{ij}V = [v_1, v_2, v_3], \quad h_{ij} \in \Phi, \tag{30}$$

where Φ is the array collection taking out the initial χ number of thrusters. This problem then turns into finding the appropriate v_1, v_2, and v_3 in order to reach the largest g_{ij} value.

For microsatellites with small volumes and quality such as PNS, which have semidiameter of less than 1 m or far less than 1 m, it is easy to ascertain that each element $a_{rirj}(ri \le 3, rj \le l)$ in $\mathbf{A}_\chi$ is less than 1. In addition, according to the Frobenius theorem and the Gerschgorin estimation, the singular values s_1, s_2, and s_3 of matrix $\mathbf{A}_\chi$ are all less than 1. Therefore, in (25)

$$\frac{v_1^2}{s_1^4} + \frac{v_2^2}{s_2^4} + \frac{v_3^2}{s_3^4} \gg 1 + \frac{v_1^2}{s_1^2} + \frac{v_2^2}{s_2^2} + \frac{v_3^2}{s_3^2}. \tag{31}$$

Thus, the optimization objective can be further simplified as (32) and the calculation complexity can be markedly simplified.

$$J = \max \quad \left(\frac{v_1^2}{s_1^4} + \frac{v_2^2}{s_2^4} + \frac{v_3^2}{s_3^4}\right). \tag{32}$$

For the SPMA, according to its layout collection $\Gamma^{m\times m}$ in Figure 2, (31) can be further simplified:

$$h_{ij} = \left(\mathbf{d}_{ij} + \Delta_o \mathrm{H}\right) \times \mathbf{e}_{ij} = \mathbf{d}_{ij} \times \mathbf{e}_{ij} + \Delta_o \mathrm{H} \times \mathbf{e}_{ij},$$
$$\mathbf{d}_{ij} = \left[x_{ij}, y_{ij}, z_{ij}\right], \tag{33}$$

$$\Delta H = \left[\Delta_o\delta_{ij}, \Delta_o\gamma_{ij}, \Delta_o\zeta_{ij}\right]^T, 1 \le \delta_{ij}, \gamma_{ij}, \zeta_{ij} \le m-1, \delta_{ij}, \gamma_{ij}, \zeta_{ij} \in Z. \tag{34}$$

To optimize the overall energy consumption, adding one more thruster and making the incremental effect maximum, the energy impact factor of new ignition position is $\Lambda_{\lambda+1} = \Delta_o H \times \mathbf{e}_{ij}$, $S_l = [1/s_1^4, 1/s_2^4, 1/s_3^4]^T$. Then, ring the problem into the integer optimization concerning $\delta_{ij}, \gamma_{ij}, \zeta_{ij}$

$$J = \max \quad \Lambda_{l+1} V_l S_\lambda. \tag{35}$$

Finally, the AECF optimization can be easily solved using integer optimization algorithms such as branch and bound method.

4.3. Optimization of Three-Dimensional Configuration. When SPMA is used to provide triaxial directional control output, SPMA should form a specific included angle with plane $O-xyz$ as shown in Figure 3.

Therefore, (5) can be modified as

$$\mathbf{d}_{ij} = \left[(\mathbf{d}_i + \Delta_o r_i)\cos\alpha, \mathbf{d}_{ij} + \Delta_o r_i, \left(\mathbf{d}_{ij} + \Delta_o r_i\right)\sin\alpha\right],$$
$$\mathbf{e}_i = [0, -\sin\alpha, \cos\alpha]^T. \tag{36}$$

Equation (34) is rewritten as

$$\Delta H = \left[\Delta_o\delta_{ij}\cos\alpha, \Delta_o\gamma_{ij}, \Delta_o\zeta_{ij}\sin\alpha\right]^T,$$
$$1 \le \delta_{ij}, \gamma_{ij}, \zeta_{ij} \le m-1, \delta_{ij}, \gamma_{ij}, \zeta_{ij} \in Z. \tag{37}$$

The relevant descriptions are similar to the above equations.

5. Simulation Analysis

5.1. Verification of AECF Algorithm. To verify the effectiveness of AECF and optimization algorithm, simulation analysis is conducted. The 10×10 scale and 50×50 scale of SPMA are tested in this section. Parameters of SPMA: $L = 0.04$ m; $\Delta_o = 2$ mm; the unit impulse of thruster is $I_0 = 1\times 10^{-4}$ Ns. Expected control $C_d = [1.84, -1.77, 0]\times 10^{-4}\mathrm{N\cdot s\cdot m}$ for 10×10 scale and $C_d = [19.54, -19.69, 0]\times 10^{-4}\mathrm{N\cdot s\cdot m}$ for 50×50 scale.

5.1.1. 10 × 10 Scale SPMA. By chain basement calculation, the thruster array composed of 38 thrusters is obtained and its AECF is 1125. Thus, Figures 4–6 can be obtained through AECF optimization algorithm.

The horizontal axis in Figure 4 shows the number of added actuators requiring optimization, and the vertical axis

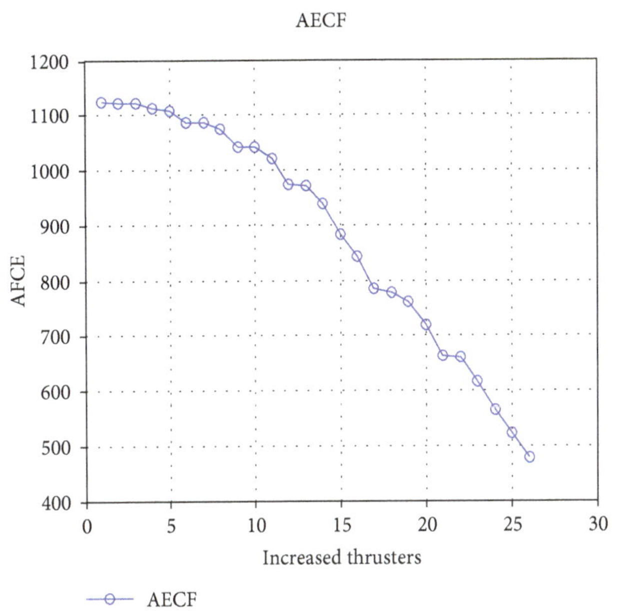

FIGURE 4: Relationship between optimization thruster and AECF.

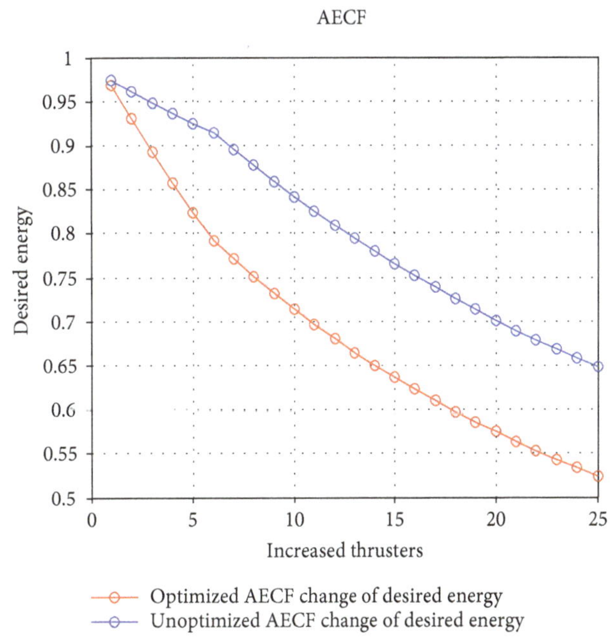

FIGURE 6: Comparison of optimization effects.

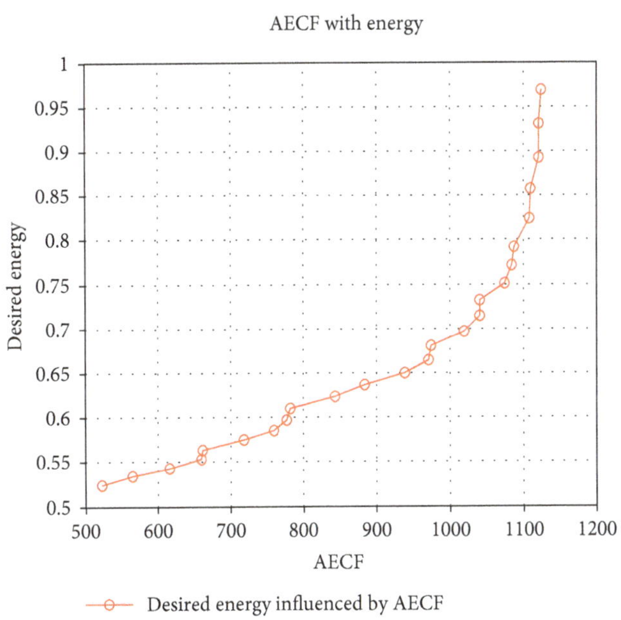

FIGURE 5: Relationship between the energy requirement and the AECF.

is the value of AECF; it can be seen from Figure 4 that as the number of optimized thrusters increases, the AECF constantly decreases, and the simulation results verify the validity of (27). According to the definition, as the AECF decreases, the thruster energy consumption required by control in Figure 4 decreases. When the optimizing thruster quantity is 26, the AECF is 478, reduced by more than half the amount compared to the initial AECF.

The horizontal axis in Figure 5 is AECF, and the vertical axis is the desired impulse of thruster. From Figure 5, it is obvious that as the AECF gradually decreases, the desired impulse of thruster significantly decreases. Eventually when AECF reaches a value of 478, the desired impulse of the thruster falls sharply, by 47.6%. This means that the overall thruster array can save nearly half of the energy consumption while achieving the same control goal, and it can also provide a higher resolution of control as well as meeting a higher precision control.

The horizontal axis in Figure 6 is the number of added optimized thrusters, and the vertical axis is the demanded thruster energy consumption. The red line is the thruster energy consumption demand using the AECF optimization algorithm proposed in this paper, and the blue line is the thruster energy consumption demand randomly optimized and obtained from the unignited available thrusters. We can see that as the number of optimized thrusters increases, the desired energy decreases, which agrees with the results of theoretical analysis. Significantly, adopting the proposed optimization algorithm can save more than 13% of the energy consumption compared to the stochastic optimization.

5.1.2. 50×50 Scale SPMA. Test results of 50×50 scale are shown in Figure 7.

From Figure 7, similar conclusions can be drawn that with the increase of the optimized number of array thrusters, the AECF decreases gradually; in the end of the AECF process, the energy consumption is only 44.72% of the original level, decreasing by 55.28%. The results prove the efficiency of proposed algorithm. By dynamically analyzing and processing the information of thrusters based on AECF, allocation problem of large-scale thrusters can also be effectively solved.

5.2. Statistical Test of AECF. Without a loss of generality, the expected control C_d randomly chooses 10 times from [0.7,

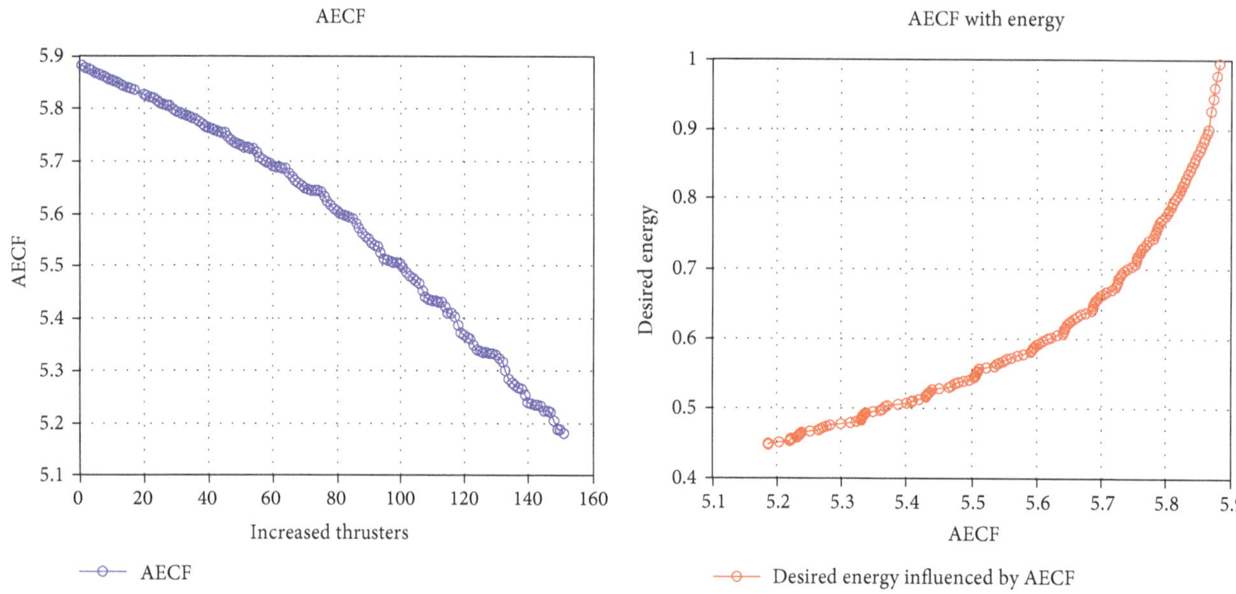

(a) (b)

FIGURE 7: Verification on 50 × 50 scale SPMA.

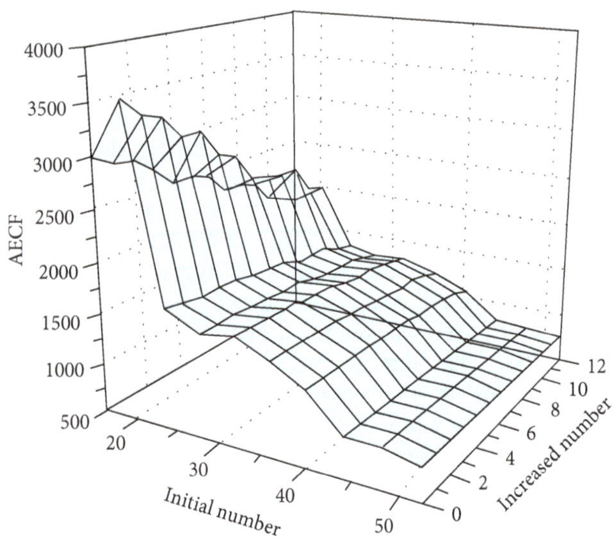

FIGURE 8: AECF varies with different control expectations and quantities of optimized thrusters.

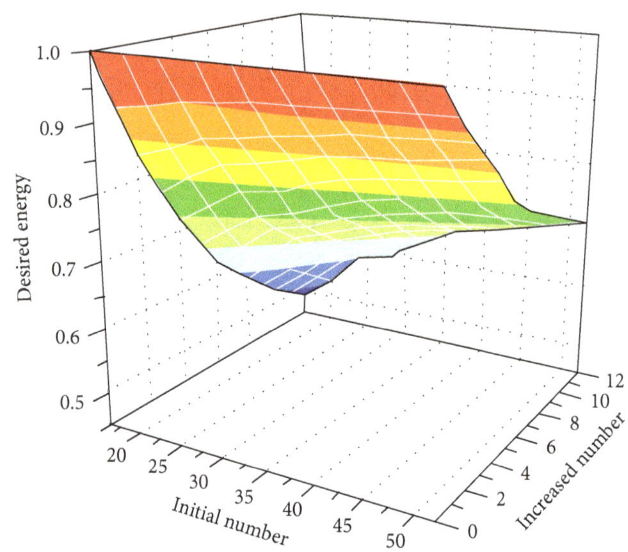

FIGURE 9: Changes in energy demand.

$-0.7,\ 0] \times 10^{-4}\mathrm{N} \cdot \mathrm{s} \cdot \mathrm{m}$ to $C_d = [2.5,\ -2.5,\ 0] \times 10^{-4}\mathrm{N} \cdot \mathrm{s} \cdot \mathrm{m}$. Using the chain basement algorithm from Section 4.1, the corresponding initial number of the execution thrusters range is [16, 52]. The simulation results are shown in Figures 8–11.

Figure 8 demonstrates the corresponding changes of AECF affected by the number of working thrusters, reflecting increases from 16 to 52, while the number of optimized thrusters increases from 0 to 12. From Figure 8, we can see that no matter how many thrusters are participating in the execution, the AECF will still decrease because of the optimization and addition of the number of the aiming thrusters in this layout. Figures 9 and 10 reflect the corresponding changes in desired energy. Taking Figure 9 as an example, with the increase of the number of working thrusters in the vertical axis, the corresponding energy demand gradually decreases. From Figure 10, we can see that when the number of the thrusters of initial participation is relatively small and then add the thrusters, the effects of energy demand will be more significantly improved.

The influence of the AECF variation on the unit impulse of a thruster can be seen more clearly in Figure 11.

The horizontal axis in Figure 11 is the change rate of AECF, and the vertical axis is the change in desired energy. It can be seen no matter how many initial thrusters there are, with the decrease of AECF, the unit thruster energy demand significantly decreases. Taking the red curve and

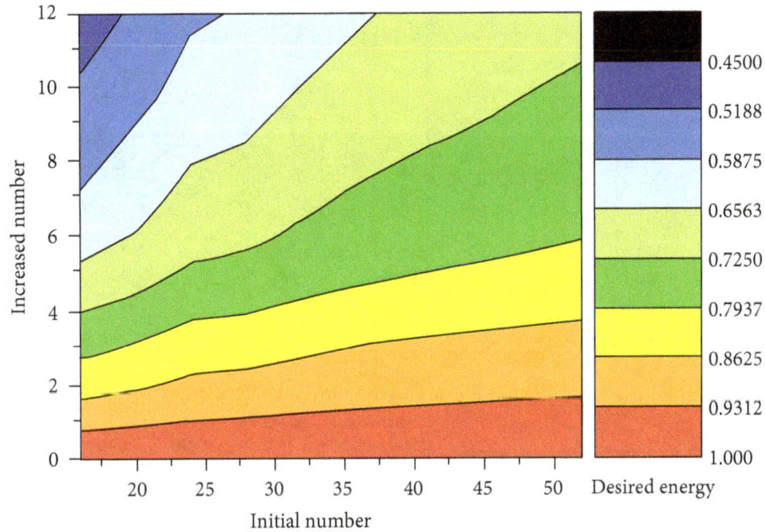

FIGURE 10: Equipotential map of energy demand.

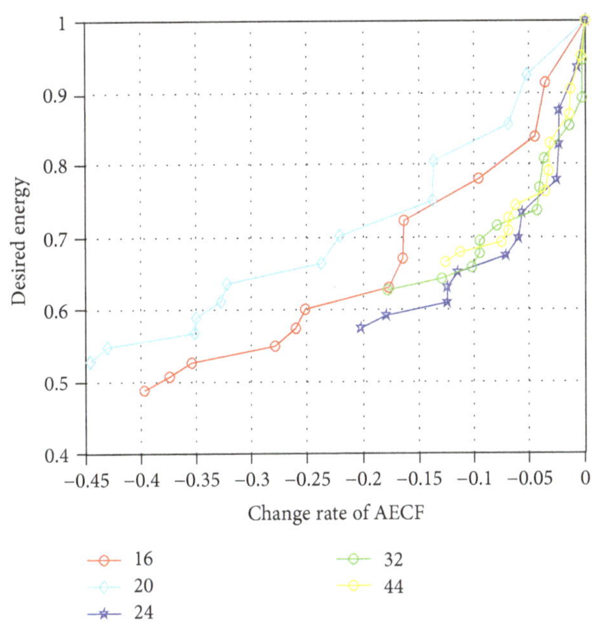

FIGURE 11: AECF variation and energy demand.

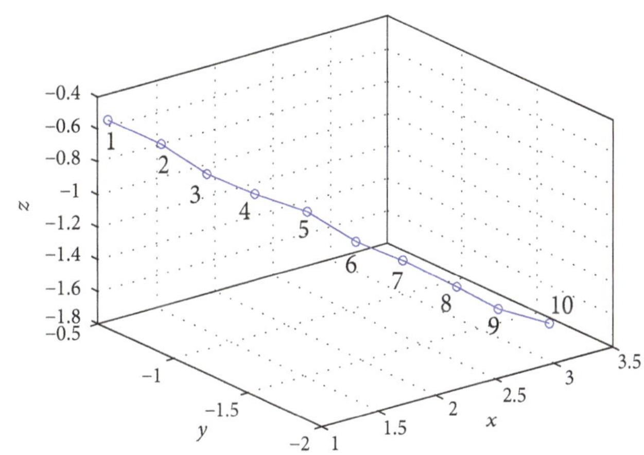

FIGURE 12: Triaxial expected control demand.

16 thrusters with initial participation as an example, when the energy consumption factor AECF decreases by 3.5%, its energy demand decreases by 8.7%, and as the AECF decreases by 39%, its energy consumption decreases by 52%.

5.3. Three-Dimensional Control Vector of AECF. Taking thruster installation angle $\alpha = 45^\circ$ as an example, there are 10 groups of control requirements in Figure 12, whose expected triaxial C_d goes from $[1, -0.5, -0.5] \times 10^{-4}\text{N} \cdot \text{s} \cdot \text{m}$ to $C_d = [3.5, -1.8, -1.8] \times 10^{-4}\text{N} \cdot \text{s} \cdot \text{m}$. The SPMA needs to, respectively, satisfy these control demands.

The horizontal axis in Figure 13 is the number of addition of optimized thrusters, and the vertical axis is the change of AECF. With the gradual increase of optimized thrusters, the AECF gradually shrinks. Taking the number 2 cyan curve as an example, $C_2 = [1.28, -0.67, -0.67] \times 10^{-4}\text{N} \cdot \text{s} \cdot \text{m}$ and the AECF is initially 2028. With the increase of the optimized of thrusters, the AECF decreases, and when the optimization increases by 19 thrusters, the AECF sharply reduces to 1144.

The horizontal axis in Figure 14 is the change rate of the AECF, and the vertical axis indicates the change in the energy demand. It can be seen that with the decrease of AECF, unit thruster energy demand decreases significantly. In addition, taking the number 2 cyan curve as an example, when the AECF optimization of this curve decreases by 13%, its energy demand decreases by 43%. Figure 15 also clearly reflects the relationship between the number of initial thrusters, the number of thrusters participating in optimization, and the thruster energy demand.

5.4. Verification of Attitude Control. To estimate the performance of the proposed approach more completely, large angle maneuver with perturbations has been tested in this

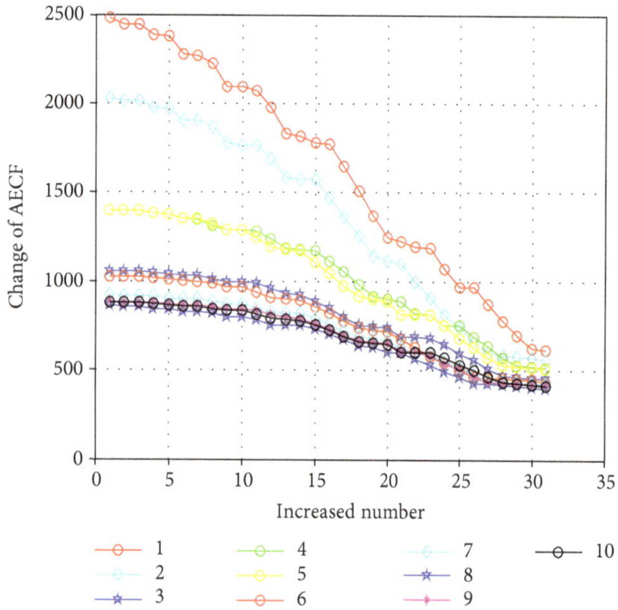

Figure 13: AECF variation.

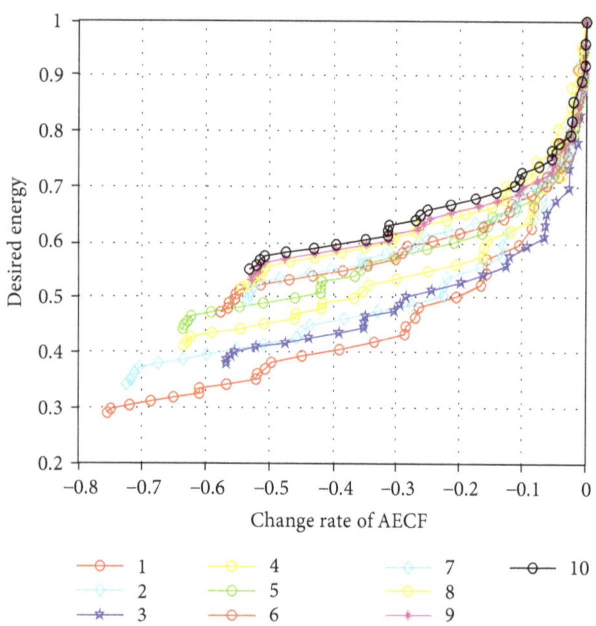

Figure 14: AECF and energy demand.

section. A sliding mode controller is designed to perform the large angle maneuver efficiently. Comparison with wheels is also been described in the verification part.

Kinetic equation and the motion equation of the PNS are expressed as follows:

$$\begin{aligned} \text{SPMA}: I_b\dot{\omega}_b + \omega_b^{\times} I_b \omega_b &= T_c + T_d, \\ \text{Wheel}: I_b\dot{\omega}_b + \omega_b^{\times}(I_b\omega_b + H_w) &= T_c + T_d, \\ \dot{q}_e &= \frac{1}{2}F(q_e)\omega_b, \end{aligned} \tag{38}$$

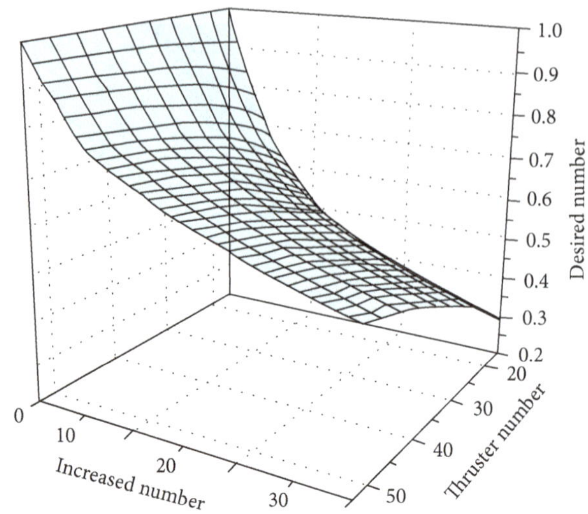

Figure 15: Energy demand.

where I_b is the rotational inertia of satellite, T_c denotes the control torque, T_d is the disturbance from space environment, $\dot{q}_e$ is the error of attitude angular rate represented by quaternion, and H_w is the angular momentum of wheel.

The torque (T_R) generated from the wheel can be obtained by the difference between the command torque (T_c) and the friction torque (T_f) as follows [31, 32]:

$$T_R = T_c - T_f. \tag{39}$$

Friction torques generated by wheels are particularly critical to PNS. For numerical analysis, the friction torques are modeled by means of a viscous torque proportional to the wheel speed, coulomb torque, and a continuous Stribeck friction as follows:

$$T_f = K_v\Omega + \left(T_k + \frac{T_s}{1 + (\Omega/\Omega_s)^2}\right)\text{sgn}(\Omega), \tag{40}$$

where T_k is Coulomb friction torque, K_v represents viscous friction coefficient, T_s denotes Stribeck friction torque, and Ω and Ω_s are angular velocity of the wheel and Stribeck velocity, respectively.

A sliding mode function is employed as

$$s = \omega_b + kq_e. \tag{41}$$

Control law for SPMA:

$$T_c = T_{eq} + T_n = \omega_b^{\times} I_b\omega_b - I_b k\dot{q}_e - I_b\alpha_1 s - I_b\alpha_2\,\text{sgn}(s). \tag{42}$$

Control law for wheel:

$$\begin{aligned} T_c &= T_{eq} + T_n \\ &= \omega_b^{\times}(I_b\omega_b + H_w) - I_b k\dot{q}_e - I_b\alpha_1 s - I_b\alpha_2\,\text{sgn}(s). \end{aligned} \tag{43}$$

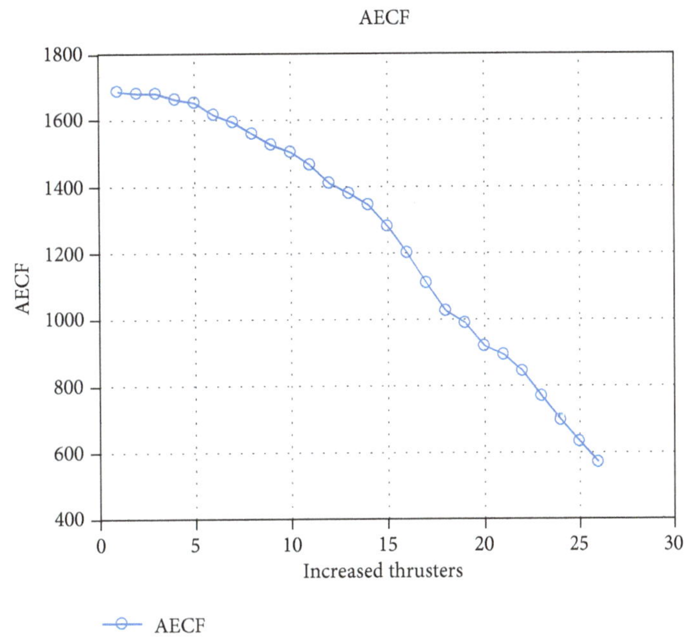

(a) Process of AECF optimization

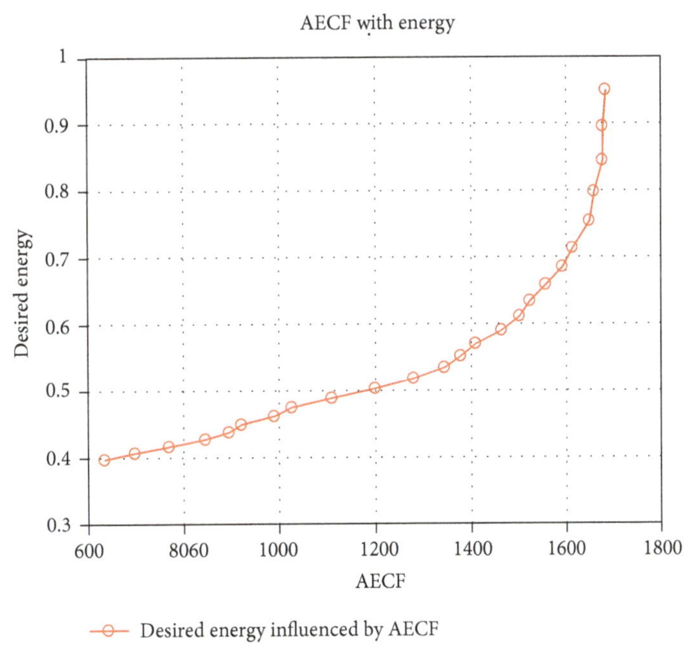

(b) Energy consumption

FIGURE 16: AECF optimization.

Assuming the Lyapunov function candidate is

$$V = \frac{1}{2} s^T s. \tag{44}$$

The derivative of V can be obtained from both SPMA and wheel as follows:

$$\dot{V} = s^T \left[I_b^{-1} T_d - \alpha_1 s - \alpha_2 \operatorname{sgn}(s) \right]. \tag{45}$$

When appropriate parameters are chosen such as α_1 and α_2, the stability of the closed loop system can be guaranteed.

Simulation parameters are set as follows:

(1) Parameters of satellite are the following:

Rotational inertia of satellite:

$$I_b = \begin{bmatrix} 0.01219 & 0 & 0 \\ 0 & 0.01406 & 0 \\ 0 & 0 & 0.009375 \end{bmatrix} \left(\text{kg} \cdot \text{m}^2\right). \tag{46}$$

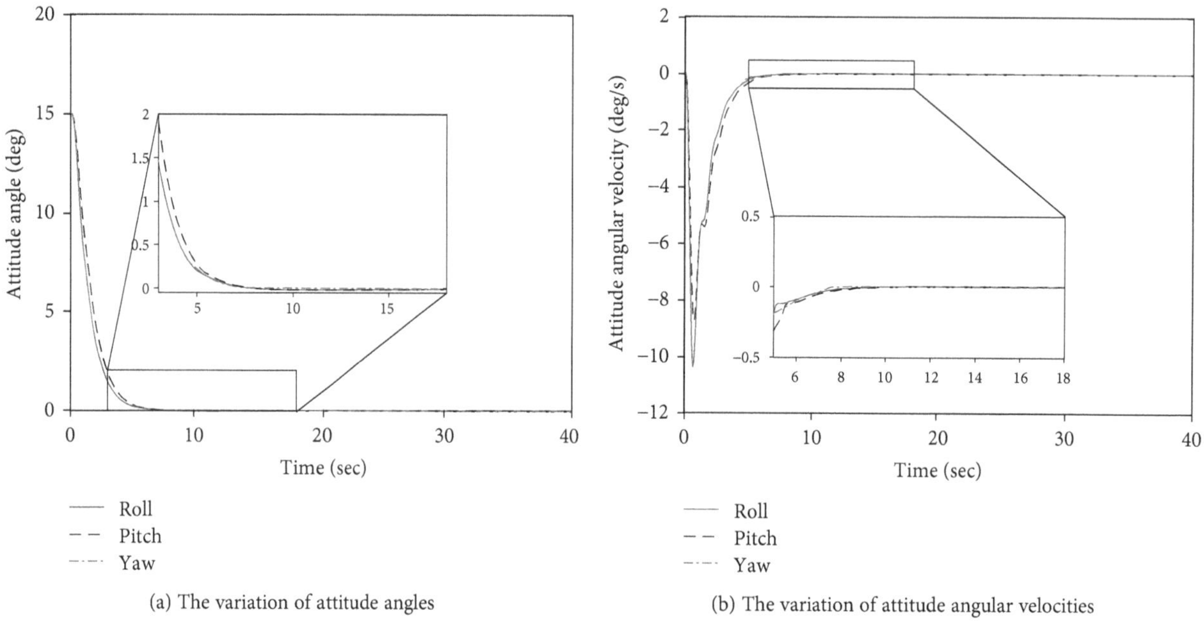

FIGURE 17: Control results of SPMA using AECF.

Initial Euler attitude angle: $\phi = 15^\circ$, $\theta = 15^\circ$, and $\psi = 15^\circ$.

Expected Euler attitude angle: $\phi = 0^\circ$, $\theta = 0^\circ$, and $\psi = 0^\circ$.

Initial attitude angular rate: $\omega_b(0) = [0.05 \quad 0.05 \quad 0.05]^T (^\circ/s)$.

Rotational inertia of wheels: $I_w = 1.6 \times 10^{-6}\,\mathrm{kg \cdot m^2}$.

Orbit angular velocity: $\Omega_0 = 1.131 \times 10^{-3}$ rad/s.

Aerodynamic disturbance moment:

$$T_a = \begin{bmatrix} 0.4 \times 10^{-9} + 1.9 \times 10^{-9} \cos(\Omega_0 t) \\ 2.0 \times 10^{-8} + 0.4 \times 10^{-8} \cos(\Omega_0 t) \\ 2.02 \times 10^{-8} + 0.18 \times 10^{-8} \cos(\Omega_0 t) \end{bmatrix}. \quad (47)$$

Magnetic disturbance moment:

$$T_m = \begin{bmatrix} 9.5 \times 10^{-8} \sin(\Omega_0 t) + 0.25 \times 10^{-8} \\ 2.6 \times 10^{-8} \cos(\Omega_0 t) - 5 \times 10^{-8} \sin(\Omega_0 t) \\ 0.5 \times 10^{-8} \sin(\Omega_0 t) - 5.2 \times 10^{-8} \cos(\Omega_0 t) - 0.25 \times 10^{-8} \end{bmatrix}. \quad (48)$$

(2) Parameters of sliding mode control are the following: $k = 0.8$, $\alpha_1 = 0.5$, and $\alpha_2 = 1 \times 10^{-5}$.

(3) Parameters of SPMA are the following: size: per square inch 36×36 microthruster array with thruster nozzle diameter of 700 μm; unit thruster impulse $I_0 = 1 \times 10^{-3}$Ns and dip angle $\alpha = 30^\circ$.

(4) Parameters of wheel are the following: $T_k = 2.5 \times 10^{-7}\,\mathrm{N \cdot m}$, $K_v = 6.4 \times 10^{-10}\,\mathrm{N \cdot m \cdot s/rad}$, $T_s = 2.5 \times 10^{-4}\,\mathrm{N \cdot m}$, and $\Omega_s = 2.5$ rad/s.

(5) The errors of sensors are zero for both SPMA and wheel.

The optimization process of the proposed ACEF algorithm is presented in Figure 16. The variations of the attitude angle and the attitude angular velocity using AECF are shown in Figure 17.

Figure 16(a) shows that the AECF before the optimization is 1686.1435, with the increase of the optimal number of array thrusters, AECF decreases gradually; eventually, AECF significantly decreases to 572.3267, reflecting the marked decrease of energy consumption demand. From the Figure 16(b), we can see that at the end of the AECF process, the energy consumption is only 39.78% of the original level, decreased by 60.22%, showing the ability to save energy consumption effectively. It can be determined from Figure 17

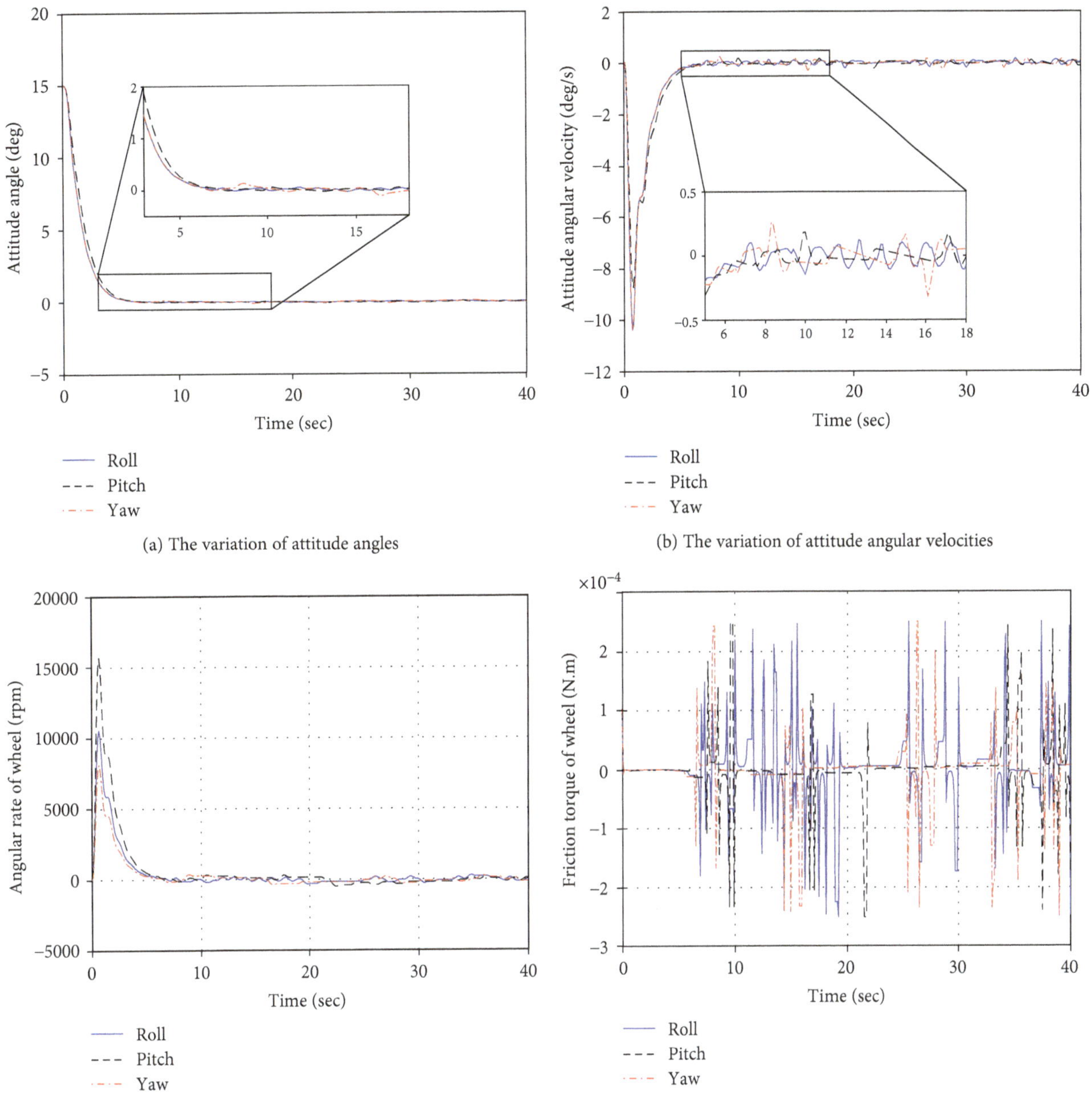

(a) The variation of attitude angles

(b) The variation of attitude angular velocities

(c) Angular rates of wheel

(d) Frictional torques produced by wheel

FIGURE 18: Control results of wheel.

that the attitude precision controls (3σ) are 0.0362°, 0.0325°, and 0.0204°, and the precisions of the attitude angular velocities (3σ) are 0.0013°/s, 0.0025°/s, and 0.0052°/s.

The variations of the attitude angle and the attitude angular velocity using wheel are shown in Figure 18.

Figure 18(a) shows the attitude angles controlled by wheel; Figure 18(b) presents the attitude angular velocities. Figures 18(c) and 18(d) show the angular rates of the wheel and the corresponding friction torque produced by wheel. As can be seen from Figures 18(c) and 18(d), during the stable state, the angular rates of wheel always change around 0. Thus, friction torques are produced by wheel as demonstrated in Figure 18(d), reducing the precision controls of PNS.

The control results of SPMA, SPMA with AECF, and the wheels are shown in Table 1.

Table 1 shows that SPMA using AECF achieves highest precision and the best performance. The errors of the attitudes obtained by SPMA using the AECF are 65.94%, 46.49%, and 80.95% of the errors obtained by SPMA. The errors of attitude angular velocity obtained by SPMA using AECF are 61.90%, 62.50%, and 50.98% of the errors obtained by SPMA. Meanwhile, the errors of attitudes obtained by SPMA using AECF are 51.49%, 43.39%, and 18.46% of the errors obtained by wheels; the errors of attitude angular velocity obtained by SPMA using AECF are 0.68%, 1.68%, and 2.34% of the errors obtained by wheels, demonstrating the feasibility and effectiveness of the proposed algorithm.

Table 1: Comparisons among SPMA, SPMA with AECF, and wheels.

Errors	SPMA	SPMA using AECF	Wheels
Roll (3σ)	0.0549°	0.0362°	0.0703°
Pitch (3σ)	0.0699°	0.0325°	0.0749°
Yaw (3σ)	0.0252°	0.0204°	0.1105°
Attitude angular velocity of roll (3σ)	0.0021°/s	0.0013°/s	0.1909°/s
Attitude angular velocity of pitch (3σ)	0.0040°/s	0.0025°/s	0.1487°/s
Attitude angular velocity of yaw (3σ)	0.0102°/s	0.0052°/s	0.2220°/s

The proposed AECF algorithm can improve control resolutions and reach higher precision controls.

6. Conclusion

Aiming at complicated large-scale control allocation problems, this paper solves the problem of previous optimal allocation methods not being able to make the overall design of the number, energy consumption, and configuration of large-scale thrusters. Through the establishment of SPMA application model, the energy consumption factor is put forward. With the increase of AECF, the energy demand of thruster increases. The decrease of AECF can reduce the thruster energy consumption and alleviate the satellite loads. With the increase of the number of optimal thrusters, the AECF will be reduced. Moreover, the relationship between the ignition thrusters and AECF is provided. The simulation results indicate that through AECF optimizing allocation, the thruster energy demand can be efficiently reduced, control resolution and precision can be improved, and energy consumption can be reduced as well. To the best of our knowledge, this is the first research to apply highly integrated actuators to control the picosatellites and nanosatellites. This work only studies the control allocation problems. As SPMA is not reproducible, to achieve higher precision and reliability, optimal compensation strategy for misalignment and inappropriate positions, heat loss phenomena will be our future work for SPMA.

Conflicts of Interest

The authors declare that there is no conflict of interest regarding the publication of this paper.

References

[1] M. Marszalek, O. Kurz, M. Drentschew, M. Schmidt, and K. Schilling, "Intersatellite links and relative navigation: preconditions for formation flights with pico- and nanosatellites," *IFAC Proceedings Volumes*, vol. 44, no. 1, pp. 3027–3032, 2011.

[2] A. Sinn, T. Riel, F. Deisl, R. Saathof, and G. Schitter, "Low-Power reflective optical communication system for Pico- and Nano-satellites," in *Proceeding Advanced Maui Optical and Space Surveillance Technologies Conference (AMOS)*, Wailea, Maui, Hawaii, September 2016.

[3] C. J. Thoemel, F. Singarayar, T. Scholz, D. Masutti, and P. Testani, "Status of the qb50 cubesat constellation mission," in *65th International Astronautical Congress*, pp. 1–22, Toronto, Canada, 2014.

[4] J. Bouwmeester and J. Guo, "Survey of worldwide pico- and nanosatellite missions, distributions and subsystem technology $," *Acta Astronautica*, vol. 67, no. 7-8, pp. 854–862, 2010.

[5] E. Buchen, "SpaceWorks'2014 Nano/Microsatellite Market Assessment," in *28th Annual AIAA/USU Conference on Small Satellites*, pp. 1–5, 2014.

[6] T. Xiang, T. Meng, H. Wang, K. Han, and Z. H. Jin, "Design and on-orbit performance of the attitude determination and control system for the ZDPS-1A pico-satellite," *Acta Astronautica*, vol. 77, pp. 182–196, 2012.

[7] Y. Zhang, Y. M. Zheng, M. Yang, H. Li, and Z. H. Jin, "Design and implementation of the highly-reliable, low-cost housekeeping system in the ZDPS-1A pico-satellite," *Journal of Zhejiang University SCIENCE C*, vol. 13, no. 2, pp. 83–89, 2012.

[8] K. Han, H. Wang, T. Xiang, and Z. Jin, "Magnetometer compensation scheme and experimental results on ZDPS-1A pico-satellite," *Chinese Journal of Aeronautics*, vol. 25, no. 3, pp. 430–436, 2012.

[9] M. Yang, H. Wang, C. Wu et al., "Space flight validation of design and engineering of the ZDPS-1A pico-satellite," *Chinese Journal of Aeronautics*, vol. 25, no. 5, pp. 725–738, 2012.

[10] D. H. Lewis Jr, S. W. Janson, R. B. Cohen, and E. K. Antonsson, "Digital micropropulsion," *Sensors and Actuators A: Physical*, vol. 80, no. 2, pp. 143–154, 2000.

[11] C. Rossi, S. Orieux, B. Larangot, T. Do Conto, and D. Esteve, "Design, fabrication and modeling of solid propellant microrocket - application to micropropulsion," *Sensors and Actuators, A: Physical*, vol. 99, no. 1-2, pp. 125–133, 2002.

[12] K. L. Zhang, S. K. Chou, S. S. Ang, and X. S. Tang, "A MEMS-based solid propellant microthruster with Au/Ti igniter," *Sensors and Actuators A: Physical*, vol. 122, no. 1, pp. 113–123, 2005.

[13] S. K. Chou, W. M. Yang, K. J. Chua, J. Li, and K. L. Zhang, "Development of micro power generators - a review," *Applied Energy*, vol. 88, no. 1, pp. 1–16, 2011.

[14] B. Yang, J. Miao, Z. Fan, J. Long, and X. Liu, "Modified cuckoo search algorithm for the optimal placement of actuators problem," *Applied Soft Computing*, vol. 67, pp. 48–60, 2018.

[15] X. Liu, T. Li, Z. Li, H. Ma, and S. Fang, "Design, fabrication and test of a solid propellant microthruster array by conventional precision machining," *Sensors and Actuators A: Physical*, vol. 236, pp. 214–227, 2015.

[16] W. C. Durham, "Constrained control allocation - three-moment problem," *Journal of Guidance, Control, and Dynamics*, vol. 17, no. 2, pp. 330–336, 1994.

[17] J. M. Buffington and D. F. Enns, "Lyapunov stability analysis of daisy chain control allocation," *Journal of Guidance, Control, and Dynamics*, vol. 19, no. 6, pp. 1226–1230, 1996.

[18] D. B. Doman and A. D. Ngo, "Dynamic inversion-based adaptive/reconfigurable control of the X-33 on ascent," *Journal of Guidance, Control, and Dynamics*, vol. 25, no. 2, pp. 275–284, 2002.

[19] A. Page and M. Steinberg, "A closed-loop comparison of control allocation methods," in *AIAA Guidance, Navigation, and Control Conference and Exhibit*, Denver, CO, USA, August 2000.

[20] J. Virnig and D. Bodden, "Multivariable control allocation and control law conditioning when control effectors limit," in *Guidance, Navigation, and Control Conference*, pp. 572–582, Scottsdale, AZ, U.S.A, August 1994.

[21] J. Jin, "Modified pseudoinverse redistribution methods for redundant controls allocation," *Journal of Guidance, Control, and Dynamics*, vol. 28, no. 5, pp. 1076–1079, 2005.

[22] H. Alwi and C. Edwards, "Propulsion control of a large civil aircraft using on-line control allocation," in *2009 American Control Conference*, pp. 4581–4586, St. Louis, MO, USA, June 2009.

[23] R. S. Sanchez Pena, R. Alonso, and P. A. Anigstein, "Robust optimal solution to the attitude/force control problem," *IEEE Transactions on Aerospace and Electronic Systems*, vol. 36, no. 3, pp. 784–792, 2000.

[24] M. Wang and Y. Xie, "Design of the optimal thruster combinations table for the real time control allocation of spacecraft thrusters," in *Proceedings of the 48h IEEE Conference on Decision and Control (CDC) held jointly with 2009 28th Chinese Control Conference*, pp. 5063–5068, Shanghai, China, December 2009.

[25] Q. Shen, D. Wang, S. Zhu, and E. K. Poh, "Robust control allocation for spacecraft attitude tracking under actuator faults," *IEEE Transactions on Control Systems Technology*, vol. 25, no. 3, pp. 1068–1075, 2017.

[26] A. Cristofaro and T. A. Johansen, "Fault tolerant control allocation using unknown input observers," *Automatica*, vol. 50, no. 7, pp. 1891–1897, 2014.

[27] M. Kirchengast, M. Steinberger, and M. Horn, "Input matrix factorizations for constrained control allocation," *IEEE Transactions on Automatic Control*, vol. 63, no. 4, pp. 1163–1170, 2018.

[28] C. Liu, C. Li, and Z. He, "Adaptive control of a morphing aircraft through updating the number of actuators," in *AIAA Guidance, Navigation, and Control Conference*, pp. 1–9, Minneapolis, Minnesota, USA, August 2012.

[29] R. Jamil, F. Salama, M. Lone, A. Savvaris, and J. Whidborne, "Multi-objective control allocation with conflict resolution," in *AIAA Guidance, Navigation, and Control Conference*, pp. 1–12, Minneapolis, Minnesota, USA, August 2012.

[30] J. Sherman and W. J. Morrison, "Adjustment of an inverse matrix corresponding to a change in one element of a given matrix," *The Annals of Mathematical Statistics*, vol. 21, no. 1, pp. 124–127, 1950.

[31] H.-U. Oh, J. Onoda, and K. Minesugi, "Semiactive isolator with liquid-crystal type ER fluid for momentum-wheel vibration isolation," *Journal of Vibration and Acoustics*, vol. 126, no. 2, p. 272, 2004.

[32] V. Carrara, A. G. Da Silva, and H. K. Kuga, "A dynamic friction model for reaction wheels," *Advances in the Astronautical Sciences*, vol. 145, pp. 343–352, 2012.

6

A Simplified Mobile Ad Hoc Network Structure for Helicopter Communication

Abdeldime Mohamed Salih Abdelgader,[1,2] Lenan Wu,[1] and Mohammed Mohsen Mohammed Nasr[1]

[1]*Southeast University, School of Information Science and Engineering, Nanjing 210096, China*
[2]*Karary University, College of Engineering, Electrical and Computer Department, 12304 Khartoum, Sudan*

Correspondence should be addressed to Abdeldime Mohamed Salih Abdelgader; abdeldime@hotmail.com

Academic Editor: Mohamed Gad-el-Hak

There are a number of volunteer and statutory organizations who are capable of conducting an emergency response using helicopters. Rescue operations require a rapidly deployable high bandwidth network to coordinate necessary relief efforts between rescue teams on the ground and helicopters. Due to massive destruction and loss of services, ordinary communication infrastructures may collapse in these situations. Consequently, information exchange becomes one of the major challenges in these circumstances. Helicopters can be also employed for providing many services in rugged environments, military applications, and aerial photography. Ad hoc network can be used to provide alternative communication link between a set of helicopters, particularly in case of significant amount of data required to be shared. This paper addresses the ability of using ad hoc networks to support the communication between a set of helicopters. A simplified network structure model is presented and extensively discussed. Furthermore, a streamlined routing algorithm is proposed. Comprehensive simulations are conducted to evaluate the proposed routing algorithm.

1. Introduction

In the recent years, Ad hoc networks emerged with the rapid development of communication technologies. Typical ad hoc approaches consider nodes with similar capabilities that move in random patterns relative to one another over time and the communication is separated from the other activity of the node. This type of networks is known as mobile ad hoc network (MANET), which is a wireless network that is formed by a collection of self-organizing mobile nodes [1]. Each node communicates with its neighbors over a shared wireless medium. In MANET, wireless nodes cooperate to relay packets over multiple relay hops from source to destination that is known as multihop network [2–4]. Due to the lack of central management, nodes in MANET are designed to act as end system and routers for other nodes. The network is connected dynamically and does not rely on any preexisting network infrastructure. In MANET, nodes are free to move and have the capability to deliver messages in a decentralized manner. MANET has been utilized in vehicular environments to form the vehicular ad hoc networks (VANETs) [3, 5]. VANETs recently emerged with great interest due to their impact in reducing traffic jams and increasing safety. It can also be used for providing many services to passengers, drivers, and authorities. VANETs are limited only to urban and highways areas and they are difficult to work in rugged environments and some extraordinary situations without special communication equipment and facilities.

Unlike planes, helicopters have the advantage of not requiring a runway for taking off or landing. Therefore, they can carry both people and goods to virtually impassable locations often not connected by road and only accessible on foot with difficulty. Helicopters are therefore indispensable tool for rescue organizations, armies, navies, and especially air forces. Recently, the ambulance service has a statutory responsibility to respond to traumatic and medical emergencies. Fire and rescue authority has a statutory duty to deal with fires, road traffic accidents, and maybe certain types of terrorist related incidents such as mass decontamination. Fire and rescue authorities with discretion require responding to events such as flooding or animal rescue [6]. Helicopters can be a valuable and versatile asset in any Search and Rescue (SAR) events [4, 7, 8]. Emergency services personnel are increasingly calling on the support of a helicopter air assistance [9]. Radio contact is still of great importance for communication between helicopters, planes, and ground stations. The operation environment is different from that of vehicular networks and traditional ad hoc networks. Helicopters outperform the performance of vehicles, particularly in cases that need fast response and in rugged or distant areas. However, it requires special and reliable communication requirements to provide reliable services. Helicopters network can also serve as alternative emergency communication system in case of natural disasters, when there is lack of ordinary ground communication systems. Furthermore, in many situations, communication between helicopters and ground vehicles or operation center is highly required.

Recently, the increased capabilities in embedded systems have enabled unmanned flight through the removal of the human pilot from the aircraft cockpit. As the applications of drone and Unmanned Aircraft System (UAS) expand, the principal issues of communication technologies are flexibility, adaptability, and controllability of the information/data flows. Future systems will be net-centric and rely on mobile ad hoc networking to provide real-time connectivity among cooperating agents and to provide backhaul of sensor and telemetry data from mobile nodes to a data repository or command center [10]. There is a current push towards the development of simple, small, low-cost UAS that work individually and in cooperative teams to accomplish complex and dangerous missions. These networks can be applied to numerous applications, including diffuse gas and plume detection [11], coordinated search [6] and reconnaissance [12], and in situ atmospheric sensing [13], as agents in the battlefield [14], aerial photography, and mass decontamination, and as components in command and control architectures. With smaller, cheaper UAS that have limited sensor capabilities, cooperative control relies heavily on communication with neighbors [6, 15–17]. That means, by forming a multihop communication network, information can be efficiently shared among helicopters or UAS. Consequently, this will increase the overall mission capabilities and extend the operational range.

Communication networks between and through aerial vehicles are the backbone of the current battlefield communications. For instance, Google recently established a research and development project with the mission of offering Internet access to rural, rugged, and remote areas. The project is denoted as Project Loon [18]. The project uses high-altitude balloons placed in the stratosphere at an altitude of about 18 km to create an aerial wireless network with up to 4G-LTE speeds. Project Loon balloons float in the stratosphere, twice as high as airplanes and the weather. In the stratosphere, there are many layers of wind, and each layer of wind fluctuates in direction and speed. Loon balloons go where they are needed by rising or descending into a layer of wind blowing in the desired direction of travel. By associating with telecommunications companies to share cellular spectrum, loon balloons can enable people to connect to the balloon network directly from their phones and other LTE-enabled devices. The signal is then passed across the balloon network and back down to the global Internet on Earth. Each balloon can provide connectivity to a ground area (*footprint*) about 80 km in diameter using LTE wireless communications technology [19]. In a similar manner, helicopter network can provide Internet\network access for a specific area and for a particular mission by connecting ground users to a backbone network in a certain time.

Present helicopters communication systems use specialized high-cost radios that operate in designated military radio and other communication bands. Small low-cost Commercial Off-the-Shelf (COTS) radio equipment combined with powerful computer processing can be mounted on helicopters or small UA and has the potential to revolutionize battlefield communications and open up many scientific and commercial applications. One of COTS technologies is the IEEE 802.1x wireless LANs that connect wireless mobile nodes to a fixed infrastructure and is being widely deployed, including in UAS applications [15, 20]. Both IEEE 802.11 and IEEE 802.16 define Peer-to-Peer (P2P) and wireless ad hoc networks, where an end user communicates to users or servers on another local area network (LAN) or other users using its access point or base station. More interesting applications are possible when swarm of helicopters connect to each other in peer-to-peer ad hoc meshed wireless networks [2, 21]. This can form directly or through multihop connected airplane, a new kind of MANET called aerial vehicular ad hoc networks (AVANETs).

The ability and performance of utilizing MANET as platform for connecting helicopters in one network have not been thoroughly investigated. MANET can be used in communicating a group of helicopters or drones working in specific mission to each other. It can also be used to communicate a helicopter or drone to the data center. Nowadays, people activities extended to deserts and other rugged environments for many purpose such as rescue, exploration, transportation, research, and mining. Helicopters can serve a lot in these areas. Improving the transmission of information about the accidents, weather, and locations in these areas will provide people with a higher degree of safety and comfort.

To the best of our knowledge, the use of ad hoc networks in helicopter communications has never been previously studied, despite their reliability, scalability, high data rate, bandwidth, diversity of application, and minimum cost compared to the existing solutions. This and the promising expected applications are the motivations of this work. This paper addresses the utilization of MANET in forming AVANET that consists of helicopters or UAS nodes. The main objective of this work is to investigate the utilization of MANET as alternative data communication link for helicopters, drones, or UAS environments. We present a simple model for the network structure and extensively discusse the different expected communication scenarios. In the proposed structure, a swarm of helicopters are clustered to one or multiple clusters which can communicate to some central servers through a wireless mesh backbone as well as Internet. Within each cluster, a self-organizing MANET is formed such that all helicopters are connected to all others within the same cluster. As the research on this topic, along with other related issues, is still in its infancy, we bring the attention to network structure formation and information routing. Accordingly, we propose a simple routing protocol. Furthermore, as many MANET routing protocols, such as Clustering Based Routing Protocol (CBRP) [22], use the ID number of the nodes to select the cluster head, this paper proposes a novel swarm head (SH) election approach which considers the communication equipment, velocity, and location of the helicopters in the election process. The proposed SH election approach uses a weighting mechanism to achieve a better utilization of all the nodes and perform optimum SH selection by using the election priority factor (EPF) of the nodes. The selection based on the location, equipment, and velocity of the nodes has never been addressed by other existing protocols which only use the ID or velocity to elect the SH.

The rest of this paper is organized as follows. Section 2 tackles ordinary helicopters' communication techniques and their limitations. This section also presents the promising communication scenarios which can be achieved by utilizing ad hoc networks for helicopter networks. In Section 3, we discuss the research opportunities which have not been considered by this work. The proposed network structure model and the SH election algorithm are presented in Section 4, while the routing algorithm is proposed in Section 5. Section 6 presents our simulations settings and their results. Section 7 concludes this work.

2. Helicopter Communications

Ordinarily, communications between air ambulance, police air support, SAR helicopter, and land SAR teams usually are via the ground to air channel as identified in the National Band Plan which is VHF High Band Land Search and Rescue Radio Channels (VHF-HBLS-RRC) [9]. Communications with coastguard rescue teams uses VHF channel 156 MHz. Communication between responding helicopters usually uses Scene of Search 123.1 MHz AM. SAR and some air ambulance helicopters are often fitted with VHF/UHF/HF and FM radios [9]. Some helicopters carry the Airwave Communications System (ACS), providing Terrestrial Trunked Radio (TETRA) communications used by emergency services and police through several talk groups [9]. In addition, the ACS provides a means to make telephone calls from the helicopter to mobile or PSTN telephone numbers. Some helicopters also carry a spare FM radio that can be deployed to incident control or to mobile ground units. Some kinds of helicopters, such as the Maritime Coastguard Agency (MCA) helicopters, have a satellite link (Satcom) and there is potential to communicate directly by phone [23]. Specific aspects of operations, such as hovering, winching, or refueling, may make it difficult to contact or get a reply from the crews. Although crew can monitor all radios throughout the copilot, these systems may not function with high efficiency, in case of urgent messages and life-threatening situations. Terrain masking can also be a factor and, in this case, emergency services may consider using local MCA units or MRTs helicopters (Mountain Rescue Teams) to enhance communications using satellite links. Emergency situations sometimes necessitate exchange of some information such as images, maps, and video that require high data rate and bandwidth. Besides, in some special missions an immediate exchange of information between helicopters themselves is quite needed. Conventional VHF/UHF communication systems are not the efficient way to carry this type of information.

Nowadays, Aircraft Communications Addressing and Reporting networks (ACARs) are under operation in the aeronautical world for aircraft-ground communications to offer a cockpit/maintenance/cabin operation. The ACARs data link service uses VHF radio channels. The VHF ground stations coverage is "line-of-sight." This means that a station at an airport can provide coverage to aircrafts located at the airport and above the airport within a cone limited by the curvature of the earth.

ACARs usually share the VHF band with voice communications, because most channels had been assigned to voice when ACARs was implemented. Thus, there were not enough VHF channels to assign a dedicated frequency to each ACARs ground station. This required the ACARs VHF design to allow service providers to have all their ground stations across a region use the same base frequency. Furthermore, by using the traditional VHF voice systems, the data rate in ACARs system is limited to only 2.4 kilobits per second. At a higher data rate, the rate of errors in decoding the received signal would rise and communications would become impossible. Moreover, there are already no VHF frequency channels available for allocation to new VHF ACARs services in Europe. This will soon be true in certain parts of USA. Therefore, to increase VHF data link capacity further, it will be quite essential to implement a system providing more capacity per channel. Although some helicopters have satellite communication or ACS Links, these solutions have limited bandwidth, are very expensive, and are only used for necessary voice calls [23]. For instance, in surveillance, monitoring, or rescue operations, the image or video of the target area must be relayed from the helicopter to the control center with a very strict delay bound, and it requires high bandwidth. In addition, by the help of the technological advancements on sensor technologies, it is possible to collect data with very high resolution, and this makes the bandwidth requirement much higher. The collaboration and coordination of multiple helicopters also need additional bandwidth resources. Sometimes, a type of communication to provide comfortable applications, such as entertainments, Internet services, and others, for the working team is also needed. In these cases, ad hoc network possibly is one of the alternative cheap solutions.

Ad hoc networks can be used to handle the communications between helicopter nodes using different scenarios:

(1) They can be used as internal communication between helicopters. Using this scenario a group of helicopters, drones, UA, or aircrafts can form together a full or half mesh topology LAN by which they can communicate to each other as shown in Figure 1(a). Forming this

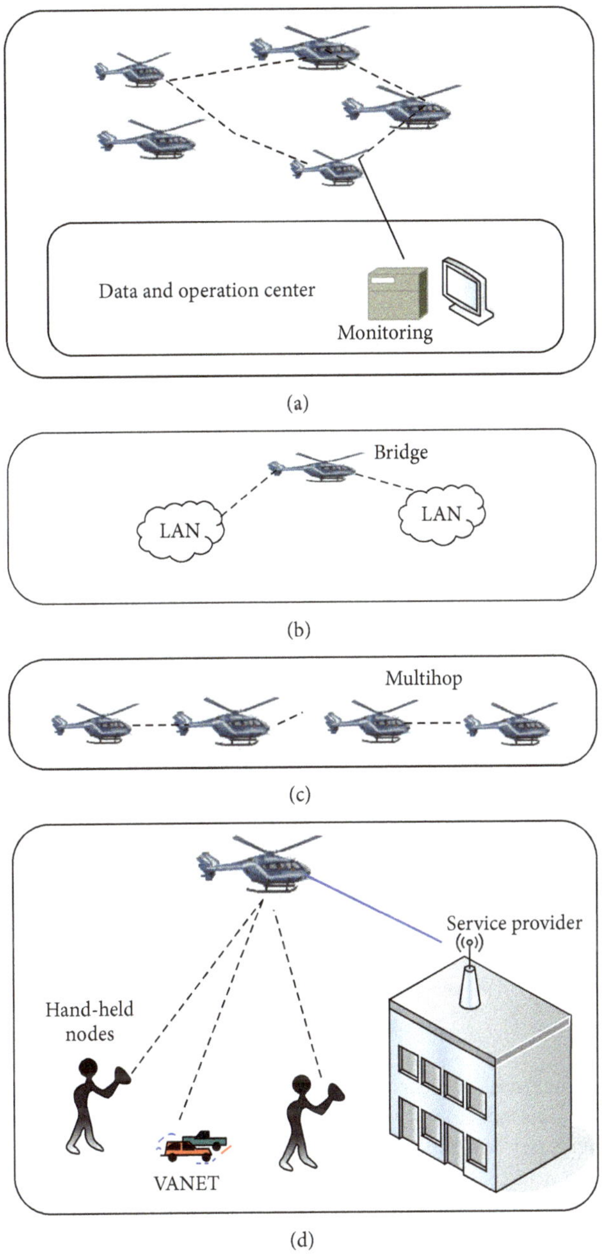

Figure 1: Helicopter network scenarios. (a) LAN and WAN. (b) Helicopter as bridge. (c) Extending network coverage range through multihops. (d) Providing Internet/network access for some users in specific area.

network can enable them to share and exchange resources and information between each other.

(2) They can be usedas a bridge in case of natural disasters or in special scenarios in which a communication between two networks is needed when the ordinary communication systems are lacking. This can be done using a rescue helicopter or small UA, to connect two nodes or networks as shown in Figure 1(b). This scenario can be temporarily implemented for a particular mission, when there is some natural obstacles which prevent installing ordinary network infrastructure, such as in seas (to connect two ships), across rivers, and over mountains.

(3) They can be used to extend the communication range: The range of the network can be extended beyond the ordinary distance of *Wi-Fi* or any other wireless technology, using the ad hoc multihop characteristic, as shown in Figure 1(c). This scenario has many applications using helicopter nodes. It can be used to provide temporary link to certain location in some circumstances. This characteristic is very useful in case of natural obstacles, natural disasters, or loss of ordinary communication infrastructure in specific location. The environment also affects the communication coverage of the ordinary VANETs. Obstacles, such as mountains, walls, or buildings and constructions, may block the radio signals. AVANET can help to operate behind the obstacles, and it can extend the scalability of multihelicopter applications [24].

(4) They can be used to connect to a central point: A group of helicopters or air crafts can be connected to form a local area network (LAN) and then select one of them to act as gateway to connect to the central operation network. That means that they can exchange and share resources and information with the operation center using this topology, as shown in Figure 1(a).

(5) *Helicopter or drone networks* can be an active part of the upcoming technologies Internet of Vehicle (IoV) and Internet of Things (IoT). Providing Internet connection and network access to helicopters would provide a promising applications for many activities such as that of Figure 1(d).

(6) *Helicopter networks* can also serve as a large-scale wireless sensor network (WSN) for future ITS and many other systems, particularly in natural disasters situations and rugged environments, because every helicopter can be regarded as a super sensor node. Helicopters are commonly equipped with thermometers, microphones, one or more cameras, ultrasound radar, GPS, and other sensory features. A WSN of such capabilities is unprecedented and can extend perceptive computer systems to every point of the world. That means that, without the need for restrictive infrastructure, information can be shared in a peer-to-peer manner.

3. Important Consideration

The use of MANET in helicopters and drones networks is a new wide field of research. Up to our knowledge, this work seems to be the first one that studies the ability of using MANET to provide communication between helicopters. Therefore, many works require to be conducted in order to provide a complete investigation and implementation of MANET in such kind of environments. Many researches are required to consider the physical, MAC, and application layers in more details and study the suitability of the existing MANET and VANET solutions for AVANETs. Considering that helicopters can be a part of intelligent transport system (ITS), the use of wireless access in vehicular environment (WAVE) standard is also need to be investigated, as this standard is particularly designed for vehicular networks, and some AVANETs' characteristics may not be similar to that of VANET. Moreover, security is one of the most important issues, as this paper does not include a security solution for communication between helicopters in different scenarios. The existing solutions need to be investigated to ensure their suitability, reliability, latency, and scalability for these types of networks.

Furthermore, a general model for modeling the speed and the distribution of helicopters is much needed. Many other investigations and studies, which are related to helicopter network applications, routing, MAC, and physical layers can also be conducted. The implementation of OSI model in such type of networks may not fit well due to some constraints imposed by the nodes itself and the environment. Therefore, cross layering techniques are much needed to facilitate the interaction between layers and make it possible to tune layer parameters so that the overall performance can be improved. Many MAC and routing protocols [21, 25–27] can be used in this type of network but their impact, performance, reliability, and suitability need to be intensively examined. Additionally, the effect of these communication systems on the safety of the helicopter is one of the most important issues that is required to be carefully studied. There are many challenges for helicopter networks such as modeling mobility, connectivity, broadcast method, and routing protocols. The above stated points and many others are the next phase of research in AVANET. This paper proposes a general structure for the helicopters network and presents a routing mechanism for information exchange. The communication problem initially can be solved by using other systems' technologies, or by utilizing other researches' results [10, 15, 23, 28], but the high mobility and the dynamic change in topology and environments remain the biggest challenges, which require special attention. Hence, what set this work apart are the exploration and the use of MANET technologies and their induction into the scope of the proposed structure model, although not all technical difficulties are completely eliminated.

4. The Proposed Network Model

The network scenarios adopted throughout this paper are rescue environments or a group of helicopters/drones working in specific mission. The network structure of the proposed model is shown in Figure 2. We assume a number of helicopters (N). The maximum numbers are not more than $N_{\max}$. Each of these helicopters is moving with velocity equal to v_I. When the number of nodes in one group exceeds $N_{\max}$, it will be divided into $g = K/N_{\max}$ neighbor groups; $K > N_{\max}$ is the total number of helicopters. In this paper, we use the notation *swarm* to represent a group of N helicopters moving in the same direction and within a communication coverage range. The basic coverage range (L) between two

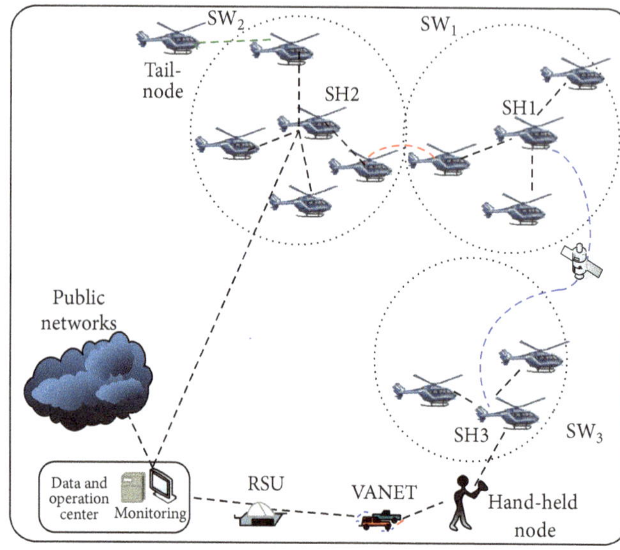

FIGURE 2: Simple structure for helicopters communication network.

helicopter nodes (I, J) is according to (1), where x, y, z are the coordinates of nodes. The presence of multihop operation can extend the coverage range to $L_{\max}$. The multihop operation can be conducted between at least three helicopters. The maximum distance between the first and the third one does not exceed $2 \times L$. That means that the maximum coverage distance in one swarm can be up to $L_{\max} \leq (N - 1) \times L$. The maximum value of L depends on the type of the physical layer communication technology, such as IEEE 802.11p and IEEE 802.16. Conisder

$$L \leq \sqrt{\left(x_i - x_j\right)^2 + \left(y_i - y_j\right)^2 + \left(z_i - z_j\right)^2}. \qquad (1)$$

The mission of the swarm can be rescue operation, aerial reconnaissance, or military mission as well as alternative emergency communication system in case of natural disasters when there is lack of ordinary communication systems. The communication conditions can be according to three scenarios shown in Figure 2. The first scenario is the communication among the swarm members. The second one is the communication between two or more swarms, in case of very large number of helicopter nodes or when two or more swarms are moving in different directions. The third scenario is the communication between the swarm members and the ground operation center, the ground VANET nodes, or the communication to ground hand-held node which is a soldier or a victim that requires rescue. Practically, the number of helicopters moving in one direction for specific mission usually is very small. The small number of nodes eases swarm management and provides good performance. However, in some cases the large number of nodes may offer more alternative communication routes and facilities.

In this model, we assume that all helicopters are equipped with ad hoc radio equipment in addition to the traditional VHF/UHF systems used for voice communications. The ad hoc uses the specification of IEEE 802.11p to form an ad hoc network with a coverage range about 1 Km. Using

802.11 with high-gain antenna can extend the coverage up to 20 km; however, this requires more power, while the utilization of IEEE 802.16 can increase the coverage to more than 40 km. However, some nodes may have other types of direct communication links. Therefore, this paper classifies helicopters according to their equipment into two types. The first type is that equipped with satellite link or any direct data communication link to ground terminals. We will use the notation S_1 to represent this type. The second type is the helicopters equipped with ad hoc network onboard units in addition to the ordinary voice communication links. We use the notation S_2 to represent this type of helicopters.

4.1. Swarm Formation. Helicopter/drone initiates the communication by broadcasting a *HELLO* message to discover the surrounding nodes. The *HELLO* message contains the information of the helicopter. These pieces of information include the ID number, velocity, and location and direction in degree and communication equipment type. We define a transitional state called *S_UNDECIDED* for smoother operation of swarm formation. "*Undecided*" means that a node is still in search for its host *swarm*. Other nodes within the coverage region upon receiving the message will respond by sending a *REPLY* message. In case of the responding node already joining an existing swarm, *REPLY* message will include its ID number, velocity, location, direction, and the ID number of the swarm. If it is not a member of a swarm it will send *FREE* message that includes its direction, ID number, velocity, and location. In case of the fact that the responding node is the SH of an existing swarm, it will send its information and the swarm information which include the amount of current nodes (N) in the swarm. If $N < N_{\max}$, the SH will include an invitation message (*INVITE*) for the *Undecided* helicopter to join the swarm. The helicopter will directly become a swarm member and has all the information of the swarm and it can use the swarm facilities. It will change its state to "*S_member.*" The direction is one of the important factors that qualify any helicopter to join a certain swarm. The direction should be not more than θ degree of the SH direction.

If $N \geq N_{\max}$, SH will send to the new node a JOIN_GUEST message. The *JOIN_GUEST* msg includes the information of the nearest neighbor node to the guest one. The SH finds the nearest nodes using the information of its swarm and the location of the guest node. The new node will form a temporary network with the nearest neighbor node. The nearest node will work temporarily as relay node to carry the information of the guest node through SH to the rest of the network. The *Undecided* node will be treated as *Tail-Node* (TN) of the nearest node as shown in Figure 2. This situation will continue until the *Undecided* helicopter joins a swarm. In case of the responding node of type *Undecided*, the two nodes will store the information and directly start the SH election process according to the SH election procedure described in Section 4.2. After formation of the new swarm, the nearest node and its TN will act as relay between the two swarms. The swarm formation is not limited to the nodes that located within one MANET coverage; a set of nodes moving in the same direction can form a swarm using multihop operation until the number of nodes equals $N_{\max}$.

As a rule, SH and the neighbor SH can communicate directly if they are in the same ad hoc coverage. Nonetheless, in the actual situation of the physical structure of the network, SH1 and its neighbor SH2 may not necessarily to be within the same ad hoc communication range. Therefore, SH selects boundary members to act as a relay between the neighbor SHs and forward information. Therefore, swarm formation not only is the process of building the swarm structure and electing the SH, but also includes the operation of determining the boundary nodes used to communicate between the adjacent swarms as shown in Figure 2.

4.2. Swarm Head Election. SH is responsible for route finding and organizing of interswarm communication. It is also responsible for finding the best route to distant destinations by exchanging its swarm information with other SHs. In many situations, SH participates in information forwarding process. Therefore, helicopter with best resources and facilities, located in a suitable location and having an average velocity among other nodes, is qualified to serve as a SH. The helicopters of any swarm need to select a SH. The selection of the SH depends on different factors, such as communication equipment, velocity, and location within the swarm. Helicopter with satellite link has the highest equipment priority. Helicopter located in the center of the swarm also has the highest location priority.

In this paper for the purpose of SH election process, we make the following assumptions. The notation P_i is the SH election priority factor (EPF). Each helicopter node has an ID number which is unique and known to the whole network. Each helicopter is equipped with a GPS used to periodically collect location, velocity, and direction. The maximum speed for each swarm member is $v_{\max}$. If the number of swarm members is too small, number of helicopter nodes that are able to act as a SH will increase and the network will become more isolated. If swarm members are too high the burden on the SH will increase. Consequently, the intraswarm communication cost increases, because the SH is responsible for communication between both its nodes and the neighbor SHs. Therefore, the swarm size needs certain constraints. In this paper, we define $M_{\max}$ to represent the maximum number of hops between the swarm members and the SH. We also define $N_{\max}$ as the maximum number of nodes in each swarm. Furthermore, if the movement direction of the SH is opposite or different to most of the nodes within the swarm, the node quickly leaves the range of swarm and joins other swarms, which increases the swarm maintenance burden. Therefore, movement direction of the nodes needs to be taken into account when selecting SH. In actual scenarios, it is very difficult to find two nodes moving exactly with the same orientation. To solve this problem, we used a threshold (θ) to determine whether the directions are approximately the same or not. Two helicopters will be considered as moving in the same direction if the angle between them does not exceed the maximum thresholds. In this paper, helicopters are moving to the same direction and the maximum threshold of difference between two helicopters is 15 degrees. Helicopters moving in the opposite direction combine another swarm. SHs moving in different directions are able to connect as long as they were in the communication range. When they are outside the ad hoc coverage, they can communicate using satellite communication or through other SH. In order to avoid frequent changes in swarm structure, the speed of the nodes also is one of the important considerations. A SH should be the node relatively near to the average rate of other nodes velocity. Based on the above assumptions, a node is elected as a SH according to its EPF value; the following expresses the mathematical calculation of the EPF (P_i).

Assume that I represents the intended helicopter node, J is another node in the same swarm, and each I node has $N-1$ neighbors, $J_1, J_2, J_3, \ldots, J_{(N-1)}$, with coordinates $(x_{J_1}, y_{J_1}), (x_{J_2}, y_{J_2}), \ldots, (x_{J_{N-1}}, y_{J_{N-1}})$. Φ_I is the group of nodes within the same swarm which are neighbors of node I. Intended helicopter coordinate is x_I, y_I. Helicopters velocities at instance election time are $v_1, v_2, \ldots, v_N$. Helicopters are classified according to their communication equipment into S_1 and S_2 as mentioned above.

Firstly, according to the equipment of the helicopter, the equipment priority value P_E is as follows:

$$P_E = \begin{cases} 0 & \text{if type } S_1 \\ 1 & \text{if type } S_2. \end{cases} \tag{2}$$

Helicopter with $P_E = 0$ has highest equipment priority.

Secondly, to determine the helicopter which has the best location within the swarm, we used the summation of distances for each helicopter from all others. Distance between I and J nodes is obtained by

$$D_{I,J} = \sqrt{(x_I - x_J)^2 + (y_I - y_J)^2 + (z_I - z_j)^2}. \tag{3}$$

Each helicopter in the swarm has $(N-1)$ distances toward the others, which form the $\mathbf{D}_I$ distance set:

$$\mathbf{D}_I = [D_{I1} \ \ D_{I2} \ \ D_{I3} \ \ \cdots \ \ D_{I(N-1)}]. \tag{4}$$

Summation of these distances for each node is obtained by

$$S_I = \sum_{J=1}^{N-1} D_{IJ}. \tag{5}$$

Obtaining S_I $\forall J \in \emptyset_I$ yields

$$\mathbf{S}_i = [S_1 \ \ S_2 \ \ S_3 \ \ \cdots \ \ S_N], \tag{6}$$

where S_i is distance values of all nodes in the swarm. The minimum value of the set S_i is the best location. Helicopter with less value is the nearest one to all others and will take a high location priority in SH election. This means that the selected helicopter is not usually located in the centre of swarm, but it is the closest one to all swarm members. As the election process depends on many factors such as ID, equipment, and velocity priorities, we define another notation which is the location priority deviation factor (P_L), where P_L is associated with other priority factors to calculate the EPF (P_i) of a helicopter. It is a value used to represent the location priority deviation of a helicopter from that of best location priority. The following equations are used to calculate (P_L):

$$\begin{aligned} &S_\mu = \min(\mathbf{S}_i), \\ &P_L = S_I - S_\mu \quad \text{For node with minimum } \mathbf{S}_I,\ P_L = 0. \end{aligned} \tag{7}$$

P_L of all nodes can be represented by

$$\mathbf{P}_{\mathbf{L}_\mathbf{I}} = [P_{L_1}, P_{L_2}, P_{L_3}, \ldots, P_{L_N}]. \tag{8}$$

Simply, it can be normalized by scaling between 0 and 1 using

$$\text{Normlized}(P_L) = \frac{P_L - \min(P_{L_I})}{\max(\mathbf{P}_{\mathbf{L}_\mathbf{I}}) - \min(\mathbf{P}_{\mathbf{L}_\mathbf{I}})}. \tag{9}$$

Thirdly, to calculate the helicopter velocity priority, first simply calculate the mean velocity v_μ:

$$v_\mu = \frac{1}{N}\sum_{i=1}^{N} v_I. \tag{10}$$

The velocity priority factor P_v for each helicopter (I) can be calculated as follows:

$$\begin{aligned} P_v &= \frac{1}{N}(v - v_\mu)^2, \\ \mathbf{P}_{\mathbf{v}_\mathbf{I}} &= [P_{v_1} \ \ P_{v_2} \ \ P_{v_3} \ \ \cdots \ \ P_{v_N}]. \end{aligned} \tag{11}$$

The set $\mathbf{P}_{\mathbf{v}_\mathbf{I}}$ represents the velocity priority factor of all helicopters. Node with the lowest P_{v_I} has the highest velocity priority. The normalized velocity priority factor for each node can be

$$\text{Normlized}(P_v) = \frac{P_v - \min(\mathbf{P}_{\mathbf{v}_\mathbf{I}})}{\max(\mathbf{P}_{\mathbf{v}_\mathbf{I}}) - \min(\mathbf{P}_{\mathbf{v}_\mathbf{I}})}. \tag{12}$$

Finally, the EPF (P_i) is a function of the normalized version of P_E, P_L, P_v which are calculated in (2), (9), and (12), respectively. Every node calculates these values and then calculates the EPF (P_i) according to the following:

$$P_i = aP_E + bP_L + cP_v, \tag{13}$$

where a, b, and c are weighting values that satisfy $a+b+c=1$. The EPFs of all nodes are expressed by $\mathbf{P}_{\mathbf{i}_\mathbf{I}}$:

$$\mathbf{P}_{\mathbf{i}_\mathbf{I}} = [P_{i_1} \ \ P_{i_2} \ \ P_{i_3} \ \ \cdots \ \ P_{i_N}] \tag{14}$$

$$\text{Normlized}(\mathbf{P}_i) = \frac{\mathbf{P}_i - \min(\mathbf{P}_{\mathbf{i}_\mathbf{I}})}{\max(\mathbf{P}_{\mathbf{i}_\mathbf{I}}) - \min(\mathbf{P}_{\mathbf{i}_\mathbf{I}})}. \tag{15}$$

(1) Each helicopter periodically sends a *HELLO* message (`msg, hop_cnt`) at the same time it forwards the *HELLO* messages of the other. Upon doing k hops the node will be

capable of processing other nodes' information, such as the basic specification and condition of the nodes, which include the ID number, the swarm election priority (P_i), location information, velocity (v), and hop count (m). Every node upon receiving the HELLO message from others stores the information of the other nodes and subsequently uses this information to determine the hop count (m) of each node and saves it for further usage. During the information collection process, the receiving node always determines the hop count of the other node, stores the original m, adds a number 1 to the count, and then forwards the information to the next node. Through this process, the next node will be able to determine the m hop of that node; also it ensures that the number of hops does not exceed M_{max} and the receiving node avoids the looping. After a number of transmissions, the nodes collect their neighbors' information, including m hops.

(2) After the collection of the information, each node (I) calculates their value $P_i(I)$ according to (13) and (15) and compares the result with $P_i(J)$ for all other nodes in the group Φ_I, if

$$P_i(I) = \min\left(P_{i_I}\right). \quad (16)$$

The node with a lower P_i value has the highest priority. Thus it will upgrade itself as a SH. If two or more nodes share the least value of P_i, hence

$$\begin{aligned} P_i(I) &= P_i(J) = \min\left(P_{i_I}\right) \\ P_E(I) &< P_E(J). \end{aligned} \quad (17)$$

Node with least P_E value is selected as a SH, if

$$\begin{aligned} P_E(I) &= P_E(J) \\ \mathrm{ID}(I) &< \mathrm{ID}(J). \end{aligned} \quad (18)$$

Node with the least ID should be elected as the SH.

(3) The helicopter, after upgrading itself as a SH, changes its state to *S_head* and broadcasts a message, *HEAD (Swarm_ID, head_msg, route, hop_cnt)*, propagating swarm information, where *Swarm_ID* is a swarm label, *head_msg* is the SH node information, *route* is the route information, and *hop_cnt* is the number of hops. In the initial transmission, the value of *route* is set to null and *hop_cnt* is set to zero. Other nodes within the swarm, upon receiving the HEAD message, store all values and judge the value of *hop_cnt*. If *hop_cnt* is less than M_{max}, then they will add one to this value and then forward the information to the next hop. If *hop_cnt* is equal to M_{max}, then there is no need for further treatments. When other nodes receive the same *Swarm_ID*, they investigate the m values and select the lowest one and then use this value to determine the value of their *hop_cnt* return. When they receive a message from a different *Swarm_ID*, they analyze *head_msg* of the SHs and the information of the high-priority node is retained.

(4) Nodes send apply message *APPLY ((Swarm_ID), msg, route, hop_cnt)* to the SH which has been selected according to steps (2) and (3). Other nodes on the route interrupt this message and update the route information by adding +1 to the *hop_cnt* and then forward the new route information to the SH. This way SH can determine the hop count of its member.

(5) SH will add the node to its member list just after verifying that the number of nodes is less than N_{max}. In this case, SH sends back an *ACCP* message to the intended node but otherwise sends back a *JOIN_GUEST* message.

(6) Intended node after receiving the ACCP message broadcasts *MEMR (Swarm_ID, msg, route, dis_cnt, hop_cnt)* message and then is not needed to be involved in other election processes. It will send *dis_cnt* to other SHs.

(7) Repeat the process until all nodes become SH or swarm members.

(8) Border swarm members search for the adjacent node of the other swarm using the *hop_cnt* and *Swarm_ID* information making a list of all neighbor nodes.

(9) Each node broadcasts its neighbor table information periodically using *HELLO* packets. Therefore, by examining the neighbor table from its neighbors, a node is able to gather complete information about the network topology that is at most two hops away from itself. This two-hop topology information is kept in a small database in each node.

(10) Send a *BUND (msg, route, dis_cnt, hop_cnt)* message *hop_cnt* message to its SH indicating that it became a boundary node.

The SH and the neighbor's SH can communicate directly if they are located in the same coverage, but in the actual situation of the physical structure of the network the SH and the neighbor SH are not necessarily within the same communication range. For that reason SH selects one of the boundary swarm members to act as a gateway between the neighbor SHs. Therefore, SH election process not only is the election of the SH but also includes the operation of determining boundary nodes, used for communication between adjacent swarms, as shown in Figure 2.

After finishing SH election, the movement of nodes could change swarm structure. Therefore, swarm structure continuously needs maintenance. SH periodically checks member list and neighbors swarm listing status. In case of adding a new member or one of the member lefts, swarm structure will not change; SH only updates the members list. Sometimes due to the movement, the total number of nodes in two or more neighbor's swarms may detract from N_{max}. In this case, the two swarms merge to form a new swarm. This situation leads to competition on SH using the same SH election process. Loser SH sends messages to inform its member about the new SH information. SH also updates the boundary nodes periodically. When a boundary node leaves the swarm, SH will select new border nodes. Nodes also periodically send message to ensure the presence of SH and update their information, particularly location and velocity. A new SH should be elected in case of loss or leaving of SH due to abnormal situation. Moreover, SH periodically calculates its P_i and sends it to all members; if one or more members find that their P_i value became less than that of the SH, they will notify the SH. SH immediately notifies its members about the new SH. Other members respond accordingly by updating their SH information. The objective is to maintain the stability of the swarm structure, avoid exchanging many messages, and reduce performing many calculations. Therefore, the structure of the swarm will not be changed unless for the above mentioned situation or in case of deterioration in the SH performance.

5. Information Routing

Due to node mobility, a major challenge is how to route data packets over such network that changes its structure dynamically, especially when the source and the destination helicopters are out of MANET transmission range. Most of the MANET routing algorithms are not ideal for AVANETs, because of the helicopter specific issues such as rapid changes in the link quality and very high node mobility. Maintaining a routing table, as in proactive methods, also is not an optimal solution. However, repetitive path finding before each packet delivery, as in reactive routing, can also be exhaustive. Therefore, specific routing solutions are needed. A routing strategy only based on the location information of the nodes can satisfy the requirements of helicopter networks. Although some VANET routing protocols can work in these types of networks, majority of these protocols mainly rely on road side units (RSU) [29], road intersections [30], and other assumptions which may not be available in helicopters networks [31]. In this paper, we propose a simple routing algorithm which fits the special requirements of helicopter networks.

5.1. Route Finding. As a rule, helicopters located in the same swarm and satisfying (1) can directly use ad hoc networks to communicate with each other. In order to establish a communication between two nodes beyond the ad hoc coverage (L), source node issues a communication request to SH. SH after receipt of request verifies whether the node is one of the swarm members. After verification SH finds from its storage the location and other information of the destination node and then starts the process of best route selection, which is subject to destination and source locations. SH is not only responsible of the route selection process; in many situations SH can participate in information forwarding process.

In case of the fact that the intended node is one of the swarm members, SH searches the best route according to the following. The presence of multiple nodes between source and destination helicopters may lead to many paths (P). Each path has its own amount of hops M_P. d_m is the distance of a specific hop m, and d_m satisfies (1), where $M_P \leq M_{\max}$. Therefore, the general formula for short path selection is

$$P_s = \min\left[\sum_{m=1}^{M_1} d_m, \sum_{m=1}^{M_2} d_m, \sum_{m=1}^{M_3} d_m, \ldots, \sum_{m=1}^{M_P} d_m\right]. \quad (19)$$

For simplicity, assume that we have five nodes SH, I, K, J, and D, as shown in Figure 3. Assume that their locations coordinates are $(x_{\mathrm{SH}}, y_{\mathrm{SH}}, z_{\mathrm{SH}})$, (x_I, y_I, z_I), (x_k, y_K, z_K), (x_J, y_J, z_J), and (x_d, y_d, z_d), respectively, where I is the source node, D is the destination node, and J, k are nodes in the same coverage (L) of SH. J and K act as a next hop. Distance (K, SH), (I, SH), (K, I), and (K, D) satisfy (1). SH finds

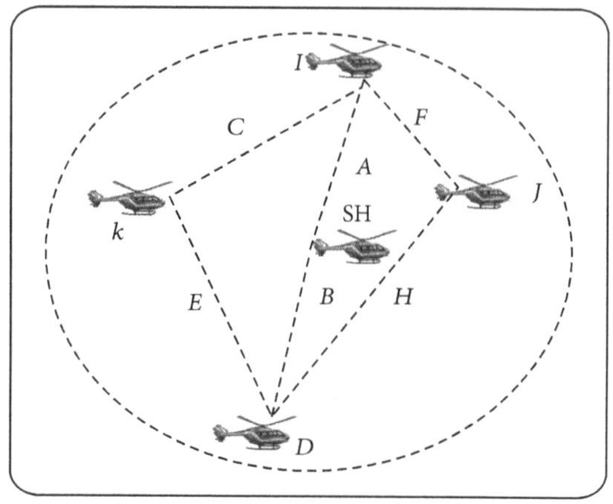

FIGURE 3: Information routing inside swarm.

the information of these nodes and starts calculating the distance according to the following:

$$\begin{aligned}
A &= \sqrt{(x_I - x_{\mathrm{SH}})^2 + (y_I - y_{\mathrm{SH}})^2 + (z_I + z_{\mathrm{SH}})^2},\\
B &= \sqrt{(x_{\mathrm{SH}} - x_D)^2 + (y_{\mathrm{SH}} - y_D)^2 + (z_{\mathrm{SH}} - z_D)^2},\\
C &= \sqrt{(x_I - x_K)^2 + (y_I - y_K)^2 + (z_I - z_K)^2},\\
E &= \sqrt{(x_K - x_D)^2 + (y_K - y_D)^2 + (z_K - z_D)^2},\\
F &= \sqrt{(x_I - x_J)^2 + (y_I - y_J)^2 + (z_I - z_J)^2},\\
H &= \sqrt{(x_J - x_D)^2 + (y_J - y_D)^2 + (z_J - z_D)^2},
\end{aligned} \quad (20)$$

where A, B, C, E, F, and H are the distances between nodes as shown in Figure 3. Shorter path (P_s) to the destination node is

$$P_s = \min\{A + B, C + E, F + H\}. \quad (21)$$

In case of the fact that the SH does not find the destination information on its storage, it requests the closest SHs for the destination information and waits for their responses. If the waiting time exceeds a threshold (t_r) and the route response has not yet been received, then *RREQ* is resent. If retransmission exceeds the maximum retransmission limit ($r_{\max}$), the route search process is terminated. To reduce network congestion, not all neighbor SHs receiving *RREQ* will respond. Only those located on the route toward the destination and having the ability to serve can participate in the routing process.

Therefore, neighbor SHs after having received *RREQ* message firstly determine if they are located on the route of the SH request; if not, the request is discarded. Otherwise the procedure is as follows:

(1) A neighbor SH checks the request; if it is the first time received, then it continues the procedure; if not then the request is discarded.

(2) It checks whether the destination is in the same swarm. If not, then it applies step (4).

(3) It forwards route request to the intended destination and waits for the routing response RREP; it also goes to (5).

(4) SH adds its own msg to (*REEQ* msg) message and then forwards the updated message to the SHs of the next hope neighbours and waits for the route response.

(5) If the waiting time exceeds t_r and it does not receive the route response, then it retransmits request and adds one to request number; otherwise it goes to (7).

(6) If the number of retransmissions exceeds the limit r_{max}, then it ends route request process. Alternatively, in accordance to intended destination situation it directly applies step (3) or (4).

(7) If the SH received more than one route response, then it chooses the node with fewest number of hops and minimum distance to destination and adds its msg to (*REEP* msg) message. Then, it forwards the message to the previous hop SH. If the routing request failed, probably because the destination swarm is far away from the source and the routing and communication request cannot be established through ad hoc or through neighbours' SHs, the SH can establish a connection via satellite or other direct links.

(8) In some rare situations and according to SH election process, SH may not have a satellite equipment to communicate with the far away swarm's heads; in this case the SH finds one of its swarm members which are equipped with satellite, to forward request to neighbors SH.

(9) If still not able to establish connection and routing through satellite of the members, it forwards request to neighbors SH and it sends notification *WAIT* indicating that the communication needs to be completed through satellite link of neighbors SH.

Nodes usually prefer the use of ad hoc connections, but if the destination is out of the coverage, a node needs to communicate with operation center or public networks. In these cases, the source SH firstly searches its members to find a member that is equipped with an appropriate communication link to the intended network. If all swarm members are not equipped with suitable equipment, then SH forwards the request to the neighbor SH that is equipped with the appropriate communication equipment. Only helicopters type S_1 can communicate with the outside networks. Upon receiving a communication request to outside networks, neighbor SH will proceed as follows:

(1) It check the request. If it is the first time received, it continues the procedure; if not, then it is discarded.

(2) It checks whether its equipment is of type S_2; then it will skip to step (4).

(3) It checks whether this swarm has a free available S_1 and then forwards the routing request, waits for response, and then skips to step (5).

(4) It forwards the routing request to the neighboring SH and waits for a response.

(5) If the waiting time exceeds t_r and it does not receive the route response, then it will retransmit the request and add 1 to the request number; otherwise it will skip to step (7).

(6) If the number of retransmissions exceeds the limit r_{max}, then it will end the route request process.

(7) If the SH receives more than one route response, then it chooses the helicopter with fewest numbers of routing hops and the minimum distance and cost to the destination. The route through this node will be marked as the primary route. At the same time, it stores other routes as a backup (*Store Route as Backup (SRB)*) to be used in the future if needed. To ovoid looping multiple links to the same destination cannot be used at the same time. Any redundant link that is not as preferred is blocked until the primary link goes down.

(8) It adds its own msg to (REEP msg) message and forwards the message to the previous hop SH in the source direction. If the routing process fails to find a helicopter type S_1, it will send a REER notification to the source helicopter indicating a routing failure.

5.2. Route Maintenance (RM). According to nodes' movement the established route may be lost. This time, the communication will temporarily be disconnected. At the same time, the node stores the destination routing information for a while and tries to resend the request. If it is successful in sending a new route request to the intended destination, that means that the route has been recovered; if not, then it will search its backup routing information and try to use one of the best backup routes. If it fails to send the message it will start a new route finding procedure. The route maintenance also can be accomplished by using the knowledge of the nodes about their neighbor status. A forwarding node can aid in the route maintenance process by repairing a broken route using its 2-hop-topology information and modifies the source route header accordingly. The destination node sends a gratuitous route reply to inform source of the new working route.

6. Simulation and Results

6.1. Simulation Configurations and Settings. To evaluate the proposed work, we conducted two different types of simulations. Firstly, we evaluate the SH election approach using a Matlab program. Secondly, to evaluate the performance of the proposed routing protocol, we used the well-known simulation program OPNET [15, 32, 33], which is a discrete event network simulator that includes a rich set of detailed models for ad hoc networks. As this work does not consider the physical and data link layer the simulation program is built on the specification of IEEE 802.11p standard. The main parameters of the simulations are summarized in Table 1.

In the first simulation, we implemented a simulation program to demonstrate the velocity, equipment, location,

Table 1: Simulation settings and configurations.

Parameters	Values
Packet size	1024
Channel sensing	0.1
Reception power thresholds	0.5 W
Traffic type	Poison
Simulation time	1200 s
Coverage area	5000 × 5000 (m)
L	1000 m
M_{max}	5 hop
N	15 node
Simulation period	20 minutes

and movement direction of the helicopters. The purpose of the program is to show the SH election process according to the proposed election approach. In this simulation, we assume a number of helicopters in one swarm moving toward the same direction with different velocities, location, and equipment values. We assume that the velocity and location of all helicopters are randomly distributed and the starting point is randomly selected. In addition, we supposed that the number of helicopters of types S_1 and S_2 is 20% and 80%, respectively. The simulation is executed several times using different number of helicopter nodes to calculate the different value of priority factors. The priority factors versus the ID numbers of helicopters are plotted in different situations.

In the second simulation, we used OPNET to compare the performance of the proposed routing algorithm with both CBRP [22] and Dynamic Source Routing (DSR) [25] protocols. As the proposed routing protocol has both the on-demand and clustering characteristics, we select these protocols for compression because of the on-demand properties of DSR and the clustering characteristics of CBRP. These protocols also demonstrated acceptable performance in vehicular networks. All protocols are operated at 12 Mbps data rate. The power transmit level of 1 mw was used for all scenarios. The nodes are moving in the simulation area according to two scenarios. In the first one, helicopters are moving according to a random waypoint model. Each node chooses a random location destination and moves toward it at a speed randomly distributed between 0 and 500 km/h. In the second scenario a group of helicopters are moving in regular swarm with a constant speed 200 Km/h. The regular swarm is constructed by following three rules, which are move in the same direction as your neighbors, remain close to your neighbors, and avoid collisions with your neighbors. All helicopter nodes in the network are configured to run the proposed routing algorithm during the first round. Then they used DSR and CBRP during the second and third rounds, respectively. In case of CBRP, we treated the helicopter swarm as a cluster and helicopter with the least ID number as a cluster head. Since the elected SH plays a major role during the routing process, we used it during the evaluation of the proposed routing algorithm. The performance of the proposed algorithm is evaluated against DSR and CBRP in terms of swarm structure reconstructions, Packet Delivery Ratio (PDR), average end-to-end delay (AD), and routing cost.

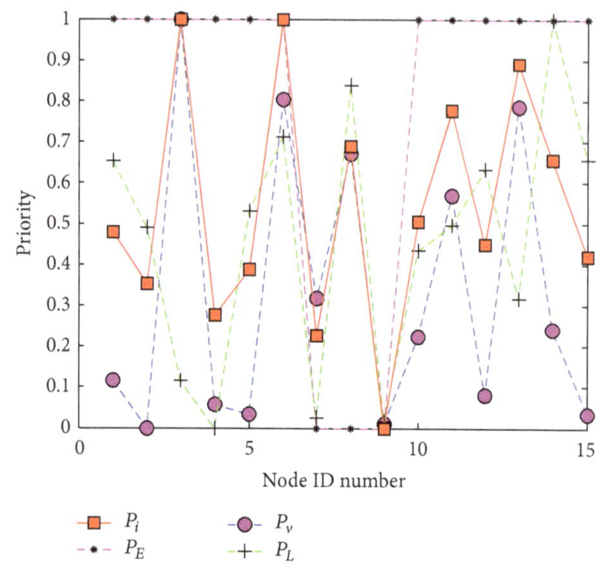

Figure 4: SH election priority. Node with the least EPF (P_i) is elected as a SH (ID = 9).

6.2. Results and Analysis. Simulation results are shown in Figures 4–9. The graphs show the election procedure result and the comparison between the three mentioned protocols using different parameter values. Figure 4 shows the priority factors (location, velocity priority equipment, and EPF) of helicopter nodes. From Figure 4, node ID number nine is elected as a SH according to EPF. Obviously, in case of using only velocity to elect the SH, as many other routing protocols do [34], node ID number two will be elected as SH though it is type S_2. Likewise, in case of using the location to elect the SH, there are three nodes (4, 7, and 9) with the same location priority; therefore, it will be difficult to elect SH among them. In order to achieve a better utilization of all nodes and optimum selection the use of EPF will solve the problem by using a weighted mechanism. This problem has not been addressed by other existing protocols that used location or velocity to elect the SH.

Furthermore, to demonstrate that the proper election of the SH has a great effect on the stability of the network, we run the simulation using the proposed routing algorithm with SH elected using one priority factor (velocity, location, equipment, or EPF), and then we monitor the swarm structure every 100 s. The accumulated number of the swarm reconstruction is plotted in Figure 5. It is realized that, using SH which is elected according to P_E, the network structure is better than when using velocity or location to elect SH. However, sometimes the variation in the velocity of the well-equipped nodes affects its performance and consequently affects the network structure. The effect of P_L and P_v seems to be convergent because the velocity affects the location of the nodes. When EPF is used, the swarm structure is more stable, because the selection process considers all factors.

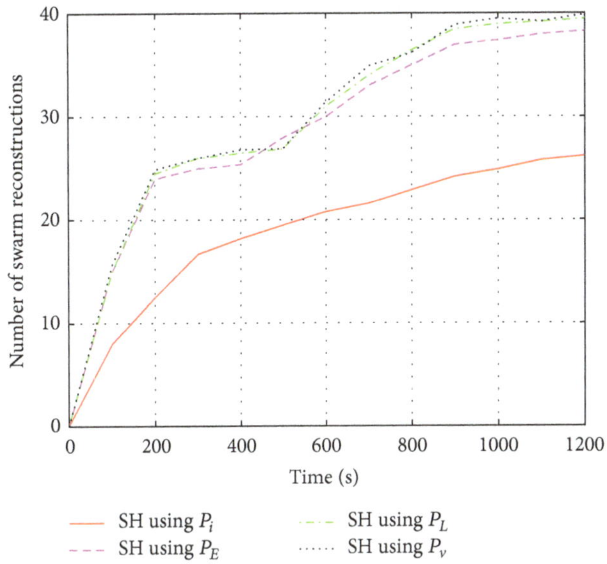

FIGURE 5: The effect of SH election process in swarm stability.

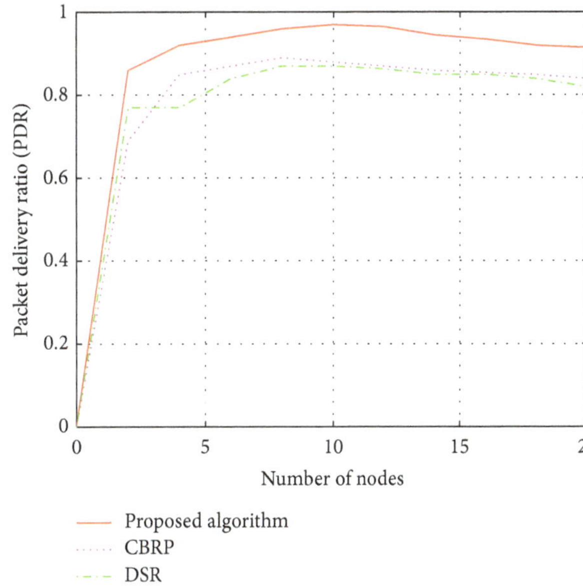

FIGURE 7: The PDR of the DSR, CBRP, and proposed algorithms.

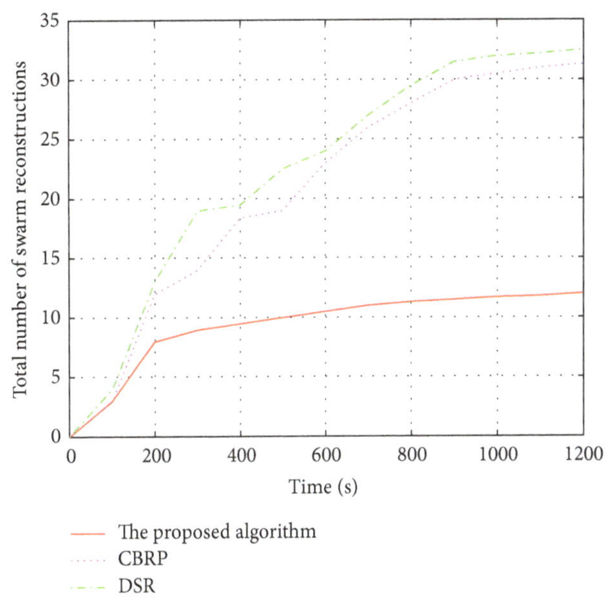

FIGURE 6: Swarm structure stability over time for the proposed, CBRP, and DSR protocols.

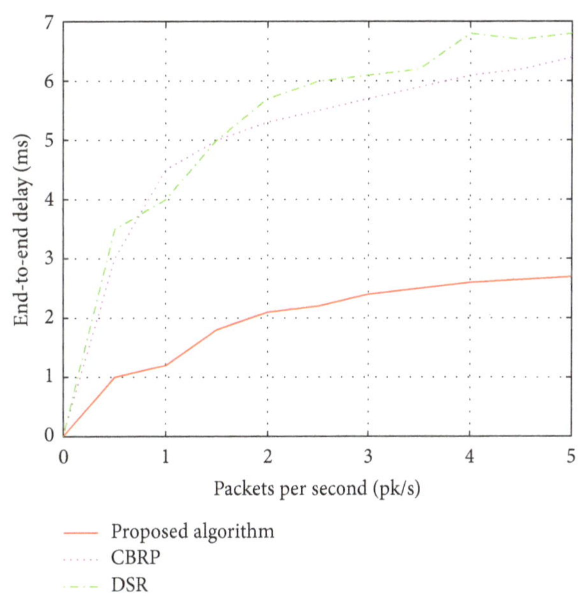

FIGURE 8: The proposed algorithm, CBRP, and DSR number of sent packets versus end-to-end delay for 15 helicopters.

We also investigate the swarm reconstruction using the CBRP, DSR, and the proposed algorithm. As shown in Figure 6, the swarm structure stability of both CBRP and DSR is poor. This is because CBRP algorithm uses only the IDs of the nodes as a basis for SH election. The smallest ID is elected as a SH. A swarm structure may change rapidly in case of the fact that the elected node moves with high velocity. On the other hand, using DSR, a separate periodic algorithm must be implemented, to support propagating the SH advertisements across the swarm. The response of the algorithm may affect the network structure. In contrast, the proposed election process takes into consideration the movement, location, and equipment of neighbouring helicopters, to reduce the probability of swarm change over time, so as to maintain higher stability.

It is observed from Figure 7 that the PDR of the proposed algorithm remains high when the number of nodes is increasing. The increase of the number of nodes does not affect the PDR because of the high efficiency and swarm structure stability. PDR of CBRP and DSR is less than that obtained by the proposed algorithm, because the source node and the intermediate nodes store the next hop information corresponding to each flow for data packet transmission. DSR and CBRP use a source routing in which a data packet carries

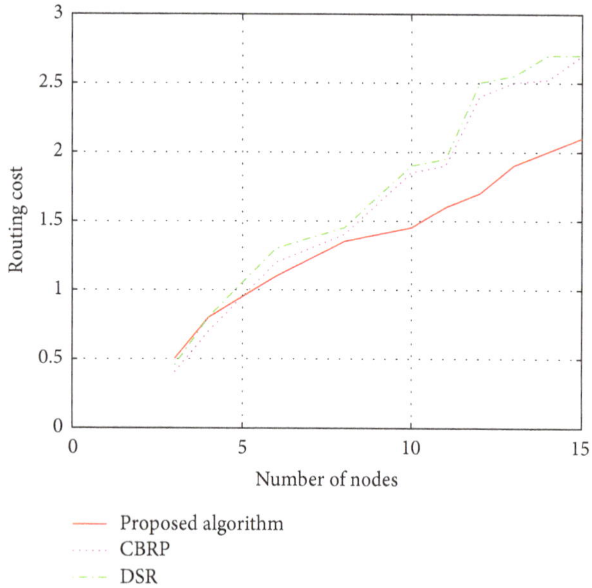

FIGURE 9: Proposed algorithm, CBRP, and DSR routing cost versus number of helicopters.

the complete path to be traversed. The PDR is too low in the beginning of the curve due to the random initiation of the simulation program.

Concerning Figure 8, we observed that the average end-to-end delay of the proposed algorithm is the best among others. The proposed algorithm end-to-end delay is smaller than the others because the route search is done only once and remains stable until the swarm structure changes. This decreases the time needed for the whole process of packet delivery and reduces the number of routing messages. Other protocols have a longer delay because the route finding process takes more time as every intermediate node tries to extract information before forwarding the reply while the protocol tries to search for a new route. The proposed algorithm also outperforms others because of the built-in RM and SRB characteristics. While using other protocols, the routing mechanism searches for a new route at every request and when a route failure occurs. The packet delivery delay time is affected by both the route search algorithm and the packet delivery process itself.

According to Figure 9, it is observed that the routing cost of the proposed algorithm, CBRP and DSR, increases with respect to the increase of the number of nodes. This is because the increase in nodes quantity increases the amount of hops, thereby increasing the congestion in the routing process. The initiation of the route finding process of the proposed algorithm takes more time. However, this calculation is done once and remains steady until the swarm structure changes. Hopefully, the increase of swarm members will increase the opportunity of more backup route in the proposed algorithm that enhances the route maintenance process. Moreover, the number of nodes per swarm and the number of hops in the proposed algorithm are limited by N_{max} and M_{max}, respectively. The amount of N_{max} and M_{max} is specified according to certain threshold of the routing cost.

The problem of other routing protocols is that the increase of intermediate nodes can lead to inconsistent routes when the source sequence number is very old and the intermediate nodes have a higher but not the latest destination sequence number, thereby having no existing or stale entries. CBRP and DSR have a higher routing cost, because they use a flooding approach to all nodes to determine the destination route, while in the proposed routing algorithm only SHs in the requested swarm that have the capability of finding the destination are able to participate in the routing process.

7. Conclusions

This paper studied the possibility of using MANET as a platform for data communication between helicopters working in certain mission and special scenarios. A simple model for the network structure is proposed and intensively explained. A simple routing protocol is also presented and comprehensively evaluated using simulation. Theoretical approach for selecting the SH is proposed and extensively discussed. The investigation results show that ad hoc network can be involved in providing communication between helicopters with high degree of reliability and minimum expense. Our comprehensive simulation results show that the proposed SH election approach can assist in obtaining best results of SH selection, so as to maintain high network stability. The proper selection has great impact in the structure stability of the swarm and consequently affects the whole performance of the network. Simulation results also show that the proposed routing algorithm can be used in these types of networks with high delivery rate and less amount of delay and routing overhead. Generally this work sets the path for many researches that are related to utilization of MANET in helicopter networks. Many researches related to the PHY, data link, routing, and application of AVANET are desired. Although the general aspects are considered in this paper, there is still a lot of work and further detailed research can be conducted in more depth. Future researches can focus on studying the capability of using bidirectional routing and enhancing the SH election process by adding more priority factors such as signal strength and the movement stability of nodes as well as considering security issues. Many researches and efforts can focus on improving the physical link performance in order to increase the coverage range of a single hop. Future work can also consider obtaining actual results by implementing a kind of real practical deployment.

Conflict of Interests

The authors declare that there is no conflict of interests regarding the publication of this paper.

Acknowledgments

This work is supported by National Natural Science Foundation of China (no. 610011024) and the National Key Technology R&D Program under Grant 2012BAH15B00.

References

[1] E. C. Eze, S. Zhang, and E. Liu, "Vehicular ad hoc networks (VANETs): current state, challenges, potentials and way forward," in *Proceedings of the 20th International Conference on Automation and Computing (ICAC '14)*, pp. 176–181, Cranfield, UK, September 2014.

[2] C. Elliott and B. Heile, "Self-organizing, self-healing wireless networks," in *Proceedings of the IEEE Aerospace Conference*, pp. 149–156, Big Sky, Mont, USA, March 2000.

[3] C. Prabha, D. S. Kumar, and D. R. Khanna, "Wireless multi-hop ad-hoc networks: a review," *IOSR Journal of Computer Engineering*, vol. 16, no. 2, pp. 54–62, 2014.

[4] N. Gondaliya and D. Kathiriya, "An application of ad hoc networks in disaster area for search and rescue operation: a survery and challenges," *International Journal of Advanced Research in Computer and Communication Engineering*, vol. 3, no. 3, pp. 5711–5714, 2014.

[5] A. M. Abdelgader and W. Lenan, "The physical layer of the IEEE 802.11 p WAVE communication standard: the specifications and challenges," in *Proceedings of the World Congress on Engineering and Computer Science (WCECS '14)*, San Francisco, Calif, USA, October 2014.

[6] J. Schlecht, K. Altenburg, B. M. Ahmed, and K. E. Nygard, "Decentralized search by unmanned air vehicles using local communication," in *Proceedings of the International Conference on Artificial Intelligence (IC-AI '03)*, pp. 757–762, June 2003.

[7] R. Haagensen, K.-Å. Sjøborg, A. Rossing, H. Ingilæ, L. Markengbakken, and P.-A. Steen, "Long-range rescue helicopter missions in the arctic," *Prehospital and Disaster Medicine*, vol. 19, no. 2, pp. 158–163, 2004.

[8] K. Jusoff, "Search and rescue (SAR) operations for the missing Bell 206 Long Ranger helicopter in Sarawak, Malaysia using near real-time airborne hyperspectral imaging system," *Disaster Prevention and Management*, vol. 17, no. 1, pp. 94–103, 2008.

[9] R. A. F. Master Aircrew Duncan Tripp RAF, *Working with Search & Rescue Helicopters, RAF SAR Force Version 1.0*, vol. 1, 2011.

[10] I. Bekmezci, O. K. Sahingoz, and Ş. Temel, "Flying ad-hoc networks (FANETs): a survey," *Ad Hoc Networks*, vol. 11, no. 3, pp. 1254–1270, 2013.

[11] Y. Chen, K. L. Moore, and Z. Song, "Diffusion boundary determination and zone control via mobile actuator-sensor networks (MAS-net): challenges and opportunities," in *Intelligent Computing: Theory and Applications II*, vol. 5421 of *Proceedings of SPIE*, pp. 102–113, Orlando, Fla, USA, April 2004.

[12] P. K. Branch, "Unmanned vehicles: mainstays of future airborne reconnaissance," in *Airborne Reconnaissance XXIII*, vol. 3751 of *Proceedings of SPIE*, pp. 36–40, Denver, Colo, USA, December 1999.

[13] B. Argrow, D. Lawrence, and E. Rasmussen, "UAV systems for sensor dispersal, telemetry, and visualization in hazardous environments," in *Proceedings of the 43rd AIAA Aerospace Sciences Meeting and Exhibit*, pp. 15097–15107, Reno, Nev, USA, January 2005.

[14] Y.-J. Eun and H. Bang, "Cooperative control of multiple UCAVs for suppression of enemy air defense," in *Proceedings of the AIAA 3rd Unmanned Unlimited Technical Conference, Workshop and Exhibit*, vol. 6529, Chicago, Ill, USA, September 2004.

[15] A. I. Alshbatat and L. Dong, "Performance analysis of mobile ad hoc unmanned aerial vehicle communication networks with directional antennas," *International Journal of Aerospace Engineering*, vol. 2010, Article ID 874586, 14 pages, 2010.

[16] P. Gaudiano, B. Shargel, E. Bonabeau, and B. T. Clough, "Swarm intelligence: a new c2 paradigm with an application to control swarms of uavs," DTIC Document, 2003.

[17] D. L. Gu, G. Pei, H. Ly, M. Gerla, B. Zhang, and X. Hong, "UAV aided intelligent routing for ad-hoc wireless network in single-area theater," in *Proceedings of the IEEE Wireless Communications and Networking Conference (WCNC '00)*, pp. 1220–1225, Chicago, Ill, USA, September 2000.

[18] S. Levy, "How Google will use high-flying balloons to deliver internet to the hinterlands," *Wired*, 2013.

[19] F. Lardinois, *Google X Announces Project Loon: Balloon-Powered Internet for Rural, Remote and Underserved Areas*, vol. 14, TechCrunch/AOL, 2013.

[20] K. Schulze and J. Buescher, "A scalable, economic autonomous flight control and guidance package for UAVs," in *Proceedings of the 2nd AIAA "Unmanned Unlimited" Systems, Technologies and Operations—Aerospace*, pp. 15–18, San Diego, Calif, USA, September 2003.

[21] K. Xu, X. Hong, M. Gerla, H. Ly, and D. L. Gu, "Landmark routing in large wireless battlefield networks using UAVs," in *Proceedings of the IEEE Communications for Network-Centric Operations: Creating the Information Force (MILCOM '01)*, pp. 230–234, IEEE, October 2001.

[22] M. Jian, J. Li, and Y. C. Tay, "Cluster based routing protocol (CBRP) functional specification," Internet Draft draft-ietf-manet-cbrp-spec-00.txt, 1998.

[23] W. Chujo, "Helicopter satellite communication system for disaster control operations," in *Proceedings of the 6th Asia-Pacific Seminar on Next Generation Mobile Communications*, 2005.

[24] T. Samad, J. S. Bay, and D. Godbole, "Network-centric systems for military operations in urban terrain: the role of UAVs," *Proceedings of the IEEE*, vol. 95, no. 1, pp. 92–107, 2007.

[25] D. Johnson, Y. Hu, and D. Maltz, "The dynamic source routing protocol (DSR) for mobile ad hoc networks for IPv4," RFC 47282007, 2007.

[26] T. Taleb, E. Sakhaee, A. Jamalipour, K. Hashimoto, N. Kato, and Y. Nemoto, "A stable routing protocol to support ITS services in VANET networks," *IEEE Transactions on Vehicular Technology*, vol. 56, no. 6, pp. 3337–3347, 2007.

[27] O. Abedi, M. Fathy, and J. Taghiloo, "Enhancing AODV routing protocol using mobility parameters in VANET," in *Proceedings of the IEEE/ACS International Conference on Computer Systems and Applications (AICCSA '08)*, pp. 229–235, Doha, Qatar, April 2008.

[28] B. Kate, J. Waterman, K. Dantu, and M. Welsh, "Simbeeotic: a simulator and testbed for micro-aerial vehicle swarm experiments," in *Proceedings of the 11th ACM/IEEE Conference on Information Processing in Sensing Networks (IPSN '12)*, pp. 49–60, Beijing, China, April 2012.

[29] K. Mershad, H. Artail, and M. Gerla, "ROAMER: roadside Units as message routers in VANETs," *Ad Hoc Networks*, vol. 10, no. 3, pp. 479–496, 2012.

[30] J. Nzouonta, N. Rajgure, G. Wang, and C. Borcea, "VANET routing on city roads using real-time vehicular traffic information," *IEEE Transactions on Vehicular Technology*, vol. 58, no. 7, pp. 3609–3626, 2009.

[31] K. C. Lee, U. Lee, and M. Gerla, "Survey of routing protocols in vehicular ad hoc networks," *Advances in Vehicular Ad-Hoc Networks: Developments and Challenges*, pp. 149–170, 2010.

[32] W. Wang and J. Zhang, *OPNET Modeler and Network Simulation*, Posts & Telecom Press, Beijing, China, 2003.

[33] G. F. Lucio, M. Paredes-Farrera, E. Jammeh, M. Fleury, and M. J. Reed, "OPNET modeler and NS-2: comparing the accuracy of network simulators for packet-level analysis using a network testbed," *WSEAS Transactions on Computers*, vol. 2, pp. 700–707, 2003.

[34] K.-T. Feng, C.-H. Hsu, and T.-E. Lu, "Velocity-assisted predictive mobility and location-aware routing protocols for mobile Ad Hoc networks," *IEEE Transactions on Vehicular Technology*, vol. 57, no. 1, pp. 448–464, 2008.

7

Deformation Measuring Methods based on Inertial Sensors for Airborne Distributed POS

Xiaolin Gong,[1,2] Haojie Liu,[1,2] and Xing-Gang Yan[3]

[1]*School of Instrumentation Science and Optoelectronics Engineering, Beihang University, Beijing 100191, China*
[2]*Science and Technology on Inertial Laboratory, Beijing 100191, China*
[3]*School of Engineering and Digital Arts, University of Kent, Canterbury CT2 7NT, UK*

Correspondence should be addressed to Xiaolin Gong; gongxiaolin@buaa.edu.cn

Academic Editor: Nicolas Avdelidis

This paper is focused on deformation measuring methods based on inertial sensors, which are used to achieve high accuracy motion parameters and the spatial distribution optimization of multiple slave systems in the airborne distributed Position and Orientation System or other purposes. In practical application, the installation difficulty, cost, and accuracy of measuring equipment are the key factors that need to be considered synthetically. Motivated by these, deformation measuring methods based on gyros and accelerometers are proposed, respectively, and compared with the traditional method based on the inertial measurement unit (IMU). The mathematical models of these proposed methods are built, and the detailed derivations of them are given. Based on the Kalman filtering estimation, simulation and semiphysical simulation based on vehicle experiment show that the method based on gyros can obtain a similar estimation accuracy to the method based on IMU, and the method based on accelerometers has an advantage in y-axis deformation estimation.

1. Introduction

The airborne distributed Position and Orientation System (POS) has been proposed to achieve multipoint spatiotemporal motion parameters for synthetical earth observation systems with multiple remote sensing loads [1–3]. Distributed POS can be composed of a few high precision master systems, some low precision slave systems, POS Computer System (PCS), and postprocessing software. Usually, the master system is a high precision integrated system of Strapdown Inertial Navigation System and Global Navigation Satellite System [4] (also called the main POS). The slave system is only an inertial measurement unit (IMU), which consists of three orthogonally mounted gyros and accelerometers, respectively, and is placed as close as possible to the location of the load. The slave systems, also called the sub-IMUs, depend on the master system to transfer alignment to achieve their high accuracy motion parameters. Due to the deformation of aircraft caused by gust, turbulence, and other factors, there is a time-varying and complex flexure angle between the main POS and each sub-IMU besides the rigid misalignment angle. The schematic diagram of the measuring system and the cross section of aircraft with deformation at a certain moment are shown in Figure 1, where the grey part with dotted line is the ideal state of the wings without any deformation. It is clear to see that the premise and key of high accuracy transfer alignment is the attitude transformation, determined by flexure and misalignment angle between the master system and slave system, which can be estimated and compensated with high accuracy.

Furthermore, when there are many remote sensing loads working simultaneously, airborne distributed array antenna Synthetic Aperture Radar (SAR) is a typical example which has many subantennas on both sides of the wing; the high accuracy motion parameters of each load must be measured [5, 6]. Since the bearing capacity of aircraft is limited, especially the wing section, there are very stringent requirements on the weight and size of the measurement equipment, while the measurement accuracy of sub-IMU is proportional to the weight and size. It should be noted that a high accuracy sub-IMU may not be available at the location of each load, and the positions of sub-IMU and load are not

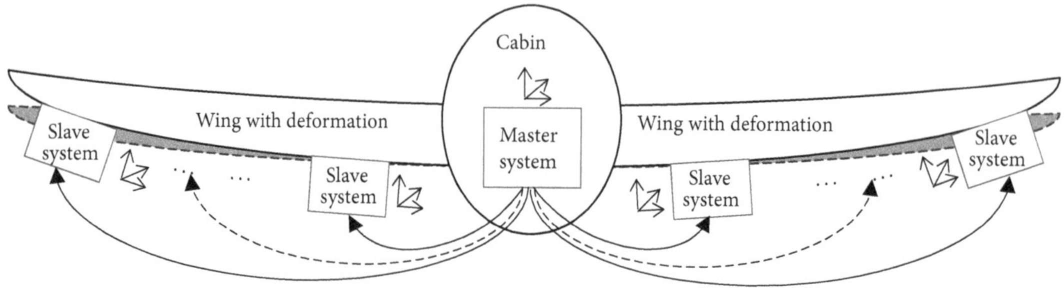

FIGURE 1: The schematic diagram of the measuring system and the cross section of aircraft.

always matched with each other. Therefore, it is necessary to consider the arrangement optimization of the distributed POS, such that the high precision motion parameters of all loads can be obtained using the minimum sub-IMUs. And the arrangement optimization also requires the attitude transformation between each node [7–9].

At present, aircraft deformation measuring methods can be summarized into three types: strain sensor measurement, optical measurement, and inertial measurement. The strain sensor measurement can be traced back to the 1940s and later it was improved by Skopinski et al. [10, 11]. It is a kind of mechanical measuring method which is widely used because of its convenient operation. However, it has limitation to the aircraft material and needs many wires which will increase the load of the aircraft. Besides, the strain sensor is easily affected by the physical abrasion, temperature, and so forth [12–14]. As for optical measurement, the Dutch National Aerospace Laboratory (NLR) used a camera to record the black and white striped pattern on the wing surface to estimate the flexible deformation [15]. Then, NLR presented a noncontact optical measurement which can obtain the deformation rule [16]. In addition, there are other optical measurements using visual sensors, optical fiber sensors, and bionic optical sensors to measure the flexible deformation [17–19]. All those optical measurements need the external measurement components and the beam transceivers must be intervisible, which make them not only complex to be installed, but also prone to be affected by the weather conditions. Inertial measurement is mainly based on the IMUs which are installed at the places of both main and subnodes. The difference of the navigation results between the main node and subnode, such as attitude difference and velocity difference, is utilized to estimate the flexible deformation. This procedure is known as the transfer alignment [20, 21].

Compared with IMU, three gyros or three accelerometers will reduce the weight, cost, and size of the measurement equipment. Particularly, high precision accelerometers have significant advantages of small size, light weight, low cost, and convenient installation when compared with the IMUs and gyros. It is necessary and significant to study the deformation measurement using gyros or accelerometers only. Therefore, this paper deduces the formulations of deformation measurement based on gyros (Gs) and accelerometers (As), respectively, and provides the mathematical modeling and algorithm design of deformation measuring method based on IMU, Gs, and As. Finally, taking the deformation measurement of the wing as an example, the measurement accuracies of these three methods are compared and analyzed by the simulation experiment based on the Kalman filter (KF).

The rest of this paper is organized as follows. In Section 2, the mathematical modeling of deformation measuring algorithms based on IMU, Gs, and As is given. In Section 3, detailed numerical simulation and semiphysical simulation are performed. Section 4 concludes this paper.

2. Mathematical Modeling

The detailed mathematical modeling of the three deformation measuring algorithms based on inertial sensors is given in this section. The details of KF can be found in [22, 23].

The measurement systems at the main node and the subnode can be called the master system and slave system, respectively. The coordinate frames used in this paper are defined as follows: i and e denote the earth-centered inertial frame and the earth-centered earth-fixed frame, respectively. The navigation frames of the master system and slave system (an IMU or three gyros or three accelerometers) are defined with x-y-z-axes pointing to east-north-up (E-N-U), represented by n_m and n_s, respectively. The body frames of the master system and slave system are defined on the rigid body of the inertial measurement unit and are denoted by b_m and b_s, respectively. A detailed description of these coordinate frames is available in [21, 24, 25].

2.1. *Mathematical Model of Deformation Measuring Algorithm Based on IMU.*

This method needs three orthogonally mounted gyros and accelerometers, respectively, on each note. For distributed POS, the main POS can be used as the master system and the sub-IMU can be used as the slave system. The mathematical model for KF includes the state equation and measurement equation.

2.1.1. *Error Equation.*

The state equation is based on the linear inertial navigation error equation of the sub-IMU, and the differential equation of the flexure angle and rigid misalignment angle is added.

Due to the existence of both flexure angle and rigid misalignment angle, the attitude transfer relationship between the master system and the slave system is shown as follows:

$$\mathbf{C}_{b_s}^{b_m} = \begin{bmatrix} 1 & -u_z & u_y \\ u_z & 1 & -u_x \\ -u_y & u_x & 1 \end{bmatrix} = I + \mathbf{u}\times, \tag{1}$$

where $\mathbf{C}_{b_s}^{b_m}$ is the coordinate transformation matrix from b_s-frame to b_m-frame, and it is an orthogonal matrix; system error angle $\mathbf{u} = \boldsymbol{\rho} + \boldsymbol{\theta}$, where $\boldsymbol{\rho} = [\rho_x\ \rho_y\ \rho_z]^{\mathrm{T}}$ and $\boldsymbol{\theta} = [\theta_x\ \theta_y\ \theta_z]^{\mathrm{T}}$ represent rigid misalignment angle and flexure angle between the master system and slave system, respectively.

The following are all the error equations based on the defined coordinate frames. Firstly, the inertial navigation error equation of the sub-IMU is given in (2), which includes attitude error equation, velocity error equation, position error equation, and inertial sensor constant error equation [26, 27]:

$$\begin{aligned}
\dot{\boldsymbol{\phi}}^{n_s} &= -\boldsymbol{\omega}_{in_s}^{n_s} \times \boldsymbol{\phi}^{n_s} + \delta\boldsymbol{\omega}_{in_s}^{n_s} + \mathbf{C}_{b_s}^{n_s}\boldsymbol{\varepsilon}^{b_s}, \\
\delta\dot{\mathbf{V}}^{n_s} &= \mathbf{f}^{n_s} \times \boldsymbol{\phi}^{n_s} - \left(2\delta\boldsymbol{\omega}_{ie}^{n_s} + \delta\boldsymbol{\omega}_{en_s}^{n_s}\right) \times \mathbf{V}^{n_s} \\
&\quad - \left(2\boldsymbol{\omega}_{ie}^{n_s} + \boldsymbol{\omega}_{en_s}^{n_s}\right) \times \delta\mathbf{V}^{n_s} + \mathbf{C}_{b_s}^{n_s}\boldsymbol{\nabla}^{b_s}, \\
\delta\dot{L} &= \frac{\delta V_{\mathrm{N}}}{R_{\mathrm{M}} + H} - \frac{V_{\mathrm{N}}}{(R_{\mathrm{M}} + H)^2}\delta H, \\
\delta\dot{\lambda} &= \frac{\sec L}{R_{\mathrm{N}} + H}\delta V_{\mathrm{E}} + \frac{V_{\mathrm{E}}\sec L\tan L}{R_{\mathrm{N}} + H}\delta L \\
&\quad - \frac{V_{\mathrm{E}}\sec L}{(R_{\mathrm{N}} + H)^2}\delta H, \\
\delta\dot{H} &= \delta V_{\mathrm{U}}, \\
\dot{\boldsymbol{\varepsilon}}_c^{b_s} &= 0, \\
\dot{\boldsymbol{\nabla}}_c^{b_s} &= 0,
\end{aligned} \tag{2}$$

where the subscripts E, N, and U represent east, north, and up, respectively; $\boldsymbol{\phi}^{n_s} = [\phi_{\mathrm{E}}\ \phi_{\mathrm{N}}\ \phi_{\mathrm{U}}]^{\mathrm{T}}$ is the attitude error vector in n_s-frame; $\boldsymbol{\omega}_{in_s}^{n_s}$ is the rotation velocity of the n_s-frame relative to the i-frame expressed in n_s-frame with error $\delta\boldsymbol{\omega}_{in_s}^{n_s}$; $\mathbf{C}_{b_s}^{n_s}$ is the coordinate transformation matrix from b_s-frame to n_s-frame; $\boldsymbol{\varepsilon}^{b_s}$ is the gyro random drift of the slave system in b_s-frame, which consists of random constant drift $\boldsymbol{\varepsilon}_c^{b_s}$ and Gaussian white noise $\boldsymbol{\omega}_{\varepsilon}^{b_s}$ [28, 29] with $\boldsymbol{\varepsilon}_c^{b_s} = \left[\varepsilon_x^{b_s}\ \varepsilon_y^{b_s}\ \varepsilon_z^{b_s}\right]^{\mathrm{T}}$ and $\boldsymbol{\omega}_{\varepsilon}^{b_s} = \left[\omega_{\varepsilon_x}^{b_s}\ \omega_{\varepsilon_y}^{b_s}\ \omega_{\varepsilon_z}^{b_s}\right]^{\mathrm{T}}$; $\mathbf{V}^{n_s} = [V_{\mathrm{E}}\ V_{\mathrm{N}}\ V_{\mathrm{U}}]^{\mathrm{T}}$ is the velocity in n_s-frame with error $\delta\mathbf{V}^{n_s} = [\delta V_{\mathrm{E}}\ \delta V_{\mathrm{N}}\ \delta V_{\mathrm{U}}]^{\mathrm{T}}$; $\mathbf{f}^{n_s} = [f_{\mathrm{E}}\ f_{\mathrm{N}}\ f_{\mathrm{U}}]^{\mathrm{T}}$ is the specific force measured by the accelerometers of sub-IMU expressed in n_s-frame; $\boldsymbol{\omega}_{ie}^{n_s}$ is the rotation velocity of the e-frame relative to the i-frame expressed in n_s-frame with error $\delta\boldsymbol{\omega}_{ie}^{n_s}$; $\boldsymbol{\omega}_{en_s}^{n_s}$ is the rotation velocity of the n_s-frame relative to the e-frame expressed in n_s-frame with error $\delta\boldsymbol{\omega}_{en_s}^{n_s}$; $\boldsymbol{\nabla}^{b_s}$ is the accelerometer random bias of slave system in b_s-frame, which consists of random constant bias $\boldsymbol{\nabla}_c^{b_s}$ and Gaussian white noise $\boldsymbol{\omega}_{\nabla}^{b_s}$ [28, 29] with $\boldsymbol{\nabla}_c^{b_s} = \left[\nabla_x^{b_s}\ \nabla_y^{b_s}\ \nabla_z^{b_s}\right]^{\mathrm{T}}$ and $\boldsymbol{\omega}_{\nabla}^{b_s} = \left[\omega_{\nabla_x}^{b_s}\ \omega_{\nabla_y}^{b_s}\ \omega_{\nabla_z}^{b_s}\right]^{\mathrm{T}}$; R_{M} and R_{N} denote the meridian and transverse radius of curvature, respectively; L and H denote the latitude and altitude, respectively. The symbols δL, $\delta\lambda$, and δH denote the error of the latitude, longitude, and altitude, respectively. $\sec L = 1/\cos L$.

Secondly, the differential equation of rigid misalignment angle and flexure angle is shown in the following equations:

$$\begin{gathered}
\dot{\rho}_j = 0, \\
\ddot{\theta}_j + 2\beta_j\dot{\theta}_j + \beta_j^2\theta_j = \eta_j, \\
j = x, y, z,
\end{gathered} \tag{3}$$

where the flexure angle θ_j is described by the second-order Markov process [30]; $\beta_j = 2.146/\tau_j$ and τ_j is the correlation time; x, y, and z represent the axis of b_s-frame; η_j is the Gaussian white noise with covariance $Q_{\eta_j} = 4\beta_j^3\sigma_j^2$, and σ_j^2 is the covariance of θ_j; $\mathbf{Q}_{\eta} = \left[Q_{\eta_x}\ Q_{\eta_y}\ Q_{\eta_z}\right]^{\mathrm{T}}$.

2.1.2. System State Equation. The system state equation can be described as follows:

$$\dot{X}_I = \mathbf{F}_I X_I + \mathbf{G}_I W_I, \tag{4}$$

where the subscript I represents the deformation measuring algorithm based on IMU; $\mathbf{X}_I$ is a 24×1 error state vector defined by $\mathbf{X}_I = [\phi_{\mathrm{E}}\ \phi_{\mathrm{N}}\ \phi_{\mathrm{U}}\ \delta V_{\mathrm{E}}\ \delta V_{\mathrm{N}}\ \delta V_{\mathrm{U}}\ \delta L\ \delta\lambda\ \delta H\ \varepsilon_x^{b_s}\ \varepsilon_y^{b_s}\ \varepsilon_z^{b_s}\ \nabla_x^{b_s}\ \nabla_y^{b_s}\ \nabla_z^{b_s}\ \rho_x\ \rho_y\ \rho_z\ \theta_x\ \theta_y\ \theta_z\ \dot{\theta}_x\ \dot{\theta}_y\ \dot{\theta}_z]^{\mathrm{T}}$; $\mathbf{F}_I$ and $\mathbf{G}_I$ are system transition matrix and system noise distribution matrix, respectively; the elements of $\mathbf{F}_I$ and $\mathbf{G}_I$ can be obtained from (2) and (3); system noise vector $W_I = \left[\omega_{\varepsilon_x}^{b_s}\ \omega_{\varepsilon_y}^{b_s}\ \omega_{\varepsilon_z}^{b_s}\ \omega_{\nabla_x}^{b_s}\ \omega_{\nabla_y}^{b_s}\ \omega_{\nabla_z}^{b_s}\ \eta_x\ \eta_y\ \eta_z\right]^{\mathrm{T}}$ is the zero-mean Gaussian white noise vector with covariance $\mathbf{Q}_I$ which consists of covariance $\mathbf{Q}_{\varepsilon}$ of gyro random drift, covariance $\mathbf{Q}_{\nabla}$ of accelerometer random bias, and $\mathbf{Q}_{\eta}$.

2.1.3. Measurement Equation. Based on the velocity plus attitude matching method, the measurement equation is given by

$$\mathbf{Z}_I = \mathbf{H}_I\mathbf{X}_I + \mathbf{v}_I, \tag{5}$$

where measurement vector $Z_I = [\delta V_{\mathrm{E}}'\ \delta V_{\mathrm{N}}'\ \delta V_{\mathrm{U}}'\ \delta\psi\ \delta\theta\ \delta\gamma]^{\mathrm{T}}$; $\delta\mathbf{V}' = [\delta V_{\mathrm{E}}'\ \delta V_{\mathrm{N}}'\ \delta V_{\mathrm{U}}']^{\mathrm{T}}$ denotes the differences of velocity between the slave system and master system after compensation for lever arm velocity; $\delta\psi$, $\delta\theta$, and $\delta\gamma$ denote the differences of heading, pitch, and roll between the slave system and master system, respectively; measurement noise $\mathbf{v}_I = \left[v_{\delta V_{\mathrm{E}}}\ v_{\delta V_{\mathrm{N}}}\ v_{\delta V_{\mathrm{U}}}\ v_{\delta\psi}\ v_{\delta\theta}\ v_{\delta\gamma}\right]^{\mathrm{T}}$ is the zero-mean Gaussian white noise sequence with covariance $\mathbf{R}_I$; measurement matrix $\mathbf{H}_I$ is given in (6); the details of $\mathbf{H}_1$, $\mathbf{H}_2$, and $\mathbf{H}_3$ can be found in [31, 32], and

$$\mathbf{H}_I = \begin{bmatrix} \mathbf{0}_{3\times3} & \mathbf{I}_{3\times3} & \mathbf{0}_{3\times9} & \mathbf{0}_{3\times3} & \mathbf{0}_{3\times3} & \mathbf{0}_{3\times3} \\ \mathbf{H}_1 & \mathbf{0}_{3\times3} & \mathbf{0}_{3\times9} & \mathbf{H}_2 & \mathbf{H}_3 & \mathbf{0}_{3\times3} \end{bmatrix}. \quad (6)$$

$\delta\mathbf{V}'$ is computed by

$$\delta\mathbf{V}' = \mathbf{V}^{n_s} - \mathbf{V}^{n_m} - \mathbf{V}_r^{n_m}, \quad (7)$$

where $\mathbf{V}^{n_m}$ is the velocity of the master system in n_m-frame and the lever arm velocity $\mathbf{V}_r^{n_m}$ can be calculated by [30]

$$\mathbf{V}_r^{n_m} = \mathbf{C}_{b_m}^{n_m}\left(\boldsymbol{\omega}_{ib_m}^{b_m} \times \mathbf{r}^{b_m}\right) + \mathbf{C}_{b_m}^{n_m}\dot{\mathbf{r}}^{b_m}, \quad (8)$$

where $\mathbf{C}_{b_m}^{n_m}$ is the coordinate transformation matrix from b_m-frame to n_m-frame; $\boldsymbol{\omega}_{ib_m}^{b_m}$ is the rotation velocity of the b_m-frame relative to the i-frame expressed in the b_m-frame; $\mathbf{r}^{b_m}$ is the lever arm between the master system and slave system expressed in the b_m-frame.

2.2. Mathematical Model of Deformation Measuring Algorithm Based on Gs. This method needs three orthogonally mounted gyros as the slave system and main POS as the master system. The deformation is estimated by using the difference of gyros between the master system and slave system as the measurement of KF.

2.2.1. State Vector Selection. The rigid misalignment angle, flexure angle and its derivative, and gyro constant drifts of the master system and slave system are selected as the state vector:

$$\mathbf{X}_G = \left[\rho_x, \rho_y, \rho_z, \theta_x, \theta_y, \theta_z, \dot{\theta}_x, \dot{\theta}_y, \dot{\theta}_z, \varepsilon_x^{b_m}, \varepsilon_y^{b_m}, \varepsilon_z^{b_m}, \varepsilon_x^{b_s}, \varepsilon_y^{b_s}, \varepsilon_z^{b_s}\right]^{\mathrm{T}}, \quad (9)$$

where subscript G represents the deformation measuring algorithm based on Gs; the symbols $\varepsilon_j^{b_m}$ and $\varepsilon_j^{b_s}$ $(j = x, y, z)$ denote gyro random constant drift of the master system and slave system, respectively.

2.2.2. System State Equation. The system state equation is given by

$$\dot{\mathbf{X}}_G = \mathbf{F}_G\mathbf{X}_G + \mathbf{G}_G\mathbf{W}_G, \quad (10)$$

where $\mathbf{F}_G$ and $\mathbf{G}_G$ are system transition matrix and system noise distribution matrix, respectively; system noise vector $\mathbf{W}_G = [\eta_x \ \eta_y \ \eta_z]^{\mathrm{T}}$ is a zero-mean Gaussian white noise vector with covariance $\mathbf{Q}_G = \mathbf{Q}_\eta$.

The differential equation of rigid misalignment angle and flexure angle is the same as in (3). The differential equation of gyro random constant drift is given by

$$\dot{\varepsilon}_j^{b_m} = 0,$$
$$\dot{\varepsilon}_j^{b_s} = 0, \quad (11)$$
$$j = x, y, z.$$

According to (3) and (11), $\mathbf{F}_G$ and $\mathbf{G}_G$ are given as follows:

$$\mathbf{F}_G = \begin{bmatrix} \mathbf{0}_{3\times3} & \mathbf{0}_{3\times3} & \mathbf{0}_{3\times3} & \mathbf{0}_{3\times3} & \mathbf{0}_{3\times3} \\ \mathbf{0}_{3\times3} & \mathbf{0}_{3\times3} & \mathbf{I}_{3\times3} & \mathbf{0}_{3\times3} & \mathbf{0}_{3\times3} \\ \mathbf{0}_{3\times3} & \mathbf{B}_1 & \mathbf{B}_2 & \mathbf{0}_{3\times3} & \mathbf{0}_{3\times3} \\ \mathbf{0}_{3\times3} & \mathbf{0}_{3\times3} & \mathbf{0}_{3\times3} & \mathbf{0}_{3\times3} & \mathbf{0}_{3\times3} \\ \mathbf{0}_{3\times3} & \mathbf{0}_{3\times3} & \mathbf{0}_{3\times3} & \mathbf{0}_{3\times3} & \mathbf{0}_{3\times3} \end{bmatrix},$$
$$\mathbf{G}_G = \left[\mathbf{0}_{3\times3} \ \mathbf{0}_{3\times3} \ \mathbf{I}_{3\times3} \ \mathbf{0}_{3\times3} \ \mathbf{0}_{3\times3}\right]^{\mathrm{T}}, \quad (12)$$

where

$$\mathbf{B}_1 = \begin{bmatrix} -\beta_x^2 & 0 & 0 \\ 0 & -\beta_y^2 & 0 \\ 0 & 0 & -\beta_z^2 \end{bmatrix},$$
$$\mathbf{B}_2 = \begin{bmatrix} -2\beta_x & 0 & 0 \\ 0 & -2\beta_y & 0 \\ 0 & 0 & -2\beta_z \end{bmatrix}. \quad (13)$$

2.2.3. Measurement Equation. The relationship between gyro outputs of the master system and slave system can be expressed by

$$\boldsymbol{\Omega}^{b_s} + \boldsymbol{\varepsilon}^{b_s} = \mathbf{C}_{b_m}^{b_s}\left(\boldsymbol{\Omega}^{b_m} + \boldsymbol{\varepsilon}^{b_m}\right) + \dot{\boldsymbol{\theta}}, \quad (14)$$

where $\boldsymbol{\Omega}^{b_m} = \left[\omega_x^{b_m} \ \omega_y^{b_m} \ \omega_z^{b_m}\right]^{\mathrm{T}}$ and $\boldsymbol{\Omega}^{b_s} = \left[\omega_x^{b_s} \ \omega_y^{b_s} \ \omega_z^{b_s}\right]^{\mathrm{T}}$ are gyro angular velocity of the master system and slave system, respectively; $\dot{\boldsymbol{\theta}} = \left[\dot{\theta}_x \ \dot{\theta}_y \ \dot{\theta}_z\right]^{\mathrm{T}}$; $\boldsymbol{\varepsilon}^{b_m}$ is the gyro random drift of the master system in b_m-frame, which consists of random constant drift $\boldsymbol{\varepsilon}_c^{b_m}$ and Gaussian white noise $\boldsymbol{\omega}_\varepsilon^{b_m}$.

According to (1) and (14), the difference of gyro output between the master and slave systems is

$$\begin{aligned} \Delta\boldsymbol{\Omega} &= \boldsymbol{\Omega}^{b_s} - \boldsymbol{\Omega}^{b_m} = \mathbf{C}_{b_m}^{b_s}\boldsymbol{\Omega}^{b_m} + \mathbf{C}_{b_m}^{b_s}\boldsymbol{\varepsilon}^{b_m} + \dot{\boldsymbol{\theta}} - \boldsymbol{\varepsilon}^{b_s} - \boldsymbol{\Omega}^{b_m} \\ &= \left(\mathbf{C}_{b_m}^{b_s} - \mathbf{I}\right)\boldsymbol{\Omega}^{b_s} + \dot{\boldsymbol{\theta}} + \mathbf{C}_{b_m}^{b_s}\boldsymbol{\varepsilon}^{b_m} - \boldsymbol{\varepsilon}_c^{b_s} \\ &= \left[-\left(\boldsymbol{\rho} + \boldsymbol{\theta}\right)\times\right]\boldsymbol{\Omega}^{b_s} + \dot{\boldsymbol{\theta}} + \mathbf{C}_{b_m}^{b_s}\boldsymbol{\varepsilon}^{b_m} - \boldsymbol{\varepsilon}_c^{b_s} \\ &= \boldsymbol{\Omega}^{b_s} \times \left(\boldsymbol{\rho} + \boldsymbol{\theta}\right) + \dot{\boldsymbol{\theta}} + \mathbf{C}_{b_m}^{b_s}\boldsymbol{\varepsilon}^{b_m} - \boldsymbol{\varepsilon}_c^{b_s} \\ &= \widehat{\boldsymbol{\Omega}}\left(\boldsymbol{\rho} + \boldsymbol{\theta}\right) + \dot{\boldsymbol{\theta}} + \mathbf{C}_{b_m}^{b_s}\boldsymbol{\varepsilon}^{b_m} - \boldsymbol{\varepsilon}_c^{b_s} \\ &\approx \widehat{\boldsymbol{\Omega}}\left(\boldsymbol{\rho} + \boldsymbol{\theta}\right) + \dot{\boldsymbol{\theta}} + \boldsymbol{\varepsilon}^{b_m} - \boldsymbol{\varepsilon}^{b_s}, \end{aligned} \quad (15)$$

where

$$\widehat{\boldsymbol{\Omega}} = \begin{bmatrix} 0 & -\omega_z^{b_m} & \omega_y^{b_m} \\ \omega_z^{b_m} & 0 & -\omega_x^{b_m} \\ -\omega_y^{b_m} & \omega_x^{b_m} & 0 \end{bmatrix}. \quad (16)$$

Substituting $\boldsymbol{\varepsilon}^{b_m} = \boldsymbol{\varepsilon}_c^{b_m} + \boldsymbol{\omega}_\varepsilon^{b_m}$ and $\boldsymbol{\varepsilon}^{b_s} = \boldsymbol{\varepsilon}_c^{b_s} + \boldsymbol{\omega}_\varepsilon^{b_s}$ into (15) yields

$$\Delta\boldsymbol{\Omega} = \widehat{\boldsymbol{\Omega}}\left(\boldsymbol{\rho} + \boldsymbol{\theta}\right) + \dot{\boldsymbol{\theta}} + \boldsymbol{\varepsilon}_c^{b_m} - \boldsymbol{\varepsilon}_c^{b_s} + \boldsymbol{\omega}_\varepsilon^{b_m} - \boldsymbol{\omega}_\varepsilon^{b_s}. \quad (17)$$

Based on (17), the measurement equation is given as follows:

$$\mathbf{Z}_G = \mathbf{H}_G\mathbf{X}_G + \mathbf{v}_G, \tag{18}$$

where measurement vector $\mathbf{Z}_G = \Delta\boldsymbol{\Omega}$; measurement matrix $\mathbf{H}_G = \left[\widehat{\boldsymbol{\Omega}}\ \widehat{\boldsymbol{\Omega}}\ \mathbf{I}_{3\times3}\ \mathbf{I}_{3\times3}\ -\mathbf{I}_{3\times3}\right]$; measurement noise $\mathbf{v}_G = \left[v_{\delta\omega_x}\ v_{\delta\omega_y}\ v_{\delta\omega_z}\right]^{\mathrm{T}}$ is a zero-mean Gaussian white noise sequence with covariance $\mathbf{R}_G$.

2.3. Mathematical Modeling of Deformation Measuring Algorithm Based on As. In this method, taking three orthogonally mounted accelerometers as the slave system and main POS as the master system, the difference of accelerometers between the master system and slave system is selected as the measurement of KF to estimate the deformation.

2.3.1. State Vector Selection. The state vector is defined by

$$\mathbf{X}_A = \left[\rho_x, \rho_y, \rho_z, \theta_x, \theta_y, \theta_z, \dot{\theta}_x, \dot{\theta}_y, \dot{\theta}_z, \nabla_x^{b_m}, \nabla_y^{b_m}, \nabla_z^{b_m}, \nabla_x^{b_s}, \nabla_y^{b_s}, \nabla_z^{b_s}\right]^{\mathrm{T}}, \tag{19}$$

where subscript A represents the deformation measuring algorithm based on As; $\nabla_j^{b_m}$ and $\nabla_j^{b_s}$ $(j = x, y, z)$ are accelerometer random constant bias of the master system and slave system, respectively.

2.3.2. System State Equation. The system state equation is given by

$$\dot{\mathbf{X}}_A = \mathbf{F}_A\mathbf{X}_A + \mathbf{G}_A\mathbf{W}_A, \tag{20}$$

where $\mathbf{F}_A$ and $\mathbf{G}_A$ are system transition matrix and system noise distribution matrix, respectively; system noise vector $\mathbf{W}_A = \left[\eta_x\ \eta_y\ \eta_z\right]^{\mathrm{T}}$ is a zero-mean Gaussian white noise vector with covariance $\mathbf{Q}_A = \mathbf{Q}_\eta$.

The differential equation of rigid misalignment angle and flexure angle is the same as in (3). The differential equation of accelerometer random constant bias is given by

$$\begin{aligned} \dot{\nabla}_j^{b_m} &= 0, \\ \dot{\nabla}_j^{b_s} &= 0, \\ j &= x, y, z. \end{aligned} \tag{21}$$

The expression of $\mathbf{F}_A$ and $\mathbf{G}_A$ can be obtained from (3) and (21). Furthermore, $\mathbf{F}_A = \mathbf{F}_G$ and $\mathbf{G}_A = \mathbf{G}_G$.

2.3.3. Measurement Equation. The relationship between the accelerometer outputs of the master system and slave system can be expressed by

$$\mathbf{f}^{b_s} + \nabla^{b_s} = \mathbf{C}_{b_m}^{b_s}\left(\mathbf{f}^{b_m} + \nabla^{b_m} + \mathbf{a}_r^{b_m}\right), \tag{22}$$

where $\mathbf{f}^{b_m} = \left[f_x^{b_m}\ f_y^{b_m}\ f_z^{b_m}\right]^{\mathrm{T}}$ and $\mathbf{f}^{b_s} = \left[f_x^{b_s}\ f_y^{b_s}\ f_z^{b_s}\right]^{\mathrm{T}}$ are the specific forces measured by the accelerometers of the master system and slave system, respectively; ∇^{b_m} is the accelerometer random bias of the master system in b_m-frame, which consists of random constant bias $\nabla_c^{b_m}$ and Gaussian white noise $\boldsymbol{\omega}_\nabla^{b_m}$; $\mathbf{a}_r^{b_m}$ is the relative acceleration between the master system and slave system caused by lever arm; $\mathbf{a}_r^{b_m}$ is given by

$$\mathbf{a}_r^{b_m} = \boldsymbol{\Omega}^{b_m} \times \boldsymbol{\Omega}^{b_m} \times \mathbf{r}^{b_m} + \dot{\boldsymbol{\Omega}}^{b_m} \times \mathbf{r}^{b_m} + 2\left(\boldsymbol{\Omega}^{b_m} \times \dot{\mathbf{r}}^{b_m}\right) + \ddot{\mathbf{r}}^{b_m}. \tag{23}$$

According to (1) and (22), the difference of accelerometer output between the master system and slave system is

$$\begin{aligned} \Delta\mathbf{f} &= \left(\mathbf{f}^{b_s} - \mathbf{C}_{b_m}^{b_s}\mathbf{a}_r^{b_m}\right) - \mathbf{f}^{b_m} \\ &= \mathbf{C}_{b_m}^{b_s}\mathbf{f}^{b_m} + \mathbf{C}_{b_m}^{b_s}\nabla^{b_m} - \nabla^{b_s} - \mathbf{f}^{b_m} \\ &= \left(\mathbf{C}_{b_m}^{b_s} - \mathbf{I}\right)\mathbf{f}^{b_m} + \mathbf{C}_{b_m}^{b_s}\nabla^{b_m} - \nabla^{b_s} \\ &= \left[-\left(\boldsymbol{\rho} + \boldsymbol{\theta}\right)\times\right]\mathbf{f}^{b_m} + \mathbf{C}_{b_m}^{b_s}\nabla^{b_m} - \nabla^{b_s} \\ &= \mathbf{f}^{b_m} \times \left(\boldsymbol{\rho} + \boldsymbol{\theta}\right) + \mathbf{C}_{b_m}^{b_s}\nabla^{b_m} - \nabla^{b_s} \\ &= \widehat{\mathbf{f}}\left(\boldsymbol{\rho} + \boldsymbol{\theta}\right) + \mathbf{C}_{b_m}^{b_s}\nabla^{b_m} - \nabla^{b_s} \\ &\approx \widehat{\mathbf{f}}\left(\boldsymbol{\rho} + \boldsymbol{\theta}\right) + \nabla^{b_m} - \nabla^{b_s}, \end{aligned} \tag{24}$$

where

$$\widehat{\mathbf{f}} = \begin{bmatrix} 0 & -f_z^{b_m} & f_y^{b_m} \\ f_z^{b_m} & 0 & -f_x^{b_m} \\ -f_y^{b_m} & f_x^{b_m} & 0 \end{bmatrix}. \tag{25}$$

Substituting $\nabla^{b_m} = \nabla_c^{b_m} + \boldsymbol{\omega}_\nabla^{b_m}$ and $\nabla^{b_s} = \nabla_c^{b_s} + \boldsymbol{\omega}_\nabla^{b_s}$ into (24) yields

$$\Delta\mathbf{f} = \widehat{\mathbf{f}}\left(\boldsymbol{\rho} + \boldsymbol{\theta}\right) + \nabla_c^{b_m} - \nabla_c^{b_s} + \boldsymbol{\omega}_\nabla^{b_m} - \boldsymbol{\omega}_\nabla^{b_s}. \tag{26}$$

Based on (26), the measurement equation is

$$\mathbf{Z}_A = \mathbf{H}_A\mathbf{X}_A + \mathbf{v}_A, \tag{27}$$

where measurement vector $\mathbf{Z}_A = \Delta\mathbf{f}$; measurement matrix $\mathbf{H}_A = \left[\widehat{\mathbf{f}}\ \widehat{\mathbf{f}}\ \mathbf{0}_{3\times3}\ \mathbf{I}_{3\times3}\ -\mathbf{I}_{3\times3}\right]$; measurement noise $\mathbf{v}_A = \left[v_{\delta f_x}\ v_{\delta f_y}\ v_{\delta f_z}\right]^{\mathrm{T}}$ is a zero-mean Gaussian white noise sequence with covariance $\mathbf{R}_A$.

3. Simulation and Semiphysical Simulation

In order to verify the estimation effect of methods based on Gs and As, respectively, proposed in Section 2, a long rod with a master system and a slave system installed on both ends is a better way. But in this case, it is not possible to know the exact value of deformation between the master system and slave system. Therefore, flight simulation and vehicle semiphysical simulation are provided and compared with the method based on IMU in this section.

TABLE 1: Parameters setting of simulation trajectory.

Time (s)	Motion state
0–100	Uniform linear motion
100–200	Turn 60° clockwise
200–300	Turn 60° anticlockwise
300–700	Uniform linear motion
700–900	Turn 180° clockwise
900–1300	Uniform linear motion

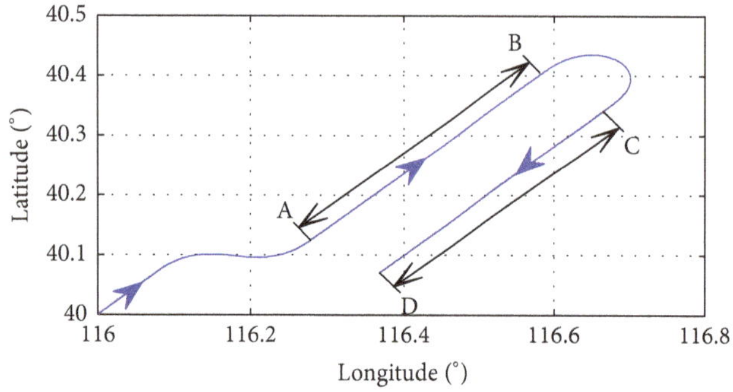

FIGURE 2: Plane trajectory with S-shaped maneuver and U-shaped flight.

3.1. Flight Simulation and Analysis. From the attitude transfer relationship shown in (1), it can be seen that the attitude difference between the master system and slave system, also called the system error angle, is determined by the sum of rigid misalignment angle and flexure angle. Therefore, not only should the estimation accuracy of the flexure angle or the rigid misalignment angle be evaluated, but also the estimation accuracy of the system angle error should be evaluated. In connection with this, the estimation error curves of flexure angle, rigid misalignment angle, and system error angle are given in the simulation, and the estimation error of the system error angle is used to evaluate the measuring precision of each method.

3.1.1. Design of Simulation. In this paper, the typical "S + U"-shaped trajectory of airborne earth observation is simulated. The plane trajectory and trajectory parameters are shown in Figure 2 and Table 1, respectively. Total flight time is 1300 s. AB and CD section can be regarded as the imaging section in Figure 2. Initial heading angle, pitch angle, and roll angle are 40°, 0°, and 0°, respectively. The flight velocity is 100 m/s and the altitude is 500 m. This simulation has been performed ten times.

The measurement noise of the main POS (as the master system) at heading, pitch, roll, and velocity are 0.02° (1σ), 0.005° (1σ), 0.005° (1σ), and 0.03 m/s (1σ), respectively. Both gyro constant drift and random drift of the main POS are 0.01°/h. Both accelerometer constant bias and random bias of the main POS and slave system are 50 μg. Both gyro constant and random drift of the slave system are 0.1°/h. The misalignment angle of the slave system relative to the main POS is given as $\boldsymbol{\rho} = \left[0.5^\circ\ \ 0.5^\circ\ \ 0.5^\circ\right]^{\mathrm{T}}$, and the lever arm between the main POS and slave system is $\mathbf{r}^{b_m} = \left[5\,\mathrm{m}\ \ 0.1\,\mathrm{m}\ \ 0.1\,\mathrm{m}\right]^{\mathrm{T}}$. The data update rate of the main system and slave system is 100 Hz. For the deformation measurement of the wing, the flexure angle rotated around the vertical axis is big, while the flexure angles around the other two axes are small. Accordingly, the correlation times of the second-order Markov processes are selected as 2, 5, and 2, and the covariances of flexure angle are 0.01, 0.15, and 0.01, respectively. The curves of flexure angle and flexure angle rate are shown in Figures 3 and 4.

Data Generation. A trajectory generator is used to generate the theoretical data of the scheduled flight trajectory, which include position, velocity, attitude, and the output data of gyros and accelerometers. The real outputs of the main POS are obtained by adding the corresponding measurement noise to the theoretical position, velocity, and attitude. Then, the theoretical outputs of gyros and accelerometers are converted by rigid misalignment angles and flexure angles, and the constant noise and random noise are added to be the inertial sensor outputs of the slave system.

3.1.2. Simulation Results Analysis. Figures 5–7 show the estimate error curves of deformation measuring method based on IMU, Gs, and As, respectively, including the estimate errors of rigid misalignment angle, flexure angle, and system error angle. For improving the estimate accuracy, a maneuver is added in the S-shape of the flight trajectory and the system error angle estimate errors are shown in Figure 8. The details of the maneuver are as follows: the flight velocity of S-shape is increased equably from 100 m/s to 300 m/s between 100 s and 200 s and then decreased equably to 100 m/s between 200 s and 300 s. This kind of maneuver is very easy to implement for earth observation aircraft, because it is usually necessary to make S or other types of maneuver to improve the estimation precision of POS before the aircraft enters the observation area.

Besides, the Root Mean Square Error (RMSE) and Standard Deviation (STD) values of system error angle estimate errors in imaging segments AB and CD are counted and shown in Table 2, where all values are the average of ten simulations. Since the maneuver mentioned above only affects the method based on Gs (it can be seen from Table 2), only the system error angle estimate errors of method based on IMU, Gs without this maneuver, and method based on Gs with this maneuver are shown in Figure 8.

Figures 5 and 6 show that the estimation accuracies of flexure angle and rigid misalignment angle are not good at the same time and coupled with each other, while the sum of two estimation errors of system error angle shown in Figure 7 is relatively stable on each axis.

Figures 7 and 8 and Table 2 show that the deformation measurement based on IMU has the highest estimation accuracy without any additional maneuver, followed by the method based on Gs, and the method based on As is the worst. The velocity maneuvers in S-shape can improve the estimation effect of the method based on Gs, especially

TABLE 2: Estimate errors of system error angle in simulation test (′).

Errors	No variation at velocity in S-shape			Variation at velocity in S-shape		
	IMU	Gs	As	IMU	Gs	As
STD						
x-axis	0.4108	0.4593	17.0632	0.4086	0.4333	17.0642
y-axis	0.3832	0.2834	2.6423	0.3882	0.2403	2.6825
z-axis	1.1755	0.4152	435.9763	1.0088	0.3902	437.4117
RMSE						
x-axis	0.6643	1.7323	24.2849	0.6546	0.8323	24.3184
y-axis	0.5720	1.3091	3.9038	0.5943	0.6703	3.9360
z-axis	2.1673	7.7563	1152.3785	2.1943	2.2219	1761.8890

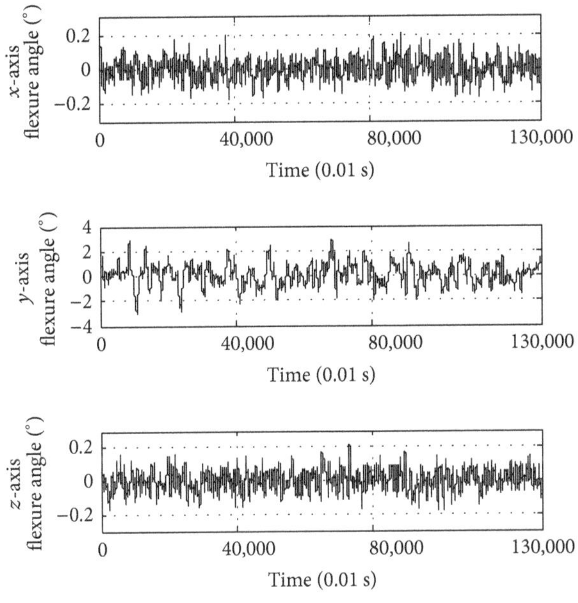

FIGURE 3: Flexure angle.

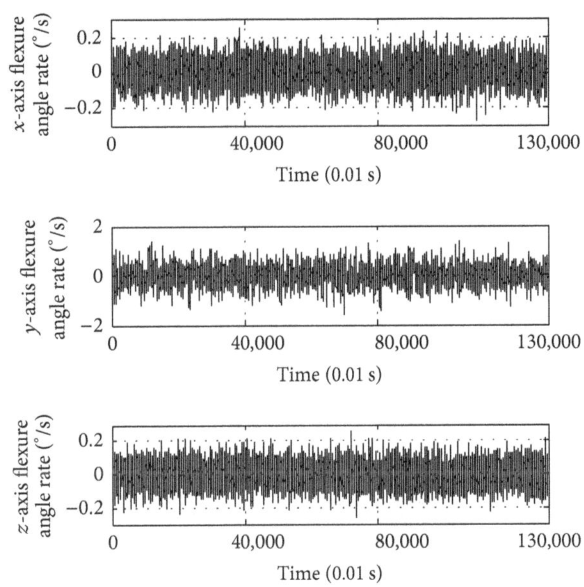

FIGURE 4: Flexure angle rate.

the RMSE, and obtain the estimation precision close to the method based on IMU. It is worth mentioning that although the method based on As has very poor estimation accuracy on x-axis and z-axis, the STD and RMES of system error angle estimation error on y-axis are only 3.3′ and 5′, respectively, which is very small relative to the flexure angle on y-axis shown in Figure 3.

3.2. Semiphysical Simulation and Analysis. A real road experiment is carried out in Shahe Town, Changping District, Beijing, China. In this vehicle experiment, a high precision POS is used as the main system, whose gyros, accelerometers, position, velocity, and attitude output are recorded. Based on these data, the flexure and misalignment sets in Section 3 are added, and then the theoretical data of the slave system can be obtained. After considering the error of the gyro and accelerometer in the slave system, the real output of the gyros and accelerometers in the slave system can be simulated.

3.2.1. Hardware Configuration. The van and the sensors installation are shown in Figures 9 and 10, respectively. The high-precision POS, developed by the Integrated System Research Group at Beihang University, Beijing, China, consists of a laser gyro-based IMU, a PCS, and Novatel DLV-3 GPS receiver (based on Novatel OEMV-3 receiver board) and has mobile station and base station equipment with 20 Hz output rate [33]. The trajectory of the test is shown in Figure 11 and the total test time is 1500 s. Figure 12 shows the van's velocity and acceleration.

The specifications of the high-precision POS inertial sensors applied in the laser gyro-based IMU and position, velocity, and attitude postprocessing output are listed in Table 3. The inertial sensors' errors of the simulated slave system are the same as these used in Section 3. The output rate of the main POS is 100 Hz.

3.2.2. Semiphysical Simulation Results Analysis. Figure 13 shows the system error angle estimate error curves of deformation measuring method based on IMU, Gs, and As. In

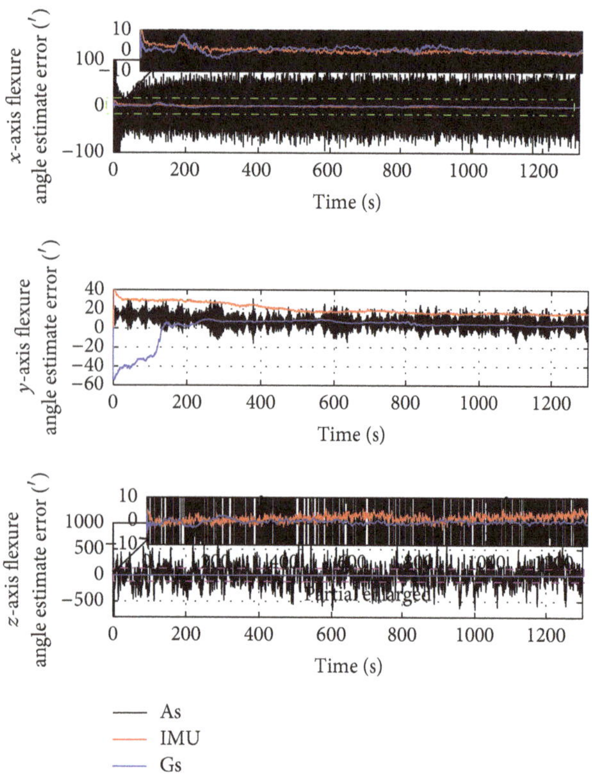

FIGURE 5: Flexure angle estimate error in simulation test.

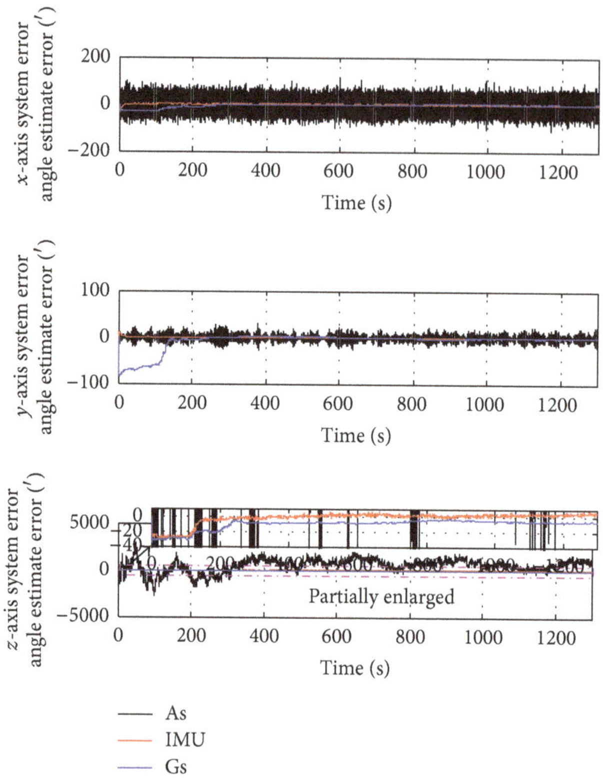

FIGURE 7: System error angle estimate error in simulation test.

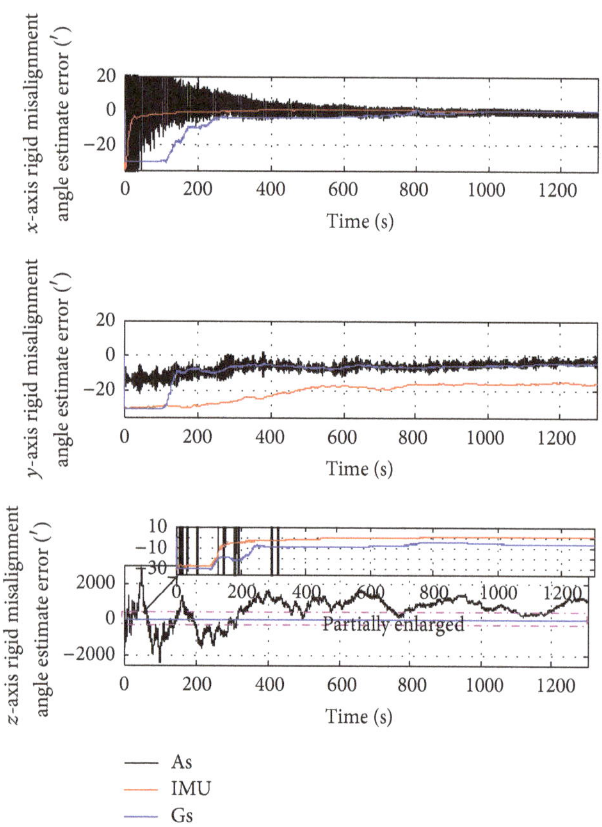

FIGURE 6: Rigid misalignment angle estimate error in simulation test.

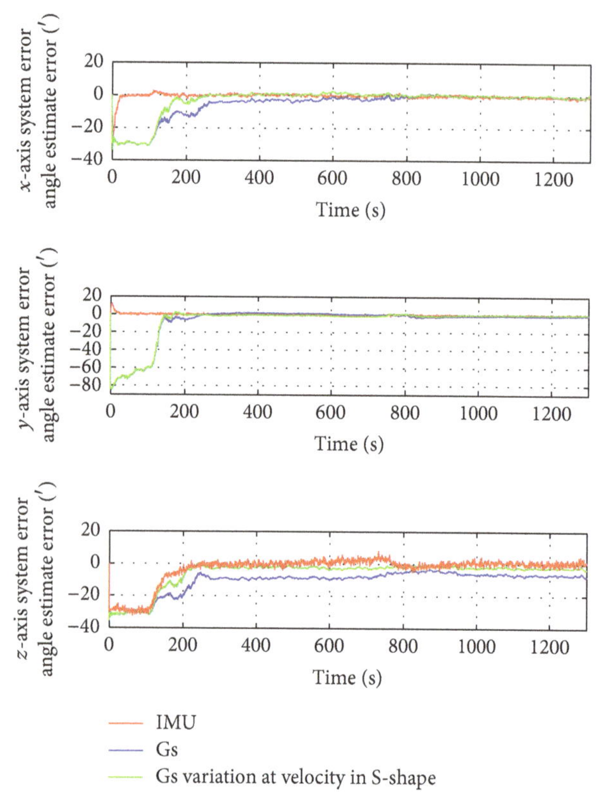

FIGURE 8: System error angle estimate error of method based on IMU and Gs in simulation test.

FIGURE 9: The van and sensors installed on the top of the van.

FIGURE 10: Sensors installed in the van and the base station equipment of GPS.

TABLE 3: Specifications of the high-precision POS (1δ).

Parameters	Values
Laser gyro	
Constant drift	0.01°/h
Random drift	0.01°/h
Accelerometer	
Constant bias	50 μg
Random bias	50 μg
Position	
Horizontal	<0.1 m
Vertical	<0.2 m
Velocity	0.005 m/s
Heading	0.005°
Pitch and roll	0.0025°

order to see the changing trends of errors more clearly, the error curves of the first two methods are shown in Figure 14, where L1 and L2 are two long straight segments in the trajectory. The statistics of system error angle estimate errors of L1 and L2 are given in Table 4.

Figures 13 and 14 and Table 4 show that the estimation accuracy of the method based on Gs is close to that of IMU method. In the horizontal direction, the estimation accuracy of the IMU method is more stable and the Gs method is greatly influenced by the turning of the vehicle; in the vertical direction, since there is not any maneuver and the vehicle began to move in a straight line, the estimation accuracy of the IMU method is gradually improved with the turn maneuver, and filtering is also gradually stabilized, while the Gs method can converge quickly. The method based on As has the largest estimation error, but its STD and RMES of system error angle estimation error on y-axis are no more than $30'$. Overall, the vehicle experiment results and the simulation results are basically similar and consistent.

4. Conclusion

In this paper, the mathematical models of deformation measurement based on Gs and As are derived, respectively, and compared with the deformation measuring method based on IMU. The results of simulation and semiphysical simulation show that the measuring method based on Gs can achieve similar estimation accuracy to the method based on IMU. Since the method based on IMU has the disadvantages of large size, high cost, and heavy weight, when the accuracy requirement is not very high and can be met by the method based on Gs or As, the measurement equipment will be greatly simplified. In particular, the method based on As has a good estimation accuracy on y-axis, which is a better choice for the cases with large deformation only on y-axis and strict

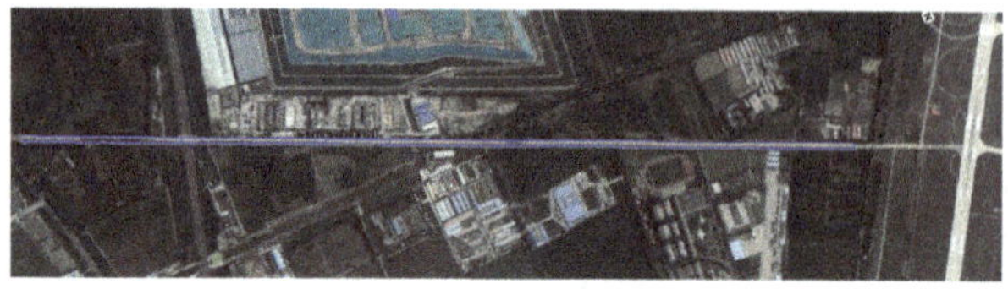

Figure 11: Trajectory of the experiment.

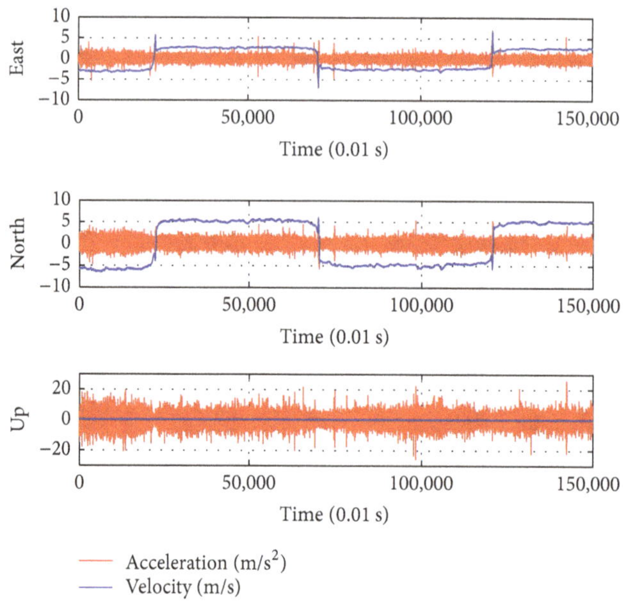

Figure 12: The van's velocity and acceleration.

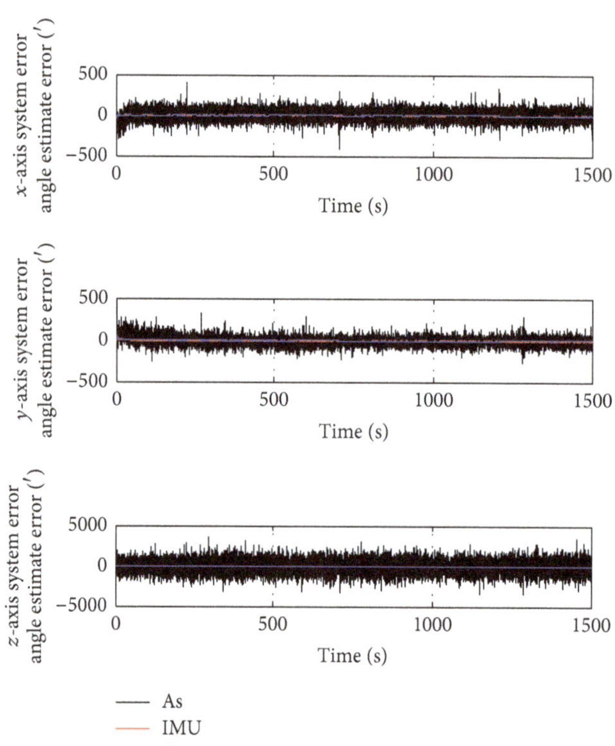

Figure 13: System error angle estimate error in semiphysical simulation.

TABLE 4: Estimate errors of system error angle in experiment test (′).

Errors	L1 (300 s–700 s)			L2 (800 s–1200 s)		
	IMU	Gs	As	IMU	Gs	As
STD						
x-axis	0.5469	0.8500	42.9655	0.5124	0.5549	42.7889
y-axis	0.2256	0.4333	27.5946	0.2340	0.3020	27.4013
z-axis	1.5704	0.5979	647.4333	0.5919	0.8776	642.9184
RMSE						
x-axis	0.6640	0.9841	43.1235	0.5625	1.0232	42.9072
y-axis	0.2401	0.5429	29.7259	0.2556	0.9420	29.2839
z-axis	5.7615	2.0966	648.8067	0.6068	1.3364	644.4869

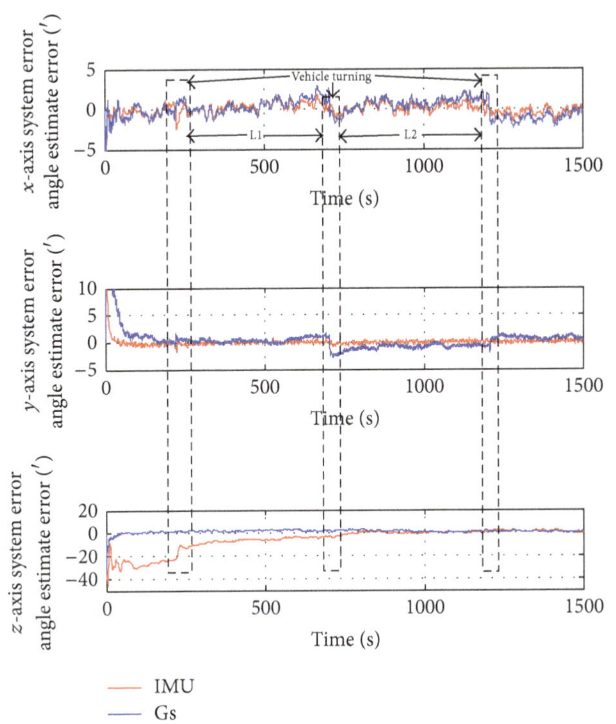

FIGURE 14: System error angle estimate error of methods based on IMU and Gs, respectively, in semiphysical simulation.

limitation on the weight, size, and cost of the measurement equipment. In view of the deformation measuring method based on Gs and As, the intrinsic relationship between the maneuver and the estimation accuracy needs further analysis and research in the future.

Conflicts of Interest

The authors declare that there are no conflicts of interest regarding the publication of this paper.

Acknowledgments

This work was supported in part by the National Natural Science Foundation of China (Grants nos. 61473020, 61421063, and 61233005), by the National High Technology Research and Development Program of China (863 Program) (Grants nos. 2015AA124001 and 2015AA124002), and by the International (Regional) Cooperation and Communication Project (Grant no. 61661136007).

References

[1] W. Xu, Y. Deng, and R. Wang, "Multichannel synthetic aperture radar systems with a planar antenna for future spaceborne microwave remote sensing," *IEEE Aerospace and Electronic Systems Magazine*, vol. 27, no. 12, pp. 26–30, 2012.

[2] P. Scherbaum, J. Brauchle, T. Kraft, and S. Pless, "MACS-Mar - A Real-Time Capable Multisensor Remote Sensing System for Maritime Applications: Current State of Development," in *Proceedings of the IEEE International Conference on Aerospace Electronics and Remote Sensing Technology, ICARES 2015*, Bali, Indonesia, 2015.

[3] Q. Zhang, X. J. Niu, Q. J. Chen, H. P. Zhang, and C. Shi, "Using Allan Variance to Evaluate the Relative Accuracy on Different Time Scales of GNSS/INS Systems," *Measurement Science and Technology*, vol. 24, no. 8, 2013.

[4] L.-Z. Chen and J. Fang, "A hybrid prediction method for bridging GPS outages in high-precision POS application," *IEEE Transactions on Instrumentation and Measurement*, vol. 63, no. 6, pp. 1656–1665, 2014.

[5] J. T. Case, M. T. Ghasr, and R. Zoughi, "Correcting mutual coupling and poor isolation for real-time 2-D microwave imaging systems," *IEEE Transactions on Instrumentation and Measurement*, vol. 63, no. 5, pp. 1310–1319, 2014.

[6] J. Wang, L. Y. Chen, X. D. Liang et al., "Multi-input Multi-output Frequency-modulated Continuous Wave Synthetic Aperture Radar System using Beat-frequency Division Waveforms," *Measurement Science and Technology*, vol. 24, no. 7, Article ID 074016, 2013.

[7] A. Liu and H. X. Ma, "Notice of Retraction Optimization Method for the Layout of IMUs in Deformation Detection System of Warship Based on Genetic Algorithm," in *Proceedings of the 6th International Conference on Natural Computation (ICNC)*, vol. 8, 155, pp. 4014–4017, Shandong, China, 2010.

[8] R. C. Zheng, G. L. Yang, C. Y. Chen, and H. N. Weng, "Layout Optimization of Shipborne IMU Based on Genetic Algorithm," *Journal of Chinese Inertial Technology*, vol. 15, no. 2, pp. 177–280, 2007.

[9] H. Dai, A. Liu, J. Lu, S. Dai, X. Wu, and Y. Sun, "Optimization about the layout of IMUs in large ship based on fruit fly optimization algorithm," *International Journal for Light and Electron Optics*, vol. 126, no. 4, pp. 490–493, 2015.

[10] T. H. Skopinski, W. S. Aiken, and W. B. Huston, "Calibration of Strain-gage Installations in Aircraft Structures for the Measurement of Flight Loads," NACA-Report-1178, Technical Report Archive and Image Library, 1954, pp. 505–536.

[11] J. M. Jenkins and V. M. DeAngelis, "A Aummary of Numerous strain-gage Load Calibrations on Aircraft Wings and Tails in A Technology Format," Nasa Technical Memorandum 4804, Dryden Flight Research Center, Edwards, Calif, UAS, 1997.

[12] A. Lizotte and W. Lokos, "Deflection-based aircraft structural loads estimation with comparison to flight," in *Proceedings of the 46th AIAA/ASME/ASCE/AHS/ASC Structures, Structural Dynamics and Materials Conference*, pp. 1–18, Austin, Tex, USA, 2005.

[13] V. De Angelis, "HiMAT flight deflection measurements," in *Proceedings of the NASA HiMAT Symposium Paper*, 1984.

[14] V. M. De Angelis, "In-flight deflection measurement of the HiMAT aeroelastically tailored wing," *Journal of Aircraft*, vol. 19, no. 12, pp. 1088–1094, 1982.

[15] H. Kannemans, "Flight testing of a wing deflection measurement method," in *Proceedings of the 20th Atmospheric Flight Mechanics Conference*, Baltimore, Md, USA, 1995.

[16] H. P. J. Veerman, H. Kannemans, and H. W. Jentink, "High accuracy in-flight wing deformation measurements based on optical correlation technique," in *Proceedings of the 19th SFTE (EC) Symposium*, Manching, Germany, 2008.

[17] J. T. Black, N. A. Pitcher, M. F. Reeder, and R. C. Maple, "Videogrammetry dynamics measurements of a lightweight flexible wing in a wind tunnel," *Journal of Aircraft*, vol. 47, no. 1, pp. 172–180, 2010.

[18] S. A. Frost, C. H. G. Wright, R. W. Streeter, M. A. Khan, and S. F. Barrett, "Bio-mimetic optical sensor for structural deflection measurement," in *Proceedings of the Bioinspiration, Biomimetics, and Bioreplication*, San Diego, Calif, USA, March 2014, 9055:90550A.

[19] M. Scheerer, Z. Djinovic, and M. Schüller, "Fiber optic system for deflection and damage detection in morphing wing structures," in *Proceedings of the 2013 SPIE Conference on Sensors and Smart Structures Technologies for Civil, Mechanical, and Aerospace Systems 2013*, March 2013.

[20] W. Wu, S. Chen, and S. Qin, "Online estimation of ship dynamic flexure model parameters for transfer alignment," *IEEE Transactions on Control Systems Technology*, vol. 21, no. 5, pp. 1666–1678, 2013.

[21] J. H. Cheng, T. D. Wang, D. X. Guan, and M. L. Li, "Polar transfer alignment of shipborne SINS with a large misalignment angle," *Measurement Science and Technology*, vol. 27, no. 3, Article ID 035101, 2016.

[22] R. E. Kalman, "A new approach to linear filtering and prediction problem," *Journal of Basic Engineering*, vol. 82, no. D, pp. 35–45, 1960.

[23] S. Eichstädt, N. Makarava, and C. Elster, "On the evaluation of uncertainties for state estimation with the kalman filter," *Measurement Science and Technology*, vol. 27, no. 12, 1968.

[24] X. Gong, R. Zhang, and J. Fang, "Application of unscented R-T-S smoothing on INS/GPS integration system post processing for airborne earth observation," *Measurement: Journal of the International Measurement Confederation*, vol. 46, no. 3, pp. 1074–1083, 2013.

[25] M. S. Grewal, L. R. Weill, and A. P. Andrews, *Global Positioning Systems, Inertial Navigation, and Integration*, John Wiley & Sons, Inc., Hoboken, NJ, USA, 2007.

[26] S. Nassar, *Improving the Inertial Navigation System (INS) Error Model for INS and INS/DGPS Applications [Ph.D. thesis]*, Department of Geomatics Engineering, University of Calgary, Canada, 2003.

[27] M. Y. Zhong, Q. Cao, J. Guo, and D. H. Zhou, "Simultaneous lever-arm compensation and disturbance attenuation of POS for a UAV surveying system," *IEEE Transations on Instrumentation Measurement*, vol. 65, no. 12, pp. 2828–2839, 2016.

[28] X. M. Wang and W. B. Ni, "An Improved Particle Filter and Its Application to an INS/GPS Integrated Navigation System in a Serious Noisy Scenario," *Measurement Science and Technology*, vol. 27, no. 9, p. 2016, 2016.

[29] M. Y. Zhong, J. Guo, D. F. Guo, and Z. H. Yang, "An Extended H_i/H_∞ Optimization Approach to Fault Detection of INS/GPS-integrated System," *IEEE Transations on Instrumentation Measurement*, vol. 65, no. 11, pp. 2495–2504, 2016.

[30] Y. Lu and X. Cheng, "Random misalignment and lever arm vector online estimation in shipborne aircraft transfer alignment," *Measurement: Journal of the International Measurement Confederation*, vol. 47, no. 1, pp. 756–764, 2014.

[31] D. Li, M. Zhong, and D. Guo, "Error Detection and Compensation in Transfer Alignment for the Distributed POS," in *Proceedings of the 25th Chinese Control and Decision Conference*, pp. 25–27, Guiyang, China, 2013.

[32] Z. Feng, W. Jiang, and S. Gao, "Research of rapid transfer alignment and observability analysis," in *Proceedings of the 2nd International Conference on Intelligent Control and Information Processing, ICICIP 2011*, pp. 204–207, Harbin, China, 2011.

[33] "Novatel OEMV-3 receiver board," http://www.novatel.com.cn/products/gnss-receivers/oem-receiver-boards/oemv-receivers/.

8

Novel Fractional Order Calculus Extended PN for Maneuvering Targets

Jikun Ye, Humin Lei, and Jiong Li

Air and Missile Defense College, Air Force Engineering University, Shaanxi, Xi'an 710051, China

Correspondence should be addressed to Jikun Ye; jikunbo@sina.com

Academic Editor: Christopher J. Damaren

Based on the theory of fractional order calculus (FOC), a novel extended proportional guidance (EPN) law for intercepting the maneuvering target is proposed. In the first part, considering the memory function and filter characteristic of FOC, the novel extended PN guidance algorithm is developed based on the conventional PN after introducing the properties and operation rules of FOC. Further, with the help of FOC theory, the average load and ballistics characteristics of proposed guidance law are analyzed. Then, using the small offset kinematic model, the robustness of the new guidance law against autopilot parameters is studied theoretically by analyzing the sensitivity of the closed loop guidance system. At last, representative numerical results show that the designed guidance law obtains a better performance than the traditional PN for maneuvering target.

1. Introduction

The proportional navigation (PN) has been widely used because of its simplicity, ease of implementation, and effectiveness in practical systems [1–5]. Due to the fact that end of the line of sight (LOS) angular rate rapidly rotating easily leads to overload saturation when PN against maneuvering target, many scholars modified traditional PN guidance law to improve the performance of guidance law [6–9]. A Lyapunov method is used by Dhananjay to analyze the effect of LOS rate delay on the performance of PN guidance law in a head-on or tail-chase missile-target engagement scenario [10]; the author proposed the way for selecting the navigation gain to overcome the problem of LOS rate delay. Ghosh et al. derived a modified PN guidance law against higher speed nonmaneuvering targets to control terminal impact angle, which uses standard pure PN and retro-PN guidance laws in a 3D engagement scenario based on the initial engagement geometry and terminal engagement requirements. And the capture region of the new guidance law was researched [11]. Jeon and Lee modified PN for an ideal pursuer with a stationary target by obtaining an optimal feedback solution minimizing a performance index of the range-weighted control energy; the authors presented the exact analyses on optimality of PN based on a nonlinear formulation [12]. The core idea of the above guidance law design is to add the compensation term against LOS rate delay, target maneuver, acceleration of gravity, and so on. The above studies centered on the integer order differential LOS rate, while the fractional calculus LOS rate has not been deeply researched and widely used.

With the development of fractional order calculus (FOC) theory, the FOC has been more and more widely used in engineering [13–15]. Considering the different application scenario, there are three common FOC definition: Riemann Liouville (RL) definition, Grumwald Letnkov (GL) definition, and Caputo definition [16, 17]. Zhang et al. discussed the order range of FOC and point out the differences of fractional order derivatives and integer order derivatives; the relations between RL definition, GL definition, and Caputo definition were studied too [18]. Unlike discontinuous change of integer order calculus, FOC expands calculus from the integer to fraction; the order of FOC can be changed continuously and reflects the nature of continuity better. The physical model established based on the FOC theory can describe physical phenomena with memory and time-dependent.

The major advantage of FOC is that it has fewer parameters and more simple forms; thus it is easy to be applied in reality. The specific memory functions and stability characteristics of fractional system have been more and more applied in the area of guidance and control [19]. Considering that the atmospheric viscosity and interactions between air vehicles makes aircraft have aerodynamic properties of FOC, thus the FOC theory can describe the characteristics of the aircraft kinematic comprehensive and truly.

Sun and Zhu developed a new fractional order tension control law for the fast and stable deployment control problem of a space tether system and proved its stability [20]. Almeida and Torres presented a method to solve fractional optimal control problems, where the dynamic control system depended on integer order and Caputo fractional derivatives [21]. Vishal et al. used the theory of fractional order derivative to discuss the local stability of the Mathieu–van der Pol hyperchaotic system and studied the analysis of control time with respect to different fractional order derivatives [22].

Reference [23] studied the application of FOC controller for guided projectiles attitude control; the author concluded that fractional order control was better than classical PID control in robustness and control performance. Li et al. studied the stability of the fractional order unified chaotic system by the equivalent passivity method. With the FOC theory, the Lyapunov function was constructed by which it was proved that the controlled fractional order system was stable [24]. The FOC theory was introduced to the guidance area in reference [25]; the author employed the PD^{λ} guidance law to carry forward the advantages of traditional PN and avoid its drawbacks based on FOC theory and concluded that the PD^{λ} guidance had the higher guidance precision and shorter interception time. Zhu et al. proposed a modified FOC pure PN guidance law by means of Lyapunov-like method, which solved the control problem for the capture of a target with random maneuvers effectively [26]. However, the conclusion of [25, 26] was obtained in the two-dimensional plane under some certain conditions, so there were some strict restrictions on the designed guidance law, and this greatly limits its practical engineering applications.

This paper presents a novel 3D extended PN guidance algorithm for maneuvering target based on FOC theory, and it overcomes the disadvantage of PN. First, the basics knowledge and relative nature of FOC are outlined. Second, the proposed guidance law is given, which is composed of traditional LOS rate term and FOC LOS rate term. And the trajectory features, the load character, and the robustness of the new guidance law are analyzed theoretically. At last, simulations show that the proposed guidance law has superior performance, especially for a maneuvering target.

2. Preliminary Knowledge

The fractional calculus differential operator can be expressed as follows [13, 25]:

$$ {}_{\alpha}D_t^p = \begin{cases} \dfrac{d^p}{dt^p} & R(p) > 0 \\ 1 & R(p) = 0 \\ \displaystyle\int_0^t (dt)^{-p} & R(p) < 0, \end{cases} \tag{1} $$

where α, t denote the upper limit and lower limit respectively. p is the calculus operator, its value can be any plurality, and $R(p)$ presents the real part of p. If $p > 0$, then ${}_{\alpha}D_t^p$ denotes fractional derivative.

Definition 1. If the function $f(t)$ can be taken n order derivation, here $m - 1 \leq p < m$ and $n \in \mathbf{N}$, then we have the definition of Caputo derivative as follows:

$$ D_t^p = \frac{1}{\Gamma(m-p)} \int_0^t \frac{f^m(\tau)}{(t-\tau)^{p-m+1}} d\tau, \tag{2} $$

where $\Gamma(x)$ denotes Gamma function, which is generalization of the factorial function $n!$, and we can take derivation of Gamma function with noninteger or even plurality. The expression of Gamma function can be denoted as

$$ \Gamma(x) = \int_0^{\infty} t^{x-t} e^{-t} dt. \tag{3} $$

The Laplace transform of Caputo fractional derivative can be shown as

$$ \begin{aligned} L\left[{}_0D_t^p f(t)\right] &= \int_0^{\infty} e^{-st} {}_0D_t^p f(t)\, dt \\ &= s^p F(s) - \sum_{k=0}^{n-1} s^{\alpha-k-1} f^k(0). \end{aligned} \tag{4} $$

In order to understand the links and differences between the integer order and fractional order calculus, we suppose that $f(t)$ is flat functions in the range $[a, t]$. Taking the derivation of $f(t)$ with n order, we have

$$ f^n(t) = \lim_{h \to 0} \frac{1}{h^n} \sum_{j=0}^{n} (-1)^j \frac{n!}{j!\,(n-j)!} f(t - jh). \tag{5} $$

When the noninteger p replaces integer n and Γ function replaces binomial coefficients, we get the definition of Grunwald-Letnikov derivative as follows:

$$ D_t^p f(t) = \lim_{h \to 0} \frac{1}{h^p} \sum_{j=0}^{n} (-1)^j \frac{\Gamma(p+1)}{j!\,\Gamma(p-j+1)} f(t - jh). \tag{6} $$

As can be seen from the above derivation, the most significant difference between integer order derivative and fractional derivative is that the former contains the recent information of several points, while the latter contains the past information of all the points. We can see that, from equation (6), the fractional derivative has the ability of memory function, and the closer the point, the stronger the impact factor [14].

In fact, the integer order calculus is a special case of fractional calculus, when the orders of fractional calculus $p \in N$, the above two are equivalent. Another point to note is that FOC uniforms the forms of the differential and integral expressions by fractional operator. When $p > 0$, it denotes fractional differential operator, while when $p < 0$, it represents fractional integral operator, so the study of integral and differential operation can be considered as an operator.

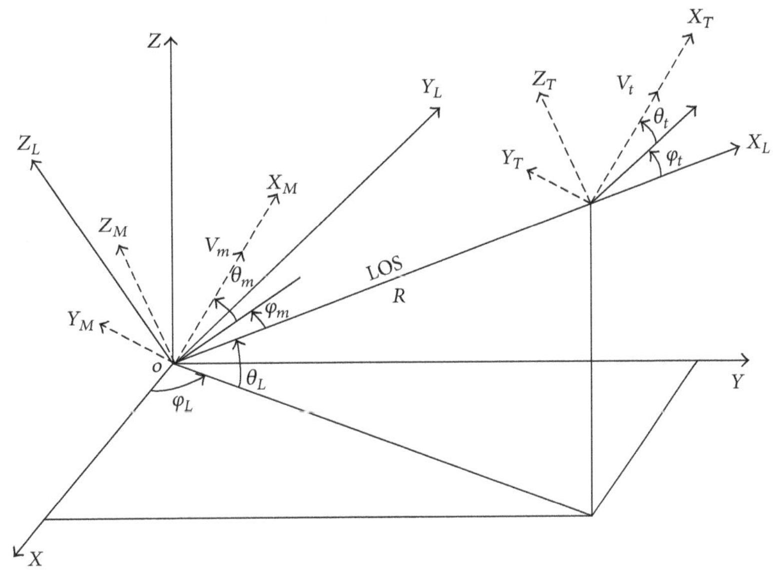

Figure 1: Pursuit geometry in 3-dimensional space.

Compared with the integer order linear system, the fractional linear system described by the FOC has the advantage of better memory function, better stability, and a greater choice of controller parameters.

3. Design of EPN Guidance Algorithm

We assume that the missile is an aerodynamic-controlled interceptor in a surface-to-air tactical missile defense scenario with a flat no rotating Earth. Consider a three-dimensional homing scenario shown in Figure 1, where M, T, and MT denote missile, target, and the missile-target LOS, respectively. The inertial frame is defined as $OXYZ$, and it is inertially fixed and centered at the launch site at the instant of launching. $OX_LY_LZ_L$ denotes the LOS coordinate, r is the distance between missile and target, the relative velocity of missile and target is represented as V_r, and the elevation and azimuth of LOS are denoted by θ_L and φ_L, respectively. V_m is the velocity of missile, and θ_m and φ_m are the elevation and azimuth of V_m, respectively. And θ_t and φ_t are the elevation and azimuth of V_t, respectively.

According to the TPN law, the guidance command is defined as

$$U_a = K_1 V_r \dot{\theta}_L,$$
$$U_b = K_1 V_r \dot{\varphi}_L \cos\theta_L, \tag{7}$$

where the guidance coefficient $K_1 > 2$ ensures convergent trajectory. $\dot{\theta}_L$ and $\dot{\varphi}_L$ denote the LOS rate, which can be obtained by detection system.

The conventional TPN consists of the relative velocity V_r, the LOS rate ($\dot{\theta}_L$ and $\dot{\varphi}_L$), and the proportional coefficient K_1. V_r can be detected by the seeker and K_1 can be set in advance before launching the missile; the derivative of LOS angle (θ_L and φ_L) contains the information of instantaneous relative motion between missile and target. The instantaneous LOS rate is used to calculate the load of TPN; however, the historical information of LOS rate is neglected, which may be useful to adjust the trajectory of missile. Considering that the derivative of FOC has unique historical memory effect, here we add the term of fractional order LOS rate to the guidance loop in order to carry forward the guidance effect of conventional TPN. In this way, the guidance command contains the more comprehensive information of relative motion to control the trajectory of missile.

Based on the theory of fractional order, here we construct a new form of EPN guidance law as follows:

$$U_a^p = K_1 V_r \dot{\theta}_L + K_1 V_r p D_t^p \theta_L, \tag{8}$$

$$U_b^p = K_1 V_r \dot{\varphi}_L \cos\theta_L + K_1 V_r D_t^p (\varphi_L) \cos\theta_L. \tag{9}$$

From (8) and (9), we find that the EPN guidance law is composed of two terms: one term is the traditional PN control, and the other term is fractional order differential of the LOS rotation rate. The fractional order term, used as compensation for the PN, contains the past information of LOS rotation rate. Note that when $p = 0$, the proposed EPN guidance law can be transformed to conventional PN guidance law.

4. Ballistic Characteristics Analysis of EPN Guidance Law

4.1. Characteristics Analysis of Average Overload. In order to simplify the expression of EPN, (7) can be rewritten as follows

$$U_a^p = K_{a1}\dot{\theta}_L + K_{a2} D_t^p (\theta_L),$$
$$U_b^p = K_{b1}\dot{\varphi}_L + K_{b2} D_t^p (\varphi_L), \tag{10}$$

where $K_{a1} = K_1 V_r$, $K_{b1} = K_1 V_r \cos\theta_L$, $K_{a2} = K_1 V_r p$, and $K_{b2} = K_1 V_r D_t^p (\varphi_L) \cos\theta_L$.

According to the EPN expression, the average overload guidance law can be expressed as

$$\overline{U} = \frac{1}{t_f}\int_0^{t_f}(U_a + U_b)\,dt = \frac{1}{t_f}\int_0^{t_f}\Big(K_{a1}\dot{\theta}_L + K_{a2}D_t^p(\theta_L) + K_{b1}\dot{\varphi}_L + K_{b2}D_t^p(\varphi_L)\Big)\,dt = \frac{1}{t_f}\cdot\int_0^{t_f}K_{a1}\dot{\theta}_L dt + \frac{1}{t_f}\int_0^{t_f}K_{b1}\dot{\varphi}_L dt + \frac{1}{t_f}\cdot\int_0^{t_f}K_{b2}D_t^p(\varphi_L)\,dt + \frac{1}{t_f}\int_0^{t_f}K_{a2}D_t^p(\theta_L)\,dt \tag{11}$$

Equation (11) can be simplified by the following term:

$$\begin{aligned} U_{a1} &= \frac{1}{t_f}\int_0^{t_f}K_{a1}\dot{\theta}_L dt,\\ U_{a2} &= \frac{1}{t_f}\int_0^{t_f}K_{a2}D_t^p(\theta_L)\,dt,\\ U_{b1} &= \frac{1}{t_f}\int_0^{t_f}K_{b1}\dot{\varphi}_L dt,\\ U_{b2} &= \frac{1}{t_f}\int_0^{t_f}K_{b2}D_t^p(\varphi_L)\,dt. \end{aligned} \tag{12}$$

The integration operation of U_{a1}, U_{a2}, U_{b1}, and U_{b2} with the interval $[0, t_f]$ (t_f represents the final time) yields

$$\begin{aligned} U_{a1} &= \frac{1}{t_f}\int_0^{t_f}K_{a1}\dot{\theta}_L dt = \frac{K_{a1}\left(\theta_{Lf}-\theta_{L0}\right)}{t_f},\\ U_{b1} &= \frac{1}{t_f}\int_0^{t_f}K_{b1}\dot{\varphi}_L dt = \frac{K_{b1}\left(\varphi_{Lf}-\varphi_{L0}\right)}{t_f},\\ U_{a2} &= \frac{1}{t_f}\int_0^{t_f}K_{a2}D_t^p(\theta_L)\,dt\\ &= \lim_{h\to 0}\frac{K_{a2}}{t_f}\int_0^{t_f}h^{-p}\sum_{j=0}^{[t/h]}\omega_j^p\dot{\theta}_L(t-jh)\,dt\\ &= \lim_{h\to 0}\frac{K_{a2}}{t_f}h^{-p}\sum_{j=0}^{[t/h]}\omega_j^p\int_0^{t_f}\dot{\theta}_L(t-jh)\,dt\\ &\approx \frac{K_{a2}}{t_f}h^{-p}\sum_{j=0}^{[t/h]}\omega_j^p\left(\theta_{Lf}-\theta_{L0}\right). \end{aligned} \tag{13}$$

The following equation can be obtained according to operation of U_{a2},

$$U_{a2} = \widetilde{K}_{a2}\frac{\left(\theta_{Lf}-\theta_{L0}\right)}{t_f}, \tag{14}$$

where $\widetilde{K}_{a2} = (K_{a2}/t_f)h^{-p}\sum_{j=0}^{[t/h]}\omega_j^p$ and the subscripts 0 and f denote the initial state and final state, respectively.

The process of solving U_{b2} is similar to U_{a2}, so we have $U_{b2} = (\widetilde{K}_{b2}\varphi_{Lf} - \varphi_{L0})/t_f$, where $\widetilde{K}_{b2} = (K_{b2}/t_f)h^{-p}\sum_{j=0}^{[t/h]}\omega_j^p$.

Let $\overline{U}_a$, $\overline{U}_b$ denote the average pitch and lateral load, respectively; they can be expressed as

$$\begin{aligned} \overline{U}_a &= U_{a1} + U_{a2} = \left(K_{a1} + \widetilde{K}_{a2}\right)\frac{\left(\theta_{Lf}-\theta_{L0}\right)}{t_f}\\ \overline{U}_b &= U_{b1} + U_{b2} = \left(K_{b1} + \widetilde{K}_{b2}\right)\frac{\left(\theta_{Lf}-\theta_{L0}\right)}{t_f}. \end{aligned} \tag{15}$$

From the above operation, we get the average overload of EPN in the vertical plane as follows:

$$\begin{aligned} \overline{U} &= \overline{U}_a + \overline{U}_b\\ &= \left(K_{a1} + K_{a2}\right)\frac{\left(\theta_{Lf}-\theta_{L0}\right)}{t_f}\\ &\quad + \left(K_{b1} + K_{b2}\right)\frac{\left(\varphi_{Lf}-\varphi_{L0}\right)}{t_f}. \end{aligned} \tag{16}$$

Consider that the average overload of traditional TPN can be expressed as

$$\overline{U}_{\mathrm{TPN}} = K_{\mathrm{TPN1}}\frac{\left(\theta_{Lf}-\theta_{L0}\right)}{t_f} + K_{\mathrm{TPN1}}\frac{\left(\varphi_{Lf}-\varphi_{L0}\right)}{t_f}, \tag{17}$$

where K_{TPN1} denotes the proportional coefficient of conventional PN. According to (16) and (17), we find that the average overload is closely related to change of LOS angle. The absolute value of average overload will increase as the LOS rotation rate becomes bigger. And calculations demonstrate that the average overload of EPN is closer to PN; it means that the ballistic controller's quantity of EPN is closer to PN.

4.2. Analysis of Required Overload. If we want to keep the trajectory of designed guidance law placidly and reduce the overload requirements of interceptor, the following conditions must be met:

$$\begin{aligned} \left|\theta_L - \theta_{L0}\right| &\longrightarrow \min,\\ \left|\theta_{Lf} - \theta_{L0}\right| &\longrightarrow \min,\\ \left|\theta_{Lf} - \theta_L\right| &\longrightarrow \min,\\ \left|\varphi_L - \varphi_{L0}\right| &\longrightarrow \min,\\ \left|\varphi_{Lf} - \varphi_{L0}\right| &\longrightarrow \min,\\ \left|\varphi_{Lf} - \varphi_L\right| &\longrightarrow \min. \end{aligned} \tag{18}$$

From (8), (9), and (18), the absolute load of EPN meets the following condition:

$$\begin{aligned} \left|U_a^p\right| &\le K_{a1}\left|\dot{\theta}_L\right| + K_{a2}\left|D_t^p(\theta_L)\right|,\\ \left|U_b^p\right| &\le K_{b1}\left|\dot{\varphi}_L\right| + K_{b2}\left|D_t^p(\varphi_L)\right|. \end{aligned} \tag{19}$$

From (19), we conclude that the maximum load of EPN can be denoted as follows:

$$\max\{U_a^p\} \le \max\{K_{a1}\dot{\theta}_L + K_{a2}D_t^p(\theta_L)\},$$
$$\max\{U_b^p\} \le \max\{K_{b1}\dot{\varphi}_L + K_{b2}D_t^p(\varphi_L)\}. \quad (20)$$

Note that, as the variation range of LOS angle is small, the trajectory of missile is closer to the straight line. When the LOS angle approaches zero, the acceleration of EPN $U_a^p \to 0$ and $U_b^p \to 0$. The extreme requirement loads in vertical and horizontal direction can be denoted as $N_{my} = \max\{|U_a|\}/g$ and $N_{mz} = \max\{|U_b|\}/g$. When $U_a^p \to 0$ and $U_b^p \to 0$, we have $N_{my} \to 0$ and $N_{mz} \to 0$. Thus, the proposed guidance EPN guidance law has the features of minimizing energy control and keeping trajectory straightly.

4.3. Characteristics Analysis of Ballistics. Considering the ups and downs of trajectory are closely related to the increment of LOS angle, if the LOS angles are controlled in a small range, the more flat trajectory can be obtained. Using this idea, compare the trajectory characteristic of EPN and PN by LOS angle changes during the process of flying. The lead angle of missile in pitch and lateral direction can be represented as follows:

$$\eta_m = \theta_L - \theta_m, \quad (21)$$

$$\varphi_m = \varphi_L - \varphi_m. \quad (22)$$

Considering the analysis process of trajectory in vertical plane is similar to the horizontal plane, here, we analyze the changes of lead angle in vertical plane only.

Taking the derivation of (21), we have

$$\dot{\eta}_m = \dot{\theta}_L - \dot{\theta}_m. \quad (23)$$

The ballistic inclination rate of missile can be denoted as

$$\dot{\theta}_m = \frac{K_1 V_r \dot{\theta}_L + K_1 V_r p D_t^p(\theta_L)}{V_m} = K_{\theta 1}\dot{\theta}_L + K_{\theta 2} D_t^p(\theta_L). \quad (24)$$

The integration of (23) for $t \in [0, t_f]$ yields

$$\begin{aligned}\eta_{mf} &= \eta_{m0} + (1 - K_{\theta 1})(\theta_{Lf} - \theta_{L0}) - K_{\theta 2}\int_0^{t_f} D_t^p(\dot{\theta}_L)\,dt \\ &= \eta_{m0} + (1 - K_{\theta 1})(\theta_{Lf} - \theta_{L0}) - K_{\theta 2} h^{-p}\sum_{j=0}^{[t/h]} \omega_j^p (\theta_{Lf} - \theta_{L0}).\end{aligned} \quad (25)$$

Let $\widetilde{K}_{\theta 2} = K_{\theta 2} h^{-p}\sum_{j=0}^{[t/h]}\omega_j^p$. Substituting $\widetilde{K}_{\theta 2}$ to (25) results in

$$\eta_{mf} = \eta_{m0} + (1 - K_{\theta 1})(\theta_{Lf} - \theta_{L0}) - \widetilde{K}_{\theta 2}(\theta_{Lf} - \theta_{L0}). \quad (26)$$

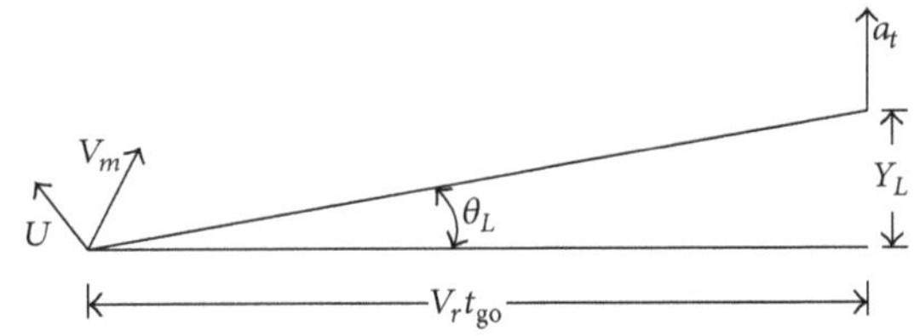

FIGURE 2: Illustration of small offset motion.

When $\eta_{m0} = 0$, the following equations can be obtained based on (26):

$$\theta_{Lf} - \theta_{L0} = \frac{\eta_{mf}}{K_{\theta 1} + \widetilde{K}_{\theta 2} - 1}. \quad (27)$$

For the traditional PN guidance law, the lead angle changes in vertical plane can be denoted as follows:

$$\theta_{Lf} - \theta_{L0} = \frac{\eta_{mf}}{K_{\theta 1} - 1} \quad (28)$$

According to (27) and (28), we find that the introduced fractional order term of EPN further increases the trajectory smoothness by decreasing the variation range of ballistic elevation. So the trajectory of EPN can be controlled easily and the requirements of actuator design reduced greatly. Considering the analysis in horizontal is similar to the vertical, therefore we no longer study it.

4.4. Robustness Analysis of EPN against Autopilot. Figure 2 shows the kinematic relation of small offset.

By using the approximation $\sin\theta_L \approx \theta_L$ (this approximation is rational because Y_L is small offset), we have

$$\theta_L = \frac{Y_L}{V_r t_{go}}, \quad (29)$$

where t_{go} denotes time-to-go; the solution of the differential equation (29) can be obtained as

$$\dot{\theta}_L = Y_L + \frac{\dot{Y}_L t_{go}}{V_r t_{go}^2}. \quad (30)$$

The transform function of autopilot can be expressed as follows:

$$G(s) = \frac{b_m s^m + b_{m-1}s^{m-1} + \cdots + b_1 s + 1}{a_n s^n + a_{n-1}s^{n-1} + \cdots + a_1 s + 1} \quad (n > m) \quad (31)$$

Taking the terminal homing guidance as an example, the guidance loop of small offset model can be denoted as Figure 3. Note that, comparing with traditional PN guidance loop, the EPN increases the fractional order term which is parallel with the PN term in the homing guidance loop.

Let $U_y(s)/Y(s) = G_y(s)$; according to the properties of fractional order, we have

$$L\left[D_t^p f(t)\right] = s^p L[f(t)] - \sum_{k=1}^{n-1} s^k \left[D_t^{p-k-1} f(t)\right]_{t=0}$$

$$\simeq s^{\lambda} \frac{1+st_{\text{go}}}{V_c t_{\text{go}}^2} - s^0 h^{-p} \sum_{j=0}^{[t/h]} \omega_j^{-1} f(0) \tag{32}$$

$$- s^1 h^{-p} \sum_{j=0}^{[t/h]} \omega_j^{p-2} f(0).$$

The transfer function $G_y(s)$ can be obtained as

$$G_y(s) = K_{a1} \frac{1+st_{\text{go}}}{V_r t_{\text{go}}^2} + s^p \frac{1+st_{\text{go}}}{V_r t_{\text{go}}^2}$$

$$+ h^{-p} \sum_{j=0}^{[t/h]} \omega_j^{p-1} \frac{2+st_{\text{go}}}{V_r t_{\text{go}}^3} \tag{33}$$

$$+ h^{-p} \sum_{j=0}^{[t/h]} \omega_j^{p-2} \frac{s+s^2 t_{\text{go}}}{V_r t_{\text{go}}^3}.$$

The guidance loop transfer coefficient of EPN can be expressed as

$$K_{\text{FOCPN}}^y = \lim_{s\to 0} s^2 \frac{1}{s^2} G_y(s) G(s)$$

$$= \frac{1}{V_r t_{\text{go}}^3} \left(K_{a1} t_{\text{go}} + 2\widetilde{K}_{a2}\right), \tag{34}$$

where $\widetilde{K}_{a2} = K_{a2} h^{-p} \sum_{j=0}^{[t/h]} \omega_j^p$.

The guidance loop transfer coefficient of PN and PD (proportional differential) guidance law can be denoted, respectively, as follows:

$$K_{\text{PN}}^y = \frac{K_1 t_{\text{go}}}{V_r t_{\text{go}}^3},$$

$$K_{\text{PD}}^y = \frac{\left(K_{a1} t_{\text{go}} + 2K_{a2}\right)}{V_r t_{\text{go}}^3}. \tag{35}$$

Comparing the transfer coefficients of EPN with PN and PD, we have

$$K_{\text{FOCPN}}^y > K_{\text{PD}}^y > K_{\text{PN}}^y \tag{36}$$

The conclusion of transfer coefficients in horizontal plane is similar to the vertical plane; the relationship can be expressed as

$$K_{\text{FOCPN}}^x > K_{\text{PD}}^x > K_{\text{PN}}^x. \tag{37}$$

According to (36) and (37), we can see the guidance ability against maneuvering target strengthened as a result of increasing the gain coefficients of transfer function by the introduction of fractional order term. So the guidance systematic errors will be reduced and the guidance accuracy will be improved in some sense.

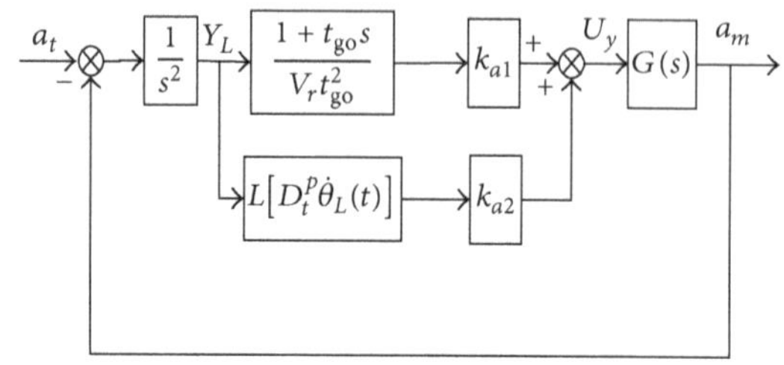

Figure 3: Illustration of the closed loop EPN guidance system.

Definition 2. The sensitivity of systems T against coefficient K is defined as follows:

$$S_K^T = \frac{d(\ln(T))}{d(\ln(K))}. \tag{38}$$

The sensitivity of stability guidance system against the coefficients K_{a1} and K_{a2} can be calculated as follows:

$$S_{K_{a1}}^e = -\frac{K_{a1} t_{\text{go}}}{K_{a1} t_{\text{go}} + 2K_{a1}},$$

$$S_{K_{a2}}^e = -\frac{K_{a2}}{K_{a1} t_{\text{go}} + 2K_{a1}}. \tag{39}$$

When using traditional PN guidance, the sensitivity $S_{K_1}^e = -1$. In contrast with formula (39), we have $|S_{K_1}^e| > |S_{K_{a1}}^e|$, $|S_K^e| > |S_{K_{a2}}^e|$; this means that the sensitivity of EPN guidance against each coefficient (K_{a1}, K_{a2}) is smaller than PN guidance. Thus, the proposed EPN guidance law has a stronger response against parameter changes.

According to the homing guidance system loop in Figure 3, the differential functions of system transmission error can be obtained

$$\frac{dG_e(s)}{dG(s)} = -\frac{s^2 G_1(s)}{\left[s^2 + G_1(s) G(s)\right]^2},$$

$$\frac{dG(s)}{db_i} = \frac{s^i}{P(s)}, \quad i = 1, 2, \ldots, m, \tag{40}$$

$$\frac{dG(s)}{da_j} = -\frac{G(s) s^j}{P(s)}, \quad j = 1, 2, \ldots, n,$$

where $P(s)$ is the intermediate computing term; therefore, according to (40), we obtain

$$S_{b_i}^e = E(s) \frac{b_i s^i}{P(s)}, \quad i = 1, 2, \ldots, m,$$

$$S_{a_j}^e = -E(s) G(s) \frac{a_j s^j}{P(s)}, \quad j = 1, 2, \ldots, n, \tag{41}$$

where $E(s) = -G(s)/(s^2 + G_1(s)G(s))$; as $s = 0$, we have

$$\left.\frac{S_{b_i}^e}{s^i}\right|_{s=0} = -\frac{b_i}{K_{a1} t_{\text{go}} + 2K_1}, \quad i = 1, 2, \ldots, m,$$

$$\left.\frac{S_{a_j}^e}{s^j}\right|_{s=0} = -\frac{a_j}{K_{a1} t_{\text{go}} + 2K_1}, \quad j = 1, 2, \ldots, n,$$

TABLE 1: Engagement specifications.

Parameters of missile and target	
$(X_{m_0}, Y_{m_0}, Z_{m_0})$	(0, 0, 0) km
$(X_{t_0}, Y_{t_0}, Z_{t_0})$	(10, 10, 10) km
θ_t	0
φ_t	180
θ_m	36
φ_m	45
V_m	800
V_t	400

$$\left.\frac{S_{b_i}^e}{s^i}\right|_{s=0}^{PN} = -\frac{b_i}{K_{a1} t_{go}}, \quad i = 1, 2, \ldots, m,$$
$$\left.\frac{S_{a_j}^e}{s^j}\right|_{s=0}^{PN} = -\frac{a_j}{K_{a1} t_{go}}, \quad j = 1, 2, \ldots, n. \tag{42}$$

It can be seen from (42) that the absolute sensitivity of EPN against autopilot parameter decreases because of the increase in denominator by introducing the fractional term. Therefore, the robustness of EPN against autopilot parameter changes is more robust than PN guidance.

5. Simulation Results and Analysis

In this section, some results of simulation are presented for 3D engagement to verify the performance of the designed EPN, simulation initial step is 0.01 s, and the proportional coefficient is taken as 3.5. Some other initial parameters of missile and target are shown in Table 1. The airframe dynamics including autopilot has been modeled as a second-order system with time lag as described by [27]. The total autopilot lag is modeled as

$$T = \frac{\omega_n^2}{(\tau s + 1)(s^2 + 2\zeta\omega_n s + \omega_n^2)}, \tag{43}$$

where the autopilot parameters are $\tau = 0.5$ s, $\xi = 0.6$, and $\omega_n = 18$ rad/s. Due to the order of FOC in the range $[0, 1]$, according to traverse optimization methods, the optimum value of differential order p is taken as 0.6. Consider that the LOS rate is contaminated with uncertainty during the terminal guidance phase, and we assume that the measurement error of the onboard seeker is 0.01 deg/s.

To investigate the performance of the proposed guidance law efficiently, three cases of simulations are carried out: (1) the target with no maneuver; (2) the target maneuver in the inclined plane with the acceleration 40 m/s^2; (3) the target maneuver in a snake-like mode with the acceleration 20 m/s^2 in both the vertical and horizontal planes.

The comparison of the interception performances between the EPN guidance law and PN guidance law is presented in Table 2, where MD is the mean miss distance, defined as the closest distance between the missile and the target before divergence, and where ET denotes the engagement time. All performance results are based on 200 run Monte-Carlo simulations.

TABLE 2: Comparison of the interception performance.

	Guidance law	MD/m	ET/s
Case 1	PN	1.2129	35.55
	EPN	0.6272	35.05
Case 2	PN	3.9193	33.34
	EPN	1.978	32.89
Case 3	PN	2.3398	40.55
	EPN	1.178	40.32

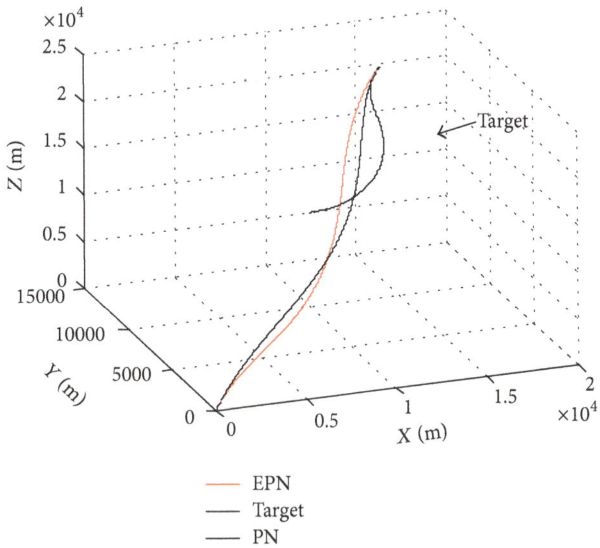

FIGURE 4: Trajectories of missile and target.

Small miss distance is the most important in the homing guidance. According to the simulation results shown in Table 2, both of the conventional PN and the proposed EPN guidance law are able to intercept targets effectively. However, the guidance precision of EPN is nearly 50% better than the PN guidance, which indicates that the EPN has greatly improved the guidance accuracy. It is also interesting to note that the engagement time of EPN is always shorter than PN, which means that the EPN will shot down the target earlier than PN.

In order to describe the guidance character of acceleration and ballistic trajectory, we give the simulation figures of case 3 only, which are shown as Figures 4–8.

As can be seen from Figure 4, the trajectory of the EPN is smoother than the PN. The curvature of the missile's trajectory guided by EPN is smaller than PN, especially in the end of engagement, which is in favor of ballistic control. As illustrated in Figures 5 and 6, the main difference between EPN and PN is the adjustment rate of elevation and azimuth. In the initial phase of engagement, the proposed EPN adjusts its attitude quicker than PN and this makes the trajectory of EPN smoother in the later stage of interception. Figures 7 and 8 denote that the acceleration of the PN guidance law tends

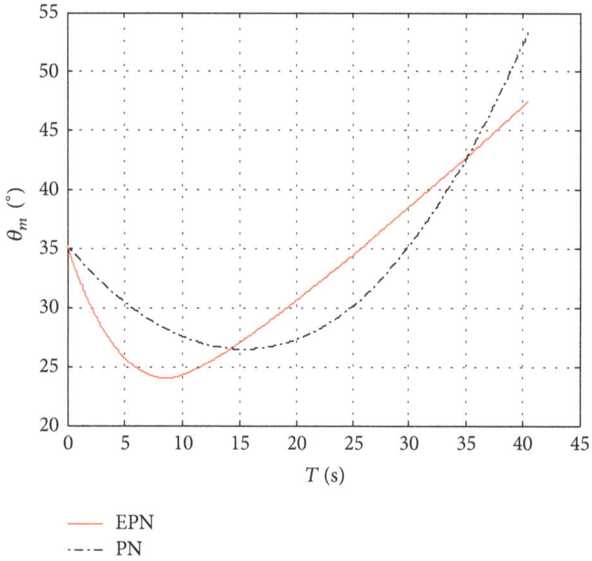

FIGURE 5: Time history of trajectory elevation.

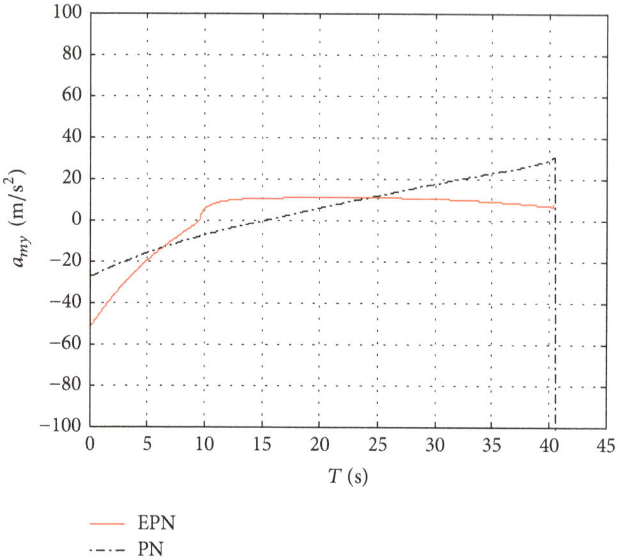

FIGURE 7: Time history of normal acceleration.

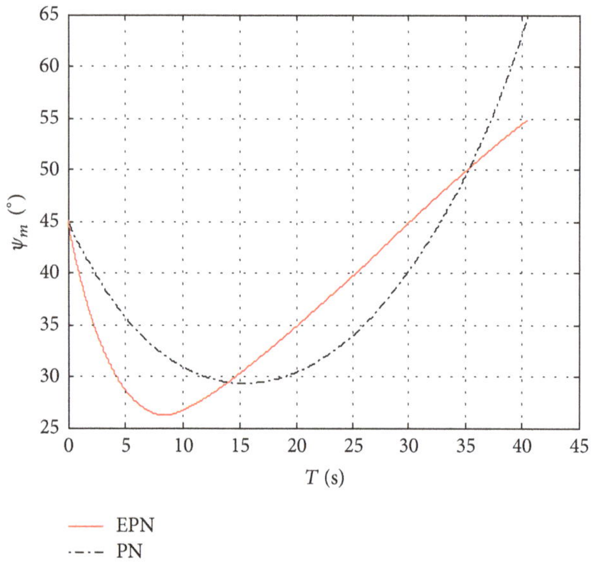

FIGURE 6: Time history of trajectory azimuth.

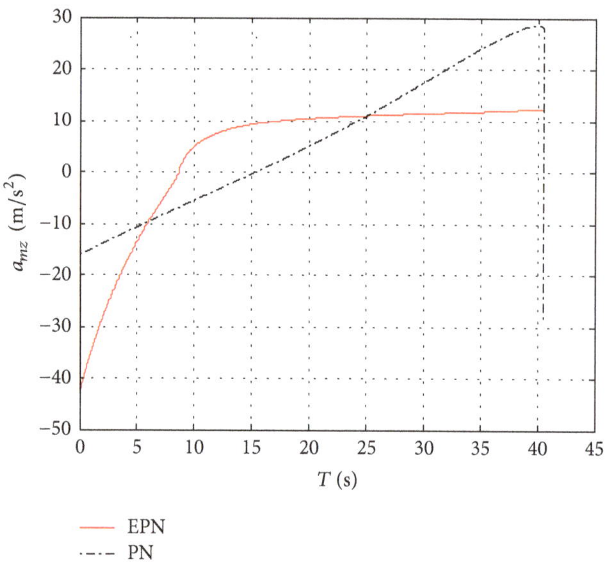

FIGURE 8: Time history of lateral acceleration.

to achieve its saturation which compensates for the target's maneuver at the end, whereas the loads of EPN undergoes a relatively stable tendency during the engagement, which positively adjusts the missile attitude at the expense of loads in the initial phase of interception without dramatic increase at the end. So the load of EPN undergoes a relatively stable tendency during the flight envelope, while the load of PN diverges at the point of impact caused by the divergence of LOS rate.

With respect to the PN guidance law, the proposed EPN guidance law maintains the tracking performance and exhibit more favorable guidance performance in intercepting maneuvering target. In fact, EPN solves the problem of LOS rate estimation by the fractional order differential treatment of LOS rate. In other words, the fractional order differential controller plays the role of filter, which improves the control precision and stability of the guidance system. In a word, the EPN extends the guidance performance of missile by carrying forward the advantages of PN guidance law and avoids its shortcomings in some sense.

6. Conclusions

To improve the performance of conventional PN guidance, this paper proposed a novel EPN guidance law based on the fractional order calculus theory. The guidance characteristics of EPN, such as average load, trajectory stability, and robustness, are analyzed theoretical. The inducement of fractional order term extended the performance of PN greatly especially in the guidance precision and interception time and overcome the load saturation caused by LOS rate divergence in the end of engagement. And the EPN guidance law has more flexible structure and better robustness. Extensive simulations of various engagements demonstrate that the proposed guidance law provides satisfactory performance against maneuvering targets.

The proposed guidance law is designed with the consideration of maneuvering target in the atmosphere; however, the guidance law needs further analysis, especially to achieve direct collision interception in the near space against hypersonic target.

Competing Interests

The authors declare that there is no conflict of interests regarding the publication of this paper.

Acknowledgments

The first author would like to thank the Air and Missile Defense College of Air Force Engineering University of China for giving permission for conducting guidance research at the laboratory. This work was supported by National Natural Science Foundation of China (nos. 61573374, 61503408, and 61603410) and Aviation Science Innovation Foundation of China (no. 20150196006).

References

[1] L. Chen and B. Zhang, "Novel TPN control algorithm for exoatmospheric intercept," *Journal of Systems Engineering and Electronics*, vol. 20, no. 6, pp. 1290–1295, 2009.

[2] Q. Z. Zhang, Z. B. Wang, and F. Tao, "Optimal guidance law design for impact with terminal angle of attack constraint," *Optik*, vol. 125, no. 1, pp. 243–251, 2014.

[3] K. Li, L. Chen, and X. Bai, "Differential geometric modeling of guidance problem for interceptors," *Science China Technological Sciences*, vol. 54, no. 9, pp. 2283–2295, 2011.

[4] J. Ye, H. Lei, D. Xue, J. Li, and L. Shao, "Nonlinear differential geometric guidance for maneuvering target," *Journal of Systems Engineering and Electronics*, vol. 23, no. 5, pp. 752–760, 2012.

[5] F. Tyan, "Unified approach to missile guidance laws: a 3D extension," *IEEE Transactions on Aerospace and Electronic Systems*, vol. 41, no. 4, pp. 1178–1199, 2005.

[6] K. Li, T. Zhang, and L. Chen, "Ideal proportional navigation for exoatmospheric interception," *Chinese Journal of Aeronautics*, vol. 26, no. 4, pp. 976–985, 2013.

[7] S.-C. Han, H. Bang, and C.-S. Yoo, "Proportional navigation-based collision avoidance for UAVs," *International Journal of Control, Automation and Systems*, vol. 7, no. 4, pp. 553–565, 2009.

[8] C.-Y. Li and W.-X. Jing, "Geometric approach to capture analysis of PN guidance law," *Aerospace Science and Technology*, vol. 12, no. 2, pp. 177–183, 2008.

[9] Y. Xia and M. Fu, "Missile guidance laws based on SMC and FTC techniques," in *Compound Control Methodology for Flight Vehicles*, vol. 438 of *Lecture Notes in Control and Information Sciences*, pp. 211–224, Springer, Berlin, Heidelberg, 2013.

[10] N. Hananjay, K.-Y. Lum, and J.-X. Xu, "Proportional navigation with delayed line-of-sight rate," *IEEE Transactions on Control Systems Technology*, vol. 21, no. 1, pp. 247–253, 2013.

[11] S. Ghosh, D. Ghose, and S. Raha, "Three dimensional PN based impact angle control for higher speed nonmaneuvering targets," in *Proceedings of the American Control Conference (ACC '13)*, pp. 31–36, Washington, DC, USA, June 2013.

[12] I.-S. Jeon and J.-I. Lee, "Optimality of proportional navigation based on nonlinear formulation," *IEEE Transactions on Aerospace and Electronic Systems*, vol. 46, no. 4, pp. 2051–2055, 2010.

[13] R. Caponetto, G. Dongola, and L. Fortuna, *Fractional Order Systems: Modeling and Control Applications*, World Scientific, Singapore, 2010.

[14] Q. Zeng, G. Cao, and X. Zhu, "Research on controllability for a class of fractional-order linear control systems," *Journal of Systems Engineering and Electronics*, vol. 16, no. 2, pp. 376–381, 2005.

[15] I. P. Cristina, I. Clara, D. K. Robain, and E. H. Dulf, "Robustness evaluation of fractional order control for varying time delay processes," *Signal, Image and Video Processing*, vol. 6, no. 3, pp. 453–461, 2012.

[16] K. S. Miller and B. Ross, *An Introduction to the Fractional Calculus and Fractional Differential Equations*, A Wiley-Interscience Publication, John Wiley & Sons, New York, NY, USA, 1993.

[17] S. G. Smako, A. A. Kilbas, and O. I. Marichev, *Fractional Integrals and Derivitives, Theory and Applications*, Gorden and Breach Science Publishers, New York, NY, USA, 1987.

[18] X.-X. Zhang, T.-S. Qiu, and H. Sheng, "A physical interpretation of fractional calculus and fractional calculus with constant extent of integral," *Acta Electronica Sinica*, vol. 41, no. 3, pp. 508–512, 2013.

[19] A. A. Kilbas, H. M. Srivastava, and J. J. Trujillo, *Theory and applications of fractional differential equations*, vol. 204 of *North-Holland Mathematics Studies*, Elsevier Science B.V., Amsterdam, Netherlands, 2006.

[20] G. Sun and Z. H. Zhu, "Fractional-order tension control law for deployment of space tether system," *Journal of Guidance, Control, and Dynamics*, vol. 37, no. 6, pp. 2057–2061, 2014.

[21] R. Almeida and D. F. M. Torres, "A discrete method to solve fractional optimal control problems," *Nonlinear Dynamics*, vol. 80, no. 4, pp. 1811–1816, 2015.

[22] K. Vishal, S. K. Agrawal, and S. Das, "Hyperchaos control and adaptive synchronization with uncertain parameter for fractional-order Mathieu-van der Pol systems," *Pramana*, vol. 86, no. 1, pp. 59–75, 2016.

[23] J.-G. Shi, Z.-Y. Wang, S.-J. Chang, and B.-S. Zhang, "Fractional-order control systems for guided projectiles," *Journal of Nanjing University of Science and Technology (Natural Science)*, vol. 35, no. 1, pp. 52–56, 2011.

[24] T.-Z. Li, Y. Wang, and M.-K. Luo, "Control of fractional chaotic and hyperchaotic systems based on a fractional order controller," *Chinese Physics B*, vol. 23, no. 8, Article ID 080501, 2014.

[25] F. Wang and H. M. Lei, "PD^{λ} guidance law based on fractional calculus," *Control Theory & Applications*, vol. 27, no. 1, pp. 126–130, 2010.

[26] Z.-T. Zhu, Z. Liao, C. Peng, and Y. Wang, "A fractional-order modified proportional navigation law," *Control Theory & Applications*, vol. 29, no. 7, pp. 945–949, 2012.

[27] S. Rituraj, N. Prabhakar, A. K. Sarkar et al., "Three dimensional nonlinear inverse dynamics guidance law for parallel navigation," in *Proceedings of the AIAA Guidance, Navigation, and Control Conference and Exhibit*, pp. 1–8, Providence, RI, USA, 2004.

9

Research on the Reliability Analysis of the Integrated Modular Avionics System based on the AADL Error Model

Peng Wang,[1,2] **Changxiao Zhao**,[1,3] **and Fang Yan**[1]

[1]*Civil Aircraft Airworthiness and Repair Key Laboratory of Tianjin, Civil Aviation University of China, Tianjin 300300, China*
[2]*Key Laboratory of Civil Aircraft Airworthiness Technology, Civil Aviation Administration of China, Tianjin 300300, China*
[3]*School of Airworthiness, Civil Aviation University of China, Tianjin 300300, China*

Correspondence should be addressed to Changxiao Zhao; cxzhao@cauc.edu.cn

Academic Editor: Kenneth M. Sobel

In recent years, the integrated modular avionics (IMA) concept has been introduced to replace the traditional federated avionics. Different avionics functions are hosted in a shared IMA platform, and IMA adopts partition technologies to provide a logical isolation among different functions. The IMA architecture can provide more sophisticated and powerful avionics functionality; meanwhile, the failure propagation patterns in IMA are more complex. The feature of resource sharing introduces some unintended interconnections among different functions, which makes the failure propagation modes more complex. Therefore, this paper proposes an architecture analysis and design language- (AADL-) based method to establish the reliability model of IMA platform. The single software and hardware error behavior in IMA system is modeled. The corresponding AADL error model of failure propagation among components, between software and hardware, is given. Finally, the display function of IMA platform is taken as an example to illustrate the effectiveness of the proposed method.

1. Introduction

As an important development direction of future large aircraft avionics system, IMA completes the real-time processing and information exchange task in navigation, communication, monitoring, and flight management through the integrated technology of avionics system [1], to ensure the flight safety effectively.

IMA is a safety critical system of civil aircraft. It uses resource sharing, data fusion, and restoration reconfiguration technology, which makes IMA highly complex and brings great challenges to IMA safety and reliability assessment. In addition, the failure propagation mechanism of IMA is so complex that the traditional reliability analysis method is not applicable to solve the problems of IMA architecture reliability assessment. Therefore, it is of great significance to study the reliability assessment method and to complete the work of IMA architecture reliability assessment.

To solve these problems, this paper introduces the reliability analysis technology based on AADL error model [2]. The AADL error model is bound to the IMA platform architecture on the basis of the error behavior and error propagation modeling of IMA platform to complete the reliability modeling. We take the display function as an example to make a comparison and analysis on different IMA architectures of its availability to draw conclusions on quantitative analysis.

2. AADL Error Model Overview

AADL uses the architecture design and analysis language, which supports the text or graphical modeling in three levels of hardware, software, and system [3]. It is used to complete the description and analysis of software and hardware in the real-time system and has been widely used in the flight control system and avionics system. In order to further extend the description ability of AADL language, SAE issued the AS5506/1 in November 2006, which includes the error model and the behavior model annex, so that the AADL has its unique syntax and semantics. For the error behavior of IMA platform, error model annex is analyzed in detail.

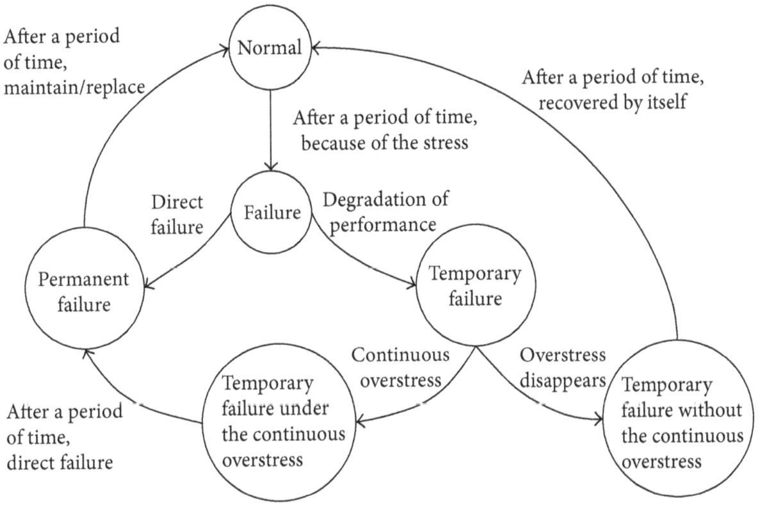

FIGURE 1: Transition behavior of IMA hardware error states.

The error model annex [4] is a state machine that can be associated with an AADL component or connection, to describe system errors, error behavior, and error propagation [5]. Including

(1) error model type: describing a set of error states, error events, and error propagation; specifying the probability of occurrence of error events and error propagation by property;

(2) error model implementation: a more detailed error model is defined, including the transition between error states and the occurrence of random events, which are used for further quantitative analysis [6].

In summary, the principle of error model can be expressed under certain conditions; components will trigger the error events due to its failure, and the error events will trigger the transition between the related error states; at the same time, the running states between the components that interact with each other will influence mutually due to the existence of the error propagation.

3. Implementation of IMA Error Model

3.1. Error Behavior Modeling of IMA Single Component

3.1.1. Error Behavior Modeling of Single Hardware. General processing modules of IMA platform can be divided into two parts: hardware and software; the software performs the related functions by loading itself into the hardware. First, considering only the single hardware error behavior model, we assume that the hardware of the general processing module is in the normal working state at the initial time; after running for some time, the hardware may fail because of the stress, such as temperature stress and electromagnetic stress. The hardware may be able to self-recover after the external stress disappears, that is, the failure lasts only a short time and only to reduce the hardware performance, which is called temporary failure [7]; however, if the external stress continues to be maintained, after a certain period of time, it will lead to permanent failure of the hardware, and it cannot recover anymore. This transition behavior of hardware error states is shown in Figure 1.

The AADL annex sublanguage is used to describe the above hardware error behavior, as shown in Figure 2. The hardware error model is divided into two parts, which are hardware error model type and error model implementation. The error model type contains 6 error states (including initial state) and 8 error events; the hardware error model implementation contains all of the transition rules between the error states, and the time occurrence or probability of the error events are described. The corresponding model meaning is shown in Figure 2 notes.

3.1.2. Error Behavior Modeling of Single Software. We assume that the software of the general processing module is in normal state at the initial time; after a period of operation, the software will be in failure due to the specific input (for the inevitable defects in the design of the complex software, the harsh design assurance level can only reduce the number of defects). Because of the fault-tolerance design of IMA software, the software will not lose its function immediately, and IMA will detect the failure and determine the failure type. If the failure is temporary, it needs to be further divided and determined whether the failure can be removed; if the failure can be removed, after a period of processing, the software can be restored to the normal state; if the failure cannot be removed, after a long time, this temporary failure will become permanent failure, and the software can be restored only through restarting operation. Figure 3 describes the transition rules between software error states.

The AADL annex sublanguage is used to describe the above software error behavior, as shown in Figure 4. The software error model is also divided into two parts, which are software error model type and error model implementation. The error model type contains 7 error states (including initial state) and 9 error events; the software error model implementation contains all of the transition rules between the error states, and the time occurrence or probability of the error

```
error model for hardware//error model type
  features//error mode features
  errorfree:initial error state;//initial state
  failed:error state;//failure state
  termporary_failed:error state;//failure state
  continual_temporary_failed:error state;//failure state
  dissappeared_temporary_failed:error state;//failure state
  overstressed:error event;//failure state
  direct_damage:error event;//failure state
  performance_degradation:error event;/failure state
  continual_overstess:error event;//failure state
  disappeared_overstess:error event;//failure state
  time_direct_damage:error event;failure state
  repaire:error event;//failure state
  recovery:error event;//failure state
end for hardware;//error model type end

error model implementation for hardware.general// error model implementation
  transitions//states transitions explanation
  errorfree-[overstress]->failed;
  failed-[direct_damage]->permanent_failed;
  failed-[performance_degradation]->temporary_failed;
  permanent_failed-[repaire]->errorfree;
  temporary_failed-[continual_overstress]->continual_temporary_failed;
  temporary_failed-[disappeared_overstress]->disappeared_temporary_failed;
  continual_temporary_failed-[time_direct_damage]->permanent_failed;
  disappeared_temporary_failed-[recovery]->errorfree;
  properties//random events properties explaination
  occurence=>poisson os applies to overstress;
  occurence=>fixed dd applieds to direct_damage;
  occurence=>fixed 1-dd applies to performance_degradation;
  occurence=>fixed co applies to continual_overstress;
  occurence=>fixed 1-co applies to disappeared_overstress;
  occurence=>poisson tdd applies to time_direct_damage;
  occurence=>poisson lambda applies to repair;
  occurence=>poisson miui applies to recovery;
end for hardware.general;//error model implementation end
```

Figure 2: Single hardware error model.

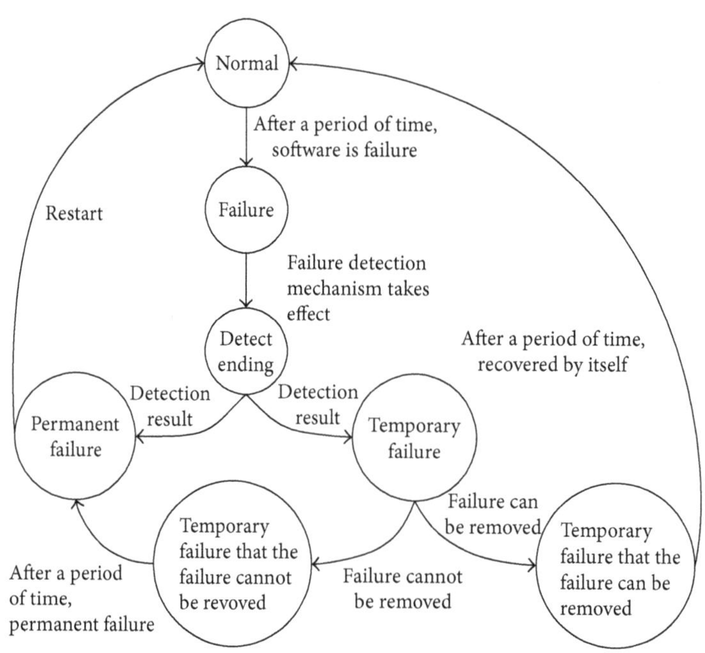

Figure 3: Transition behavior of IMA software error states.

```
error model for software//error model type
  features//error mode features
  errorfree:initial error state;//initial state
  failed:error state;//failure state
  detection_end:error state;//failure state
  permanent_failed:error state;//failure state
  temporary_failed:error state;//failure state
  unremovable_temporary_failed:error state;//failure state
  removable_temporary_failed:error state;//failure state
  fail:error event;//failure state
  detection:error event;//failure state
  direct_damage:error event;//failure state
  performance_degradation:error event;//failure state
  unremovable:error event;//failure state
  removable:error event;//failure state
  time_damage:error event;//failure state
  restart:error event;//failure state
  recovery:error event;//failure state
end for software;//error model type end

error model implementation for software.general// error model implementation
  transitions//states transitions explaination
  errorfree-[fail]->failed;
  failed-[detection]->detection_end;
  detection_end-[direct_damage]->permanent_failed;
  detection_end-[performance_degradation]->temporary_failed;
  permanent_failed-[restart]->errorfree;
  temporary_failed-[unremovable]->unremovable_temporary_failed;
  temporary_failed-[removable]->removable_temporary_failed;
  unremovable_temporary_failed-[time_damage]->permanent_failed;
  removable_temporary_failed-[recovery]->errorfree;
  properties//random events properties explaination
  occurrence=>poisson os applies to fail;
  occurrence=>poisson os applies to detection;
  occurrence=>fixed dd applieds to direct_damage;
  occurrence=>fixed 1-dd applies to performance_degradation;
  occurrence=>fixed phi applies to unremovable;
  occurrence=>fixed 1-phi applies to removable;
  occurrence=>poisson mu applies to restart;
  occurrence=>poisson td applies to time_damage;
  occurrence=>poisson theta applies to recovery;
end for software.general;//error model implementation end
```

FIGURE 4: Single software error model.

events are described. The corresponding model meaning is shown in Figure 4 notes.

3.2. Error Propagation Modeling of IMA

3.2.1. Internal Hardware/Software Error Propagation Modeling for Individual Components. The software runs on the hardware, so the hardware failure will affect the normal operation of the software. It specifically includes three cases:

(1) The stress on hardware can disappear, and the hardware is in temporary failure; it depends on whether the software is in a normal state. If the software is in normal state, after a period of operation, the software will be in failure under the influence of the hardware failure; if the software itself is in failure now, after a period of operation, the software may be in permanent failure or temporary failure because of the hardware failure, which depends entirely on the hardware failure type.

(2) Hardware is in permanent failure, which will lead the software to stop immediately, and now the software can be recovered only by restarting after the hardware is being back to normal.

(3) As for the impact of software on hardware, we only consider the permanent failure of software that will occur, which will lead the hardware to transfer from the normal state to a temporary failure state where the stress can disappear. In other cases, the software has no effect on the hardware.

The interaction between hardware and software [8] is shown in Figure 5. The AADL annex sublanguage is used to model the above error output and input behavior of software, as shown in Figures 6 and 7.

3.2.2. Error Propagation Modeling between IMA Components. IMA uses the partition method to divide each application software or function into a single partition, so that each partition has its own storage space and time gap and

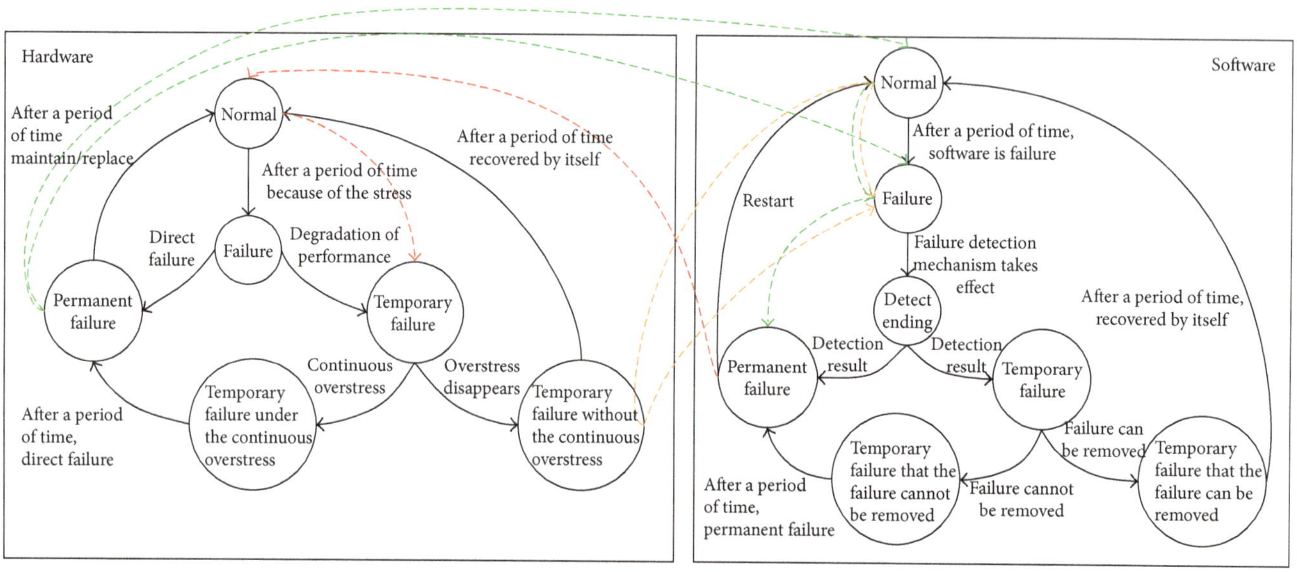

Figure 5: Interaction between software and hardware.

```
error model for software//error model type
  features//error mode features
  [...]
  (+)temporary_failed:error state;//failure state
  (+)reboot_state:error state;//failure state
  [...]
  (+)to_permanent_failed:error state;//failure state
  (+)to_temporary_failed:error state;//failure state
  (+)sw_permanent_failed_outflow:output error propagation;// output failure
  (+)disapeared_temporary_failed_outflow:input error propagation;// input failure
  (+)hw_permanent_failed_outflow:input error propagation;// input failure
  (+)reboot:input error propagation;// input failure
end for software;//error model type end
error model implementation for software.general//error model implementation
  transitions//states transitions explaination
  [...]
  (+)permanent_failed-[out_sw_permanent_failed_outflow]->permanent_failed;// output failure and state transfer
  (+)errorfree-[in disappeared_temporary_failed_flow]->failed;//input failure and state transfer
  (+)failed-[in disappeared_temporary_failed_outflow]->temporary_failed_sw;//input failure and state transfer
  (+)temporary_failed_sw-[in to_permanet_failed]->permanent_failed;//input failure and state transfer
  (+)temporary_failed_sw-[in to_temporary_failed]->temporary_failed;//input failure and state transfer
  (+)errorfree-[in hw_permanent_failed_outflow]->reboot_state;/ input failure and state transfer
  (+)failed-[in hw_permanent_failed_outflow]->reboot_state;// input failure and state transfer
  (+)permanent_failed-[in hw_permanent_failed_outflow]->reboot_state;// input failure and state transfer
  (+)unremovable_temporary_failed-[in hw_permanent_failed_outflow]->reboot_state;// input failure and state transfer
  (+)removable_temporary_failed-[in hw_permanent_failed_outflow]->reboot_state;/ /input failure and state transfer
  (+)reboot_state-[in reboot]->errorfree;//input failure and state transfer
  properties//random events properties explaination
  [...]
  (+)occurence=>poisson spfo applies to sw_permanent_failed_outflow;
  (+)occurence=>fixed 1 applieds to disappeared_temporary_failed_outflow;
  (+)occurence=>fixed tpf applies to permanent_failed;
  (+)occurence=>fixed 1-tpf applies to temporary_failed;
  (+)occurence=>fixed 1 applies to hw_permanent_failed_outflow;
  (+)occurence=>fixed 1 applies to reboot;
end for software general;//error model implementation end
```

Figure 6: Error output and input model of software.

ensures that they do not have the influence on each other. However, because of the needs for communication between applications, the errors may be transferred across different partitions during the data transmission and may damage the partition [9].

All of the message communication between partitions are completed through channels, that is, one message source transfers from a partition to one or more destination partitions through channels. The channel defines a logical connection from one source to another or multiple destinations, and the source and destination may belong to different partitions. In the initial system configuration, we need to define the message type, port, and channel. The error propagation simplified model of IMA is shown in Figure 8.

```
error model for hardware//error model type
  features//error mode features
  [...]
  (+)disappeared_temporary_failed_ovverflow:output error propagation;// output failure
  (+)hw_permanent_failed_outflow:output error propagation;// output failure
  (+)reboot:output error propagation;// output failure
  (+)sw_permanent_failed_outflow:input error propagation;// input failure
end for software;//error model type end
error model implementation for hardware.general// error model implementation
  transitions//states transitions explaination
  [...]
  (+)errorfree-[in sw_permanent_failed_outflow]->disappeared_temporary_failed;
  //input failure and state transfer
  (+)disappeared_temporary_failed-[out disappeared_temporary_failed_outflow]->disappeared_temporary_failed;
  //input failure and state transfer
  (+)permanent_failed-[out hw_permanent_failed_outflow]->permanent_failed;
  //output failure and state transfer
  (+)errorfree-[out reboot]->errorfree;
  //output failure and state transfer
  properties//random events properties explaination
  [...]
  (+)occurence=>fixed dtfo applieds to disappeared_temporary_failed_outflow;
  (+)occurence=>fixed 1 applies to hw_permanent_failed_outflow;
  (+)occurence=>poisson r applies to reboot;
  (+)occurence=>fixed 1 applies to sw_permanent_failed_outflow;
end for hardware.general;//error model implementation end
```

FIGURE 7: Error output and input model of hardware.

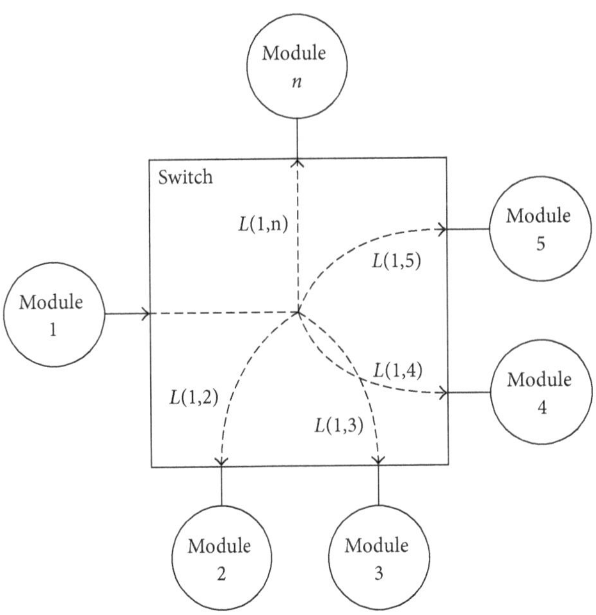

FIGURE 8: Error propagation simplified model of IMA.

After the module 1 fails, IMA will deal with it with the corresponding failure handling mechanism (such as failure detection and failure limit). We consider two factors: on the one hand, we assume that the module has 100% failure detection capability, namely, the failure must be detected after it occurred, but it does not have 100% failure limit capability, namely, the error may be transferred through the established channel after it happened; on the other hand, we consider the correctness of the channel configuration, for example, when module 1 transfers the data to module 2, the connection L(1,2) should be connected; however, because of the incorrect configuration of the switch configuration table resulting in that the connection L(1,3) and/or L(1,4) and/or L(1,n) are connected, while the connection L(1,2) is unconnected. Thus, considering these two factors, the error propagation probability from module m to module n is

$$P(m,n) = [1 - r(m)] \times L(m,n). \quad (1)$$

In the above formula, $r(m)$ is the failure limit capability of module m, that is, the ability that the module can limit the error to the area without transferring it out. $0 \le r(m) \le 1$; when $r(m) = 0$, the error must be transferred out; when $r(m) = 1$, the error cannot be transferred out. $L(m, n)$ is the probability of connecting the channel from module m to n; obviously, $0 \le L(m,n) \le 1$. Generally speaking, the reliability of AFDX network is so high that the value of $L(m, n)$ will be close to 1, when the data is transferred from module m to n.

Based on the above analysis, modeling the error propagation between IMA components, here, we only show the error transition from one component to another, and the rest is analogous. The error propagation behavior is shown in Figure 9.

The AADL error model annex language is used to describe the error propagation behavior between two components, as shown in Figures 10 and 11. Component A and component B describe their error output and input by matching the out-in names.

4. Verification Case of Safety Analysis on Display Function of IMA

The display function of IMA platform is taken as an example to illustrate the effectiveness of the proposed method. Firstly, the system architecture and error model

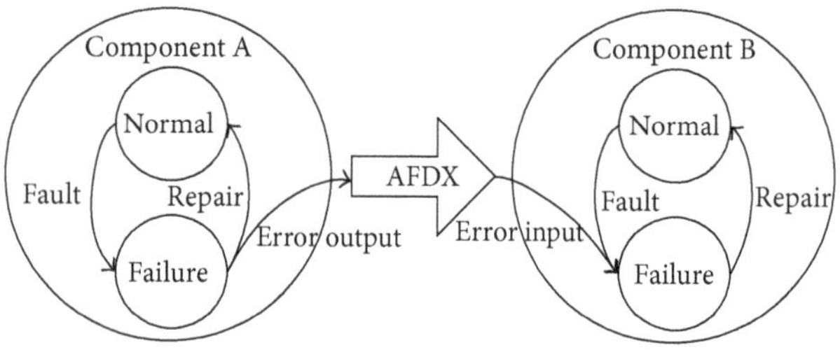

FIGURE 9: Error propagation behavior between components.

```
error model A//error model type
  features//error mode features
  [...]
  (+)error_propagation:output error propagation:// output failure
end A://error model type end

error model implementation A.general// error model implementation
  transitions//states transitions explanation
  [...]
  (+)failed-[output error_propogation]-failed;// output and transfer
  properties//random events properties explaination
  (+)occurrence=>fixed ep applieds to error_propagation;
end A.general;//error model implementation end
```

FIGURE 10: Error output model of component A.

```
error model B//error model type
  features//error mode features
  [...]
  (+)error_propagation:input error propagation;// input failure
end B;//error model type end

error model implementation B.general// error model implementation
  transitions//states transitions explanation
  [...]
  (+)errorfree-[input error_propogation]-failed:// input and transfer
  properties//random events properties explaination
  (+)occurrence=>fixed 1 applieds to error_propagation;
end B.general;//error model implementation end
```

FIGURE 11: Error input model of component B.

is constructed by AADL, and then the AADL model is transferred to GSPN model according to certain grammatical rules [10]. This method is shown in Figure 12. Finally, the availability of different architectures for display function is compared.

4.1. Reliability Modeling of Display Function. With the improvement of aircraft performance, the display control system of aircraft is gradually integrated. Cockpit display technology has undergone a long process of development from the mechanical to electronic, from dedicated devices to shared resources, and from the separated to the integrated. The typical architecture of basic integrated cockpit display system [11] is shown in Figure 13.

Sensors are distributed in various parts of the aircraft, sensing the measured information and converting the information into output signals according to certain rules, and then outputting these signals; data processing module is used to determine the display working mode and components of

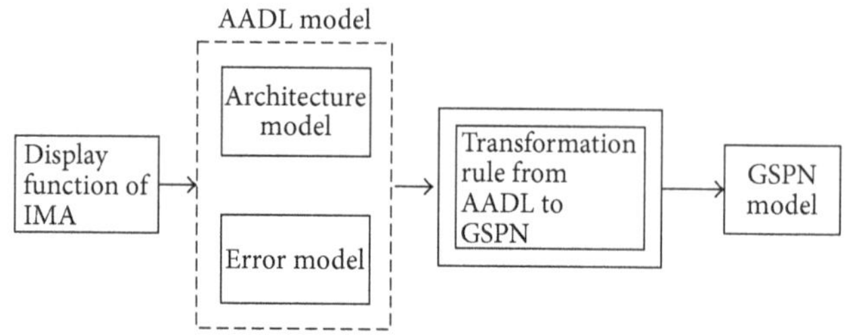

FIGURE 12: The experiment method.

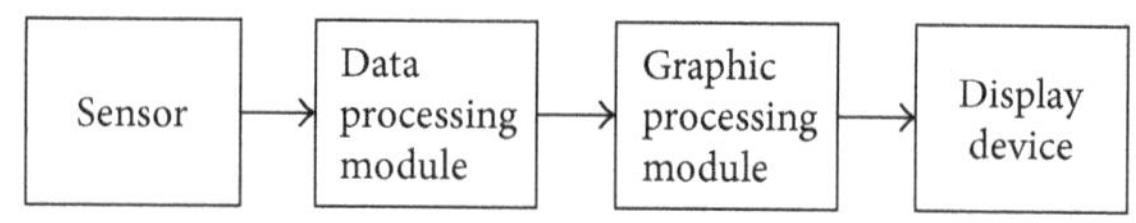

FIGURE 13: Architecture of basic integrated cockpit display system.

display function, changing the data information into graphic data and instructions, to display the control task scheduling; graphic processing module is used to construct characters and graphics; display devices include various displays and support units for displaying the flight information, engine working states, and so on.

Based on the above architecture, in fact, the integrated cockpit display system realizes the fault-tolerant property through many display devices, a lot of graphic processing modules and backup sensors. The typical architecture of it is shown in Figure 14.

There is a set of sensors on each side of the aircraft to collect the flight data, and they are backups to each other. As long as a set of sensors can work normally, data processing modules can receive the normal data. Then, data will be transferred to two graphic processing modules after being processed by the data processing module, and the two graphic processing modules can adopt the cold/hot backup configuration. After being processed by the graphic processing modules, images will be shown on the display. From the overall architecture, the same software and hardware configuration is used in both left and right display function, but tasks are different. They are working independently and backup to each other; once one side fails, the other side will take over its mission, thereby improving the reliability of the aircraft [12].

On the basis of mastering the architecture of IMA display function, the error models of basic display function and bilateral fault-tolerant display function are established, respectively, by using the AADL. Simplify every component according to the modeling method which is described in Section 3.2.2, that is, only consider the normal and failure states, regardless of the remaining error states. The error model of the components is bound to the architecture model to construct the AADL reliability model of the display function. The graphical representation of AADL reliability models is shown in Figures 15 and 16.

In Figure 15, when sensor S fails, the error will transfer along the data chain; thus, the data processing module will be influenced; once the data processing module fails, no matter how it fails (it fails by itself or affected by the failure of the sensor), the error will transfer along the data chain

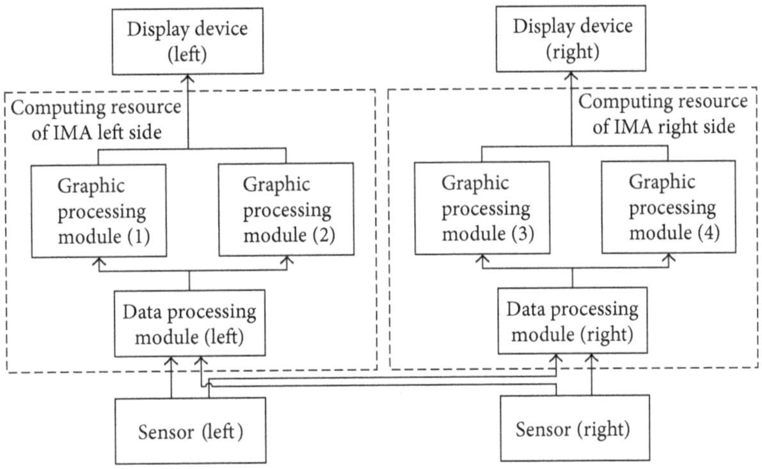

Figure 14: Architecture of bilateral tolerant display function.

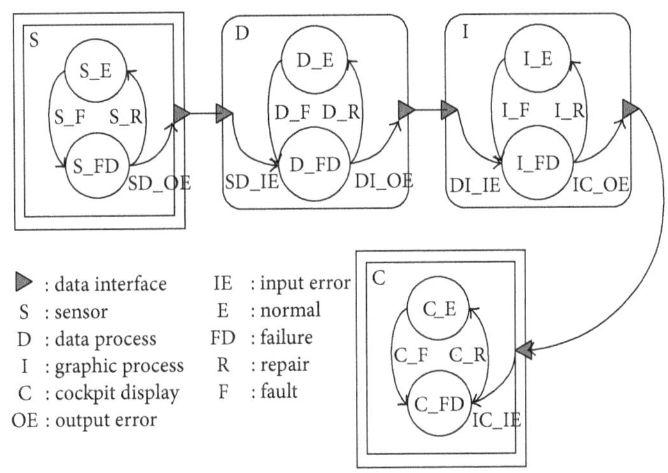

Figure 15: AADL reliability model of basic cockpit display system.

continuously and will make influence on graphic processing module; similarly, the graphic processing module and display device will be influenced by the transferred errors; finally, the cockpit display function will be influenced.

In Figure 16, sensors “S_1” and “S_2” are backups to each other; they are both in working mode and both make output to the system; only when both of them fail, the error will transfer along the data chain, then affecting the data processing module; the graphic processing module uses a hot backup architecture, namely, “I_1” and “I_3” are in working mode and make output to the system in initial conditions and “I_2” and “I_4” are also in working mode, but they do not make output to the system; once “I_1” or “I_3” fails, graphic processing module will change to hot backup mode immediately, and “I_2” or “I_4” will make output to the system. Therefore, only when both of two graphic processing modules fail, the error will transfer along the data chain and affect the cockpit display device.

4.2. Comparative Analysis of Availability of Different Architectures for Display Function. The design of IMA is an iterative process. As a necessary part of this process, the safety assessment is also an iterative process [13]. We should consider not only the cost but also the availability, so as to find a balance between them. The availability of the following ten cases is studied in detail for both the basic and the bilateral fault-tolerant display function architecture, as shown in Tables 1 and 2.

On the basis of establishing the AADL reliability model, we use the GSPN (general stochastic Petri net) [14] to analyze and compute the availability of the display function under the steady states. The availability is the proportion of time that a system is in a functioning condition. Firstly, we need to set the changeable parameters. Set failure proba bility of the display system components to 2E-3 per flight hour and complete repair once an hour and then further analyze all of the existent states and capture the availability of display function under the steady states. Results can be seen in Tables 3 and 4.

We can get the following conclusions from the above results: the availability of bilateral tolerant architectures is higher than that of the basic architectures under the steady states. When sensors are in the same architecture, the steady-state availability of the hot backup of graphic processing

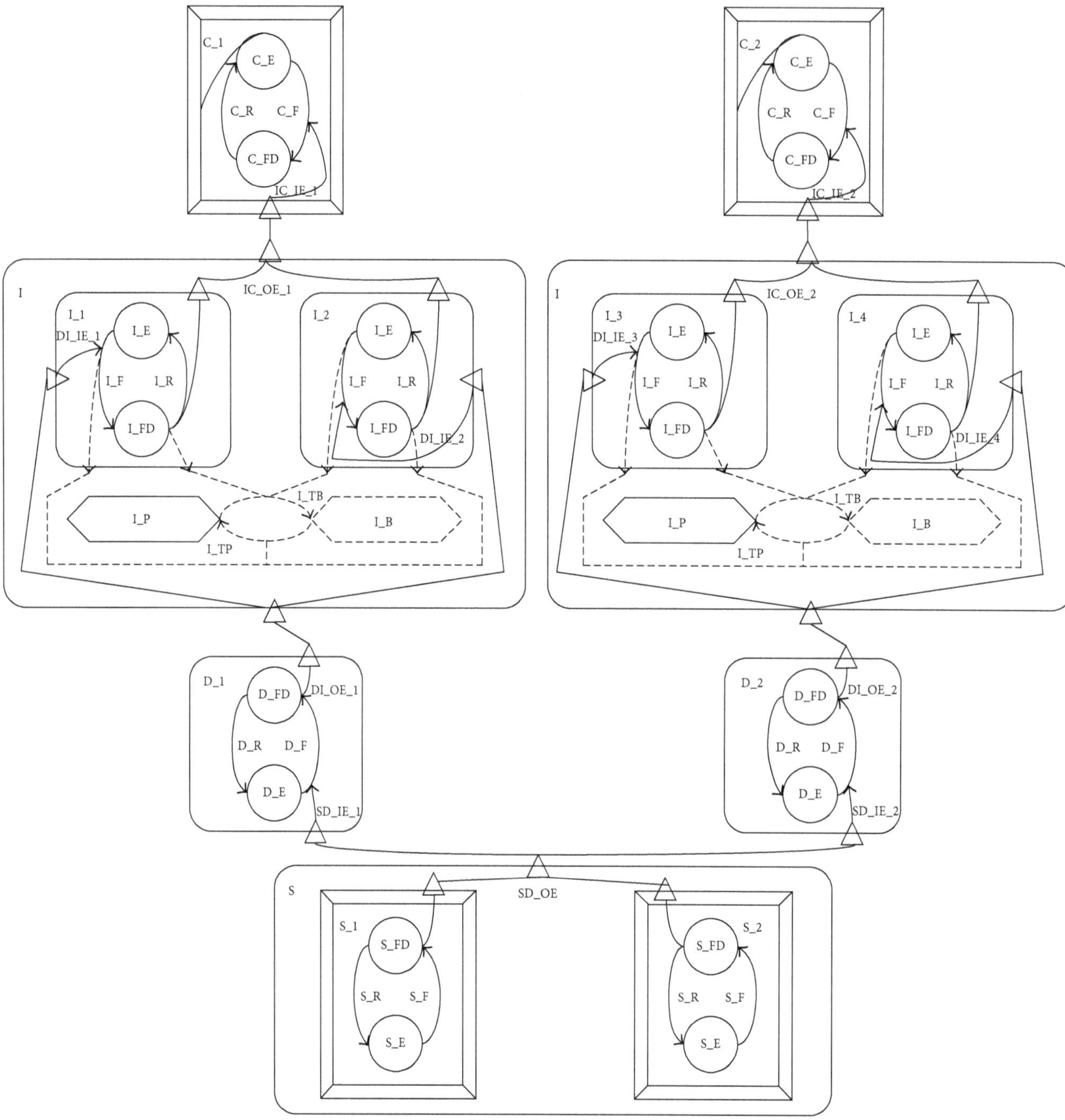

Figure 16: AADL reliability model of bilateral tolerant display function.

Table 1: Design of different architectures for basic display function.

Architecture	Sensor	Graphic processing module
1	No backup	No backup
2	No backup	Hot backup
3	Mutual backup	No backup
4	Mutual backup	Hot backup
5	Nonjudgement mutual backup	No backup
6	Nonjudgement mutual backup	Hot backup

Table 2: Design of different architectures for bilateral tolerant display function.

Architecture	Sensor	Graphic processing module
7	Mutual backup	No backup
8	Mutual backup	Hot backup
9	Nonjudgement mutual backup	No backup
10	Nonjudgement mutual backup	Hot backup

Table 3: Availability for different architectures of basic display function under the steady states.

	No backup of graphic processing module	Hot backup of graphic processing module
No backup of sensor	0.98025	0.98611
Mutual backup of sensor	0.98807	0.99300
Nonjudgement mutual backup of sensor	0.97249	0.97925

Table 4: Availability for different architectures of bilateral tolerant display function under the steady states.

	No backup of graphic processing module	Hot backup of graphic processing module
Mutual backup of sensor	0.99985	0.99995
Nonjudgement mutual backup of sensor	0.99925	0.99957

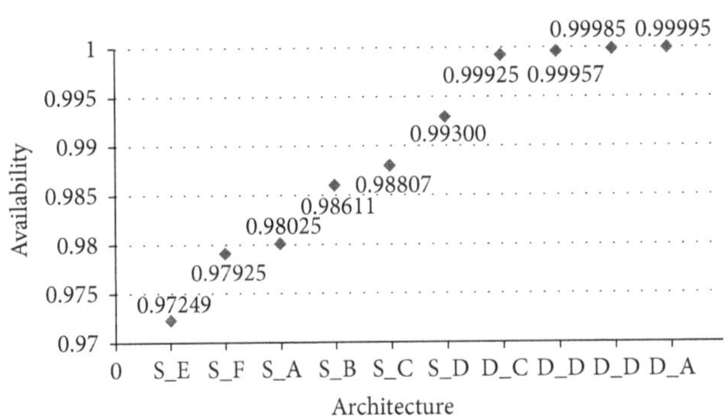

Figure 17: Availability comparison of different display function architecture.

modules is higher than that of nonbackup; when graphic processing modules are in the same architecture, the steady-state availability of the mutual backup of sensors is higher than nonjudgment mutual backup and nonbackup of sensors. The availability of ten architectures is shown in Figure 17. It should be noted that the hot backup will increase the cost; thus, the architecture of lower cost can be chosen when it meets the reliability needs.

In summary, when single component is under the pressure of failure, for the actual system which contains thousands of components, the key way to keep it operating reliably is fault limit and backup. Therefore, if the cost is limited, we should use these two ways to ensure the reliability of the system.

5. Conclusion

This paper proposed a reliability modeling method for IMA based on AADL error model, which effectively solves the problem of complex error propagation mechanism due to the highly complex and resource-sharing characteristics of IMA. It accurately describes the dynamic characteristics of IMA fault and makes up the defects of static reliability assessment method. At the same time, the establishment of AADL reliability model based on system architecture ensures the synchronization of IMA reliability model and IMA design and avoids the traditional problems of "postdesign evaluation." By analyzing the availability of display function under different architecture, we proposed that the reasonable use the fault limit and backup mechanism can improve the system reliability effectively under the condition of cost allowable. It is meaningful to safety assessment work of complex electronic hardware in avionics system.

Conflicts of Interest

The authors declare that they have no competing interests.

Acknowledgments

This paper was funded by joint fund of the National Natural Science Foundation of China and the Civil Aviation Administration of China (no. U1533105). The authors would like to thank Ms. Fan Zhang and Ms. Peipei Xing for their preliminary work.

References

[1] Q. Y. Li, "Application analysis of SMP and MPP architecture in integrated avionics system," *Electronics Word*, vol. 2013, no. 9, pp. 23-24, 2013.

[2] SAE International, "Architecture analysis and design language (AADL) annex volume," Tech. Rep. AS5506/1, 2006.

[3] P. H. Feiler and D. P. Gluch, *Model-Based Engineering with AADL-an Introduction to the SAE Architecture Analysis & Design Language*, Pearson Schweiz Ag, 2012.

[4] K. S. Kushal, M. Nanda, and J. Jayanthi, "Architecture level safety analyses for safety-critical systems," *International Journal of Aerospace Engineering*, vol. 2017, Article ID 6143727, 9 pages, 2017.

[5] P. Feiler and A. Rugina, *Dependability Modeling with the Architecture Analysis & Design Language (AADL)*, ARNEGIE-MELLON UNIV PITTSBURGH PA SOFTWARE ENGINEERING INST, 2007.

[6] P. H. Feiler, J. Hansson, and D. de Niz, "System architecture virtual integration: an industrial case study," Technical Report CMU/SEI-2009-TR-017, CMU, 2009.

[7] C. R. Spitzer, "Digital avionics technology (2): avionics development and implementation," in *Translation*, W. T. Xie, Ed., p. 119, Aviation Industry Publishing Company, Beijing, 2010.

[8] J. Shi, Y. X. Meng, and S. P. Wang, "Reliability and safety analysis of redundant vehicle management computer system," *Chinese Journal of Aeronautics*, vol. 26, no. 5, pp. 1290–1302, 2013.

[9] S. Park and G. Kwon, *Formal Verification for Inter-Partitions Communication of RTOS Supporting IMA//LNEE : Frontier and Innovation in Future Computing and Communications*, Springer, Netherlands, Berlin, 2014.

[10] P. Wang, F. Zhang, L. Dong, H. U. Jianbo, and C. Zhao, "Translation rules of fault extended model to probabilistic

checking model," *System Engineering and Electronics*, vol. 11, pp. 2501–2508, 2017.

[11] W. S. Niu, *Airborne Computer Technology*, Aviation Industry Publishing Company, Beijing, 2013.

[12] S. Liu and R. C. Lin, "Design and implementation of integrated cockpit display control system," *Modern Electronic Technology*, vol. 33, no. 15, pp. 160–162, 2010.

[13] Z. X. Xiu, *System Safety Design & Assessment in Civil Aircraft*, Shanghai Jiao Tong University, Shanghai, 2013.

[14] Y. T. Xu, Y. F. Yin, and J. Sun, "Real-time embedded software architecture modeling and reliability estimation based on time-extended petri net," *Acta Armamentarii*, vol. 36, no. 2, pp. 363–373, 2015.

10

Ionospheric Gradient Threat Mitigation in Future Dual Frequency GBAS

Michael Felux,[1] Mihaela-Simona Circiu,[1] Jiyun Lee,[2] and Florian Holzapfel[3]

[1]*German Aerospace Center (DLR), Münchener Straße 20, 82234 Weßling, Germany*
[2]*Korea Advanced Institute of Science and Technology (KAIST), 291 Daehak-ro, Yuseong-gu, Daejeon 34141, Republic of Korea*
[3]*Institute of Flight System Dynamics, Technische Universität München (TUM), Boltzmannstraße 15, 85748 Garching, Germany*

Correspondence should be addressed to Michael Felux; michael.felux@dlr.de

Academic Editor: Antonio Angrisano

The Ground Based Augmentation System (GBAS) is a landing system for aircraft based on differential corrections for the signals of Global Navigation Satellite Systems (GNSS), such as GPS or Galileo. The main impact on the availability of current single frequency systems results from the necessary protection against ionospheric gradients. With the introduction of Galileo and the latest generation of GPS satellites, a second frequency is available for aeronautical navigation. Dual frequency methods allow forming of ionospheric free combinations of the signals, eliminating a large part of the ionospheric threats to GBAS. However, the combination of several signals increases the noise in the position solution and in the calculation of error bounds. We, therefore, developed a method to base positioning algorithms on single frequency measurements and use the second frequency only for monitoring purposes. In this paper, we describe a detailed derivation of the monitoring scheme and discuss its implications for the use in an aviation context.

1. Introduction

In modern aviation, the navigation methods are increasingly based on the use of Global Navigation Satellite Systems (GNSS). This transition away from conventional terrestrial navigation aids enables a more efficient use of the airspace and provides a seamless global navigation capability. A key aspect of using GNSS is the provision of integrity, that is, a high level of trust an airborne user can put into the position solution. For different phases of flight different methods and augmentation systems can be used to achieve different levels of integrity. Receiver autonomous integrity monitoring (RAIM) is based on consistency checking of redundant GNSS measurements and is used today in support of RNP (Required Navigation Performance) operations. Advanced RAIM (ARAIM) using multiple GNSS constellations with a light integrity support will provide horizontal services soon and vertical guidance at a later stage [1]. Space based augmentation systems (SBAS) are certified for use also for precision approaches (with a certain minimum visibility still required) [2]. GBAS is currently certified for CAT I operations (or GBAS Approach Service Type (GAST) C in GBAS terminology) and is operational and in regular use at a number of locations around the world including Bremen and Frankfurt in Germany, Newark and Houston in the US, Sydney in Australia, Zurich in Switzerland, and Malaga in Spain. The standards for GBAS supporting CAT II/III operations (i.e., with lower or no limitations regarding minimum visibility, termed as GAST D in the GBAS community), were agreed upon by the International Civil Aviation Organization (ICAO) at the end of 2016. First stations supporting this new service type could become operational as soon as 2018. In GBAS a ground station with four (or potentially more) reference receivers at carefully surveyed sites at an airport generates corrections for the GNSS signals and broadcasts them to arriving aircraft. Along with those corrections, a set of integrity parameters is broadcast to enable the aircraft to calculate error bounds for the differentially corrected positions. Stations designed for GAST C and GAST D, however, only rely on signals from the GPS constellation in the L1 frequency band. A significant number of different monitors in the ground and airborne systems may result in limited availability due to frequent trips, especially in equatorial and auroral regions prone to high ionospheric activity.

A challenge for GNSS users is ionospheric scintillations. This effect leads to degraded signal tracking quality in the receiver or loss of lock to several satellites and thus results in degraded satellite geometries [3–5]. Use of signals from multiple GNSS constellations will provide significantly increased robustness against this kind of events due to the large number of satellites and thus low sensitivity to the loss of individual signals. A different threat for GBAS results from ionospheric gradients, that is, strong variations of electron density in the ionosphere over a short distance [6, 7]. This threat can be addressed by means of dual frequency positioning eliminating the ionospheric delay (to a first order which is sufficient for GBAS operations) in GNSS measurements. GPS introduced signals on a second frequency usable for safety of life applications with the latest generation of satellites (Block IIF). Galileo provides those signals on all its satellites from the beginning. Combining measurements from two frequencies, however, comes at the cost of significantly increasing the residual noise in the position solution due to the combination of the two noisy pseudorange measurements on L1/E1 and L5/E5a [8]. It is thus a likely scenario that even in future GBAS with dual frequency capability positioning in the nominal case will be based on single frequency measurements. In that case, an effective monitoring for ionospheric gradients is necessary. In this paper, we describe a method how this can be achieved and show results from testing this method in flight trials.

2. Generation of GBAS Corrections

In order to understand the difficulty of dual frequency ionospheric monitoring, we start by describing how the corrections are generated in a GBAS ground station according to [9]. In a first step, the pseudoranges $\rho(i,j)$, measured for satellite i at the reference receiver j, are passed through a low pass Hatch filter [10]. This process removes a large part of the noise and multipath errors. From the geometric range $R(i,j)$, the carrier smoothed code measurement $\rho_{csc}(i,j)$ and the correction for the satellite clock $c\Delta t_{SV,j}$ are subtracted in order to obtain a preliminary candidate for a pseudorange correction $\mathrm{PRC}_{prel}(i,j)$:

$$\mathrm{PRC}_{prel}(i,j) = R(i,j) - \rho_{csc}(i,j) - c\Delta t_{SV,j}. \quad (1)$$

The geometric range can be calculated due to precise knowledge of the reference antenna location and the known satellite position. The satellite clock correction is calculated based on parameters transmitted in the satellite navigation message. Repeating this process for each satellite in view yields a set of candidate corrections from one receiver. These correction candidates, however, all contain the receiver clock error. In order to remove the individual clock errors from the correction candidates of each receiver, a process called smooth clock adjustment is performed. In this step, the weighted average over all N correction candidates from one receiver is removed with the condition that the sum of all weights $\sum_{i=1}^{N} k_i = 1$.

$$\mathrm{PRC}_{sca}(i,j) = \mathrm{PRC}_{prel}(i,j) - \sum_{i=1}^{N} k_i \cdot \mathrm{PRC}_{prel}(i,j). \quad (2)$$

This can be done without affecting the positioning since any bias common to all satellites only projects into the user clock estimate. It is important to note, however, that in this step all biases common to all satellites are removed. This is especially true for the part of the ionospheric delay that is common to all satellites. An illustration of this process is shown in Figure 1.

An estimation of the experienced ionospheric delay at the reference receiver by analyzing the corrections is therefore not possible anymore. After repeating this process for each reference receiver, the final correction to be transmitted to the user is the average of the smoothed and clock adjusted corrections from all $M(i)$ receivers that track the particular satellite i.

$$\mathrm{PRC}_{tx}(i) = \frac{1}{M(i)} \sum_{j} \mathrm{PRC}_{sca,\tau}(i,j). \quad (3)$$

Additionally, a set of range rate corrections (RRC) is formed based on the pseudorange corrections. It is generated by dividing the difference of the PRC at the current epoch and the previous epoch by the time between the two epochs. This is done to allow the aircraft to interpolate between the reception of two sets of corrections.

In the current service types, C and D, the corrections are only formed for pseudorange measurements of GPS L1 C/A code measurements. Future processing modes will also provide corrections and integrity parameters for signals from other constellations as well as for signals on a second frequency (L5/E5a) [11, 12]. These corrections shall enable another single frequency mode based on L5 signals, should the L1 signal be unavailable, and also a dual frequency mode in order to enable ionosphere free positioning to address the ionospheric gradient threat. While the steps of forming the corrections are assumed to be the same also for the new signals, it is important to note that the smooth clock adjustment process described in (2) will result in very different averages for different frequencies. This is due to the fact that not necessarily the same set of satellites will be used (GPS satellites before Block IIF do not yet provide L5 signals) and the ionospheric delay experienced by a user on L5/E5a is about 1.8 times larger than on L1/E1.

3. Dual Frequency Monitoring Concept for Ionospheric Gradients

While it is possible to eliminate the ionospheric gradient threat using a dual frequency combination of GNSS signals, it might not be beneficial due to the significant increase in noise in the position solution and the required bounding of the residual errors in the protection level [8, 13]. A likely scenario for future service types is that positioning may still be based on single frequency methods, while a dual frequency mode is only used in case a GBAS ionospheric monitors triggers.

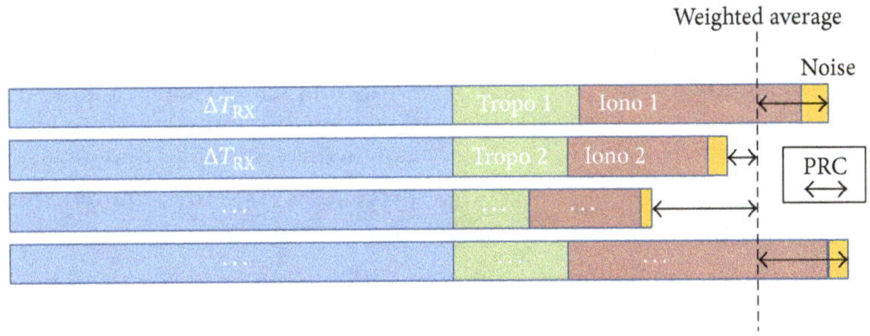

FIGURE 1: Illustration of the receiver clock adjustment process in the generation of pseudorange corrections (PRC).

In that case, however, mitigation of the ionospheric gradient threat is mandatory. Assuming that the ground station provides corrections for two frequencies and the aircraft receiver is also able to track signals on two frequencies, an effective ionospheric monitoring is possible. The proposed method marks a fundamental change where the monitoring is to be performed. While in GAST C it is the sole responsibility of the ground station to bound the errors at the airborne receiver, in GAST D the ionospheric monitoring task is shared between ground and airborne GBAS subsystems. In the method proposed here, it is the sole responsibility of the airborne systems to monitor ionospheric gradients. This allows for a more realistic error bounding because the current navigation performance (and aircraft performance) can be exploited and thus require less conservative assumptions.

The basic idea of the proposed monitoring is to simply compare the ionospheric delay estimate at the ground station with an ionospheric delay estimate at the aircraft. In the airborne system, it is possible to directly estimate the ionospheric delay $\widetilde{I}_{\text{air},i}$ experienced on $\widehat{\rho}_{\text{L1},i}$ based on the smoothed pseudorange measurements $\widehat{\rho}_{f,i}$ on frequency f and satellite i as

$$\widetilde{I}_{\text{air},i} = \frac{f_{\text{L5}}^2}{f_{\text{L1}}^2 - f_{\text{L5}}^2}\left(\widehat{\rho}_{\text{L5},i} - \widehat{\rho}_{\text{L1},i}\right). \tag{4}$$

As previously described, however, the ionospheric delay experienced at the ground station cannot be directly computed from the received corrections. Nevertheless, a pseudo-ionospheric delay estimate $\widetilde{I}_{\text{PRC},i}$ for each satellite i can be formed from the GBAS corrections in the same way as when estimating the actual ionospheric delay on the L1 frequency from pseudorange measurements. The measured pseudorange is simply replaced with the received corrections from the ground station:

$$\widetilde{I}_{\text{PRC},i} = \frac{f_{\text{L5}}^2}{f_{\text{L1}}^2 - f_{\text{L5}}^2} \cdot \left[\left(\text{PRC}_{\text{L5},i} + \Delta t \cdot \text{RRC}_{\text{L5},i}\right) - \left(\text{PRC}_{\text{L1},i} + \Delta t \cdot \text{RRC}_{\text{L1},i}\right)\right]. \tag{5}$$

In the same way as the applied correction would contain the pseudorange correction PRC and the range rate correction RRC multiplied with the time difference between the current time of measurement and the time of generation of the corrections Δt, these quantities are used also in the monitor. The indices L1 and L5 indicate the frequency for which the corrections were formed and f refers to the central frequency of the respective signals.

In a next step, these two quantities have to be made comparable. In order to do this, we remove the average of all N airborne ionospheric delay estimates per constellation to obtain a pseudo-airborne ionospheric delay measure $I_{\text{air},i}$, given by

$$I_{\text{air},i} = \widetilde{I}_{\text{air},i} - \frac{1}{N}\sum_{i=1}^{N}\widetilde{I}_{\text{air},i}. \tag{6}$$

In order to account for the fact that the airborne system may use only a subset of the satellites to which the ground system provides corrections, we repeat this process also with the pseudo-delay estimates from the ground station using only the set of satellites used in the airborne pseudo-ionospheric delay estimate. Similar to (6) the pseudo-ionospheric delay from the corrections is then

$$I_{\text{PRC},i} = \widetilde{I}_{\text{PRC},i} - \frac{1}{N}\sum_{i=1}^{N}\widetilde{I}_{\text{PRC},i}. \tag{7}$$

These two estimates are now comparable and can be used to form a test statistic $I_{\text{test},i}$ as

$$I_{\text{test},i} = \left|I_{\text{air},i} + I_{\text{PRC},i}\right|. \tag{8}$$

This test statistic is then going to be compared with a monitoring threshold that is derived in the next section. Note that the two quantities are added (a more intuitive guess would be to subtract them and compare to zero) due to the fact that the pseudo-delay estimate from the corrections would have opposite sign of the value derived from the pseudorange measurements. In the nominal case (i.e., without a significant ionospheric gradient) the test statistic is small and would mostly consist of noise. With increasing decorrelation of the ionospheric delay experienced at the ground station and the airborne user, this quantity would increase.

The previously described test statistic is only valid if the navigation is based on signals in the L1/E1 frequency band. It is, however, also possible to support positioning based on L5/E5a which may, under certain circumstances, be a desired mode. In that case, a user would experience an ionospheric delay on each measurement that is $\beta = f_{\text{L1}}^2/f_{\text{L5}}^2 \approx 1.8$ times larger than on L1/E1. Additionally, the monitoring threshold will be even tighter in that case. Thus, navigating based on L5/E5a is significantly more challenging in terms of integrity and is not considered further in this study.

4. Monitoring Threshold

After defining the test statistic above, a threshold needs to be determined, for when it is unsafe to use the signals from a certain satellite or set of satellites. A meaningful derivation of this monitoring threshold comes from the operation, namely, the requirements for a safe landing [14]. For a landing to be considered safe, the aircraft has to touch down inside the so-called "touchdown box," that is, not less than 200 ft and not more than 2700 ft behind the runway threshold and not less than 5 ft from either runway edge with a probability of not less than $1\text{–}10^{-6}$ in the nominal case [15, 16]. Two more cases are defined that also have to be considered, namely, the limit and the malfunction case. In the limit case, one parameter is kept at its most adverse value while all other influencing parameters vary according to their nominal distributions. In the malfunction case, an undetected error occurs. In both cases, the aircraft has to land within the touchdown box, where the land long limit is extended to 3000 ft behind the runway threshold.

These requirements are defined at aircraft level and thus only part of the total error budget can be attributed to the navigation system. The other main question in automatic landings, apart from how well the navigation system can determine the position of the aircraft, is how well the autopilot can land the aircraft on a desired spot. As the landing performance depends on both systems there is a possibility of trading off autopilot and navigation system performance. Note that this tradeoff is, however, only possible if the monitoring for disturbances occurs onboard the aircraft. In the case of the current service types, the ground station plays an important role in mitigating part of the threat space. Thus, it is not possible to take actual aircraft performance into account in those systems as the ground station has to support all aircraft types.

In [20], we showed the derivation of the monitoring threshold in the case a single satellite is affected by an ionospheric gradient. This was done by starting from the requirement that the aircraft has to land within the touchdown box and the land short case (i.e., landing not too close to the runway threshold) is considered to be the driving constraint. The along track error is a function of the nominal touchdown point NTDP, the nominal flight technical error FTE (i.e., how well the autopilot can place the aircraft onto the desired landing spot), the nominal navigation system error NSE (i.e., nominal errors due to noise, multipath, nominal ionosphere, and troposphere decorrelation), and an additional undetected error in form of a bias (e.g., caused by an ionospheric gradient). It is assumed that a vertical position error E_v maps into an along track position error E_{atrk} at touchdown by the simple geometric relation

$$E_{\text{atrk}} = \frac{E_v}{\tan(\text{GPA})}, \tag{9}$$

where GPA is the glide path angle (typically 3°). In the malfunction case, the aircraft has to land in the touchdown box with complete certainty. Taking the nominal NSE and FTE as Gaussian distributed random variables could of course never satisfy this condition. In that case, they are fixed at their 95th percentiles [17]. The requirement for the land short case can now be formulated as

$$200\,\text{ft} \le \text{NTDP} - \frac{\text{NSE}_{\text{vert,ff},95\%} + E_{v,\text{iono}}}{\tan(\text{GPA})} - \text{FTE}_{\text{ff},95\%}, \tag{10}$$

where $\text{NSE}_{\text{vert,ff},95\%}$ and $\text{FTE}_{\text{ff},95\%}$ are the nominal navigation and flight technical errors at the 95th percentile of their Gaussian distributions. This is illustrated in Figure 2.

$\text{NSE}_{\text{vert,ff},95\%}$ is a vertical position error and can be derived from the protection level equation [18] as

$$\sigma_{\text{NSE,vert,ff}} = \frac{\text{VPL}}{K_{\text{ffmd}}} = \frac{\text{VPL}}{5.81}, \tag{11}$$

where K_{ffmd} is the fault-free missed detection multiplier for the integrity risk of $2 \cdot 10^{-7}$ allocated to the protection level. Note that the VPL is a bound for the instantaneous nominal vertical position error and depends on both, the satellite geometry and the expected satellite ranging performance. At all times it must remain below the so-called vertical alert limit (VAL) that is limited by a value of 10 m for the final approach and landing.

The nominal touchdown point is usually considered to be located 1290 ft behind the runway threshold and σ_{FTE} = 180 ft is assumed to be a conservative value for all aircraft using GBAS [19]. With all those assumptions it is now possible to derive a limit on the largest vertical position error $E_{v,\text{iono}}$ due to an ionospheric anomaly using (10). Assuming VPL = VAL = 10 m and GPA = 3°, the resulting condition would be $E_v \le 8.4$ m. Taking a less conservative value of VPL = 5 m and leaving all other parameters constant, the resulting condition would already be relaxed to $E_v \le 10.1$ m. In the multiconstellation case VPL = 2.5 m is realistic and would further relax the required monitoring to $E_v \le 10.9$ m.

Given the test statistic per satellite described in (8), this limit in the position domain needs to be translated into a limit in the pseudorange domain. The pseudorange measurements ρ and the user position and clock estimate x are related by the weighted pseudoinverse S of the geometry matrix G. This matrix contains the normalized line-of-sight vectors between user and satellite in its rows and a vector of "1" in the fourth column for the user clock offset such that

$$x = S \cdot \rho. \tag{12}$$

A weighting matrix W containing the fault-free variances of the expected residual pseudorange errors is used to give lower weight to satellites with larger expected uncertainties. The S-matrix is defined as

$$S = \left(G^T W G\right)^{-1} G^T W \tag{13}$$

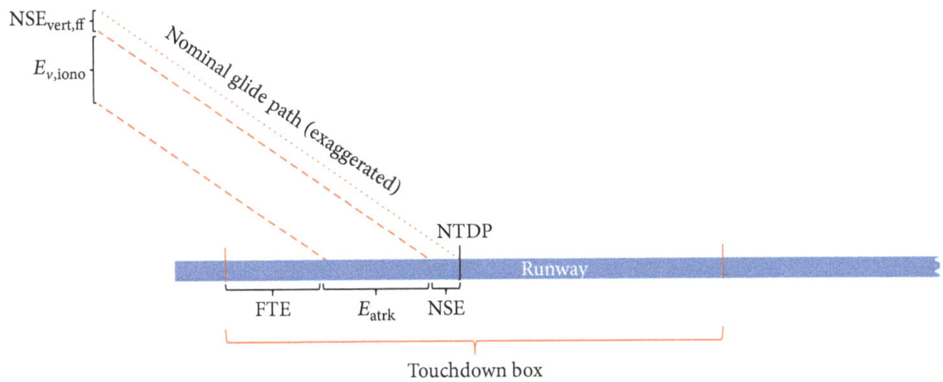

FIGURE 2: Derivation of the largest allowable vertical error $E_{v,iono}$ and illustration of the contributing error sources for automatic landings.

and the contribution of a single satellite i to the position estimate vertical to the approach track is given by

$$s_{vert,i} = s_{3,i} + s_{1,i} \cdot \tan(\text{GPA}), \tag{14}$$

where $s_{k,i}$ denotes the entry in the kth row and ith column of the S-matrix.

In the case of a single satellite affected by an undetected ionospheric gradient, a limit on the pseudorange error $E_{r,iono,i}$ for that particular satellite would be

$$E_{r,iono,i} \leq \frac{E_{v,iono}}{|s_{vert,i}|}. \tag{15}$$

It is interesting to note that, based on this equation, a pseudorange error may become arbitrarily large as long as the weight assigned to that satellite through $s_{vert,i}$ is sufficiently small. A typical behavior of values for $s_{vert,i}$ as a function of the elevation of the satellite is shown in Figure 3. Here the geometry of a combined GPS and Galileo constellation was simulated with 31 and 27 satellites, respectively, during our test flight described later in Section 5. At about 45° elevation s_{vert} of two satellites become almost zero, leading to very large pseudorange error limits.

With a missed detection probability of 10^{-9} attributed to the monitor and assuming that the noise in the test statistic follows a Gaussian distribution the monitoring condition is given as

$$I_{test,i} \leq \frac{E_{v,iono}}{|s_{vert,i}|} - 6.1 \cdot \sigma_{monitor,i}. \tag{16}$$

The test statistic $I_{test,i}$, as defined in (8), contains the airborne measurement noise $\sigma_{air,f}$ from the smoothed pseudorange measurements on both frequencies f, as well as the pseudorange correction noise $\sigma_{gnd,f}$ for both frequencies. The noise in the proposed monitor can thus be described as

$$\sigma_{monitor,i} = \frac{f_{L5}^2}{f_{L1}^2 - f_{L5}^2} \sqrt{\sigma_{gnd,L1}^2 + \sigma_{gnd,L5}^2 + \sigma_{air,L1}^2 + \sigma_{air,L5}^2}. \tag{17}$$

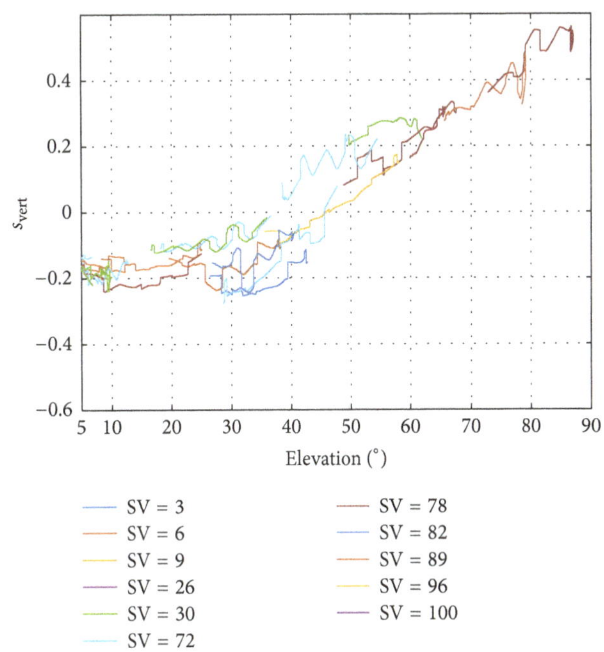

FIGURE 3: s_{vert} as a function of the elevation of the satellites for a future combined GPS (SVs 1–32)/Galileo (SVs 71–100) constellation with 31 and 27 satellites, respectively.

In [20] we showed that this monitoring condition can be fulfilled rather easily. Even with the minimum number of 5 satellites and noise from our experimental system (that is expected to be larger than that of an operational GBAS), the monitoring was possible and did not cause false alarms due to very low monitoring thresholds.

Unfortunately, assuming only a single affected satellite may not be sufficient to bound potential errors caused by ionospheric disturbances. It is possible that more than one satellite could be affected by an ionospheric front. For that reason, a scenario with two simultaneously affected satellites is investigated. In this case, the error from both affected satellites would project from the pseudorange into the position domain according to (12). A limit on the sum of both

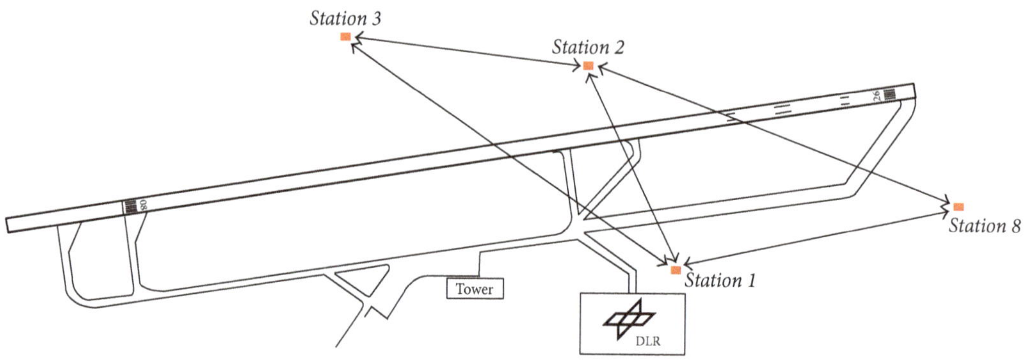

FIGURE 4: Layout of the GBAS reference stations at Braunschweig Airport as used during the flight trials.

pseudorange errors for the affected satellites i and k, using (15), then becomes

$$E_{r,\text{iono},i} + E_{r,\text{iono},k} \leq \frac{E_{v,\text{iono}}}{\left|s_{\text{vert},i} + s_{\text{vert},k}\right|}. \quad (18)$$

Finally, also the noise of both estimates needs to be considered and the monitoring condition becomes more stringent:

$$I_{\text{test},i} + I_{\text{test},k} \leq \frac{E_{v,\text{iono}}}{\left|s_{\text{vert},i} + s_{\text{vert},k}\right|} - 6.1 \cdot \sqrt{\sigma^2_{\text{monitor},i} + \sigma^2_{\text{monitor},k}}. \quad (19)$$

Note that, in this case, the test statistic can increase if the pseudo-ionospheric estimates on the two satellites have the same sign. At the same time, the noise contribution in the monitoring condition increases. The sum of s_{vert} can either increase or decrease depending on the sign of the values as shown previously in Figure 3. The worst case would occur when the satellites with the two largest s_{vert} and the same sign also have pseudo-ionospheric delays with the same sign that add up constructively.

This monitoring concept can of course also be expanded to consider more affected satellites. A generalized monitoring condition for N affected satellites would then be

$$\sum_{i=1}^{N} I_{\text{test},i} \leq \frac{E_{v,\text{iono}}}{\left|\sum_{i=1}^{N} s_{\text{vert},i}\right|} - 6.1 \cdot \sqrt{\sum_{i=1}^{N} \sigma^2_{\text{monitor},i}}. \quad (20)$$

Note that as more affected satellites are considered, the monitoring becomes more challenging. However, considering the cases with more affected satellites that are not excluded by other monitors becomes less likely with increasing N. Equation (20) does not take into account the low probability of occurrence but this can be included in the future if necessary. Furthermore, the monitoring is conservative in the way that it does not consider the actual separation of satellites in terms of azimuth angle. In the case of an ionospheric front, only satellites in certain azimuth regions would be affected; however, we consider only the s_{vert} values, irrespective of the azimuth.

The following section presents an exemplary evaluation during one of our flight trials. The contributing factors are analyzed more in detail with the example at hand to illustrate the effects inherent to this monitoring scheme.

5. Evaluation in Flight Trials and Discussion of the Monitor

The German Aerospace Center (DLR) is operating an experimental GBAS station consisting of four reference receivers tracking GPS and Galileo L1/E1 and L5/E5a signals. It is located at the Braunschweig research airport in northern Germany. The layout of the reference receiver locations is shown in Figure 4.

In November 2016 a flight test with a Dornier DO-228 aircraft (a twin-engine turboprop aircraft) was performed in the vicinity of Braunschweig. The flight lasted for about two hours and consisted of about one hour of maneuvering and flying the aircraft at different bank angles, followed by one hour of five approaches and extended traffic patterns. The ground track of the flight is shown in Figure 5. One of the purposes of the test was to evaluate the signal tracking performance in steep turns with bank angles of up to 60°. The data of that part of the flights shows frequent loss of lock events and very limited continuously tracked signals due to the banking of the aircraft [21]. That first part (shown in black in Figure 5) is therefore omitted in the following evaluations. Only the second half of the flight where normal maneuvering was performed (shown in green) is used.

Figure 6 shows a skyplot of the GPS Block IIF and Galileo satellites as observed during the test flight. There were 11 different satellites visible, at most 10 of them at the same time. This number is somewhat typical for a single full constellation. With a fully deployed Galileo constellation and all GPS satellites providing also L5 signals in the future, the number of visible satellites in Braunschweig would vary between 18 and 21.

We start the evaluations with assumptions on σ_{air} and σ_{gnd} which were experimentally derived from our ground and airborne systems for each of the two frequencies and each of the two constellations individually as previously presented in [8, 13]. For the whole flight and all satellites in the single affected satellite case, the test statistic (the monitoring condition for $i = 1$ in (20)) remains below the threshold, except one short instance that will be discussed later.

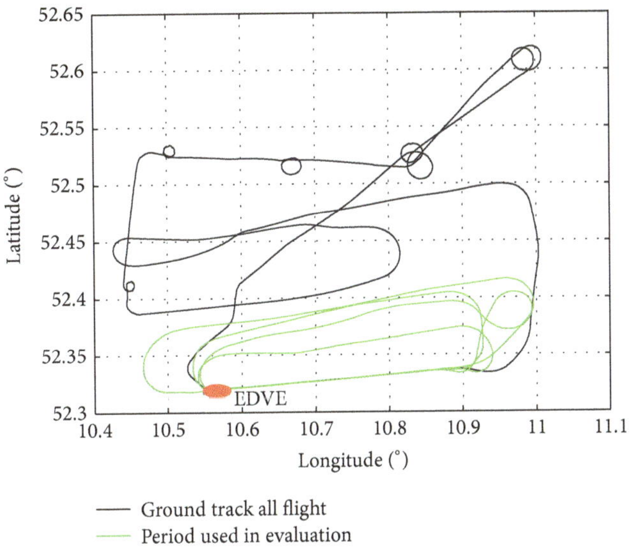

FIGURE 5: Ground track of the test flight around Braunschweig (ICAO identifier EDVE). The part used for the evaluation is shown in green.

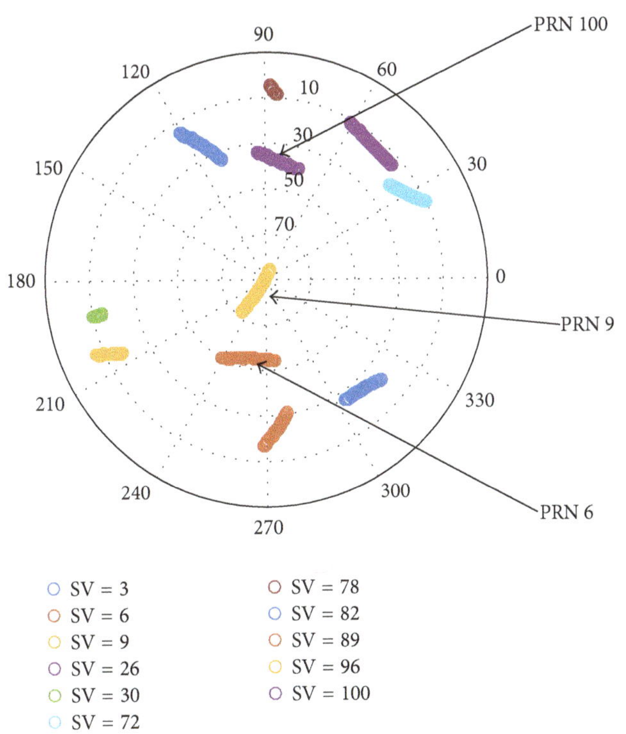

FIGURE 6: Skyplot of the combined GPS (SVs 1–32)/Galileo (SVs 71–100) constellation providing L1/E1 and L5/E5a signals as observed during the flight trial.

Figure 7 shows the number of used satellites in subplot (a) and the test statistics with the corresponding monitoring thresholds for L1/E1-based navigation for the three selected satellites marked in the skyplot in Figure 6. The drops in the number of satellites occur during turns where tracking of low elevation satellites is lost due to signal blockage by parts of the aircraft. Subplot (b) shows the results for PRN 6, a satellite at elevations between 50° and 55°. In the particular geometry, this satellite had a very small s_{vert} leading to an extremely large monitoring threshold mostly in the range of 100 m to 200 m (note the different scale of the y-axis for this subplot). As the satellite plays almost no role in the determination of the vertical position a potential error could become very large and not affect the user much. The threshold decreases to a value of about 3.7 m shortly before 16 h at the short period of time when the number of used satellites went down to 7. At that moment PRN 6 became a rather important satellite leading to a small threshold. The sharp drop in the monitoring threshold for all satellites at that time indicates a rather strong dependence on the number of satellites available in case that number is small. If many satellites are available, losing one or two of them for the position solution does not have a large impact anymore. Shortly before 15.8 h and at 16 h there are small spikes in the test statistic. They appear small due to the large scale of the y-axis but reach values of up to 7 m. These spikes result from excessive multipath or a cycle slip and were detected by the onboard GBAS monitor. However, for illustration purposes, we did not exclude the data from the plot. Due to the average removal in the test statistic as described in (6) an effect on one satellite would be visible in the test statistic of all other satellites as well.

This is the case as can be seen in subplot (c) showing the results for PRN 9 where spikes occur at the same times. PRN 9 is a very high elevation satellite between 72° and almost 90° elevation. At all times it is the satellite with the highest elevation. From a geometrical perspective, PRN 9 is a very important satellite and thus has a rather large s_{vert}. According to (15) this results in a small tolerable pseudorange error for that satellite and thus a low monitoring threshold throughout the flight. Due to the average removal in forming the test statistic the threshold for that satellite is exceeded twice just before 16 h; however, that effect by a cycle slip would have been excluded by other monitors and is again in the data just for illustration purposes.

The results for PRN 100, a Galileo satellite, are shown in subplot (d). Note that the average removal is performed per constellation and thus the spikes from subplots (b) and (c) do not occur in subplot (d). The elevation of the satellite varies between 35° and 40°. While it has quite a large weight in the beginning, it quickly becomes smaller and thus the monitoring threshold increases fast to larger values.

Figure 7 showed exemplary results for the monitoring condition for a single affected satellite. The case of more than one affected satellite will probably have to be considered for monitoring as well. We therefore also considered the case of two affected satellites; that is, $i = 2$ in (20).

Figure 8 shows the minimum over all thresholds for all possible subsets of two affected satellites during the flight test in red, together with the test statistic in black. The solid lines of the monitoring thresholds are using our own experimentally derived characterization of noise in the corrections and airborne measurements, while the dashed lines assume the performance that can be expected from an operational GBAS station with multipath limiting antennas in carefully protected sites and from airborne equipment with a reasonably

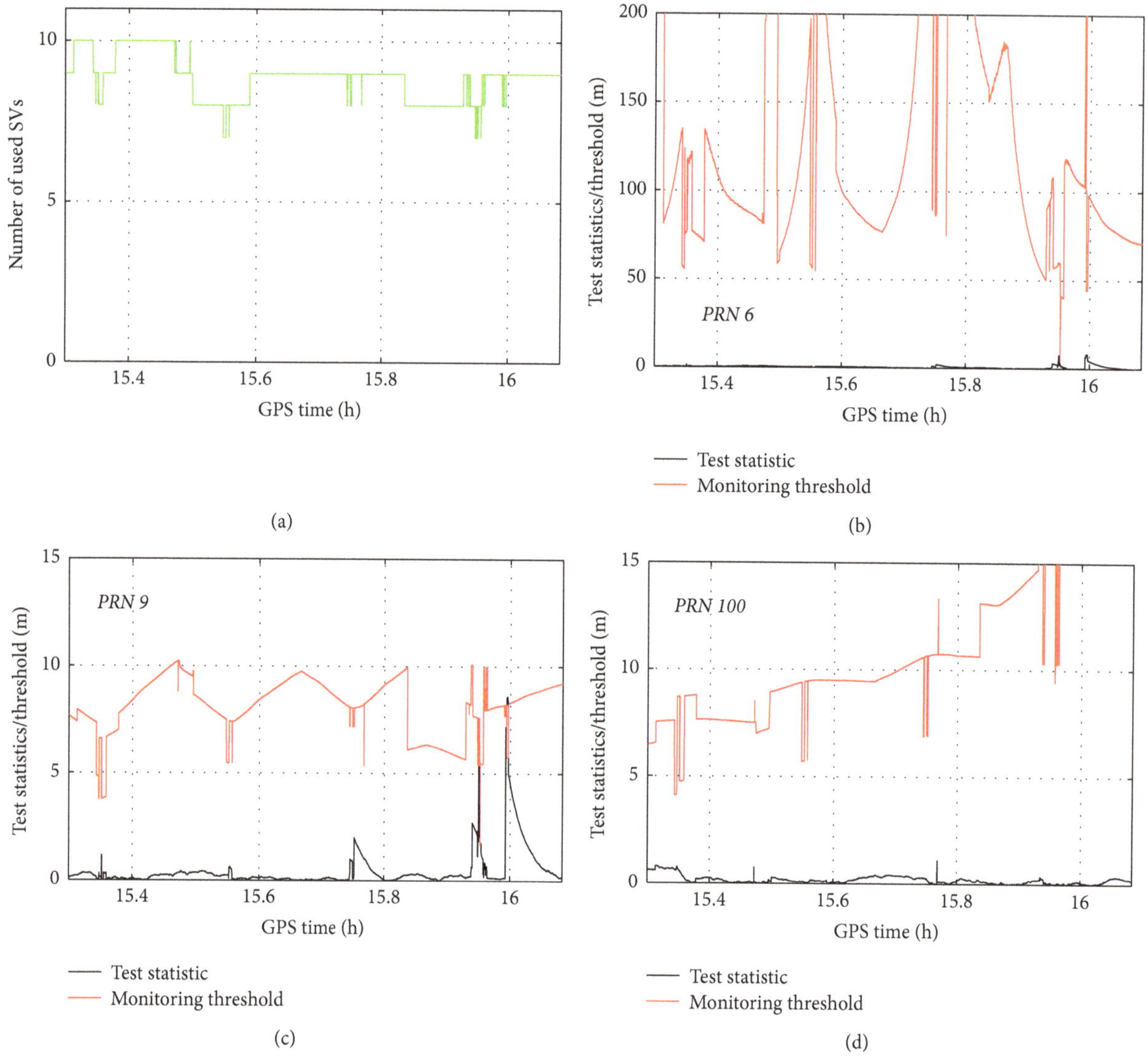

FIGURE 7: Test statistic and monitoring threshold for three example satellites.

good airborne antenna performance as presented in [9, 18]. These models can be considered realistic, while the performance of our own equipment is slightly worse and thus yields conservative results. It can be noted again that the monitoring threshold has sharp drops whenever we lose satellites in turns. At about 15.35 h and 15.95 h the monitoring threshold even becomes negative and thus, of course, makes the monitoring completely impossible assuming our own noise models. The black curve is the corresponding test statistic that is regularly exceeding the threshold whenever a drop in the number of used satellites and thus in the monitoring threshold occurs. However, recall that the experimental thresholds in red were derived from the actual measurements during the flight with a maximum of 10 satellites. Looking into the future and simulating complete dual frequency GPS and Galileo constellations with 31 and 27 satellites, respectively, the situation looks very much different. This is shown in the same plot in green. The monitoring threshold now reaches a minimum of 7.8 m at about 15.75 h. With two full dual frequency capable constellations, the monitoring thresholds will be large enough to provide sufficient margin to prevent false alarms due to noise effects for the case of two affected satellites.

This concept can be further extended to three and four affected satellites. This, of course, places a greater burden on the monitoring and yields again decreasing monitoring thresholds. Figure 9 shows simulated monitoring thresholds for the aircraft positions during the test flight assuming two, three, and four affected satellites in black, red, and green, respectively. As expected, the monitoring threshold keeps decreasing while more satellites are assumed to be affected.

But even in the case of four affected satellites the threshold just reaches values of around 4 m and thus remains above the threshold for two affected satellites and for the limited constellation we observed during the flight test.

Now the results so far only showed an example of a short time period in mid-latitudes with a rather good satellite geometry. For that reason, we also examined the effect of the satellite geometry in less favorable locations and chose as an

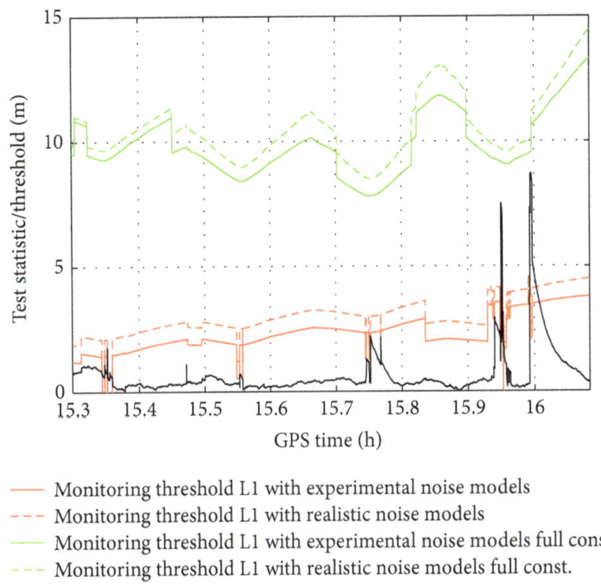

FIGURE 8: Monitoring thresholds for the case of two affected satellites with the constellation as seen during the flight trials (red) and simulated full GPS + Galileo constellation (green).

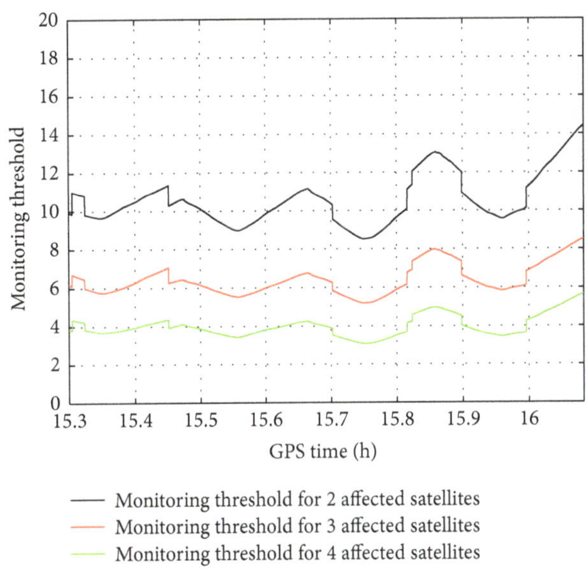

FIGURE 9: Monitoring thresholds assuming two full dual frequency capable constellations and 2, 3, and 4 simultaneously affected satellites.

example Troll Research Station in Antarctica located at 72° southern latitude. The results are shown in Figure 10. Due to the lack of high elevation satellites, the geometries become weaker in the vertical domain and thus lead to somewhat larger s_{vert} values. However, the monitoring for the 2 and 3 affected satellite cases still seems to be possible, for the 4 affected cases there are times, for example, around midnight, where the threshold is rather small. At all times the threshold remains positive and thus makes the monitoring possible in principle depending on the assumed noise in the measurement.

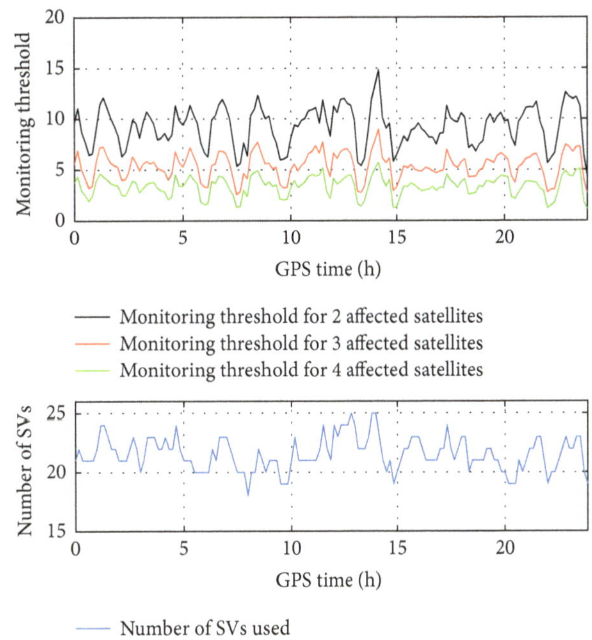

FIGURE 10: Monitoring threshold for Troll (Antarctica at 72° Southern Latitude) as example for a location with less favorable satellite geometry and the effect on the monitoring.

6. Discussion and Conclusions

In this paper, we presented a possibility for using dual frequency corrections and airborne measurements to monitor ionospheric gradients between a GBAS ground station and an airborne user. We showed the definition and calculation of a test statistic based on pseudo-ionospheric delay estimates. In a next step, we derived a monitoring threshold from autoland requirements for the case of one or more satellites that are potentially affected by an ionospheric gradient. Then we used flight test data where we had up to 10 dual frequency satellites of the latest generation of GPS and Galileo available to evaluate the performance of the monitor experimentally. The main findings showed that the monitoring is feasible in principle. By design, the monitor is sensitive to the satellite geometry at a given time. Thus, loss of lock of satellites decreases the monitoring performance. However, the tests were carried out in a scenario with a limited number of satellites. Assuming two full dual frequency constellations greatly simplifies the monitoring task and will enable safe and reliable monitoring in the case of single frequency positioning using dual frequency monitoring which is one of the likely scenarios for a future GBAS service.

The proposed monitoring follows a new philosophy where the monitoring for ionospheric gradients should occur. In the currently operational service type C ionospheric threat mitigation is the responsibility of the ground station. In the service type D, the responsibility is shared between the ground and the airborne system. Both, however, rely on conservative assumptions and the validity of ionospheric threat models. Moving the monitoring to the airborne system as proposed in this paper is a lot less conservative since the monitoring threshold is derived based on the satellite geometry used by the airborne receiver and thus reflecting much better the current positioning performance. It furthermore provides the possibility for airframe integrators to also use realistic models for the autopilot performance (in terms of σ_{FTE}) and thus obtain even larger monitoring thresholds. At all times the monitor achieves the same level of safety and relaxes the monitoring requirement only by replacing conservative assumptions with more realistic information that is available in the aircraft.

Conflicts of Interest

The authors declare that there are no conflicts of interest regarding the publication of this paper.

References

[1] Working Group C, "ARAIM Technical Subgroup—Milestone III Report," February 2016, http://ec.europa.eu/growth/tools-databases/newsroom/cf/itemdetail.cfm?item_id=8690.

[2] T. Dautermann, "Civil air navigation using GNSS enhanced by wide area satellite based augmentation systems," *Progress in Aerospace Sciences*, vol. 67, pp. 51–62, 2014.

[3] T. Fujiwara and T. Tsujii, "GBAS availability assessment and modelling of ionospheric scintillation effects," in *Proceedings of the ION's Pacific PNT Meeting*, Honolulu, Hawaii, USA, April 2015.

[4] M. Yoon, D. Kim, J. Lee, and S. Pullen, "Multi-dimensional verification methodology of ionospheric gradient observations during plasma bubble events in the Brazilian Region," in *Proceedings of the ION's Pacific PNT Meeting*, Honolulu, Hawaii, USA, April 2015.

[5] N. Hlubek, J. Berdermann, V. Wilken et al., "Scintillations of the GPS, GLONASS, and Galileo signals at equatorial latitude," *Journal of Space Weather and Space Climate*, vol. 4, article A22, pp. 1–7, 2014.

[6] S. Jung and J. Lee, "Long-term ionospheric anomaly monitoring for ground based augmentation systems," *Radio Science*, vol. 47, no. 4, Article ID RS4006, 2012.

[7] M. Kim, Y. Choi, H.-S. Jun, and J. Lee, "GBAS ionospheric threat model assessment for category I operation in the Korean region," *GPS Solutions*, vol. 19, no. 3, pp. 443–456, 2015.

[8] M.-S. Circiu, M. Felux, B. Belabbas et al., "Evaluation of GPS L5, Galileo E1 and Galileo E5a performance in flight trials for multi frequency multi constellation GBAS," in *Proceedings of the ION GNSS*, Tampa, Fla, USA, 2015.

[9] ED-114A, "Minimum operational performance specification for global navigation satellite ground based augmentation system ground equipment to support category I operations," EUROCAE, 2013.

[10] R. Hatch, "The synergism of GPS code and carrier measurements," in *Proceedings of the International Geodetic Symposium on Satellite Doppler Positioning*, vol. 1, 1983.

[11] M. Stanisak, A. Lipp, and T. Feuerle, "Possible VDB formatting for multi-constellation / multi-frequency GBAS services," in *Proceedings of the 28th International Technical Meeting of the Satellite Division of the Institute of Navigation (ION GNSS '15)*, pp. 1507–1518, Tampa, Fla, USA, September 2015.

[12] C. Milner, A. Guilbert, and C. Macabiau, "Evolution of corrections processing for the MC/MF Ground Based Augmentation System (GBAS)," in *Proceedings of the International Technical Meeting ION (GNSS '15)*, Dana Point, Calif, USA, 2015.

[13] D. Gerbeth, M.-S. Circiu, M. Caamano, and M. Felux, "Nominal performance of future dual frequency dual constellation GBAS," *International Journal of Aerospace Engineering*, vol. 2016, Article ID 6835282, 20 pages, 2016.

[14] W. Schuster and W. Ochieng, "Harmonisation of category-III precision approach navigation system performance requirements," *Journal of Navigation*, vol. 63, no. 4, pp. 569–589, 2010.

[15] AC 120-28D, "Criteria for approval of category III weather minima for takeoff, landing, and rollout," FAA, 1999.

[16] CS-AWO, "Certification Specifications for All Weather Operations," EASA, 2003.

[17] B. Clark and B. DeCleene, "Alert limits: do we need them for CAT III? Deriving GBAS requirements for consistency with cat III operations," in *Proceedings of the Institute of Navigation—19th International Technical Meeting of the Satellite Division (ION GNSS '06)*, pp. 3070–3081, Fort Worth, Tex, USA, September 2006.

[18] DO-253C, "Minimum operational performance standards for GPS local area augmentation system airborne equipment," RTCA, 2009.

[19] "Conceptual Framework for the Proposal for GBAS to Support CAT III Operations," in *Proceedings of the ICAO NSP 3 Meeting*, Montreal, Canada, December 2016.

[20] M. Felux, M.-S. Circiu, D. Gerbeth, and M. Caamano, "Experimental validation of an ionospheric monitoring scheme for dual frequency GBAS," in *Proceedings of the ION GNSS+*, Portland, Ore, USA, 2016.

[21] M. Rippl, I. Martini, B. Belabbas, and M. Meurer, "ARAIM operational performance tested in flight," in *Proceedings of the International Technical Meeting ION*, San Diego, Calif, USA, January 2014.

11

A Modified Pareto Ant Colony Optimization Approach to Solve Biobjective Weapon-Target Assignment Problem

You Li,[1] Yingxin Kou,[1] Zhanwu Li,[1,2] An Xu,[1,2] and Yizhe Chang[1]

[1] *Aeronautics and Astronautics Engineering College, Air Force Engineering University, Xi'an, Shaanxi 710038, China*
[2] *College of Electronic Communication, Northwestern Polytechnical University, Xi'an 710072, China*

Correspondence should be addressed to You Li; reason891020@126.com

Academic Editor: Linda L. Vahala

The weapon-target assignment (WTA) problem, known as an NP-complete problem, aims at seeking a proper assignment of weapons to targets. The biobjective WTA (BOWTA) optimization model which maximizes the expected damage of the enemy and minimizes the cost of missiles is designed in this paper. A modified Pareto ant colony optimization (MPACO) algorithm is used to solve the BOWTA problem. In order to avoid defects in traditional optimization algorithms and obtain a set of Pareto solutions efficiently, MPACO algorithm based on new designed operators is proposed, including a dynamic heuristic information calculation approach, an improved movement probability rule, a dynamic evaporation rate strategy, a global updating rule of pheromone, and a boundary symmetric mutation strategy. In order to simulate real air combat, the pilot operation factor is introduced into the BOWTA model. Finally, we apply the MPACO algorithm and other algorithms to the model and compare the data. Simulation results show that the proposed algorithm is successfully applied in the field of WTA which improves the performance of the traditional P-ACO algorithm effectively and produces better solutions than the two well-known multiobjective optimization algorithms NSGA-II and SPEA-II.

1. Introduction

Among air power, land power, and sea power, air power plays the most decisive role in the outcome of any war [1]. With the development of military air combat, the problem of WTA has received considerable attention [2–6]. Because the goal of air power is to maximize the military effectiveness, it becomes imperative to find a proper assignment of weapons to targets. From the perspective of the quantity of objective functions, the WTA problem can be classified into single-objective WTA problem and multiple-objective WTA problem. When taking the time factor into account, the WTA problem can be divided into dynamic WTA problem and static WTA problem. The current research status of various WTA is summarized in Table 1. In this paper, we mainly focus on the static biobjective WTA problem which aims at finding a proper static assignment of maximizing the expected damage of the enemy and minimizing the cost of missiles.

The WTA problem belongs to NP-complete problems (NP: nondeterministic polynomial), so the increment of weapons and targets will result in an explosive growth of the searching space. Although the enumeration method ensures a global solution, its computational requirement grows exponentially as the scale of WTA problem grows. Therefore, seeking better weapons for targets distribution scheme and improving the operational efficiency especially for the large-scale WTA problem are still an important issue in the command and control problem (C^2).

The study of WTA problem can be traced back to the 1950s and 1960s when Manne [7] and Day [8] built the model of WTA problem. Hosein and Athans [9] classified the WTA problem into two classes: the single-objective WTA problem and the multiple-objective WTA problem. Genetic algorithm [10], ACO algorithm [11], auction algorithm [12], VLSN algorithm [13], Tabu search [14], and other hybrid algorithms [15–17] have been used to optimize single-objective WTA model by many scholars. In contrast to single-objective WTA, multiple-objective optimization can take different criterions into consideration and is more in line with the real combat

TABLE 1: Summary of variant metaheuristic algorithms and implementation of various WTA.

Researchers	Year	Metaheuristic algorithm	Implementation (WTA)
Lee et al. [40]	2002	IS + ACO	Static single-objective
Lee et al. [41]	2002	GA	Static single-objective
Z.-J. Lee and W.-L. Lee [15]	2003	GA + ACO	Static single-objective
Galati and Simaan [14]	2007	Tabu	Dynamic single-objective
Lee [13]	2010	VLSN	Static single-objective
Xin et al. [42]	2010	VP + Tabu	Dynamic single-objective
Li and Dong [16]	2010	DPSO + SA	Dynamic single-objective
Xin et al. [43]	2011	Rule-based heuristic	Dynamic multiobjective
Chen et al. [44]	2010	SA	Static single-objective
Bogdanowicz et al. [10]	2013	GA	Static single-objective
Fei et al. [12]	2012	Auction algorithm	Static single-objective
Liu et al. [18]	2013	MOPSO	Static multiobjective
Zhang et al. [19]	2014	MOEA/D	Static multiobjective
Ahner and Parson [45]	2015	Dynamic programming	Dynamic multiobjective
Li et al. [20]	2015	NSGA-II, MOEA/D	Static multiobjective
Dirik et al. [46]	2015	MILP	Dynamic multiobjective
Liang and Kang [47]	2016	CSA	Static single-objective
Li et al. [48]	2016	MDE	Dynamic multiobjective

decision-making. So, it has aroused wide attention from scholars. Liu et al. [18] proposed an improved multiobjective particle swarm optimization (MOPSO) algorithm to solve multiple-objective WTA problem and apply it to a simple example including 7 platforms and 10 targets. However, they did not apply the proposed algorithm to different scales WTA problems for contrastive study. Zhang et al. [19] designed a WTA mathematic model and proposed a decomposition-based evolutionary multiobjective optimization algorithm. But the algorithm has not been tested on large-scale WTA problem and it has a low convergence speed. Li et al. [20] adopted NSGA-II (domination-based) and MOEA/D (decomposition-based) to solve the multiple-objective WTA problem and carried out tests on three different BOWTA problems. They only applied the proposed adaptive mechanism to the WTA problems, but they did not verify the behavior of the proposed adaptive mechanism on standard problems.

Different from other algorithms, ACO algorithm is a class of reactive search optimization (RSO) methods adopting the principle of "learning while optimizing" [21, 22]. Since it was introduced in 1992, many variant ACO algorithms have been presented, including ant colony system (ACS) [23] and MAX-MIN ant system (MMAS) [24]. Meanwhile, ACO algorithms have been intensively investigated and successfully applied to deal with multiobjective problems such as travelling salesmen problem (TSP) [25], scheduling problem [26], vehicle routing problem [27], portfolio selection problem [28], network optimization problem [29], and some others problems [30–32]. The main applications of ACO algorithms are shown in Figure 1. In 2004, Doerner et al. proposed the P-ACO algorithm [28] which combined traditional ant system with ant colony optimization. Because P-ACO has the ability to balance local search and global search, it will avoid premature convergence during the solution construction phase. The characteristic is especially suitable for solving portfolio selection problem. Therefore, we choose P-ACO as the approach for solving the BOWTA problem. However, to the best of our knowledge, there has not been any research conducted about P-ACO for BOWTA problem.

Basic multiobjective optimization algorithms have such shortcomings as slow convergence to the Pareto front, no efficient termination criterion, and a lack of theoretical convergence proof. In this paper, we have proposed a modified P-ACO for tackling the above problems. The proposed algorithm is used to seek better compromise solutions between maximizing the expected damage and minimizing the cost. Based on the framework of the traditional P-ACO, the proposed algorithm is devised with several attractive features to enhance the optimization performance, including a dynamic heuristic information calculation approach, an improved movement probability rule, a dynamic evaporation rate strategy, a global updating rule of pheromone, and a boundary symmetric mutation strategy.

Dynamic Heuristic Information Calculation Approach. Because heuristic information seriously affects the efficiency of ACO algorithm, the problem-specific heuristic information function is one of the most important problems that should be solve. Hence, we design a dynamic heuristic information calculation approach for providing necessary guidance for an efficient search.

Improved Movement Probability Rule. In order to avoid P-ACO plunge into the local optimum which is caused by the stagnation phenomenon, we proposed a dynamic movement probability rule to enhance local exploitation and alleviate the premature convergence.

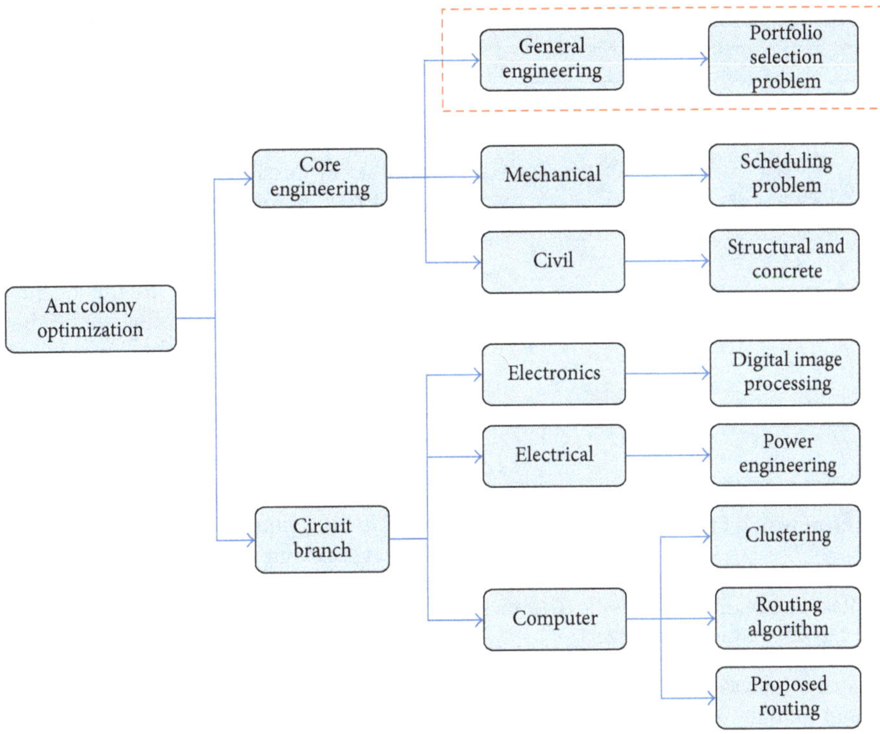

Figure 1: The main applications of ACO algorithm in recent years.

Dynamic Evaporation Rate Strategy. When the algorithm is used to solve large-scale WTA problem, a static pheromone evaporation rate will reduce global searching ability and convergence speed. Therefore, a dynamic evaporation rate strategy is proposed in this paper.

Global Updating Rule of Pheromone. The traditional static pheromone updating rule makes the algorithm easy to fall into local optimum. Hence, a global pheromone updating rule is introduced to solve the conflict between expanding search and finding optimal solution.

Boundary Symmetric Mutation Strategy. When solving large-scale problem, the distributions of targets are complex. In order to avoid the search being stuck in local optimum and diversify the solutions, a boundary symmetric mutation strategy is proposed to improve the mutation efficiency and the quality of solutions.

The rest of this paper is organized as follows. In Section 2, a mathematical model of the BOWTA problem and a problem-specific assumption description are presented. Section 3 provides algorithmic descriptions of P-ACO algorithm and presents the application of ACO algorithm for solving WTA problem. Detailed improvements of the proposed algorithm are introduced in Section 4. In Section 5, experimental evaluation is carried out and the analysis of the results is presented. Firstly, in order to validate effectiveness and show performance, the proposed algorithm is applied in a specific instance. Secondly, the proposed algorithm and others, like NSGA-II [33], SPEA-II [34], and traditional P-ACO [28], are tested on several different scales WTA problems. Section 6 provides some conclusion of the paper.

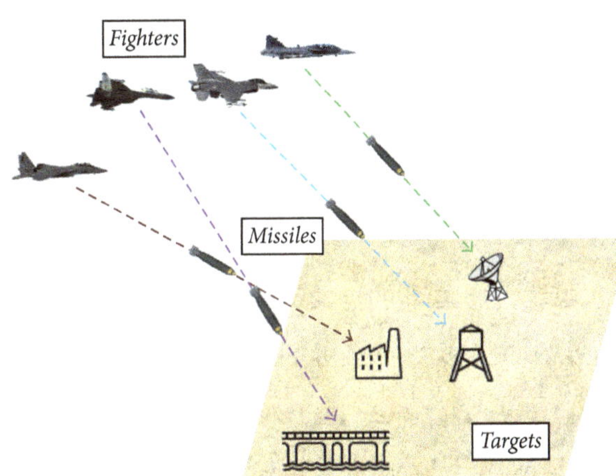

Figure 2: Illustration of the WTA problem.

2. Problem Formulation

The WTA problem focuses on how to allocate attack units to targets and can be illustrated in Figure 2. WTA assumptions and mathematical model are introduced as follows in this section.

2.1. Assumption Description. In order to establish a reasonable WTA mathematical model, the following assumptions can be defined.

Assumption 1. We assume that the mathematical model is composed of F fighters, M missiles, and T targets. The fighters

TABLE 2: The decision table of WTA.

	T_1	T_2	T_3	T_4	$\cdots$	T_n
M_1	x_{11}	x_{12}	x_{13}	x_{14}	$\cdots$	x_{1n}
M_2	x_{21}	x_{22}	x_{23}	x_{24}	$\cdots$	x_{2n}
M_3	x_{31}	x_{32}	x_{33}	x_{34}	$\cdots$	x_{3n}
M_4	x_{41}	x_{42}	x_{43}	x_{44}	$\cdots$	x_{4n}
M_5	x_{51}	x_{52}	x_{53}	x_{54}	$\cdots$	x_{5n}
M_6	x_{61}	x_{62}	x_{63}	x_{64}	$\cdots$	x_{6n}
M_7	x_{71}	x_{72}	x_{73}	x_{74}	$\cdots$	x_{7n}
M_8	x_{81}	x_{82}	x_{83}	x_{84}	$\cdots$	x_{8n}
$\vdots$	$\vdots$	$\vdots$	$\vdots$	$\vdots$	x_{ij}	$\vdots$
M_m	x_{l1}	x_{l2}	x_{l3}	x_{l4}	$\cdots$	x_{mn}

and targets are not necessarily equal in quantity (each fighter is equivalent to one platform which takes different kinds and quantities of missiles).

Assumption 2. Each fighter can use different missiles to attack one target (each missile can only attack one target).

Assumption 3. All allocated missiles cannot exceed the number of missiles resources in military air operation.

Assumption 4. We assume the probability of kill (q_{ij}) between the missile (ith unit of M) and the target (jth unit of T).

Assumption 5. If the target is in work area, then a missile can be assigned effectively. On the other hand, the missile is not.

2.2. Mathematical Model. The multiobjective optimization is to make the number of weapons to targets reach the maximum and reduce missiles consumption as much as possible. The definitions about the optimization model are shown as follows.

Definition 6. The multiobjective functions of WTA are to maximize the total effectiveness of attack and minimize the cost of missiles. The optimization mathematical model is described as

$$f^1 = \max \sum_{j=1}^{n} \left[1 - \prod_{i=1}^{m} \left(1 - \rho_k q_{ij}\right)^{x_{ij}} \right],$$
$$f^2 = \min \sum_{i=1}^{m} \sum_{j=1}^{n} c_i x_{ij}, \tag{1}$$
$$f = \min \left(\frac{1}{f^1}, f^2 \right).$$

Definition 7. $q_{ij} \in (0,1)$ represents the effectiveness of the missile unit. It is known as the probability of kill for each T target ($j = 1, 2, \ldots, n$) and M missile ($i = 1, 2, \ldots, m$).

Definition 8. The decision table $X = [x_{ij}]_{M\times T}$ can be shown as Table 2.

Here, x_{ij} is a Boolean value ($x_{ij} = \{0, 1\}$) and represents whether i missile is assigned to j target. If the i missile of k fighter is assigned to the j target, $x_{ij} = 1$; otherwise, x_{ij} is equal to 0.

Definition 9. c_i denotes the cost of i missile and is given a constant value in this paper.

Definition 10. In combat, different fighters are equipped with different numbers of missiles, so one assumes that the number of missiles carried by each fighter is not necessarily equal (the number of missiles carried by each plane is no more than 4).

Definition 11. According to the talent, training time, and operation stability of the pilot, all the above factors may affect the attack performance of the missile. One assumes that ρ_k is pilot operation factor $\rho_k \in (0, 1)$.

Definition 12. There are some constraints that the above function variables must satisfy.

Missile Assignment Constraint. For simplicity, we assume that at least one missile unit is assigned to one target. This constraint can be expressed as

$$\sum_{i=1}^{m} x_{ij} \geq 1 \quad \forall j = 1, 2, \ldots, n. \tag{2}$$

Quantity Assignment Constraint. We assume that one missile can only attack one target. The corresponding constraint is

$$\sum_{j=1}^{n} x_{ij} \leq 1 \quad \forall i = 1, 2, \ldots, m. \tag{3}$$

Parameters Constraint. In this paper, we will use some parameters like x_{ij}. The corresponding constraint is defined as follows:

$$\sum_{i=1}^{m} \sum_{j=1}^{n} x_{ij} \leq m. \tag{4}$$

3. Ant Colony Optimization Algorithm

3.1. Pareto Ant Colony Optimization Algorithm (P-ACO). The ACO is a heuristic algorithm which has emerged in recent years and has been applied to solve hard combinatorial optimization problems [35]. Scientists have developed ACO by imitating the behaviors of real ant colonies in the process of searching for food. When ants in the colony find a short path for gathering food by frequently travelling between their nest and the food source, they deposit a chemical called *pheromone trail* on the path which can be followed by other ants to search for the food source [36].

Pareto ant colony optimization (P-ACO) algorithm is one of the multiple-objective ant colony optimization (MOACO) algorithms that was proposed by Doerner et al. [28] to be originally applied to optimize the multiobjective portfolio selection. The main objective of P-ACO algorithm is to find a set of solutions ordered by fronts under the concept of Pareto front. Although it is based on the classical ACS, the update process of *global pheromone trail* is implemented by using two different ants (the best and the second-best solutions) which are generated by each objective g in current iteration.

```
(0) Pre-processing;
(1) Initialization: generate the number of ants A, initialize pheromone vectors with τ0,
        set the values of α and β;
(2) repeat until Termination conditions has not been met {
(3)   for Ant = 1 to A {
        determine the lifespan Θ of the ant randomly of the interval [1, 2, ..., N];
        create x = (0, ..., 0)/*built empty assignment solution*/;
        determine the objective weight w_g for each objective g;
        indicates the number of targets to be selected ξ = Θ;
(4)   while ξ > 0 and ∃i with η_i(x) > 0 {
        select an assignment ij using formula (5) and add it to x;
        update local pheromone information;
        decrement ξ;
        }
        check feasibility x;
(5)     if x is feasible {
          check efficiency x;
(6)     if x is efficient {
          store x and remove dominated ones;
      }}}
(7) for each objective g {
      determine best and second-best solution Δτ_ij^g for each objective g;
      update global pheromone information using best and second-best solution;
      using formula (8);
}}.
```

ALGORITHM 1

In P-ACO, τ_g denotes the pheromone matrices of objective g. During each iteration, w_g represents the weight of objective g and is used to combine the pheromone information and heuristic information. In the route, an ant selects the next node s to be visited from node r, and it uses the ACS transition rule like formula (5). ζ is a random number from 0 to 1. ζ_0 is a fixed parameter ($\zeta_0 \in (0, 1]$). When $\zeta \leq \zeta_0$, the available knowledge is exploited for selecting the best option through pheromone trail heuristic information. The formula is shown below:

$$j = \begin{cases} \arg\max_{j\in\Omega} \left(\left[\sum_{g=1}^{G} w_g \cdot \tau_{ij}^{g} \right]^{\alpha} \cdot \left[\eta_{ij} \right]^{\beta} \right), & \text{if } \zeta \leq \zeta_0 \\ \hat{j}, & \text{otherwise,} \end{cases} \quad (5)$$

where $g = \{1, 2, \dots, G\}$ is a set of objectives, η_{ij} represents the heuristic information of edge e_{ij}, and $\hat{j}$ is a probabilistic node. α is the weight coefficient of pheromone information and β is the weight coefficient of heuristic information. α and β belong to R. Ω is the current feasible neighborhood of the ant.

When $\zeta > \zeta_0$, p_{ij} is implemented for finding $\hat{j}$ and the formula is expressed as

$$p_{ij} = \begin{cases} \dfrac{\left[\sum_{g=1}^{G} w_g \cdot \tau_{ij}^{g} \right]^{\alpha} \cdot \left[\eta_{ij} \right]^{\beta}}{\sum_{u\in\Omega} \left[\sum_{g=1}^{G} w_g \cdot \tau_{iu}^{g} \right]^{\alpha} \cdot \left[\eta_{iu} \right]^{\beta}}, & \text{if } j \in \Omega \\ 0, & \text{otherwise.} \end{cases} \quad (6)$$

When an ant has travelled an edge e_{ij} in the algorithm, *local pheromone update* (online pheromone trail update) is implemented for each objective g by the rule

$$\tau_{ij}^{g} = (1 - \rho) \cdot \tau_{ij}^{g} + \rho \cdot \tau_0, \quad (7)$$

where ρ is the pheromone evaporation rate ($\rho \in (0, 1]$) and τ_0 is the value of initial pheromone trail. In this paper, we use the same method as the literature [37] to set the value of τ_0.

When each ant has constructed its solution, *global pheromone update* (offline pheromone trail update) is implemented for each objective g. The update formula is shown below:

$$\tau_{ij}^{g} = (1 - \rho) \cdot \tau_{ij}^{g} + \rho \cdot \Delta\tau_{ij}^{g}, \quad (8)$$

where $\Delta\tau_{ij}^{g}$ follows the problem-specific values:

$$\Delta\tau_{ij}^{g} = \begin{cases} 10 & \text{if edge } e_{ij} \in \text{best solution,} \\ 5 & \text{if edge } e_{ij} \in \text{second-best solution,} \\ 0 & \text{otherwise.} \end{cases} \quad (9)$$

In the process, the nondominated solutions are stored in an external set S.

The pseudocode of the P-ACO algorithm can be demonstrated as shown in Algorithm 1.

3.2. Application of ACO Algorithm for Solving WTA Problem. At the beginning, each missile has a list of possible target assignments (Figure 3). The number of assignments

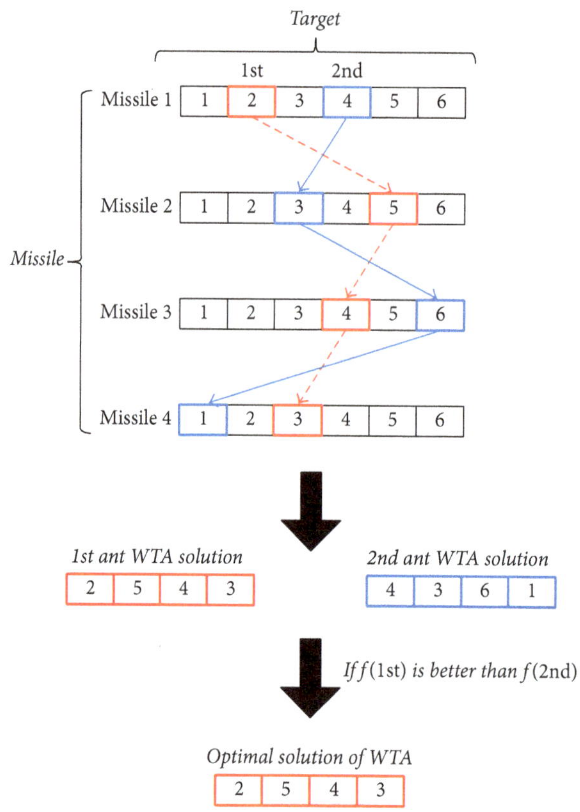

FIGURE 3: Example of WTA solutions (2 ants, 4 missiles, and 6 targets).

is independent of the conditions which are set at first. Each ant builds a tour (WTA solution) while moving from one to another list until all of them have been visited. Figure 3 gives an example with two ants, four missiles, and six targets during an iteration. $M = \{1, 2, \ldots, m\}$ is a set of missiles. The size of each list is fixed to the number of targets ($T = \{1, 2, \ldots, n\}$). To construct a WTA solution (a tour), the 1st ant visits Target 2 (missile 1), Target 5 (missile 2), Target 4 (missile 3), and Target 3 (missile 4) by using a transition rule which takes the pheromone trails of visited nodes and the heuristic information into account. On the contrary, the 2nd ant visits Target 4 (missile 1), Target 3 (missile 2), Target 6 (missile 3), and Target 1 (missile 4). If the 1st solution is better than the 2nd solution, then the 1st solution is the best solution at current iteration and the algorithm enters the next iteration.

4. Improvements of the MPACO Algorithm

4.1. Dynamic Heuristic Information Calculation Approach. The heuristic information η_{ij} denotes the probability of the ant moving from i to j and is an optional weighting function calculated by heuristic approach. Because η_{ij} must be calculated for all ants in each iteration, it may seriously affect the efficiency of the ACO algorithm. How to design the problem-specific heuristic information function is one of the important problems that should be solved. There are two main calculation approaches of heuristic information at present: (1) η_{ij} is calculated at first and remains unchanged in each iteration until the algorithm finishes [23] and (2) compared with the first approach η_{ij} is given a dynamic value according to the current iteration [38]. In the first approach, η_{ij} is more efficient but cannot completely indicate the probability. The second approach can satisfy precise estimation of the probability but will relatively reduce the efficiency of computation. In order to solve the WTA problem, an improved calculating approach for heuristic information which considers the efficiency of calculation and the quality of solution is proposed. The values of biobjective functions are considered in the approach. The formulas can be shown as follows:

$$\eta_{ij}^1 = \sum_{j=1}^{n} \left[1 - \prod_{i=1}^{m} \left(1 - \rho_k q_{ij}\right)^{x_{ij}} \right],$$
$$\eta_{ij}^2 = \sum_{i=1}^{m} \sum_{j=1}^{n} c_i x_{ij}, \tag{10}$$
$$\eta_{ij} = \frac{\vartheta \cdot \eta_{ij}^1}{\eta_{ij}^2},$$

where η_{ij}^1 and η_{ij}^2 dividedly denote the first objective function and the second one. η_{ij} is the total probability heuristic information which is represented as η_{ij}^1 multiplied by η_{ij}^2. ϑ is a positive constant.

4.2. Improved Movement Probability Rule. In practice, ACO algorithm is easy to plunge into the local optimum which is caused by stagnation phenomenon. So, the dynamic movement probability rule based on essential selection ($q \le q_0$) and random selection ($q > q_0$) is proposed to improve the strategy of global search in the process. The phenomenon τ_{ij} of the paths which have been visited is enhanced. The selected probability of the paths will increase in the next iteration. Because few ants visit the path, some better solutions cannot be stored at the beginning iterations. If more ants can visit the path less traveled in the process, it will be beneficial to the solutions space and effectively perfect the P-ACO algorithm. The probability p_{ij} can be computed as

$$p_{ij} = \begin{cases} \dfrac{\left[\sum_{g=1}^{G} p_g \cdot \tau_{ij}^{g}\right]^{\alpha} \cdot \eta_{ij}^{\beta} \cdot z_{ij}}{\sum_{u\in\Omega}\left[\sum_{g=1}^{G} p_g \cdot \tau_{iu}^{k}\right]^{\alpha} \cdot \eta_{iu}^{\beta} \cdot z_{iu}}, & \text{if } j \in \Omega \\ 0, & \text{otherwise}, \end{cases} \quad (11)$$

$$z_{ij} = \frac{A \cdot I_c}{A \cdot I_c + \delta \cdot A_{\text{sf}} \cdot \eta_{ij}/\eta_{\max}},$$

where A denotes the number of ants according to the specific problem, I_c represents the current iteration, δ is a constant, A_{sf} is the total number of ants that have visited the same node so far, and $\eta_{\max}$ is the maximum value of η_{ij}. The value of z_{ij} takes into account A_{sf} and η simultaneously. If the current iteration tends to the sub-Pareto solution, the pheromone τ_{ij} of sub-Pareto solution will be enhanced. But the number of A_{sf} increased and the value of z_{ij} decreased. This strategy can restrain the excessive growth of pheromone and escape from the local optimum.

4.3. Dynamic Evaporation Rate Strategy. In classical P-ACO algorithm, the pheromone evaporation rate ρ is a fixed parameter ($\rho \in (0,1]$). The value of ρ is one of the main factors which have an influence on global search ability and convergence speed. When P-ACO algorithm is applied to solve large-scale problem, the value of pheromone τ is close to 0 and global search ability declines. If the value of pheromone becomes very large, the selection probability of the passed nodes increases and global search ability is also directly affected. How to design the value of pheromone is crucial to the release and evaporation of pheromone. In order to improve the global search ability, the common strategy is to set a large value of ρ at the beginning of the algorithm. But the evaporation rate will also decay in the process, so that the P-ACO algorithm can quickly fall into local optimum. Two advantages are brought about by the dynamic evaporation rate strategy: (1) better global search ability and (2) faster convergence speed. The mathematical expression of decay model is shown below:

$$\rho_{I_c} = \frac{I_{\max} \cdot I_c \cdot \left[\sum_{g=1}^{G} w_g \left(\tau_{\max}^{g} - \tau_{\min}^{g}\right)\right]}{I_{\max} - 1} + \frac{\sum_{g=1}^{G} w_g \left(I_{\max} \cdot \tau_{\min}^{g} - \tau_{\max}^{g}\right)}{I_{\max} - 1}, \quad (12)$$

where $I_{\max}$ is the maximum iteration and is set at the beginning and I_c is the current iteration. $\tau_{\max}^{g}$ denotes the maximum of pheromone and $\tau_{\min}^{g}$ denotes the minimum of pheromone.

4.4. Global Updating Rule of Pheromone. A fixed pheromone update rule is mainly used in the classical P-ACO algorithm. The traditional strategy may flung the algorithm into local optimum and has three disadvantages: (1) the distribution features of solution are ignored; (2) there is the trend of stagnation; (3) the convergence rate is slow. In addition to the pheromone local updating rule, the proposed MPACO adopts an adaptive adjustment global updating rule which can effectively deal with the conflict between expanding search and finding Pareto solution. In this rule, the pheromone information of each objective is updated by the best and second-best solution. The rule also provides some instructive guidance to improve the quality of the solutions. The pheromone $\Delta\tau_{ij}^{g}$ can be computed as

$$\Delta\tau_{ij}^{g} = \begin{cases} \dfrac{1}{f_b^{g}} & \text{if } x_{\text{best},ij}^{g} = 1 \\ \dfrac{1}{f_{sb}^{g}} & \text{if } x_{\text{second-best},ij}^{g} = 1 \\ 0 & \text{otherwise}, \end{cases} \quad (13)$$

where x_{ij} is an assignment of a population best (or second-best) ant according to objective g. The strategy maintains the Pareto exploitation and balances the random search and the function about pheromone evaporation or augmentation. In order to reduce the otherness of pheromone on the local Pareto solution and expand the search scope, the positive feedback of P-ACO algorithm is restrained by adding a little amount of negative feedback pheromone in the process. If the algorithm falls into a local Pareto point during a period of iterations, then the adaptive adjustment strategy will be adopted to decrease the amount of pheromone for escaping from local Pareto front. The two expressions are similar in format, but the objective function is different.

4.5. Boundary Symmetric Mutation Strategy. Based on the mathematical statistics theory, the distribution of the samples will be close to the normal distribution when the number of samples tends to infinity. In solving WTA, targets coordinates according to the values are sorted. When a larger scale problem is solved, the distributions of targets are complex and a majority of targets are on the central region. The path of ants is boundary-center-boundary in the initial stage, so that all ants will follow the original track mode.

In order to solve the above problem, the mutation strategy is used for boundary mutation which can improve the mutation efficiency and the quality of solutions. Based on the quartile idea in probability theory, the mutation strategy is applied to mutate 1/4 solutions in both sides of the path. Due to the symmetry of the boundary, crossover and mutation only take place in the selected boundary which did not happen among others.

According to the above improvements, a modified Pareto ant colony optimization (MPACO) algorithm based on dynamic calculation approach of heuristic information, improved movement probability rules, improvement strategy of evaporation factor, improved global updating rules of pheromone, and boundary symmetric mutation strategy is proposed in this paper.

4.6. The Structure of MPACO Algorithm. According to the idea of the proposed MPACO algorithm, the structure of MPACO algorithm in solving BOWTA problem can be shown as follows.

Begin

Step 0 (preprocessing). These nodes of decision table are divided into center nodes and boundary nodes according to q_{ij} which represents the effectiveness of the missile. The classification result is saved and sorted in order to prepare for the subsequent run of the MPACO algorithm.

Step 1 (initialization of parameters for the proposed MPACO). These parameters include the colony of ants A, the maximum iteration $I_{\max}$, the pheromone information factor α, heuristic information factor β, and initial uniform probability ζ_0.

While (Termination conditions has not been met) *do*

For $I = 1, 2, \ldots, I_{\max}$.

For $A = 1, 2, \ldots, A_{\max}$.

Step 2. The A ants randomly visit targets of each missile by constraints (2), (3), and (4), and these assignments are added into the decision list of the ant $X_a = \{x_{ij}^1, x_{ij}^2, \ldots, x_{ij}^m\}$.

Step 3. Due to the dynamic calculation approach of heuristic information, formulas (10) are used to calculate coupling η_{ij} before Step 4.

Step 4. According to (11) of improved movement probability rules, the probability that the ant selects the next node is calculated (Section 4.2). Then, this node is added into the decision list, and the pheromone is locally updated.

Step 5. After the ants have completed a selection, the fitness function f will be calculated. Then, the decision list is modified. Repeat Step 4 until the ant completed the tour (namely, all missiles are allocated). The current best and second-best solutions are saved and the global best solution is updated for the current iteration.

End for

Step 6. Based on (8) and (13), the pheromone is globally updated.

Step 7. Update Pareto optimal front set.

Step 8. Set the iterative counter $I = I + 1$ for the iteration control. If $I < I_{\max}$, return to Step 2.

End while

Step 9. The proposed MPACO algorithm is terminated; output Pareto solutions.

Table 3: Parameters of MPACO algorithm (part).

Parameter	Value
A	20
α	1
β	2
$w_{1,2}$	0.5
τ_0	Formula (14)
ζ_0	0.98

Table 4: Parameters of the simulation environment.

Computer specifications	Intel(R) Core(TM) i5-4460T CPU @ 1.90 GHz with 8.00 GB RAM
Operating system	Windows 7 (x64) Operation System
Language	C++
Software	Microsoft Visual C++ 6.0

5. Experiment Results and Analysis

5.1. Parameter Settings. In order to select the initial value of pheromone τ_0, we use the formula in the literature [39], which can be shown as follows:

$$\tau_0 = \frac{G}{\sum_{g=1}^{G} f^g(X_g)}, \tag{14}$$

where X_g is a solution of the gth objective computed by a greedy algorithm, G is the number of objectives in problem-specific function, and $f^g(X_g)$ is the fitness function of the solution X_g according to the gth objective.

Based on the numbers of targets and missiles in a different BOWTA scale, the maximum number of iterations $I_{\max}$ is set. Some parameters of MPACO algorithm are shown in Table 3.

We can see from Table 4 that the numerical experiments have been performed on a PC with Intel(R) Core(TM) i5-4460T CPU @ 1.90 GHz, with 8.00 GB of RAM, running Windows 7 64 bits as the operating system. The codes of all algorithms and optimization models have been written in C++ under Microsoft Visual C++ 6.0.

5.2. Numerical Experiments and Analysis

5.2.1. Computation Results on Specific Example. A specific example is designed to verify performances of the algorithm. Assume that there are 4 fighters that carry different numbers of missiles (12 missile units in total) and 10 targets in the simulation. Missiles damage probability p_{ij} ($i = 1, \ldots, 12$; $j = 1, \ldots, 10$), pilot operation factor ρ_k ($k = 1, \ldots, 4$), and cost of missiles c_i are shown in Tables 5–7.

According to the literature [1], we present the missiles cost, as shown in Table 7.

TABLE 5: The table of missiles damage probability.

Missile unit	Target									
	1	2	3	4	5	6	7	8	9	10
1	0.82	0.85	0.78	0.75	0.52	0.88	0.44	0.76	0.72	0.56
2	0.56	0.72	0.88	0.46	0.64	0.47	0.68	0.45	0.48	0.75
3	0.45	0.62	0.54	0.73	0.84	0.76	0.78	0.42	0.53	0.65
4	0.56	0.42	0.76	0.84	0.73	0.83	0.86	0.62	0.78	0.82
5	0.45	0.58	0.81	0.44	0.63	0.59	0.78	0.77	0.65	0.70
6	0.46	0.61	0.55	0.68	0.75	0.83	0.73	0.66	0.82	0.48
7	0.66	0.71	0.65	0.44	0.86	0.79	0.59	0.85	0.53	0.56
8	0.56	0.42	0.76	0.84	0.73	0.72	0.44	0.75	0.48	0.47
9	0.88	0.78	0.44	0.67	0.56	0.86	0.58	0.65	0.73	0.42
10	0.56	0.88	0.68	0.45	0.75	0.73	0.61	0.76	0.84	0.78
11	0.84	0.54	0.44	0.42	0.65	0.56	0.71	0.55	0.45	0.88
12	0.83	0.76	0.84	0.62	0.82	0.75	0.42	0.68	0.57	0.54

TABLE 6: Pilot operation factor table.

Fighter	Missile unit	p_k
F_1	M_1, M_2, M_3	0.96
F_2	M_4, M_5	0.95
F_3	M_6, M_7, M_8, M_9	0.98
F_4	M_{10}, M_{11}, M_{12}	0.93

Firstly, an enumeration method is employed to the specific example instance and a set of evenly distributed true Pareto solutions (PSs) are obtained. Secondly, the proposed algorithm is used to find PSs.

Figure 4 presents the distribution of the true PSs solved by enumeration method and final PSs obtained by the MPACO in the objective space. We can see from the figure that the MPACO can find close and even PSs in the objective space. It is evident that MPACO can guarantee the quality of solutions. So, the MPACO for optimizing the BOWTA is well verified to be feasible.

We can get some results from Figure 5 and Table 8 as follows: when funds of national defense are enough, we can choose Scheme 1 which cost the most money and obtain the greatest expected damage of the enemy. When funds are insufficient, we should choose one scheme among Schemes 34, 35, and 36 on the basis of threat coefficient of each target. Considering that all targets must be allocated and the cost of missiles should be minimized, we can only choose Scheme 4.

5.2.2. Comparison Study. In order to verify the applicability and feasibility of the proposed algorithm, we test MPACO, P-ACO, NSGA-II, and SPEA-II on Scale 1, Scale 2, and Scale 3 instances. Three examples represent, respectively, small-, medium-, and large-scale BOWTA instances under the same condition. The quantitative indexes, such as Pareto figure and time performance, are chosen to evaluate the performance on C++. The value of three instances can be seen from Table 9.

In what follows, Scale 1 instance is firstly taken as an example to illustrate the performance of MPACO, P-ACO, NSGA-II, and SPEA-II. Then, Scale 2 and Scale 3 instances are introduced to further demonstrate the performances of both algorithms. The statistical results are shown in Figures 5–7.

Figures 5–7 present the distribution of PSs obtained in 30 runs with the best performance of each algorithm for each test instance in objective space. When especially tested on Scale 2 and Scale 3, the MPACO can obtain the best solutions among the four algorithms. Due to increments of dynamic calculation approach of heuristic information and improved movement probability rules, the calculation accuracy of MPACO algorithm is improved. Boundary symmetric mutation strategy is also good for the algorithm to solve large-scale BOWTA problem.

The primary aim of solving BOWTA problems is to lay the foundation for the DWTA problems whose goal is to provide a set of near-optimal or acceptable real-time decisions in air combat. So, in practice, the time performance of algorithms is also an important index. In the end, we test four algorithms on three instances in 30 runs and record the literation time of each algorithm. The statistical results of time performances are shown in Figures 8–10.

Although MPACO does not have time advantage in solving Scale 1 and Scale 2 compared with other algorithms, it runs as fast as SPEA-II with the same number of function evaluations on Scale 3. Boundary symmetric mutation strategy improves the mutation efficiency and the quality of solutions and new evaporation factor is good for solving large-scale problems, but dynamic heuristic information and

TABLE 7: The cost of each missile table.

	Missile unit											
	M_1	M_2	M_3	M_4	M_5	M_6	M_7	M_8	M_9	M_{10}	M_{11}	M_{12}
c_i	0.62	0.63	0.69	0.80	0.72	0.90	0.96	0.68	0.72	0.65	0.66	0.65

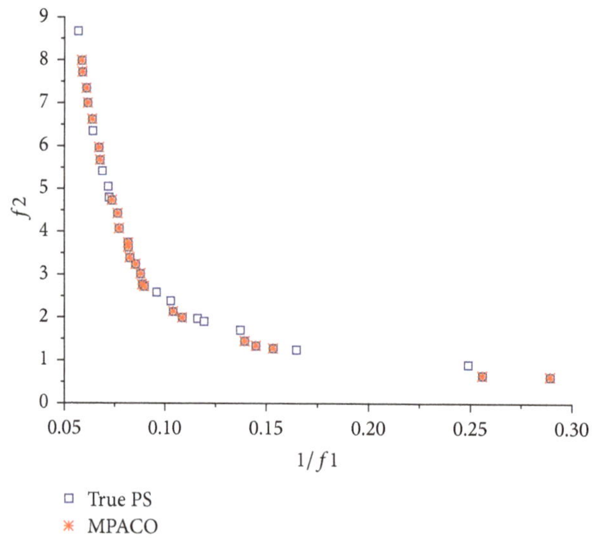

FIGURE 4: Plots of the true PS and final PSs found by MPACO on the specific example instance.

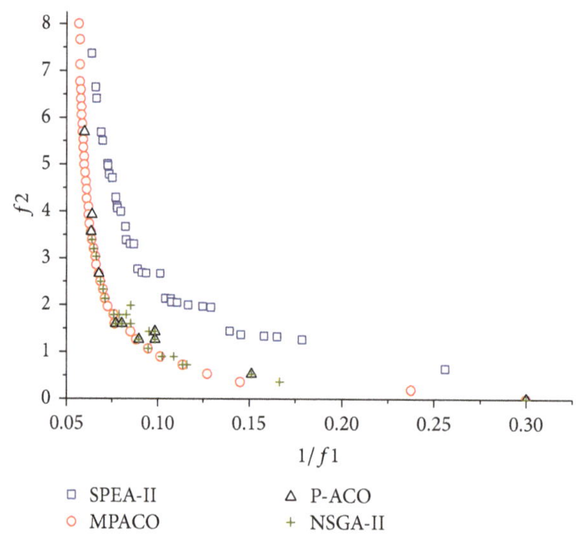

FIGURE 5: Plots of PSs best performance in 30 runs of four algorithms on Scale 1.

probability rules affect the iteration rate. In real air combat situations, pilots often make deadly decisions within seconds or even within milliseconds. Although the algorithm structure and implementation process are relatively complex and increase the computational complexity, high-performance hardware platform can be adopted to reduce the processing time and satisfy the real-time requirement.

In this section, we do some work as follows. Firstly, we test the MPACO on a specific example. We verify the feasibility after comparing two kinds of PSs and show the corresponding distribution results. Secondly, to give a comparison study among the four algorithms, three different scale instances are introduced to evaluate each algorithm. Plots of the best PSs in 30 runs of each algorithm are presented to have an intuitive comparison. Lastly, we show the time performance of the four algorithms on each instance. To summarize, MPACO has been proven to be an effective technique for BOWTA optimization problem and obviously is the best among the four algorithms.

6. Conclusion

We apply ACO to WTA problem and propose the MPACO to solve BOWTA in this paper. The main contributions of

TABLE 8: The value of PSs and results of WTA distribution.

	$1/f1$	$f2$	M_1	M_2	M_3	M_4	M_5	M_6	M_7	M_8	M_9	M_{10}	M_{11}	M_{12}
1	0.056798	8.68	6	3	7	7	8	9	5	4	1	9	10	2
2	0.058623	7.99	6	3	0	7	8	9	5	4	1	9	10	2
3	0.059	7.72	6	3	5	7	8	9	0	4	1	9	10	2
4	0.060966	7.34	6	3	0	7	8	9	5	4	1	2	10	0
5	0.061662	7.00	6	3	5	7	0	9	0	4	1	9	10	2
6	0.063811	6.62	6	3	0	7	0	9	5	4	1	2	10	0
7	0.064259	6.35	6	3	5	7	0	9	0	4	1	2	10	0
8	0.067327	5.96	6	3	0	7	0	9	5	4	1	2	0	0
9	0.067848	5.67	6	3	5	7	0	9	0	0	1	2	10	0
10	0.068893	5.42	6	3	5	7	0	0	0	0	1	9	10	2
11	0.071974	5.05	0	3	5	7	0	9	0	0	1	2	10	0
12	0.072395	4.79	6	3	5	7	0	0	0	4	1	9	0	0
13	0.073822	4.73	6	3	0	7	8	0	0	0	0	9	10	5
14	0.07645	4.42	0	3	5	7	0	9	0	4	1	0	0	0
15	0.077215	4.07	6	3	5	7	0	0	0	4	0	9	0	0
16	0.081585	3.74	0	3	5	7	0	9	0	0	1	0	0	0
17	0.081702	3.64	6	3	5	7	0	9	0	0	0	0	0	0
18	0.082457	3.39	6	3	5	7	0	0	0	0	0	9	0	0
19	0.085299	3.24	6	3	7	0	0	0	0	0	0	9	0	5
20	0.087759	3.02	0	3	5	7	0	9	0	0	0	0	0	0
21	0.08863	2.77	0	3	5	7	0	0	0	0	0	9	0	0
22	0.089675	2.73	0	3	0	7	0	0	0	0	0	9	0	5
23	0.09589	2.59	9	3	7	0	0	0	0	0	0	0	0	5
24	0.103038	2.39	0	0	5	7	0	9	0	0	0	0	0	0
25	0.104241	2.14	0	0	5	7	0	0	0	0	0	9	0	0
26	0.108826	1.99	0	0	7	0	0	0	0	0	0	9	0	5
27	0.116404	1.97	0	3	7	0	0	0	0	0	0	9	0	0
28	0.119689	1.90	9	7	0	0	0	0	0	0	0	0	0	5
29	0.13725	1.70	0	0	0	7	0	9	0	0	0	0	0	0
30	0.139392	1.45	0	0	0	7	0	0	0	0	0	9	0	0
31	0.144902	1.34	0	0	7	0	0	0	0	0	0	9	0	0
32	0.15344	1.28	0	7	0	0	0	0	0	0	0	9	0	0
33	0.164821	1.25	9	7	0	0	0	0	0	0	0	0	0	0
34	0.24888	0.90	0	0	0	0	0	9	0	0	0	0	0	0
35	0.256016	0.65	0	0	0	0	0	0	0	0	0	9	0	0
36	0.289352	0.62	9	0	0	0	0	0	0	0	0	0	0	0

TABLE 9: The values of Scale 1, Scale 2, and Scale 3 instances.

	Scale 1	Scale 2	Scale 3
Target	10	48	98
Missile	10	53	100

the thesis are summarized as follows. On the one hand, the expected damage of enemy and the cost of missiles are taken into account from a practical viewpoint; on the other hand, pilot operation factor and missiles constraints are introduced to make the model in line with the real air combat. In this aspect, five strategies are proposed and embedded into traditional P-ACO algorithm to improve performances of MPACO algorithm. The experiments have shown that MPACO can find better Pareto solutions than the other three algorithms for all instances. More importantly, MPACO is more suitable for solving large-scale problems from the time performance viewpoint.

However, we mainly have studied the static WTA problem; few researches focused on dynamic problems which are

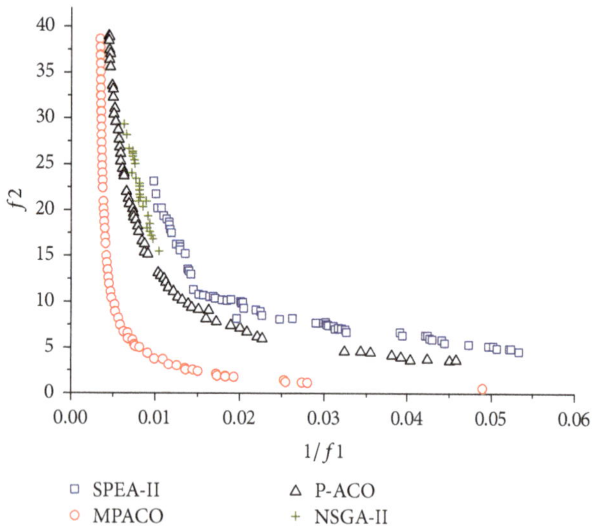

FIGURE 6: Plots of PSs best performance in 30 runs of four algorithms on Scale 2.

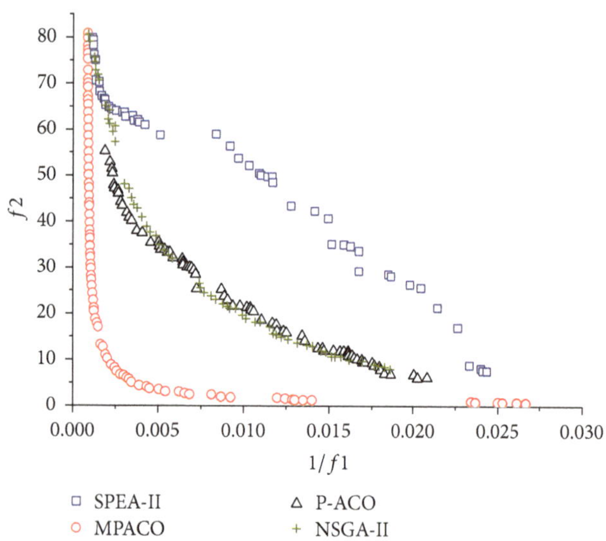

FIGURE 7: Plots of PSs best performance in 30 runs of four algorithms on Scale 3.

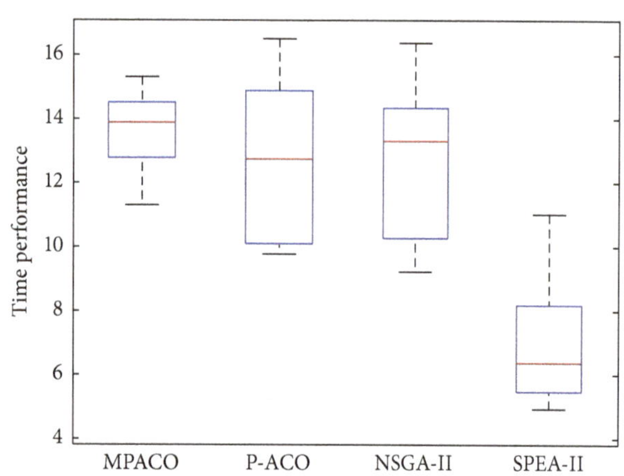

FIGURE 8: Boxplot of time performance on Scale 1 (in secs).

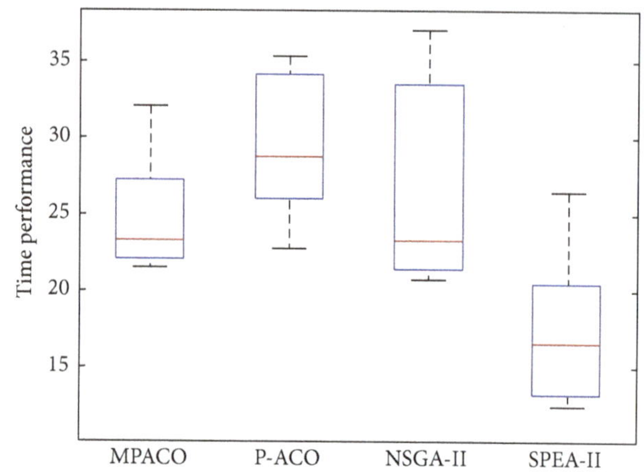

FIGURE 9: Boxplot of time performance on Scale 2 (in secs).

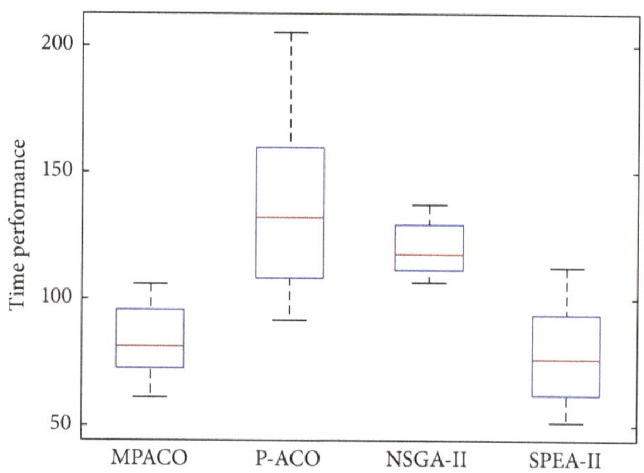

FIGURE 10: Boxplot of time performance on Scale 3 (in secs).

more instructive to real combat. In recent years, more and more researches began to pay attention to DWTA problems. A further study on this topic is one of our future tasks.

Conflicts of Interest

The authors declare that there are no conflicts of interest regarding the publication of this paper.

References

[1] R. N. Rai and N. Bolia, "Optimal decision support for air power potential," *IEEE Transactions on Engineering Management*, vol. 61, no. 2, pp. 310–322, 2014.

[2] S. Matlin, "A review of the literature on the missile-allocation problem," *Operations Research*, vol. 18, no. 2, pp. 334–373, 1970.

[3] Z.-J. Lee, S.-F. Su, and C.-Y. Lee, "Efficiently solving general weapon-target assignment problem by genetic algorithms with greedy eugenics," *IEEE Transactions on Systems, Man, and Cybernetics, Part B: Cybernetics*, vol. 33, no. 1, pp. 113–121, 2003.

[4] E. Çetin and S. T. Esen, "A weapon-target assignment approach to media allocation," *Applied Mathematics and Computation*, vol. 175, no. 2, pp. 1266–1275, 2006.

[5] X.-Y. Wang, C.-Z. Hou, J.-M. Yuan, F. Guo, and W. Hao, "Modeling and optimization method on antiaircraft firepower allocation," *Control & Decision*, vol. 21, no. 8, pp. 913–917, 2006.

[6] B. Alidaee, H. Wang, and F. Landram, "A note on integer programming formulations of the real-time optimal scheduling and flight path selection of UAVs," *IEEE Transactions on Control Systems Technology*, vol. 17, no. 4, pp. 839–843, 2009.

[7] A. S. Manne, "A target-assignment problem," *Operations Research*, vol. 6, pp. 346–351, 1958.

[8] R. H. Day, "Allocating weapons to target complexes by means of nonlinear programming," *Operations Research*, vol. 14, no. 6, pp. 992–1013, 1966.

[9] P. A. Hosein and M. Athans, *Preferential Defense Strategies. Part I: The Static Case*, MIT Laboratory for Information and Decision Systems, Cambridge, Mass, USA, 1990.

[10] Z. R. Bogdanowicz, A. Tolano, K. Patel, and N. P. Coleman, "Optimization of weapon-target pairings based on kill probabilities," *IEEE Transactions on Cybernetics*, vol. 43, no. 6, pp. 1835–1844, 2013.

[11] S. Chen, J. He, and H. Liu, "Realization and simulation of parallel ant colony algorithm to solve WTA problem," in *Proceedings of the International Conference on Systems and Informatics (ICSAI '12)*, May 2012.

[12] A.-G. Fei, L.-Y. Zhang, and Q.-J. Ding, "Multi-aircraft cooperative fire assignment based on auction algorithm," *Systems Engineering & Electronics*, vol. 34, no. 9, pp. 1829–1833, 2012.

[13] M.-Z. Lee, "Constrained weapon–target assignment: enhanced very large scale neighborhood search algorithm," *IEEE Transactions on Systems, Man, and Cybernetics Part A:Systems and Humans*, vol. 40, no. 1, pp. 198–204, 2010.

[14] D. G. Galati and M. A. Simaan, "Effectiveness of the Nash strategies in competitivemulti-team target assignment problems," *IEEE Transactions on Aerospace & Electronics System*, vol. 43, no. 1, pp. 126–134, 2007.

[15] Z.-J. Lee and W.-L. Lee, "A Hybrid search algorithm of ant colony optimization and genetic algorithm applied to weapon-target assignment problems," in *Intelligent Data Engineering and Automated Learning*, vol. 2690 of *Lecture Notes in Computer Science*, pp. 278–285, Springer, Berlin, Germany, 2003.

[16] Y. Li and Y. Dong, "Weapon-target assignment based on simulated annealing and discrete particle swarm optimization in cooperative air combat," *Acta Aeronautica et Astronautica Sinica*, vol. 31, no. 3, pp. 626–631, 2010.

[17] H.-D. Chen, S.-Z. Wang, and H.-Y. Wang, "Research of firepower assignment with multi-launcher and multi-weapon based on a hybrid particle swarm optimization," *Systems Engineering & Electronics*, vol. 30, no. 5, pp. 880–883, 2008.

[18] X. Liu, Z. Liu, W.-S. Hou, and J.-H. Xu, "Improved MOPSO algorithm for multi-objective programming model of weapon-target assignment," *Systems Engineering & Electronics*, vol. 35, no. 2, pp. 326–330, 2013.

[19] Y. Zhang, R.-N. Yang, J.-L. Zuo, and X.-N. Jing, "Weapon-target assignment based on decomposition-based evolutionary multi-objective optimization algorithms," *Systems Engineering & Electronics*, vol. 36, no. 12, pp. 2435–2441, 2014.

[20] J. Li, J. Chen, B. Xin, and L. Dou, "Solving multi-objective multi-stage weapon target assignment problem via adaptive NSGAII and adaptive MOEA/D: A Comparison Study," in *Proceedings of the IEEE Congress on Evolutionary Computation (CEC '15)*, pp. 3132–3139, May 2015.

[21] R. Battiti, M. Brunato, and F. Mascia, "Reactive search and intelligent optimization," *Operations Research*, vol. 45, no. 1, pp. 74–89, 2008.

[22] R. Battiti and G. Tecchiolli, "The reactive tabu search," *ORSA Journal on Computing*, vol. 6, no. 2, pp. 126–140, 1994.

[23] M. Dorigo and L. M. Gambardella, "Ant colony system: a cooperative learning approach to the traveling salesman problem," *IEEE Transactions on Evolutionary Computation*, vol. 1, no. 1, pp. 53–66, 1997.

[24] T. Stützle and H. H. Hoos, "Max-min ant system," *Future Generation Computer Systems*, vol. 16, no. 9, pp. 889–914, 1999.

[25] S. Iredi, D. Merkle, and M. Middendorf, "Bi-criterion optimization with multi colony ant algorithms," in *Evolutionary Multi-Criterion Optimization: First International Conference, EMO 2001 Zurich, Switzerland, March 7–9, 2001 Proceedings*, vol. 1993 of *Lecture Notes in Computer Science*, pp. 359–372, Springer, Berlin, Germany, 2001.

[26] D. Merkle, M. Middendorf, and H. Schmeck, "Ant colony optimization for resource-constrained project scheduling," *IEEE Transactions on Evolutionary Computation*, vol. 6, no. 4, pp. 333–346, 2002.

[27] J. E. Bell and P. R. McMullen, "Ant colony optimization techniques for the vehicle routing problem," *Advanced Engineering Informatics*, vol. 18, no. 1, pp. 41–48, 2004.

[28] K. Doerner, W. J. Gutjahr, R. F. Hartl, C. Strauss, and C. Stummer, "Pareto ant colony optimization: a metaheuristic approach to multiobjective portfolio selection," *Annals of Operations Research*, vol. 131, pp. 79–99, 2004.

[29] P. Cardoso, M. Jesus, and A. Márquez, "MONACO-multi-objective network optimisation based on an ACO," in *Proceedings of the X Encuentros de Geometrıa Computacional*, vol. 1, no. 1, pp. 1–10, Seville, Spain, 2013.

[30] P. R. McMullen, "An ant colony optimization approach to addressing a JIT sequencing problem with multiple objectives," *Artificial Intelligence in Engineering*, vol. 15, no. 3, pp. 309–317, 2001.

[31] T. Stützle and M. Dorigo, "ACO algorithms for the quadratic assignment problem," *New Ideas in Optimization*, vol. 3, no. 1, pp. 33–50, 2000.

[32] A. Baykasoglu, T. Dereli, and I. Sabuncu, "A multiple objective and colony ant colony optimization approach to assembly line balancing problems," in *Proceedings of the 35th International Conference on Computers and Industrial Engineering*, pp. 263–268, IEEE, Istanbul, Turkey, June 2005.

[33] K. Deb, A. Pratap, S. Agarwal, and T. Meyarivan, "A fast and elitist multiobjective genetic algorithm: NSGA-II," *IEEE Transactions on Evolutionary Computation*, vol. 6, no. 2, pp. 182–197, 2002.

[34] E. Zitzler, M. Laumanns, and L. Thiele, "SPEA2: improving the strength pareto evolutionary algorithm," in *Proceedings of the Conference on Evolutionary Methods for Design, Optimisation and Control with Applications to Industrial Problems*, pp. 95–100, Athens, Greece, September 2001.

[35] M. Dorigo and T. Tzle, "Ant colony optimization," in *Wiley Encyclopedia of Operations Research and Management Science*, pp. 1155–1173, John Wiley & Sons, New York, NY, USA, 2004.

[36] I. D. I. D. Ariyasingha and T. G. I. Fernando, "Performance analysis of the multi-objective ant colony optimization algorithms for the traveling salesman problem," *Swarm & Evolutionary Computation*, vol. 23, pp. 11–26, 2015.

[37] B. Baran and M. Schaerer, "A multiobjective ant colony system for vehicle routing problem with time windows," in *Proceedings of the 21th IASTED International Conference on Applied Informatics*, pp. 97–102, DBLP, February 2003.

[38] V. Maniezzo and A. Colorni, "The ant system applied to the quadratic assignment problem," *IEEE Transactions on Knowledge and Data Engineering*, vol. 11, no. 5, pp. 769–778, 1999.

[39] C. García-Martínez, O. Cordón, and F. Herrera, "A taxonomy and an empirical analysis of multiple objective ant colony optimization algorithms for the bi-criteria TSP," *European Journal of Operational Research*, vol. 180, no. 1, pp. 116–148, 2007.

[40] Z.-J. Lee, C.-Y. Lee, and S.-F. Su, "An immunity-based ant colony optimization algorithm for solving weapon-target assignment problem," *Applied Soft Computing Journal*, vol. 2, no. 1, pp. 39–47, 2002.

[41] Z.-J. Lee, S.-F. Su, and C.-Y. Lee, "A genetic algorithm with domain knowledge for weapon–target assignment problems," *Journal of the Chinese Institute of Engineers*, vol. 25, no. 3, pp. 287–295, 2002.

[42] B. Xin, J. Chen, J. Zhang, L. Dou, and Z. Peng, "Efficient decision makings for dynamic weapon-target assignment by virtual permutation and tabu search heuristics," *IEEE Transactions on Systems, Man and Cybernetics Part C: Applications and Reviews*, vol. 40, no. 6, pp. 649–662, 2010.

[43] B. Xin, J. Chen, Z. Peng et al., "An efficient rule-based constructive heuristic to solve dynamic weapon-target assignment problem," *IEEE Transactions on Systems Man & Cybernetics, Part A: Systems & Humans*, vol. 41, no. 3, pp. 598–606, 2011.

[44] P. Chen, Y. Zheng, and W. Zhu, "Optimized simulated annealing algorithm for thinning and weighting large planar arrays in both far-field and near-field," *Frontiers of Information Technology & Electronic Engineering*, vol. 11, no. 4, pp. 261–269, 2010.

[45] D. K. Ahner and C. R. Parson, "Optimal multi-stage allocation of weapons to targets using adaptive dynamic programming," *Optimization Letters*, vol. 9, no. 8, pp. 1689–1701, 2015.

[46] N. Dirik, S. N. Hall, and J. T. Moore, "Maximizing strike aircraft planning efficiency for a given class of ground targets," *Optimization Letters*, vol. 9, no. 8, pp. 1729–1748, 2015.

[47] H. Liang and F. Kang, "Adaptive chaos parallel clonal selection algorithm for objective optimization in WTA application," *Optik*, vol. 127, no. 6, pp. 3459–3465, 2016.

[48] N. Li, W. Huai, and S. Wang, "The solution of target assignment problem in command and control decision-making behavior simulation," *Enterprise Information Systems*, vol. 1, no. 1, pp. 1–19, 2016.

12

Robust Smooth Sliding-Mode-Based Controller with Fixed-Time Convergence for Missiles considering Aerodynamic Uncertainty

Yuhang Yun, Jie Guo, and Shengjing Tang

School of Aerospace Engineering, Beijing Institute of Technology, Beijing 100081, China

Correspondence should be addressed to Jie Guo; guojie1981@bit.edu.cn

Academic Editor: Linda L. Vahala

This paper designed a smooth fixed-time-convergent sliding mode controller for a missile flight system considering aerodynamic uncertainties. Fixed-time convergence theory is incorporated with the sliding mode control technique to ensure that the system tracks desired commands within uniform bounded time under different initial conditions. Unlike previous terminal sliding mode approaches, not only is the bound of settling time independent of initial state, indicating that performance metrics like convergence rate can be predicted beforehand, but the control input is designed to be smooth based on adaptive estimations and some mathematical results without introducing any discontinuous items like the signum function, which avoids the problem of chattering consequently. A cascade control structure is employed with the derived control algorithm, and therein, the control input signal is obtained. Finally, a number of simulations are carried out and demonstrate the effectiveness of the designed controller.

1. Introduction

Increasing demand for high accuracy and system reliability has stimulated the development of control techniques in the past decades. For the missile flight system, a well-behaved controller can not only track the command signal swiftly but also exhibit appropriate robustness against disturbances and uncertainties which exist throughout the fickle flight regime. From the perspective of overall design, it is also preferable to evaluate the control performances as much as possible once the preliminary controller design is accomplished.

The conventional missile autopilot focused on improving the poor underdamped dynamics of the missile airframe. By adding feedback loops, the PI regulator [1], root locus and frequency domain technique [2], and pole placement method [3] can be applied conveniently, all constructing a new control circuit with moderate dynamics, wherein the missile body itself becomes part of the control loop. However, these methods fail to interact directly with uncertainties or disturbances. They normally designed the controller without caring for uncertainties at the beginning. After independent tuning of designed parameters, the influence of uncertainties are alleviated through adding the conservativeness of control gains. To improve this, some modern control techniques are introduced. Reference [4] divided the missile attitude dynamic and kinematic model into fast and slow loops according to time scalar separations, which is the common idea that is widely adopted by almost all the control works for missile controller design. Variable structure approach is then applied in the loop design. In [5], an acceleration autopilot design methodology was developed based on optimal control theory, which revealed better performance and stability robustness compared with the classic three-loop controller. Reference [6] utilized the dynamic inversion method to design the control input, together with the extended-mean technique dealing with system uncertainties. References [7, 8] finished a similar work through incorporating fuzzy logic to PID control, constituting an adaptive hybrid control strategy, so that online self-tuning of the controller gains was achieved. Further, [9, 10] deliberated the fusion of a sliding mode controller and the fuzzy logic networks, comprising the advantages of both systems. Simulations revealed the effectiveness of the designed method.

Among the myriad control approaches for missile controller design, the sliding mode control (SMC) technique is widely used and performs remarkably against uncertainties and disturbances without exact modeling of the uncertainties or disturbance estimation [11–13]. This enables the system with good robustness and interference immunity. SMC method involves a sliding manifold that should be chosen based on control objectives and errors, and then appropriate reaching laws are designed to maintain the surface "on sliding mode." Conventional SMC algorithm usually prescribes the switching manifolds as linear combinations of system state and desired command with linear reaching law [11, 14, 15], which could only achieve the asymptotic or exponential convergence in state movement towards the equilibrium point. Then, the concept of "terminal sliding mode control" (TSMC) is put forward, which indicates the finite convergence property of the sliding phase [16–18]. As the sliding manifolds and reaching law of TSMC are nonlinear functions with higher order, a singularity problem is possibly found and prohibits the practical application when the error being controlled becomes very small. In [19], the sliding manifold of TSMC has been improved through substitutions of state and its derivative variables, together with the change of exponential coefficient range. Therefore, the singularity problem has been eliminated when taking the derivatives of the sliding states. Similarly, [20] applied a nonsingular TSMC method to the integrated guidance and control system design incorporated with adaptive method estimating the bound of disturbances, so that not only system states converge to the desired values in finite time but also the missile can finally hit the target in finite time.

However, in the abovementioned works, where finite-time convergence characteristic is incorporated, exhaustive estimations of settling time are not addressed. Only theoretical results are derived in the citation of lemmas. From the expression of theoretical settling time, prior knowledge of the initial system state must be known to estimate the convergence rate. Different initial conditions correspond to various settling time, of which the maximum value depends on the maximum initial errors related to the Lyapunov function. In the preliminary design of a missile control system, it would be very helpful if the settling time can be predicted without information of the initial conditions. Thus, in contrast with existing finite-time controllers, the upper bound of the settling time could be estimated in the sense of the fixed-time convergence concept [21], of which the bounded convergence time is independent of system initial conditions. With the newly constructed sliding manifold, [22] achieved a nonsingular fixed-time consensus tracking for second-order multiagent networks. Carrying the idea further, [23] expanded the fixed-time terminal sliding mode control methods for a class of second-order nonlinear systems. In [24], a controller in the sense of the fixed-time concept had been constructed for rigid spacecraft and the attitude of the spacecraft converged to the equilibrium within the defined fixed time even with actuator saturation and faults.

Further, variable structure items, which comprise discontinuous parts like the signum function, are normally introduced in the SMC control methodologies mentioned above to suppress the disturbances. The discontinuous signum function induces the well-known chattering problem as the switching gain must be chosen larger than the bound of uncertainty, which results in a degradation of system performance to some considerable extent under specific occasions. Some works replaced the discontinuous item with saturation function or sigmoid function [25], whereas control effectiveness is sacrificed to a certain degree. References [26, 27] utilized the extended state observer (ESO) to estimate uncertainties. Despite the control burden being reduced with ESO's output incorporated into the control algorithm as a feed forward scheme, the discontinuous variable structure item is unavoidable and the system suffers from the chattering phenomenon inevitably. To alleviate chattering effectively, [28] proposed a smooth terminal sliding mode control algorithm for robotic manipulators, where a kind of continuous reaching condition was elaborately devised.

Inspired by the previous discussion, this paper designed a new robust smooth sliding-mode-based controller with fixed-time convergence. An adaptive estimation and some primary mathematical results are utilized to alleviate the effect of uncertainties, as well as deducing the control algorithm. Unlike existing works, the designed smooth fixed-time-convergent sliding mode controller motivates missile control variables to converge to the equilibrium point before the uniform bounded settling time in the presence of aerodynamic uncertainties with its input inherently continuous without using any discrete items, like the signum function. The upper bound of settling time here is simply a function of designed parameters, and therefore, prior knowledge of the convergence rate can be evaluated in advance without any information of system initial conditions. With a uniform bounded convergence time, the controller can track the desired command in the presence of aerodynamic uncertainties under different initial state situations, which is very helpful for both preliminary design and performance evaluation. The smooth input avoids the problem of singularity and chattering, which ensures better performance of the missile control system and exhibits superior availability for practical application. Comprehensive simulations considering a nominal missile dynamic model with aerodynamic uncertainties are carried out to demonstrate the effectiveness of the designed attitude controller. The fixed-time convergence characteristic is fully reflected under the cases of various controller gains and initial states, and the smooth input exhibits nice continuity through comparison with the conventional terminal sliding mode method.

The rest of this paper is organized as follows. In the forthcoming section, a missile dynamic model with aerodynamic uncertainties is presented. Next, some preliminaries involving primary mathematical results and the fixed-time convergence theory are introduced, and then a smooth fixed-time-convergent sliding mode controller is designed in Section 3. In Section 4, detailed control property and system stability analysis are given. Finally, various simulations are performed in the last section and demonstrate the satisfactory performance of the designed controller.

2. Problem Formulation

Typical missile control embodies the tracking of commanded attitude angle. To begin with, some fundamental assumptions are introduced.

(a) The variation of static parameters like mass and moment of inertial through entire flight is not taken into consideration.

(b) Trustworthy measurements of attitude angles and angular velocities are available with high precision sensors.

(c) Aerodynamic uncertainties and their derivatives are bounded.

(d) In the controller design, missile velocity and flight altitude are assumed to be constant.

Regardless of gravity, the longitudinal model of the missile in the presence of aerodynamic uncertainties is considered here:

$$\begin{aligned}\dot{\alpha} &= \omega_z - b_\alpha(1 + d_{\alpha\alpha})\alpha - b_{\delta_z}(1 + d_{\alpha\delta_z})\delta_z,\\ \dot{\omega}_z &= -a_{\omega_z}(1 + d_{\omega_z\omega_z})\omega_z - a_\alpha(1 + d_{\omega_z\alpha})\alpha - a_{\delta_z}(1 + d_{\omega_z\delta_z})\delta_z.\end{aligned} \tag{1}$$

In the above equation, α denotes the angle of attack, ω_z stands for the pitch rate, and δ_z is the elevator deflection of the missile. Denotations like b_α, $b_{\delta z}$, a_α, $a_{\delta z}$, and $a_{\omega z}$ are defined as

$$\begin{aligned}b_\alpha &= \frac{P + C_y{}^\alpha qs}{mV},\\ b_{\delta^z} &= \frac{C_y{}^{\delta^z} qs}{mV},\\ a_\alpha &= \frac{-m_z{}^\alpha qsL}{J_z},\\ a_{\delta^z} &= \frac{-m_z{}^{\delta^z} qsL}{J_z},\\ a_{\omega_z} &= \frac{-m_z{}^{\bar{\omega}_z} qsL}{J_z},\end{aligned} \tag{2}$$

where m, V, P, s, and L represent the missile weight, velocity, thrust, reference area, and length, respectively. q denotes the dynamic pressure and can be calculated by $q = 0.5\rho V2$. Cy^* and mz^* are coefficients of aerodynamic force and moment, respectively. Thus, in (1), $d_{\alpha\alpha}$ and $d_{\alpha\delta_z}$ describe the aerodynamic uncertainties in terms of lift coefficient generated by angle of attack and elevator deflection. $d_{\omega_z\omega_z}$, $d_{\omega_z\alpha}$, and $d_{\omega_z\delta_z}$ depict the aerodynamic uncertainties of moment coefficient in pitch direction resulting from pitch rate, angle of attack, and elevator deflection, respectively. As assumed at the beginning, all of the coefficients representing aerodynamic perturbations are bounded.

For our convenience, (1) can be rearranged to form the following dynamic system:

$$\dot{\mathbf{x}} = \mathbf{f}_\mathrm{x} + \mathbf{g}_\mathrm{x}\mathbf{u} + \mathbf{d}, \tag{3}$$

where

$$\begin{aligned}\mathbf{x} &= [x_1, x_2]^T = [\alpha, \omega_z]^T,\\ \mathbf{f}_\mathrm{x} &= \begin{bmatrix}\omega_z - b_\alpha\alpha - b_{\delta_z}\delta_z\\ -a_\alpha\alpha - a_{\omega_z}\omega_z\end{bmatrix},\\ \mathbf{g}_\mathrm{x} &= \begin{bmatrix}0\\ -a_{\delta_z}\end{bmatrix},\\ \mathbf{d} &= \begin{bmatrix}d_1\\ d_2\end{bmatrix} = \begin{bmatrix}-b_\alpha d_{\alpha\alpha}\alpha - b_{\delta_z}d_{\alpha\delta_z}\delta_z\\ -a_{\omega_z}d_{\omega_z\omega_z}\omega_z - a_\alpha d_{\omega_z\alpha}\alpha - a_{\delta_z}d_{\omega_z\delta_z}\delta_z\end{bmatrix}.\end{aligned} \tag{4}$$

System (3) well represents the complex behavior of missiles steering by elevator deflection in the presence of aerodynamic uncertainties. With measurable state variables x_1 and x_2, items like $\mathbf{f}_\mathrm{x}$ and $\mathbf{g}_\mathrm{x}$ can be calculated.

3. Controller Design

The dynamic model developed in Section 2 is suitable to design a controller based on robust stabilizing and tracking control law. In this section, some primary mathematical results and the concept of fixed-time stability are first introduced. Afterwards, a robust smooth adaptive sliding-mode-based controller has been designed with fixed-time convergence characteristics, wherein the influence of uncertainties is alleviated without using any discontinuous items (like the signum function).

3.1. Preliminary. Two fundamental equations are presented here. For any a, b, c, $d \in \mathbf{R}$ and arbitrary positive value μ, $\zeta > 0$, one can get

$$\frac{(ab)^2}{4\mu} + \mu \geq ab, \tag{5}$$

$$\zeta c^2 + \frac{d^2}{4\zeta} \geq cd. \tag{6}$$

Equations (5) and (6) are mathematically correct and are helpful to prove the stability of the designed controller in the subsequent section.

Another important synthesis in this paper is the concept of fixed-time stability, which can be assumed as the extension of finite-time stability.

Consider the nonlinear system [17]

$$\dot{\mathbf{x}} = \mathbf{f}(\mathbf{x}, t), \quad \mathbf{f}(\mathbf{0}, t) = \mathbf{0}, \mathbf{x} \in \mathbf{R}^n, \tag{7}$$

where $\mathbf{f} : \mathbf{U}_0 \times \mathbf{R} \rightarrow \mathbf{R}^n$ is continuous on $\mathbf{U}_0 \times \mathbf{R}$ and $\mathbf{U}_0$ is an open set of the origin.

The origin is a finite-time stable equilibrium if it is Lyapunov stable and for any given initial time t_0 and initial state $\mathbf{x}(t_0) = \mathbf{x}_0 \in \mathbf{U} \subset \mathbf{U}_0$, there exists a settling time $T \geq 0$ which is dependent on $\mathbf{x}_0$, such that for every solution of system (7), $\mathbf{x}(t) = \boldsymbol{\varphi}(t : t_0, \mathbf{x}_0) \in \mathbf{U}/\{\mathbf{0}\}$ satisfies

$$\begin{aligned} &\lim_{t \to T(x_0)} \boldsymbol{\varphi}(t : t_0, \mathbf{x}_0) = \mathbf{0}, \quad t \in [t_0, T(x_0)], \\ &\boldsymbol{\varphi}(t : t_0, \mathbf{x}_0) = \mathbf{0}, \quad t > T(x_0). \end{aligned} \tag{8}$$

Moreover, if the origin is finite-time stable with $\mathbf{U} = \mathbf{R}^n$, then the origin is a global finite-time stable equilibrium. The following lemma [16] provides sufficient condition for the finite convergence control design.

Lemma 1. *Consider the differential equation of system (7). Suppose there exists a continuous positive-definite function $V(\mathbf{x}, t)$: $\widehat{\mathbf{U}} \to \mathbf{R}$, where $\widehat{\mathbf{U}}$ is a neighborhood of the origin, such that there are real numbers $c > 0$ and $0 < a < 1$ satisfying $\dot{V}(\mathbf{x}, t) + cV^{\alpha}(\mathbf{x}, t) \leq 0$ on $\widehat{\mathbf{U}}$. Then the zero solution of system (7) is finite-time stable. The settling time is given by*

$$T(\mathbf{x}_0) \leq \frac{V^{1-\alpha}(\mathbf{x}_0)}{c(1-\alpha)}. \tag{9}$$

In addition, if $\widehat{\mathbf{U}} = \mathbf{R}^n$, and V is radially unbounded, then the origin is globally finite-time stable.

From Lemma 1, the finite convergence time depends on the system initial condition. Diverse initial states result in varied settling times, which means no uniform bound of the convergence time can be obtained in advance without the knowledge of maximum $V(\mathbf{x}_0)$. If prior information of system initial states is not available, preevaluations of controller performance may be restricted to some extent. For preliminary design, it would be much helpful if system performance metrics, for example, the bound of settling time, can be estimated in advance regardless of initial system states. Moreover, for those with strict positive dwelling time, for example, a hybrid system control, it is preferable to stabilize the controlled system exactly before the next switching occurs [29]. Thus, moving further on, the concept of fixed-time convergence is addressed.

Considering system (7), the origin is said to be a "fixed-time stable" equilibrium point if it is globally finite-time stable and the settling time function $T(\mathbf{x}_0)$ is bounded, that is, there exists $T_{\max} > 0$ such that $T(\mathbf{x}_0) < T_{\max}$, $\forall \mathbf{x}_0 \in \mathbf{R}^n$.The following lemma [21] presents a helpful result for the design of a controller with fixed-time convergence.

Lemma 2. *Consider a scalar system*

$$\dot{y} = -\alpha y^{m/n} - \beta y^{p/q}, \quad y(0) = y_0, \tag{10}$$

where $\alpha > 0$ and $\beta > 0$. m, n, p, and q are positive odd integers satisfying $m > n$ and $p < q$. Then the origin of system (10) is fixed-time stable, and the settling time T is bounded by

$$T < T_{\max} \triangleq \frac{1}{\alpha}\frac{n}{m-n} + \frac{1}{\beta}\frac{q}{q-p}. \tag{11}$$

As denoted in (11), the upper bound of the settling time is independent of system initial condition and is basically determined by designed parameters, namely, m, n, p, q, α, and β. Therefore, prior knowledge of "fixed" settling time can be known in advance with appropriate selection of parameters on the basis of system performance metrics.

3.2. Smooth Adaptive Fixed-Time-Convergent Sliding Mode Controller. This part derived a robust smooth sliding-mode-based controller with fixed-time convergence for system (5). In the light of time scale separation principal and cascade control strategy, system (3) is decomposed into outer loop and inner loop architecture. The former accounts for generating proper angular rate command to trace the reference attitude command, while the latter for operating on the tail deflection to track the reference angular rate command obtained from the angle tracking loop. System (3) is expressed in a scalar form as

$$\begin{aligned} \dot{x}_1 &= f_{x1} + x_2 + d_1, \\ \dot{x}_2 &= f_{x2} + g_{x2}u + d_2, \end{aligned} \tag{12}$$

where $f_{x1} = -b_\alpha \alpha - b_{\delta_z}\delta_z$ and $f_{x2} = -a_\alpha \alpha - a_{\omega_z}\omega_z$. d_1 and d_2 are bounded by $|d_1| \leq \sigma_1$ and $|d_2| \leq \sigma_2$, respectively.

3.2.1. Outer Loop. Suppose the desired command is x_c. The objective of the outer loop control is to design the angular rate defined as a virtual control input x_{2c} to make x_1 converge to x_c within a fixed bounded time. The first equation of system (12) is rewritten here:

$$\dot{x}_1 = f_{x1} + x_{2c} + d_1. \tag{13}$$

Define a sliding manifold as

$$s_1 = x_1 - x_c = \alpha - \alpha_c. \tag{14}$$

Differentiating s_1 with respect to time yields

$$\dot{s}_1 = f_{x1} + x_{1c} + d_1 - \dot{x}_c. \tag{15}$$

To achieve fixed-time convergence, the reaching law is designed in the form

$$\dot{s}_1 = -\beta_{11}s_1^{m_1/n_1} - \beta_{21}s_1^{p_1/q_1} - \beta_{31}s_1 - \frac{\widehat{\chi}_1 s_1}{4\mu_1} + d_1, \tag{16}$$

where $\beta_{11}, \beta_{21}, \beta_{31}$, and μ_1 are nonnegative. m_1, n_1, p_1, and q_1 are positive odd constants satisfying $m_1 > n_1$ and $p_1 < q_1$. $\widehat{\chi}_1$ is determined by the updating law

$$\dot{\widehat{\chi}}_1 = -\beta_{41}\widehat{\chi} + \frac{s_1^2}{4\mu_1}, \tag{17}$$

where $\beta_{41} > 0$. $\widehat{\chi}_1$ is the estimation of χ_1 and $\chi_1 = {\sigma_1}^2$.

Substituting (16) into (15), the virtual control signal for outer loop can be designed as

$$x_{2c} = -f_{x1} - \beta_{11}s_1^{m_1/n_1} - \beta_{21}s_1^{p_1/q_1} - \beta_{31}s_1 - \frac{\widehat{\chi}_1 s_1}{4\mu_1} + \dot{\alpha}_c. \tag{18}$$

Next, with the obtained virtual control input, the control law for the inner loop can be designed under the same baseline.

3.2.2. Inner Loop. Similar procedures are conducted in the inner loop. The objective is to deduce the actual control input such that tracking error between the virtual control signal x_{2c} and the actual angular rate variable x_2 converges to zero within a fixed bounded time. Rewrite the second equation of system (12)

$$\dot{x}_2 = f_{x2} + g_{x2}u + d_2. \tag{19}$$

Another sliding manifold is defined as

$$s_2 = x_2 - x_{2c}. \tag{20}$$

Taking the derivative of s_2 gives

$$\dot{s}_2 = f_{x2} + g_{x2}u + d_2 - \dot{x}_{2c}. \tag{21}$$

In a similar way, the reaching law for inner loop is designed as

$$\dot{s}_2 = -\beta_{12}s_2^{m_2/n_2} - \beta_{22}s_2^{p_2/q_2} - \beta_{32}s_2 - \frac{\hat{\chi}_2 s_2}{4\mu_2} + d_2, \tag{22}$$

where $\beta_{12}, \beta_{22}, \beta_{32}$, and μ_2 are nonnegative. m_2, n_2, p_2, and q_2 are positive odd constants satisfying $m_1 > n_1$ and $p_1 < q_1$. $\hat{\chi}_2$ is the estimation of $\chi_2 = \sigma_2{}^2$ and is determined by

$$\dot{\hat{\chi}}_2 = -\beta_{42}\hat{\chi}_2 + \frac{s_2^2}{4\mu_2}, \tag{23}$$

where $\beta_{42} > 0$.

Considering (21) and (22), the actual control input for stabilizing the tracking error in inner loop is derived as

$$\begin{aligned} u &= \delta_z \\ &= g_{x2}^{-1}\left(-f_{x2} - \beta_{12}s_2^{m_2/n_2} - \beta_{22}s_2^{p_2/q_2} - \beta_{32}s_2 - \frac{\hat{\chi}_2 s_2}{4\mu_2} + \dot{x}_{2c}\right). \end{aligned} \tag{24}$$

It can be observed from (23) and (24) that there is no discontinuous item like the signum function, and the control input is inherently smooth and consequently avoids the problem of chattering.

4. Stability Analysis

In this section, closed form stability analysis of system (12) is given with the deduced controller. Preliminaries introduced in the last section are helpful to the proof of the main results.

Theorem 1. *For the dynamic system (12), system states can track the desired command with the designed controller within fixed bounded time and converge to a neighborhood around the sliding manifold. Further, the fixed bound of settling time is independent of initial conditions.*

Proof. For the outer loop, consider the Lyapunov function

$$V_1 = \frac{1}{2}s_1^2 + \frac{1}{2}\tilde{\chi}_1^2, \tag{25}$$

where $\tilde{\chi}_1 = \chi_1 - \hat{\chi}_1$.

Differentiating V_1 with respect to time and substituting (17) and (18) into it yield

$$\begin{aligned} \dot{V}_1 &= s_1\dot{s}_1 + \tilde{\chi}_1\dot{\tilde{\chi}}_1 = -\beta_{11}s_1{}^{(m_1+n_1)/n_1} - \beta_{21}s_1{}^{(p_1+q_1)/q_1} - \beta_{31}s_1^2 \\ &\quad + s_1 d_1 - \frac{\hat{\chi}_1 s_1^2}{4\mu_1} + \beta_{41}\hat{\chi}_1\tilde{\chi}_1 - \frac{\tilde{\chi}_1 s_1^2}{4\mu_1} \le -\beta_{31}s_1^2 + |s_1|\sigma_1 \\ &\quad - \frac{\hat{\chi}_1 s_1^2}{4\mu_1} + \beta_{41}\hat{\chi}_1\tilde{\chi}_1 - \frac{\tilde{\chi}_1 s_1^2}{4\mu_1} = -\beta_{31}s_1^2 + |s_1|\sigma_1 - \frac{\chi_1 s_1^2}{4\mu_1} \\ &\quad + \beta_{41}\chi_1\tilde{\chi}_1 - \beta_{41}\tilde{\chi}_1^2. \end{aligned} \tag{26}$$

Considering (5) and (6), one gets

$$\begin{aligned} \dot{V}_1 &\le -\beta_{31}s_1^2 + \mu_1 + \beta_{41}\zeta_1\tilde{\chi}_1^2 + \beta_{41}\frac{\chi_1^2}{4\zeta_1} - \beta_{41}\tilde{\chi}_1^2 \\ &= -\beta_{31}s_1^2 - \beta_{41}(1-\zeta_1)\tilde{\chi}_1^2 + \mu_1 + \beta_{41}\frac{\chi_1^2}{4\zeta_1} \le -\eta_1 V_1 + \varsigma_1, \end{aligned} \tag{27}$$

where $\eta_1 = \min\ \{2\beta_{31}, 2\beta_{41}(1-\zeta_1)\}, \varsigma_1 = \mu_1 + \beta_{41}\chi_1{}^2/4\zeta_1$.

According to the boundedness theorem, s_1 and $\tilde{\chi}_1$ are uniformly ultimately bounded. Assume that $|\tilde{\chi}_1| \le \xi_1$, where ξ_1 is positive.

Similarly, define another Lyapunov function of inner loop as

$$V_2 = \frac{1}{2}s_2^2 + \frac{1}{2}\tilde{\chi}_2^2, \tag{28}$$

where $\tilde{\chi}_2 = \chi_2 - \hat{\chi}_2$.

Taking the derivative of V_2, it follows from (23) and (24) that

$$\begin{aligned} \dot{V}_2 &= s_2\dot{s}_2 + \tilde{\chi}_2\dot{\tilde{\chi}}_2 = -\beta_{12}s_2{}^{(m_2+n_2)/n_2} - \beta_{22}s_2{}^{(p_2+q_2)/q_2} - \beta_{32}s_2^2 \\ &\quad + s_2 d_2 - \frac{\hat{\chi}_2 s_2^2}{4\mu_2} + \beta_{42}\hat{\chi}_2\tilde{\chi}_2 - \frac{\tilde{\chi}_2 s_2^2}{4\mu_2} \le -\beta_{32}s_2^2 + |s_2|\sigma_2 \\ &\quad - \frac{\hat{\chi}_2 s_2^2}{4\mu_2} + \beta_{42}\hat{\chi}_2\tilde{\chi}_2 - \frac{\tilde{\chi}_2 s_2^2}{4\mu_2} = -\beta_{32}s_2^2 + |s_2|\sigma_2 - \frac{\chi_2 s_2^2}{4\mu_2} \\ &\quad + \beta_{42}\chi_2\tilde{\chi}_2 - \beta_{42}\tilde{\chi}_2^2 \le -\beta_{32}s_2^2 + \mu_2 + \beta_{42}\zeta_2\tilde{\chi}_2^2 + \beta_{42}\frac{\chi_2^2}{4\zeta_2} \\ &\quad - \beta_{42}\tilde{\chi}_2^2 = -\beta_{32}s_2^2 - \beta_{42}(1-\zeta_2)\tilde{\chi}_2^2 + \mu_2 \\ &\quad + \beta_{42}\frac{\chi_2^2}{4\zeta_2} \le -\eta_2 V_2 + \varsigma_2, \end{aligned} \tag{29}$$

where $\eta_2 = \min\ \{2\beta_{32}, 2\beta_{42}(1-\zeta_2)\}$ and $\varsigma_2 = \mu_2 + \beta_{42}\chi_2{}^2/4\zeta_2$. Thus, s_2 and $\tilde{\chi}_2$ are uniformly ultimately bounded. Assume that $|\tilde{\chi}_2^*| \le \xi_2$, where ξ_2 is positive.

For the integral system, consider the Lyapunov function

$$V_3 = \frac{1}{2}s_1^2 + \frac{1}{2}s_2^2. \tag{30}$$

Combined with (17), (18), (23), and (24), the derivative of V_3 is expressed in the following form

$$\begin{aligned}\dot{V}_3 &= s_1\dot{s}_1 + s_2\dot{s}_2 = -\beta_{11}s_1^{(m_1+n_1)/n_1} - \beta_{21}s_1^{(p_1+q_1)/q_1} - \beta_{31}s_1^2 \\ &+ s_1 d_1 - \frac{\widehat{\chi}_1 s_1^2}{4\mu_1} - \beta_{12}s_2^{(m_2+n_2)/n_2} - \beta_{22}s_2^{(p_2+q_2)/q_2} - \beta_{32}s_2^2 \\ &+ s_2 d_2 - \frac{\widehat{\chi}_2 s_2^2}{4\mu_2} \le -\beta_{11}s_1^{(m_1+n_1)/n_1} - \beta_{21}s_1^{(p_1+q_1)/q_1} - \beta_{31}s_1^2 \\ &+ \frac{\xi_1 s_1^2}{4\mu_1} + \mu_1 - \beta_{12}s_2^{(m_2+n_2)/n_2} - \beta_{22}s_2^{(p_2+q_2)/q_2} - \beta_{32}s_2^2 \\ &+ \frac{\xi_2 s_2^2}{4\mu_2} + \mu_2 = -\beta_{11}s_1^{(m_1+n_1)/n_1} - \beta_{21}s_1^{(p_1+q_1)/q_1} \\ &- \left(\beta_{31} - \frac{\xi_1}{4\mu_1}\right)s_1^2 + \mu_1 - \beta_{12}s_2^{(m_2+n_2)/n_2} \\ &- \beta_{22}s_2^{(p_2+q_2)/q_2} - \left(\beta_{32} - \frac{\xi_2}{4\mu_2}\right)s_2^2 + \mu_2.\end{aligned} \tag{31}$$

Denote

$$\begin{aligned}\beta_{31}{}^* &= \beta_{31} - \frac{\xi_1}{4\mu_1}, \\ \beta_{32}{}^* &= \beta_{32} - \frac{\xi_2}{4\mu_2}.\end{aligned} \tag{32}$$

Equation (31) yields

$$\begin{aligned}\dot{V}_3 &\le -\beta_{11}s_1^{(m_1+n_1)/n_1} - \beta_{21}s_1^{(p_1+q_1)/q_1} - \left(\beta_{31}{}^* - \frac{\mu_1}{s_1^2}\right)s_1^2 \\ &- \beta_{12}s_2^{(m_2+n_2)/n_2} - \beta_{22}s_2^{(p_2+q_2)/q_2} - \left(\beta_{32}{}^* - \frac{\mu_2}{s_2^2}\right)s_2^2 \\ &\le -\beta_{11}s_1^{(m_1+n_1)/n_1} - \beta_{21}s_1^{(p_1+q_1)/q_1} - \beta_{12}s_2^{(m_2+n_2)/n_2} \\ &- \beta_{22}s_2^{(p_2+q_2)/q_2}.\end{aligned} \tag{33}$$

To ensure the stability of the integral system, designed parameters are chosen to satisfy $\beta_{31}{}^* \ge \mu_1/s_1{}^2$ and $\beta_{32}{}^* \ge \mu_2/s_2{}^2$. To simplify the design process and keep its uniformity, other parameters can be chosen conveniently in the form $\beta_{11} = \beta_{12} = \beta_1$, $\beta_{21} = \beta_{22} = \beta_2$, $m_1 = m_2 = m$, $n_1 = n_2 = n$, $p_1 = p_2 = p$, and $q_1 = q_2 = q$. Then, (33) is changed into

$$\begin{aligned}\dot{V}_3 &= \mathbf{s}_1{}^T\dot{\mathbf{s}}_1 + \mathbf{s}_2{}^T\dot{\mathbf{s}}_2 \le -\beta_1 s_1^{(m+n)/n} - \beta_2 s_1^{(p+q)/q} - \beta_1 s_2^{(m+n)/n} \\ &- \beta_2 s_2^{(p+q)/q} = -\beta_1\left[\left(s_{1i}{}^2\right)^{(m+n)/2n} + \left(s_{2i}{}^2\right)^{(m+n)/2n}\right] \\ &- \beta_2\left[\left(s_{1i}{}^2\right)^{(p+q)/2q} + \left(s_{2i}{}^2\right)^{(p+q)/2q}\right] \\ &\le -2^{(n-m)/2n}\beta_1(2V_3)^{(m+n)/2n} - \beta_2(2V_3)^{(p+q)/2q}.\end{aligned} \tag{34}$$

Considering Lemma 2, $V_3 = 0$ implies $s_1 = s_2 = 0$. If $V_3 \ne 0$, substituting $y = \sqrt{2V_3}$ into (34) yields

$$\dot{y} \le -2^{(n-m)/2n}\beta_1(y)^{m/n} - \beta_2(y)^{p/q}, \tag{35}$$

which follows from Lemma 2 that system states reach the sliding bounded region within fixed-time $t < T_{\max}$. The fixed settling time is bounded and can be expressed as follows:

$$t < T_{\max} \triangleq \frac{1}{\beta_1}\frac{2^{(m-n)/2n}n}{m-n} + \frac{1}{\beta_2}\frac{q}{q-p}. \tag{36}$$

For the control variable α, decreasing V ultimately drives the state of trajectory to converge to the small region $|s1| < (\mu_1/\beta_{31}{}^*)^{1/2}$ within fixed bounded convergence time. Eventually, the tracking of the desired command is guaranteed.

Further, (36) indicates that the uniform fixed bound of settling time is a function of designed parameters. It is independent of system initial conditions and can be known in advance.

It is noticed in (24) that the first derivative of x_{2c} should be calculated. Because of the fact that f_{x1} is a function of control input, it is difficult to deduce $\dot{x}_{2c}$ from (18) analytically. In piratical application and engineering field, the use of some low pass filters to get the value of $\dot{x}_{2c}$ is widely adopted [30–32]. In our proposed method, the following low pass filter is used.

$$\tau\dot{\bar{x}}_{2c} + \bar{x}_{2c} = x_{2c}, \quad \bar{x}_{2c}(0) := x_{2c}(0), \tag{37}$$

where τ is a small positive time constant of the filter. Denote the estimation of $\dot{x}_{2c}$ by $\dot{\bar{x}}_{2c}$. Then, the control input is revised as

$$u = \delta_z = g_{x2}^{-1}\left(-f_{x2} - \beta_{12}s_2{}^{m_2/n_2} - \beta_{22}s_2{}^{p_2/q_2} - \beta_{32}s_2 - \frac{\widehat{\chi}^*{}_2 s_2}{4\mu_2^*} + \dot{\bar{x}}_{2c}\right), \tag{38}$$

$$\dot{\widehat{\chi}}^*{}_2 = -\beta_{42}^*\widehat{\chi}^*{}_2 + \frac{s_2^2}{4\mu_2^*}, \tag{39}$$

where β_{42}^* is positive and $\widehat{\chi}_2^*$ is the estimation of χ_2^* and will be explained in the lateral analysis.

Suppose the estimate error $\dot{\bar{x}}_{2c} - \dot{x}_{2c}$ is bounded by a positive constant [31, 32], then according to the assumption at the beginning, there exists a positive constant σ_2^* such that $|\dot{\bar{x}}_{2c} - \dot{x}_{2c} + d_2| \le \sigma_2^*$. Denote $\chi_2^* = (\sigma_2^*)^2$; similarly, for the inner loop, define its Lyapunov function as

$$V_2^* = \frac{1}{2}s_2^2 + \frac{1}{2}(\tilde{\chi}_2^*)^2, \tag{40}$$

where $\tilde{\chi}_2^* = \chi_2^* - \widehat{\chi}_2^*$.

TABLE 1: The parameters for a dual-control missile.

Variable	Value
b_α	2.63
b_{δ_z}	0.45
a_{ω_z}	0.86
a_α	10.98
a_{δ_z}	7.47
δ_z	$\leq\delta_{max} = 15$ deg

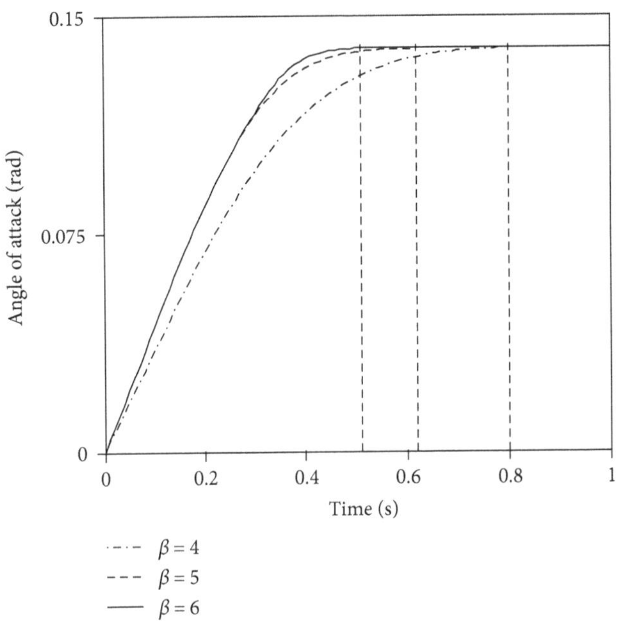

FIGURE 1: Response of the angle of attack.

Substituting (21), (38), and (39) into (40), one gets

$$\begin{aligned}\dot{V}_2^{\,*} &= s_2\dot{s}_2 + \tilde{\chi}_2^*\dot{\tilde{\chi}}_2^* \\ &= -\beta_{12}s_2^{(m_2+n_2)/n_2} - \beta_{22}s_2^{(p_2+q_2)/q_2} - \beta_{32}s_2^2 \\ &\quad + s_2\left(d_2 + \dot{\tilde{x}}_{2c} - \dot{x}_{2c}\right) - \frac{\tilde{\chi}_2^* s_2^2}{4\mu_2^*} + \beta_{42}^*\tilde{\chi}_2^*\tilde{\chi}_2^* \\ &\quad - \frac{\tilde{\chi}_2 s_2^2}{4\mu_2^*} \leq -\beta_{32}s_2^2 + |s_2|\sigma_2^* - \frac{\tilde{\chi}_2^* s_2^2}{4\mu_2^*} + \beta_{42}^*\tilde{\chi}_2^*\tilde{\chi}_2^* \\ &\quad - \frac{\tilde{\chi}_2^* s_2^2}{4\mu_2^*} \leq -\beta_{32}s_2^2 - \beta_{42}^*(1-\zeta_2)(\tilde{\chi}_2^*)^2 + \mu_2^* \\ &\quad + \beta_{42}^*\frac{(\chi_2^*)^2}{4\zeta_2} \leq -\eta_2^{\,*}V_2^* + \varsigma_2^{\,*},\end{aligned} \tag{41}$$

where $\eta_2^{\,*} = \min\{2\beta_{32}, 2\beta_{42}(1-\zeta_2)\}$, $\varsigma_2 = \mu_2 + \beta_{42}{\chi_2}^2/4\zeta_2$. Thus, s_2 and $\tilde{\chi}_2^*$ are uniformly ultimately bounded. Assume that $|\tilde{\chi}_2^*| \leq \zeta_2^*$ where ζ_2^* is positive.

Similar with the previous process, for the integral system, the derivative of Lyapunov function V_3 can be deduced in the same form with (35); thus, the fixed bounded convergence time is expressed identical with (36).

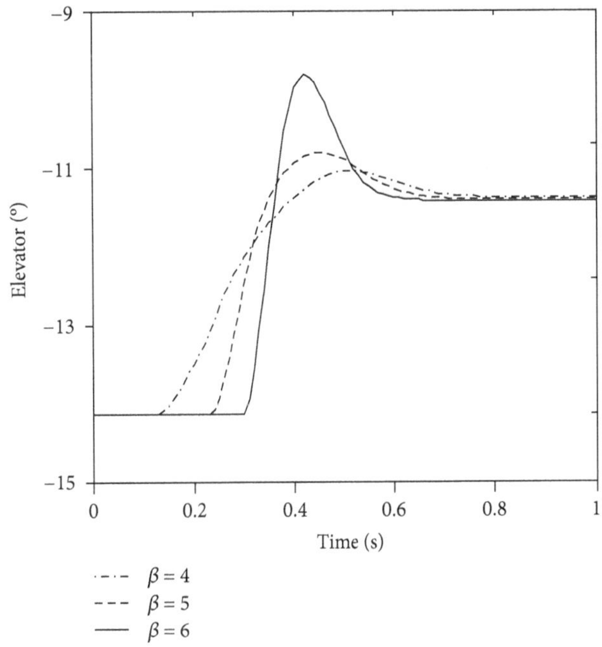

FIGURE 2: Time history of elevator deflection.

5. Simulation

In this section, a missile control problem in the longitudinal plane in the presence of aerodynamic uncertainties is considered. The dynamic model of (5) and the designed smooth fixed-time-convergent sliding mode controller of (17), (18), (38), and (39) are used in the simulation. To validate the effectiveness of the designed controller, simulations are divided into 2 subsections. Two kinds of maneuvers are pursued with varying aerodynamic uncertainties, namely, tracking fixed constant command and tracking continuous turn command. For the former, the fixed-time convergence characteristic is first validated, and afterwards, Monte Carlo simulations are carried out with different initial conditions to further demonstrate the convergence performance of the designed controller. For the latter, a continuous command is tracked under the condition of diverse aerodynamic uncertainties, which reflects the robustness of the designed controller adequately. From comparison with that using a conventional terminal sliding mode control algorithm, the superiority of the designed approach is revealed.

The flight speed and altitude are Mach 5 and 25 km, respectively. Parameters of a generic tail fin controlled missile are given in Table 1. Elevator deflection is limited to 15 degrees.

Control parameters are chosen to be $\beta_1 = \beta_2 = \beta$, $m_1 = m_2 = m = 5$, $n_1 = n_2 = n = 3$, $p_1 = p_2 = p = 3$, $q_1 = q_2 = q = 5$, $\beta_{41} = 0.03$, $\mu_1 = 0.005$, $\beta_{42}^* = 0.15$, $\mu_2^{\,*} = 0.05$, and $\tau = 0.1$

5.1. Case 1: Fixed Convergence Time with Different β and Initial Conditions.

This case simply takes the tracking ability

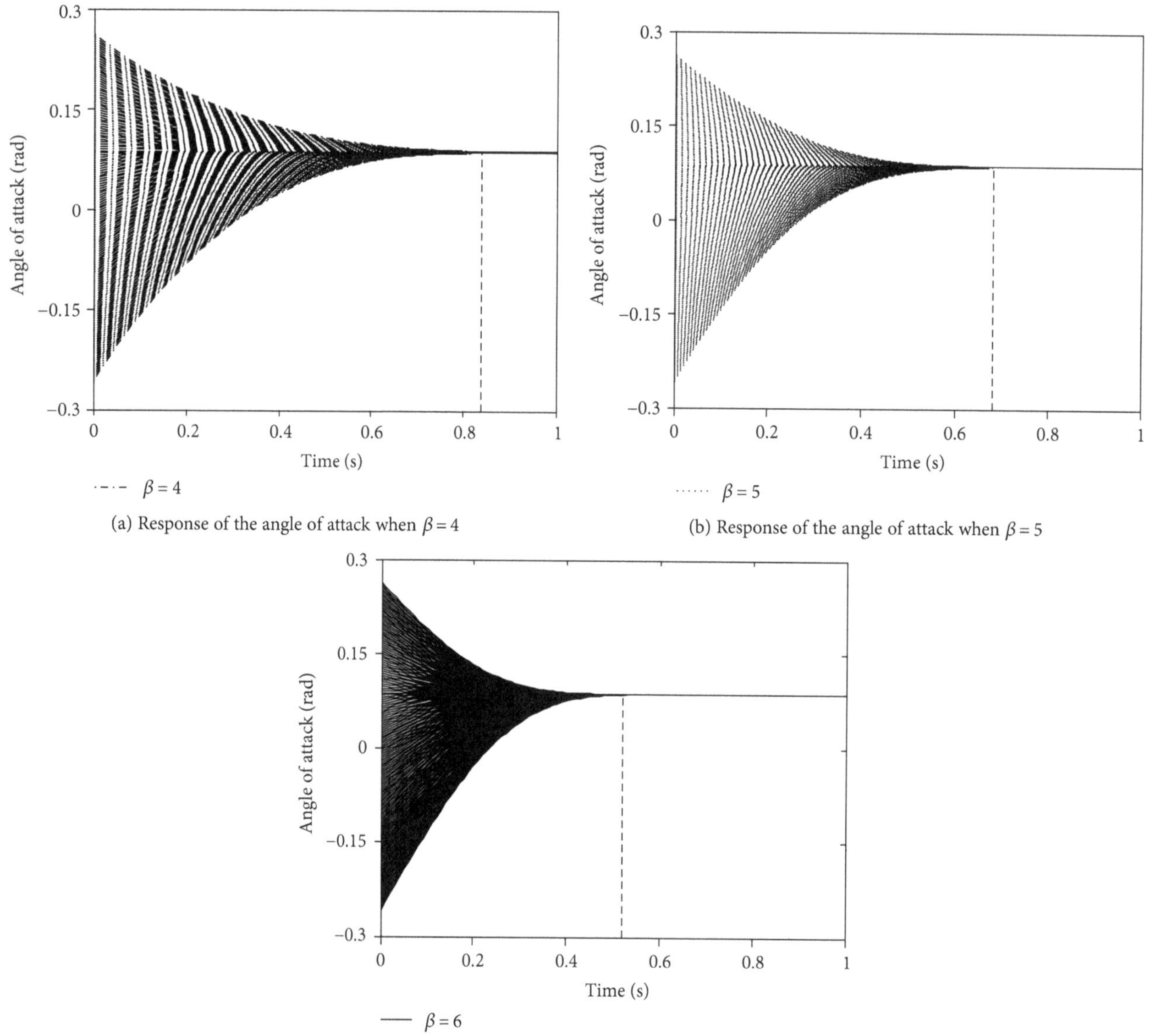

(a) Response of the angle of attack when $\beta = 4$

(b) Response of the angle of attack when $\beta = 5$

(c) Response of the angle of attack when $\beta = 6$

FIGURE 3: System response with the designed controller.

of fixed constant command into consideration. The coefficients of aerodynamic forces and moments are all increased by 30% of their respective nominal values. First, the initial system state is set to be $\alpha_0 = 0$ rad and $\alpha_c = 0.09$ rad. Figure 1 illustrates the time histories of system response in the longitudinal plane with the designed controller. Elevator deflections in different scenarios is depicted in Figure 2.

As visualized in Figure 1, with the proposed controller, the missile can track the desired command with satisfactory performance in the presence of aerodynamic uncertainties. The settling times are approximately 0.51 s, 0.62 s, and 0.8 s, which are all less than theoretical bounded fixed convergence time calculated from (36) with control parameters, namely, 0.73 s for $\beta = 6$, 0.88 s for $\beta = 5$, and 1.1 s for $\beta = 4$. At the beginning, deviations between command signal and initial state variable are considerably large, so that the tail fin goes beyond the deflection constraint promptly for the sake of tracking, as shown in Figure 2. Full aerodynamic fin deflection continuously contributes to the control efforts, propelling the required control quantities rapidly to the scope of available control load. The burden of aerodynamic actuator is reduced and gradually actuated smoothly to ensure the tracking precision. As a result, fast tracking within fixed bounded convergence time is achieved for the missile flight system with the designed controller.

Next, Monte Carlo simulations are carried out to verify the fixed-time convergence property of the designed controller, which is independent of system initial conditions. In this occasion, $\alpha_c = 0.087$ rad. The initial state varies from −0.26 to approximately 0.26 rad. Other parameters are identical with the previous occasion.

Figure 3 depicts the tracking performance of a tail controlled missile with the designed controller under different initial conditions.

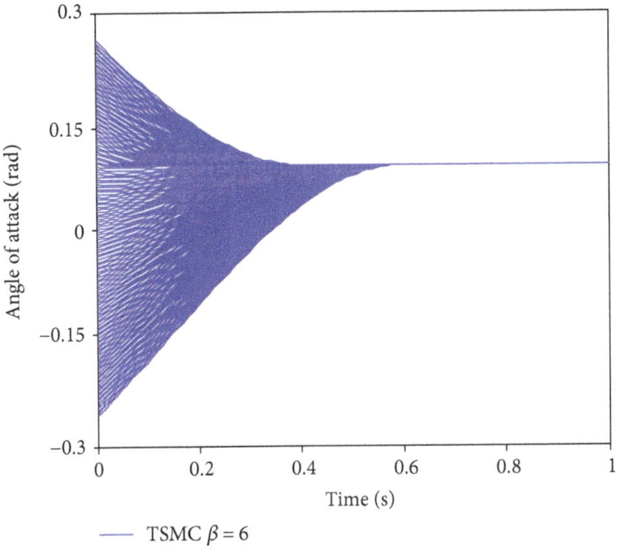

FIGURE 4: System response with TSMC controller.

As calculated above, theoretical bounds of settling time with the designed controller are 1.1 s, 0.88 s, and 0.73 s, respectively. Performance diagrams in Figure 3 demonstrate that under different initial conditions, the missile can track the command signal before the "fixed" predefined bound of convergence time. The time history of system response exhibits a nice uniform convergent characteristic when diverse initial states are given. Simulation results of this situation fully reflect the fixed-time convergence characteristic of the designed controller, which is independent of initial conditions. That is also what separates the designed method with conventional TSMC controllers.

A terminal sliding mode controller (TSMC) is used here with identical designed parameters for comparison. The expressions are given in (42) and (43).

$$x_{2c} = -f_{x1} - \beta s_1^{\eta}\,\mathrm{sgn}(s_1) - \varepsilon_1 \mathrm{sgn}(s_1), \tag{42}$$

$$u = \delta_z = g_{x2}^{-1}\left(-f_{x2} - \beta s_2^{\eta}\,\mathrm{sgn}(s_2) - \varepsilon_2\,\mathrm{sgn}(s_2) + \dot{\bar{x}}_{2c}\right), \tag{43}$$

where $\varepsilon_1 > \sigma 1$ and $\varepsilon_2 > \sigma_2$. Here, let $\beta = 6$, for various initial conditions, and time histories of system response are plotted in Figure 4.

Figure 4 depicts the fact that the missile with TSMC controller is able to track the command in an acceptable settling time. However, no uniform boundary of convergence time can be defined in advance since the settling time of the TSMC algorithm is in connection with initial states, which is very different from that in Figure 3(c). Various initial conditions correspond to diverse convergence rates, without which it is almost impossible to predict the bounded settling time when conducting preliminary design and performance evaluation. That is different with the designed method.

5.2. Case 2: Continuous Command under Different Aerodynamic Uncertainty Scenarios. In this case, it is assumed that aerodynamic uncertainty coefficients are increased by

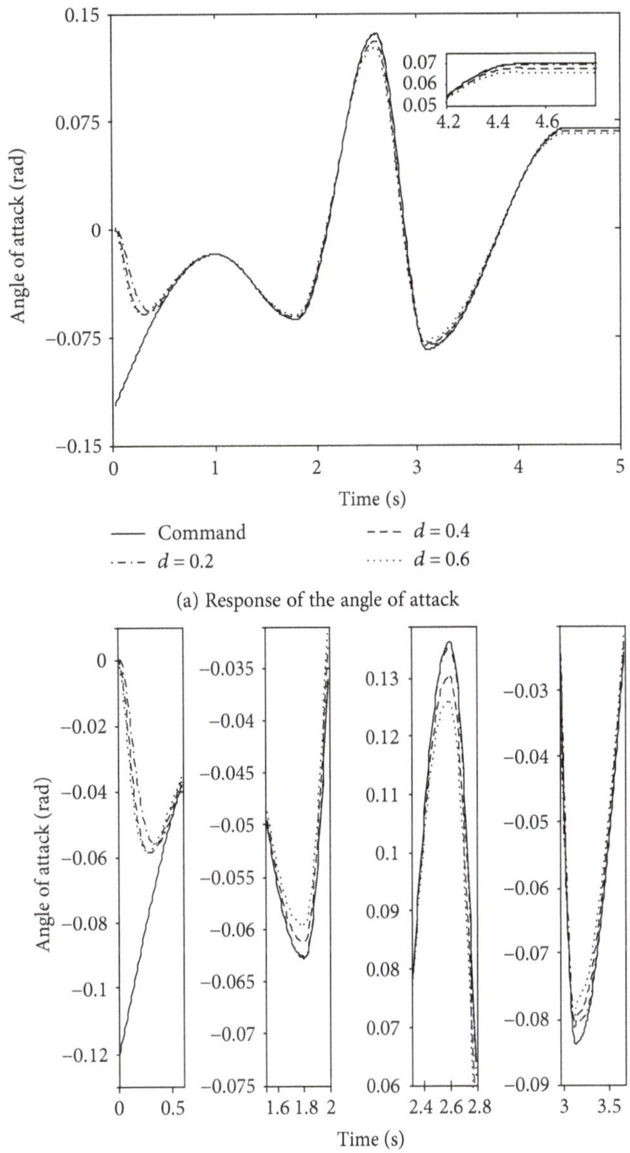

(a) Response of the angle of attack

(b) Details of response in the pitch plane

FIGURE 5: Time history of the angle of attack.

20%, 40%, and 60%. A continuous signal is generated as turn commands for the missile to maneuver, of which $\beta = 5$, and the initial angle of attack equals to zero. Time histories of system response with the designed controller in the presence of various uncertainties are depicted in Figure 5, where the continuous turn commands are illustrated as a reference. Figure 6 gives the diagram of corresponding tail deflections. Details of distinctions between diverse uncertainty conditions are supplemented in Figures 5(b) and 6(b).

It can be seen from Figure 5 that continuous command could be tracked with the proposed smooth fixed-time-convergent sliding mode controller. Within theoretical predefined settling time, which is 0.88 s, the trajectory of the missile attitude angle is driven closely to the desired value and follows the continuous command effectively afterwards. As shown in Figure 6, the deflection of elevator first exceeds

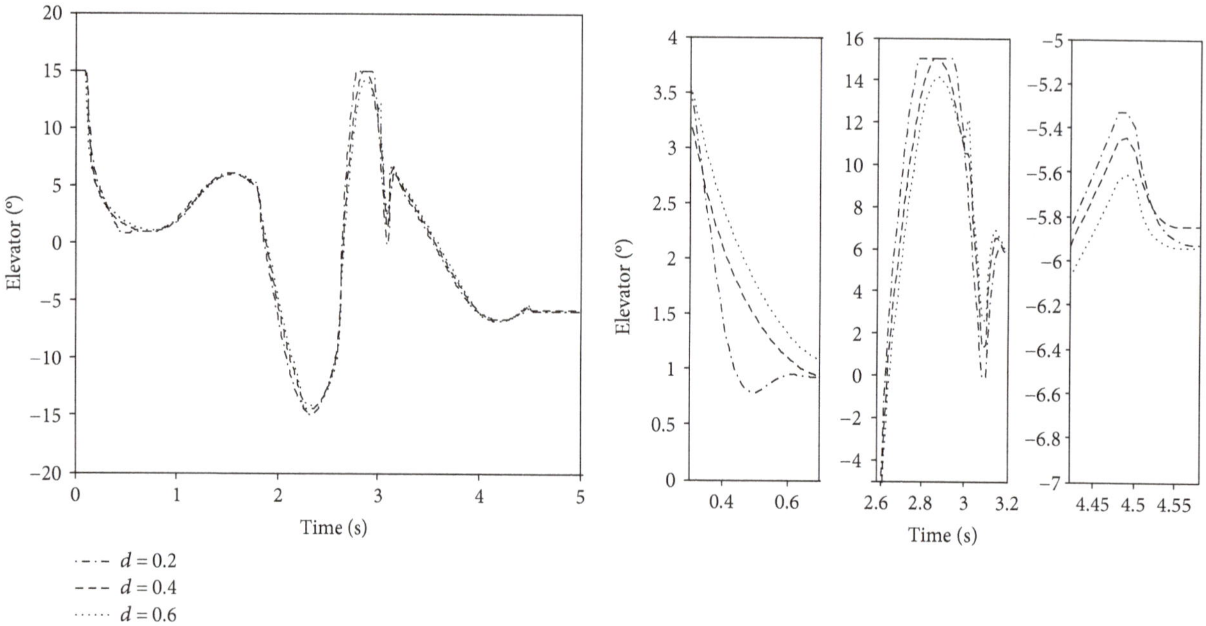

(a) Time history of elevator deflection

(b) Details of elevator deflection

FIGURE 6: Elevator deflection.

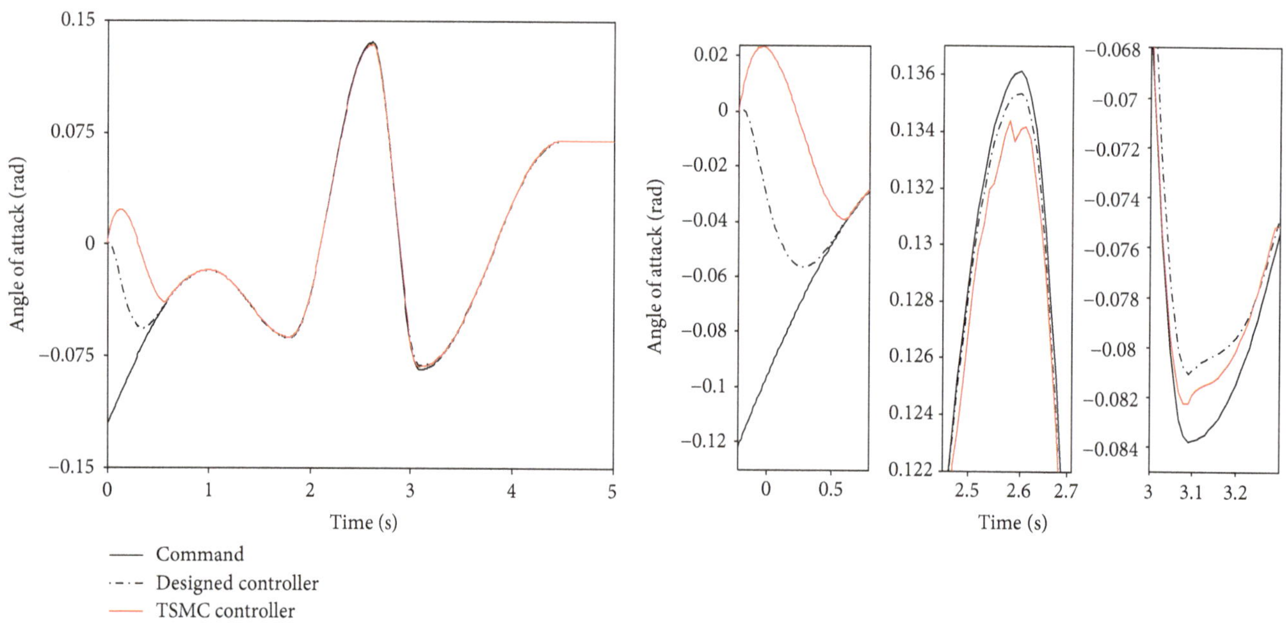

(a) Response of the angle of attack

(b) Details of response in the pitch plane

FIGURE 7: Time history of the angle of attack.

its physical constraint because of the large deviation between the initial condition and the desired command. When an initial error is swiped out to a small extent, effective control efforts becomes available, and consequently, smooth elevator deflection plays a crucial part consecutively. Although different degrees of aerodynamic uncertainty exist throughout the flight regime, the designed approach can ensure good tracking precision of the missile flight system and demonstrate satisfactory robustness against uncertainties.

To show the effectiveness of the proposed smooth controller further, the TSMC controller is simulated with identical designed parameters for comparison.

On this occasion, aerodynamic uncertainty is assumed to increase by 20%. Figures 7 and 8 illustrate the system response and control input, respectively.

Although it seems both controllers can track the continuous command, the control quality of the designed controller is superior to that of the TSMC method. It can be observed

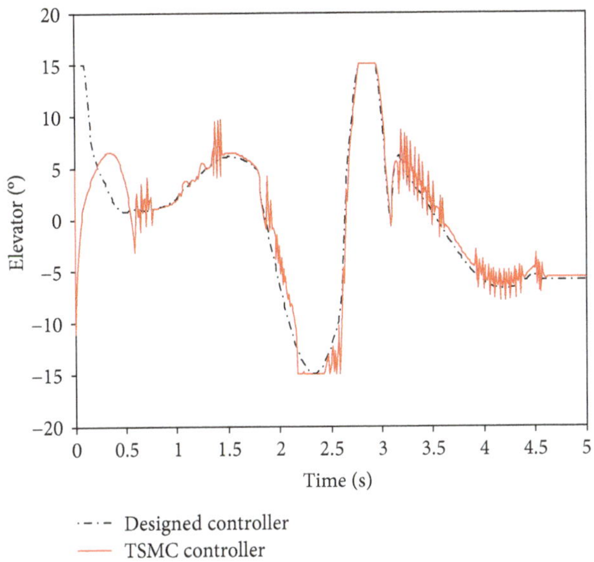

FIGURE 8: Elevator deflection.

from the above figures that the chattering problem exists in the performance of the TSMC controller. With the designed smooth controller, the input signal is smooth without singularity and chattering, which illustrates better tracking performance than the TSMC approach.

6. Conclusions

This paper designed a smooth fixed-time-convergent sliding mode controller for a missile flight system in the presence of aerodynamic uncertainties. Based on fixed-time stability theorem, some primary mathematical results, and adaptive estimations, the derived control algorithm guarantees that control variables converge to the desired command within fixed uniform bounded time regardless of initial conditions with a smooth control input. The bounded convergence rate can be predicted in advance without knowledge of system original states, which is of significant value in system preliminary design, as well as performance evaluation. Meanwhile, the input signal is designed to be continuous without introducing any discrete items like the signum function, thus eliminating the chattering phenomenon and facilitating practical application significantly. In the end, extensive simulations are carried out and validate the effectiveness and robustness of the designed controller.

Conflicts of Interest

The authors declare that there is no conflict of interest.

References

[1] C. F. Lin and W. R. Yueh, "Coordinated bank-to-turn autopilot design," in *1985 American Control Conference*, pp. 922–926, Boston, MA, USA, USA, June 1985.

[2] M. J. Kovach, T. R. Stevens, and A. Arrow, "A bank-to-turn autopilot design for an advanced air-to-air interceptor," in *Guidance, Navigation and Control Conference*, pp. 1346–1353, Monterey, CA, USA, August 1987.

[3] B. SUN and Z. QI, "Study of pole placement method for state feedback constrained autopilot design," *Journal of System Simulation*, vol. 18, Supplement 2, pp. 892–896, 2006.

[4] W. Peng, C. Wanchun, and Y. I. N. Xing-liang, "Large angle attitude reaction jet control for an air-to-air missile," *Acta Aeronautica Et Astronautica Sinica*, vol. 26, pp. 263–267, 2005.

[5] J.-f. Fan, D.-f. Lin, Z.-k. Qi, and J. Wang, "Missile autopilot design using optimal control and frequency constraint," in *2008 2nd International Symposium on Systems and Control in Aerospace and Astronautics*, pp. 1–6, Shenzhen, China, December 2008.

[6] Y. S. Choi, H. C. Lee, and J. W. Choi, "Autopilot design for agile missile with aerodynamic fin and side thruster," in *SICE 2003 Annual Conference*, vol. 2, pp. 1476–1481, Fukui, Japan, August 2003.

[7] Z. Xu and H. L. Xu, "Application of fuzzy controller to missile networked control systems," in *2008 4th International Conference on Wireless Communications, Networking and Mobile Computing*, pp. 1–4, Dalian, China, October 2008.

[8] L. Yuan, J. Cui, M. Pan, and Y. Liu, "Design of aerodynamics missile controller based on adaptive fuzzy PID," in *Measurement, Information and Control (MIC), 2012 International Conference on IEEE*, vol. 2, pp. 712–716, Harbin, China, May 2012.

[9] W. Zheng, G. Liu, J. Yang, and T. Du, "An adaptive fuzzy variable structure controller for bank-to-turn missile," *Journal of Computational Information Systems*, vol. 7, no. 2, pp. 562–569, 2011.

[10] Y. Xu, J. Yu, Y. Yuan, and W. Gu, "Adaptive fuzzy sliding-mode controller for BTT missile," in *ICARCV 2004 8th Control, Automation, Robotics and Vision Conference, 2004*, vol. 2, pp. 1222–1226, Kunming, China, China, December 2004.

[11] D. Zhou, C. Mu, and W. Xu, "Adaptive sliding-mode guidance of a homing missile," *Journal of Guidance, Control, and Dynamics*, vol. 22, no. 4, pp. 589–594, 1999.

[12] Y. X. Zhang, M. W. Sun, and Z. Q. Chen, "Sliding-mode variable structure finite-time convergence guidance law," *Control Theory & Applications*, vol. 29, no. 11, pp. 1413–1418, 2012.

[13] S. R. Kumar, S. Rao, and D. Ghose, "Sliding-mode guidance and control for all-aspect interceptors with terminal angle constraints," *Journal of Guidance, Control and Dynamics*, vol. 35, no. 4, pp. 1230–1246, 2012.

[14] Y. B. Shtessel, I. A. Shkolnikov, and A. Levant, "Guidance and control of missile interceptor using second-order sliding modes," *IEEE Transactions on Aerospace and Electronic Systems*, vol. 45, no. 1, pp. 110–124, 2009.

[15] D. Zhou, P. Qu, and S. Sun, "A guidance law with terminal impact angle constraint accounting for missile autopilot," *Journal of Dynamic Systems, Measurement, and Control*, vol. 135, no. 5, article 051009, 2013.

[16] D. Zhou, S. Sun, and K. L. Teo, "Guidance laws with finite time convergence," *Journal of Guidance, Control, and Dynamics*, vol. 32, no. 6, pp. 1838–1846, 2009.

[17] Z. Wang, S. H. Li, and S. M. Fei, "Finite-time tracking control of bank-to-turn missiles using terminal sliding mode," *ICIC Express Letters*, vol. 3, no. 3, pp. 1–6, 2009.

[18] L. Hui and J. Li, "Terminal sliding mode control for spacecraft formation flying," *IEEE Transactions on Aerospace and Electronic Systems*, vol. 45, no. 3, pp. 835–846, 2009.

[19] S. R. Kumar, S. Rao, and D. Ghose, "Non-singular terminal sliding mode guidance and control with terminal angle constraints for non-maneuvering targets," in *2012 12th International Workshop on Variable Structure Systems*, pp. 291–296, Mumbai, Maharashtra, India, January 2012.

[20] X. Wang and J. Wang, "Partial integrated missile guidance and control with finite time convergence," *Journal of Guidance, Control, and Dynamics*, vol. 36, no. 5, pp. 1399–1409, 2013.

[21] A. Polyakov, "Nonlinear feedback design for fixed-time stabilization of linear control systems," *IEEE Transactions on Automatic Control*, vol. 57, no. 8, pp. 2106–2110, 2012.

[22] Z. Zuo, "Nonsingular fixed-time consensus tracking for second-order multi-agent networks," *Automatica*, vol. 54, pp. 305–309, 2015.

[23] Z. Zuo, "Non-singular fixed-time terminal sliding mode control of non-linear systems," *IET Control Theory & Applications*, vol. 9, no. 4, pp. 545–552, 2015.

[24] B. Jiang, Q. Hu, and M. I. Friswell, "Fixed-time attitude control for rigid spacecraft with actuator saturation and faults," *IEEE Transactions on Control Systems Technology*, vol. 24, no. 5, pp. 1892–1898, 2016.

[25] Y. X. Zhang, "Research on finite-time convergent sliding-mode guidance law [PhD thesis]," NanKai University, China, 2013.

[26] F. Wang, Q. Zong, Q. Dong, and B. Tian, "Disturbance observer-based sliding mode backstepping control for a re-entry vehicle with input constraint and external disturbance," *Transactions of the Institute of Measurement and Control*, vol. 38, no. 2, pp. 165–181, 2015.

[27] S. Li and J. Yang, "Robust autopilot design for bank-to-turn missiles using disturbance observers," *IEEE Transactions on Aerospace and Electronic Systems*, vol. 49, no. 1, pp. 558–579, 2013.

[28] S. Yu, X. Yu, B. Shirinzadeh, and Z. Man, "Continuous finite-time control for robotic manipulators with terminal sliding mode," *Automatica*, vol. 41, no. 11, pp. 1957–1964, 2005.

[29] E. Cruz-Zavala, J. A. Moreno, and L. Fridman, "Uniform sliding mode controllers and uniform sliding surfaces," *IMA Journal of Mathematical Control and Information*, vol. 29, no. 4, pp. 491–505, 2012.

[30] D. Swaroop, J. K. Hedrick, P. P. Yip, and J. C. Gerdes, "Dynamic surface control for a class of nonlinear systems," *IEEE Transactions on Automatic Control*, vol. 45, no. 10, pp. 1893–1899, 2002.

[31] M. Hou, X. Liang, and G. Duan, "Adaptive block dynamic surface control for integrated missile guidance and autopilot," *Chinese Journal of Aeronautics*, vol. 26, no. 3, pp. 741–750, 2013.

[32] Y. Wei, M. Hou, and G. R. Duan, "Adaptive multiple sliding surface control for integrated missile guidance and autopilot with terminal angular constraint," in *Proceedings of the 29th Chinese Control Conference*, pp. 2162–2166, Beijing, China, July 2010.

13

Consensus Control of Time-Varying Delayed Multiagent Systems with High-Order Iterative Learning Control

Xiongfeng Deng[1] **Xiuxia Sun**[1] **Ri Liu,**[2] **and Shuguang Liu**[1]

[1]*Equipment Management and Unmanned Aerial Vehicle Engineering College, Air Force Engineering University, Xi'an 710051, China*
[2]*Theory Training Department, Air Force Harbin Flight Academy, Harbin 150001, China*

Correspondence should be addressed to Xiuxia Sun; gcxysxx@126.com

Academic Editor: Giovanni Palmerini

We address the consensus control problem of time-varying delayed multiagent systems with directed communication topology. The model of each agent includes time-varying nonlinear dynamic and external disturbance, where the time-varying nonlinear term satisfies the global Lipschitz condition and the disturbance term satisfies norm-bounded condition. An improved control protocol, that is, a high-order iterative learning control scheme, is applied to cope with consensus tracking problem, where the desired trajectory is generated by a virtual leader agent. Through theoretical analysis, the improved control protocol guarantees that the tracking errors converge asymptotically to a sufficiently small interval under the given convergence conditions. Furthermore, the bounds of initial state difference and disturbances tend to zero; the bound of tracking errors also tends to zero. In the end, some cases are provided to illustrate the effectiveness of the theoretical analysis.

1. Introduction

In the past few decades, cooperative control of multiagent systems has been attracted outstanding attention. One of the main reasons is its wide application in various fields, such as military, aerospace, and industrial [1]. The research directions include formation control, containment control, tracking control, swarm control, and flocking control [2–6]. Generally speaking, the fundamental problem for cooperative control is consensus, where the objective is to design an appropriate control law so that the whole states of a group of agents eventually reach an agreement.

In the process of studying the cooperative problem of multiagent systems, diverse control strategies are proposed in many literatures. For example, the output regulation method and distributed containment control law for the containment control of linear multiagent systems were investigated in [7, 8], while an advanced clustering with the frequency sweeping method was used for the general linear multiagent systems with communication and input delay in [9]. Nevertheless, it should be emphasized that almost all practice systems involve nonlinear dynamics. In [10, 11], the distributed adaptive control protocol and pinning control algorithm for the consensus of a class of second-order leader-following nonlinear multiagent systems with directed topology were considered. In [12], a distributed cooperative method was applied to deal with the dynamic task planning of multiple satellites. Additionally, the neural network algorithm was also introduced to study the consensus problem of nonlinear multiagent systems [13, 14].

In practice, time delays always exist due to the impact of physical factors of the sensor. Therefore, it is of great significance to study the time delay system. In [15], a distributed consensus control protocol with decaying gains for linear discrete-time multiagent systems with delays and noises was considered; and the cases of communication delay and input delay were studied in [16, 17]. For the problem of time-varying delays, Xiao and Wang [18] investigated the state control problem of multiagent systems with bounded time-varying communication delays and Chen et al. [19] presented a robust control law for a class of time-varying delayed multiagent systems with noisy environment.

As we know, iterative learning control is based upon the idea that the performance of a system that performs the same task repeatedly can be improved by learning from previous iterations [20]. In early works, iterative learning control has been considered in various issues. In [21], an iterative learning control scheme was proposed to make the tracking error trajectory converge to the prespecified error trajectory. In [22], an adaptive iterative learning updating law was applied to solve the tracking problem of high precision motion systems. In [23], the packet dropout problem of nonlinear systems with iterative learning control was developed; and Wu et al. [24] solved the high precision satellite attitude tracking control by using iterative learning control. Additionally, a high-order iterative learning identification scheme was considered to extract the projectile's aerodynamic drag coefficient curve from radar measured velocity data in [25]; and the stability analysis of a high-order iterative learning control problem for discrete-time systems and nonlinear switched systems was also studied in [26, 27], respectively. In recent years, the application of iterative learning control has been extended to multiagent systems. In [28], a D-type iterative learning control scheme was presented to deal with the tracking problem of multiagent systems. In [29], a distributed adaptive fuzzy iterative learning control algorithm for the coordination control problem of leader-following multiagent systems with unknown dynamics and nonrepeatable input disturbances was designed; and Meng and Jia [30] developed a finite-time consensus control protocol with iterative learning control for a class of multiagent systems. In addition, the iterative learning control was also considered to achieve formation control of multiagent systems [31, 32].

By analyzing the above literature, it is not difficult to find that there exist few results about the consensus problem of time-varying delayed multiagent systems with iterative learning control. Moreover, the application of high-order iterative learning control for multiagent systems is also seldom described in the existing papers. These facts inspired our current study. In this work, we divert our attention to the consensus control of time-varying delayed multiagent systems with high-order iterative learning control. The main contributions of this work are summarized as follows: (i) a class of time-varying delayed multiagent systems with directed communication topology is considered in this paper. The dynamic of each agent contains time-varying nonlinear dynamic and external disturbance, where the time-varying nonlinear term satisfies the global Lipschitz condition and the disturbance term satisfies norm-bounded condition; (ii) different from [26, 27], an improved high-order iterative learning control scheme is applied to guarantee the tracking error convergence in the iteration domain. Furthermore, we show that the bound of tracking errors also tends to zero when the bounds of initial state difference and disturbances tend to zero; and (iii) it is proven that the improved control protocol can effectively handle the consensus problem of multiagent systems with time-varying delays and external disturbances.

The rest of this paper is organized as follows. Some necessary preliminaries are introduced in Section 2. The problem formulation and main results about time-varying delayed multiagent systems with high-order iterative learning control are discussed in Sections 3 and 4, respectively. In Section 5, the effectiveness of the proposed control protocol is illustrated by simulations, and briefly, conclusions are outlined in Section 6.

2. Preliminaries

In this section, graph theory, definition, and lemma which will be utilized in this paper are introduced briefly.

2.1. Graph Theory. For a multiagent systems with n agents, the exchange information among agents can be modeled as an interaction graph with n nodes. Let $\mathcal{G} = (\mathcal{V}, \mathcal{E}, \mathcal{A})$ denote a directed graph with a node set $\mathcal{V} = (v_1, \dots, v_n)$ and a directed edge set $\mathcal{E} = \{(i, j), i, j \in \mathcal{V}, i \neq j\}$. And the adjacency matrix of the directed graph is defined by $\mathcal{A} = [a_{ij}] \in \mathbb{R}^{n\times n}$, where $a_{ij} > 0$ if and only if $(i, j) \in \mathcal{E}$ and $a_{ij} = 0$ otherwise. Moreover, it is assumed that $a_{ii} = 0$. For agent i, the set of neighbors is denoted by $\mathcal{N}_i = \{v_j : (v_j, v_i) \in \mathcal{E}\}$. The Laplacian matrix $\mathcal{L}$ is denoted by $\mathcal{L} = \mathcal{D} - \mathcal{A}$, where $\mathcal{D} = \text{diag}\{d_1, \dots, d_n\}$ with $d_i = \sum_{j=1}^{\mathcal{N}_i} a_{ij}$.

In this work, an augmented graph is described as $\bar{\mathcal{G}}$ which consists of n agents and one virtual leader agent. The communication between agents and the virtual agent is defined by the matrix $\mathcal{B} = \text{diag}\{b_1, \dots, b_n\}$. If agent i can obtain the information of the virtual agent, then $b_i > 0$ and $b_i = 0$ otherwise.

2.2. Definition and Lemma

Definition 1 (see [33]). For a function $y(t)$: $[0, T] \rightarrow \mathbb{R}^n$, the λ norm is defined by

$$\|y(t)\|_\lambda = \sup_{t\in[0,T]} \left\{e^{-\lambda t}\|y(t)\|\right\}, \quad \lambda > 0. \tag{1}$$

The following property for λ norm is held.

Property 1. For functions $\mathbf{h}(t), \mathbf{y}(t)$: $[0, T] \rightarrow \mathbb{R}^n$, if $\mathbf{y}(t) = \int_0^t \mathbf{h}(s)ds$, then have

$$\|\mathbf{y}(t)\|_\lambda \leq o(\lambda)\|\mathbf{h}(t)\|_\lambda, \tag{2}$$

where $o(\lambda) = \lambda^{-1}(1 - e^{-\lambda T})$.

Lemma 1 (see [34]). *For a real positive series* $\{a_n\}(n \in \{1, 2, \dots, \infty\})$ *satisfies*

$$a_n \leq \beta_1 a_{n-1} + \beta_2 a_{n-2} + \cdots + \beta_N a_{N-2} + \sigma, (n = N + 1, N + 2, \dots) \tag{3}$$

where $\beta_m \geq 0\ (m = 1, 2, \ldots, N)$ *and* $\sigma \geq 0$ *and exists*

$$\beta = \sum_{i=1}^{N} \beta_m < 1, \tag{4}$$

then have

$$\lim_{n\to\infty} a_n \leq \frac{\sigma}{1-\beta}. \tag{5}$$

3. Construction of the Control Protocol for the Consensus Problem

In this section, the main works are carried out to formulate the consensus problem and construct the consensus control protocol. In addition, some assumptions are also provided.

Considering the multiagent systems which consist of n identical agents with time-varying delays and nonlinear dynamics, the dynamics of the ith agent at kth iteration are described by

$$\begin{aligned}\dot{z}_i^k(t) &= f\left(z_i^k(t), z_i^k(t-\tau_{d1}), t\right) + B\left(z_i^k(t), z_i^k(t-\tau_{d2}), t\right)u_i^k(t) \\ &\quad + w_i^k(t), \\ y_i^k(t) &= g\left(z_i^k(t), t\right),\end{aligned} \tag{6}$$

where $z_i^k(t) \in \mathbb{R}^2$, $u_i^k(t) \in \mathbb{R}$, and $y_i^k(t) \in \mathbb{R}^2$ are the state, control input, and output of the system, respectively; $w_i^k(t) \in \mathbb{R}^2$ is bounded disturbance; the time $t \in [0, T]$ and T is given; $\tau_{d1}, \tau_{d2} \leq T$ are time delays; and $i \in \{0, 1, \ldots, n\}$. The functions $f(\cdot,\cdot,\cdot)$: $\mathbb{R}^2 \times \mathbb{R}^2 \times [0, T] \to \mathbb{R}^2$ and $B(\cdot,\cdot,\cdot)$: $\mathbb{R}^2 \times \mathbb{R}^2 \times [0, T] \to \mathbb{R}^2$ are piecewise continuous in t; and $g(\cdot,\cdot)$: $\mathbb{R}^2 \times [0, T] \to \mathbb{R}^2$ is differentiable in z and t, with $g_z(\cdot,\cdot)$ and $g_t(\cdot,\cdot)$. In addition, it is assumed that $z_i^k(t) = 0$ if $t < 0$.

The matrix form of (6) is expressed as

$$\begin{aligned}\dot{\mathbf{z}}^k(t) &= \mathbf{f}\left(\mathbf{z}^k(t), \mathbf{z}^k(t-\tau_{d1}), t\right) + \mathbf{B}\left(\mathbf{z}^k(t), \mathbf{z}^k(t-\tau_{d2}), t\right)\mathbf{u}^k(t) \\ &\quad + \mathbf{w}^k(t), \\ \mathbf{y}^k(t) &= \mathbf{g}\left(\mathbf{z}^k(t), t\right),\end{aligned} \tag{7}$$

where $\mathbf{z}^k(t) = [(z_1^k(t))^T, \ldots, (z_n^k(t))^T]^T$, $\mathbf{f}(\mathbf{z}^k(t), \mathbf{z}^k(t-\tau_{d1}), t) = [f^T(z_1^k(t), z_1^k(t-\tau_{d1}), t), \ldots, f^T(z_n^k(t), z_n^k(t-\tau_{d1}), t)]^T$, $\mathbf{B}(\mathbf{z}^k(t), \mathbf{z}^k(t-\tau_{d2}), t) = \mathrm{diag}\,\{B(z_1^k(t), z_1^k(t-\tau_{d2}), t), \ldots, B(z_n^k(t), z_n^k(t-\tau_{d2}), t)\}$, $\mathbf{u}^k(t) = [u_1^k(t), \ldots, u_n^k(t)]^T$, $\mathbf{w}^k(t) = [(w_1^k(t))^T, \ldots, (w_n^k(t))^T]^T$, $\mathbf{y}^k(t) = [(y_1^k(t))^T, \ldots, (y_n^k(t))^T]^T$, and $\mathbf{g}(\mathbf{z}^k(t), t) = [g^T(z_1^k(t), t), \ldots, g^T(z_n^k(t), t)]^T$.

The dynamics of virtual leader agent are given as

$$\begin{aligned}\dot{z}_d(t) &= f(z_d(t), z_d(t-\tau_{d1}), t) + B(z_d(t), z_d(t-\tau_{d2}), t)u_d(t), \\ y_d(t) &= g(z_d(t), t),\end{aligned} \tag{8}$$

which may show a given bounded desired trajectory $y_d(t) \in \mathbb{R}^2$; there exist unique bounded input $u_d(t) \in \mathbb{R}$ and $t \in [0, T]$ such that when $u(t) = u_d(t)$, the virtual agent has a unique bounded state $z_d(t) \in \mathbb{R}^2$ and output $y_d(t) = g(z_d(t), t)$.

Similar to the expression of (7), the matrix form of (8) is expressed as

$$\begin{aligned}\dot{\mathbf{z}}_d(t) &= \mathbf{f}(\mathbf{z}_d(t), \mathbf{z}_d(t-\tau_{d1}), t) + \mathbf{B}(\mathbf{z}_d(t), \mathbf{z}_d(t-\tau_{d2}), t)\mathbf{u}_d(t), \\ \mathbf{y}_d(t) &= \mathbf{g}(\mathbf{z}_d(t), t),\end{aligned} \tag{9}$$

where $\mathbf{z}_d(t) = \mathbf{1}_n \otimes z_d(t)$, $\mathbf{f}(\mathbf{z}_d(t), \mathbf{z}_d(t-\tau_{d1}), t) = \mathbf{1}_n \otimes f(z_d(t), z_d(t-\tau_{d1}), t)$, $\mathbf{B}(\mathbf{z}_d(t), \mathbf{z}_d(t-\tau_{d2}), t) = \mathbf{I}_n \otimes B(z_d(t), z_d(t-\tau_{d2}), t)$, $\mathbf{u}_d(t) = \mathbf{1}_n \otimes u_d(t)$, $\mathbf{y}_d(t) = \mathbf{1}_n \otimes y_d(t)$, and $\mathbf{g}(\mathbf{z}_d(t), t) = \mathbf{1}_n \otimes g(z_d(t), t)$; $\mathbf{1}_n = [1, \ldots, 1]^T$, $\mathbf{I}_n$ is n dimension unit matrix, and $\otimes$ represents the Kronecker product.

Given the desired output trajectory of the virtual agent $y_d(t)$, the goal is to find a control consensus $\boldsymbol{u}^k(t)$ such that when $k \to \infty$, the output $\boldsymbol{y}^k(t)$ of all agents will track the desired output trajectory $y_d(t)$ as close as possible.

According to (7) and (9), the consensus tracking error of the agent i at the kth iteration is defined as

$$\varepsilon_i^k(t) = \sum_{j\in\mathcal{N}_i} a_{ij}\left(y_i^k(t) - y_j^k(t)\right) + b_i\left(y_d(t) - y_j^k(t)\right), \tag{10}$$

and have

$$\boldsymbol{\varepsilon}^k(t) = \bar{\mathscr{H}}\mathbf{e}^k(t) \tag{11}$$

where $\bar{\mathscr{H}} = \mathscr{H} \otimes \mathbf{I}_2$, $\mathscr{H} = \mathscr{L} + \mathscr{B}$, $\varepsilon^k(t) = [\varepsilon_1^k(t), \ldots, \varepsilon_n^k(t)]^T$, $\mathbf{e}^k(t) = [e_1^k(t), \ldots, e_n^k(t)]^T$, and $\mathbf{e}^k(t) = \mathbf{y}_d(t) - \mathbf{y}^k(t)$.

Before introducing a high-order iterative learning control scheme, we give the following assumptions.

Assumption 1. The functions $\mathbf{f}(\cdot,\cdot,\cdot)$, $\mathbf{B}(\cdot,\cdot,\cdot)$, and $\mathbf{g}(\cdot,\cdot)$ and the partial derivatives $\mathbf{g}_z(\cdot,\cdot)$ and $\mathbf{g}_t(\cdot,\cdot)$ are uniformly globally Lipschitz in $\mathbf{z}(\cdot)$ on the interval $[0, T]$. That is, there exist constants $\kappa_{\boldsymbol{\alpha}}$, $\kappa_{\mathbf{f}r}$, and $\kappa_{\mathbf{B}r}$ $(r = 1, 2)$ such that

$$\begin{aligned}&\|\boldsymbol{\alpha}(\mathbf{x}_1(t), t) - \boldsymbol{\alpha}(\mathbf{x}_2(t), t)\| \\ &\quad \leq \kappa_{\boldsymbol{\alpha}}\|\mathbf{x}_1(t) - \mathbf{x}_2(t)\|, \\ &\|\mathbf{f}(\mathbf{x}_1(t), \mathbf{x}_1(t-\tau_{d1}), t) - \mathbf{f}(\mathbf{x}_2(t), \mathbf{x}_2(t-\tau_{d1}), t)\| \\ &\quad \leq \kappa_{\mathbf{f}1}\|\mathbf{x}_1(t) - \mathbf{x}_2(t)\| + \kappa_{\mathbf{f}2}\|\mathbf{x}_1(t-\tau_{d1}) - \mathbf{x}_2(t-\tau_{d1})\|, \\ &\|\mathbf{B}(\mathbf{x}_1(t), \mathbf{x}_1(t-\tau_{d2}), t) - \mathbf{B}(\mathbf{x}_2(t), \mathbf{x}_2(t-\tau_{d2}), t)\| \\ &\quad \leq \kappa_{\mathbf{B}1}\|\mathbf{x}_1(t) - \mathbf{x}_2(t)\| + \kappa_{\mathbf{B}2}\|\mathbf{x}_1(t-\tau_{d2}) - \mathbf{x}_2(t-\tau_{d2})\|,\end{aligned} \tag{12}$$

where $\boldsymbol{\alpha} \in \{\mathbf{g}, \mathbf{g}_z, \mathbf{g}_t\}$.

Assumption 2. The functions $\mathbf{g}_\mathbf{z}(\cdot,\cdot)$, $\mathbf{B}(\cdot,\cdot,\cdot)$, $\mathbf{w}^k(\cdot)$, and $\dot{\mathbf{z}}_d(\cdot)$ are uniformly bounded with bounds b_{g_z}, b_B, b_w, and $b_{\dot{z}_d}$ which are denoted by

$$\begin{aligned} b_{g_z} &= \max_{k\to\infty, t\in[0,T]} \sup \left\| \mathbf{g}_\mathbf{z}\left(\mathbf{z}^k(t), t\right) \right\|, \\ b_B &= \max_{k\to\infty, t\in[0,T]} \sup \left\| \mathbf{B}\left(\mathbf{z}^k(t), \mathbf{z}^k(t-\tau_{d2}), t\right) \right\|, \\ b_w &= \max_{k\to\infty, t\in[0,T]} \sup \left\| \mathbf{w}^k(t) \right\|, \\ b_{\dot{z}_d} &= \sup_{t\in[0,T]} \|\dot{\mathbf{z}}_d(t)\|. \end{aligned} \tag{13}$$

We now present the following improved high-order iterative learning control scheme for the multiagent systems (7) and (9):

$$\begin{aligned} \mathbf{u}^{k+1}(t) = & \sum_{m=1}^{N} (1-\mu)\mathbf{P}_m(t)\mathbf{u}^l(t) + \mu\mathbf{u}^0(t) \\ & + \sum_{m=1}^{N} \left\{ \mathbf{Q}_m\left(\mathbf{y}^l(t), t\right)\boldsymbol{\varepsilon}^l(t) + \mathbf{R}_m\left(\mathbf{y}^l(t), t\right)\dot{\boldsymbol{\varepsilon}}^l(t) \right. \\ & \left. + \mathbf{S}_m\left(\mathbf{y}^l(t), t\right)\int_0^t \boldsymbol{\varepsilon}^l(s)ds \right\}, \end{aligned} \tag{14}$$

where $l = k - m + 1$, $\boldsymbol{\varepsilon}^l(t) = \bar{\mathscr{H}}\mathbf{e}^l(t)$, $\mathbf{u}^0(t)$ is initial given control input, integer $N \geq 1$ is the order of the iterative learning control scheme, and $0 < \mu < 1$ is a weighting parameter to prevent the large fluctuation of the input at the beginning of the iterative operation. The term μ may be allowed to vary with the iteration and let it fixed for simplicity in this paper. $\mathbf{P}_m(t)$ is the time-varying matrix to regulate the rate of convergence, which needs to satisfy certain convergence condition (will be introduced below). Like the description of [25, 27], $\mathbf{Q}_m(\cdot,\cdot)$, $\mathbf{R}_m(\cdot,\cdot)$, and $\mathbf{S}_m(\cdot,\cdot)$ are learning matrices. In addition, it is assumed that $\mathbf{Q}_m(\cdot,\cdot)\bar{\mathscr{H}}$, $\mathbf{R}_m(\cdot,\cdot)\bar{\mathscr{H}}$, and $\mathbf{S}_m(\cdot,\cdot)\bar{\mathscr{H}}$ are bounded, and their bounds are defined by b_Q, b_R, and b_S, respectively, for instance,

$$b_Q = \max_{m\in\{1,\ldots,N\}, t\in[0,T]} \sup \left\| \mathbf{Q}_m\left(\mathbf{y}^l(t), t\right)\bar{\mathscr{H}} \right\|. \tag{15}$$

Remark 1. If $N = 1$ for (14), the iterative learning control scheme may be considered as a PID-type control law similar to the convention PID controller; and if $N \geq 2$, it can be seen as consisting of dual or multiple PID controllers. As more past learning errors are considered, it implies that the high-order iterative learning control scheme has better control effect than one-order. Generally speaking, the higher the orders of the control scheme, the better the iterative learning control performance is expected. However, the choice of N in practice is normally not more than 3.

Remark 2. Like the choice of parameters of the traditional PID algorithm, the choice of $\mathbf{Q}_m(\cdot,\cdot)$, $\mathbf{R}_m(\cdot,\cdot)$, and $\mathbf{S}_m(\cdot,\cdot)$ is usually based on a trial-and-error method. Considering that the purpose of $\mathbf{Q}_m(\cdot,\cdot)$ is to increase the convergence speed of the system, $\mathbf{R}_m(\cdot,\cdot)$ is to eliminate the steady-state error of the system, and $\mathbf{S}_m(\cdot,\cdot)$ is to overcome the oscillation and improve the stability of the system, their weight can be tuned according to the control requirements. In addition, the control approaches such as a fuzzy adaptive method, a neural network algorithm, and particle swarm optimization algorithm can be also applied to obtain the optimal learning matrices.

4. Convergence Analysis

In this section, we will analyze the convergence of time-varying delayed multiagent systems (7) and (9) with a high-order iterative learning control scheme (14). The convergence conditions are shown in the following theorem.

For clarification of the remaining discussion, the variable t will be omitted, and the following notations will be introduced. That is,

$$\begin{aligned} \boldsymbol{\alpha}^k &= \boldsymbol{\alpha}\left(z^k(t), t\right), \\ \boldsymbol{\alpha}_d &= \boldsymbol{\alpha}(\mathbf{z}_d(t), t), \\ \delta\boldsymbol{\alpha}^k &= \boldsymbol{\alpha}_d - \boldsymbol{\alpha}^k, \\ b_\alpha &= \max_{k\to\infty, t\in[0,T]} \sup \left\| \boldsymbol{\alpha}\left(\mathbf{z}^k(t), t\right) \right\|. \end{aligned} \tag{16}$$

where $\boldsymbol{\alpha}$ represents a function concerned.

The partial derivatives of $\boldsymbol{\alpha}(\mathbf{x}(t), t)$ are expressed by

$$\begin{aligned} \boldsymbol{\alpha}_{\mathbf{z}^k} &= \left.\frac{\partial\boldsymbol{\alpha}(\mathbf{x}(t), t)}{\partial\mathbf{x}(t)}\right|_{\mathbf{x}(t)=\mathbf{z}^k(t)}, \\ \boldsymbol{\alpha}_{t^k} &= \left.\frac{\partial\boldsymbol{\alpha}(\mathbf{x}(t), t)}{\partial t}\right|_{\mathbf{x}(t)=\mathbf{z}^k(t)}, \\ \boldsymbol{\alpha}_{\mathbf{z}_d} &= \left.\frac{\partial\boldsymbol{\alpha}(\mathbf{x}(t), t)}{\partial\mathbf{x}(t)}\right|_{\mathbf{x}(t)=\mathbf{z}_d(t)}, \\ \boldsymbol{\alpha}_{t_d} &= \left.\frac{\partial\boldsymbol{\alpha}(\mathbf{x}(t), t)}{\partial t}\right|_{\mathbf{x}(t)=\mathbf{z}_d(t)}. \end{aligned} \tag{17}$$

Hence, it is easy to find that $\|\delta\boldsymbol{\alpha}^k\| \leq \kappa_\boldsymbol{\alpha}\|\delta\mathbf{z}^k\|$, where $\kappa_\boldsymbol{\alpha}$ is the Lipschitz constant of the function $\boldsymbol{\alpha}$ on the interval $[0, T]$. We now state the main result of this paper.

Theorem 1. *Let the multiagent systems (7) and the virtual agent dynamic (9) satisfy Assumptions 1 and 2 and suppose that the initial state difference $\mathbf{z}_d(0) - \mathbf{z}^k(0)$ is bounded. If there exist*

$$\sum_{m=1}^{N} \mathbf{P}_m(t) = \mathbf{I}_n, \tag{18}$$

and positive numbers β_m satisfying

$$\begin{gathered}\|(1-\mu)\mathbf{P}_m(t)-\mathbf{R}_m(\mathbf{y}(t),t)\bar{\mathscr{H}}\mathbf{g}_\mathbf{z}(\mathbf{z}(t),t)\mathbf{B}\\ \cdot(\mathbf{z}(t),\mathbf{z}(t-\tau_{d2}),t)\|\le\beta_m,\\ \sum_{m=1}^{N}\beta_m=\beta<1.\end{gathered}\tag{19}$$

and then the bounds of tracking errors $\|z_d(t)-z_i^k(t)\|$ *and* $\|y_d(t)-y_i^k(t)\|$ *converge asymptotically to a sufficiently small interval as* $k\to\infty$ *for* $i\in\{1,\ldots,n\}$. *Furthermore, the bounds of initial state difference and disturbances tend to zero; the bound of tracking errors also tends to zero.*

Proof. According to the multiagent systems (7) and (9), we have

$$\mathbf{e}^l=\mathbf{y}_d-\mathbf{y}^l=\delta\mathbf{g}^l,\tag{20}$$

and then

$$\begin{aligned}\dot{\mathbf{e}}^l&=\dot{\mathbf{y}}_d-\dot{\mathbf{y}}^l=\mathbf{g}_{\mathbf{z}_d}\dot{\mathbf{z}}_d+\mathbf{g}_{t_d}-\left(\mathbf{g}_{\mathbf{z}^l}\dot{\mathbf{z}}^l+\mathbf{g}_{t^l}\right)\\&=\mathbf{g}_{\mathbf{z}_d}\dot{\mathbf{z}}_d-\mathbf{g}_{\mathbf{z}^l}\dot{\mathbf{z}}_d+\mathbf{g}_{\mathbf{z}^l}\dot{\mathbf{z}}_d-\mathbf{g}_{\mathbf{z}^l}\dot{\mathbf{z}}^l+\mathbf{g}_{t_d}-\mathbf{g}_{t^l}\\&=\left(\mathbf{g}_{\mathbf{z}_d}-\mathbf{g}_{\mathbf{z}^l}\right)\dot{\mathbf{z}}_d+\mathbf{g}_{\mathbf{z}^l}\left(\dot{\mathbf{z}}_d-\dot{\mathbf{z}}^l\right)+\left(\mathbf{g}_{t_d}-\mathbf{g}_{t^l}\right)\\&=\delta\mathbf{g}_{\mathbf{z}^l}\dot{\mathbf{z}}_d+\mathbf{g}_{\mathbf{z}^l}\delta\dot{\mathbf{z}}^l+\delta\mathbf{g}_{t^l},\end{aligned}\tag{21}$$

where $l=k,k-1,\ldots,k-N+1$.

Also,

$$\begin{aligned}\delta\dot{\mathbf{z}}^l&=\dot{\mathbf{z}}_d-\dot{\mathbf{z}}^l\\&=\mathbf{f}_d+\mathbf{B}_d\mathbf{u}_d-\left(\mathbf{f}^l+\mathbf{B}^l\mathbf{u}^l+\mathbf{w}^l\right)\\&=\left(\mathbf{f}_d-\mathbf{f}^l\right)+\left(\mathbf{B}_d-\mathbf{B}^l\right)\mathbf{u}_d+\mathbf{B}^l\left(\mathbf{u}_d-\mathbf{u}^l\right)-\mathbf{w}^l\\&=\delta\mathbf{f}^l+\delta\mathbf{B}^l\mathbf{u}_d+\mathbf{B}^l\delta\mathbf{u}^l-\mathbf{w}^l,\end{aligned}\tag{22}$$

where $\delta\mathbf{f}^l=\mathbf{f}(\mathbf{z}_d(t),\mathbf{z}_d(t-\tau_{d1}),t)-\mathbf{f}(\mathbf{z}^l(t),\mathbf{z}^l(t-\tau_{d1}),t)$ and $\delta\mathbf{B}^l=\mathbf{B}(\mathbf{z}_d(t),\mathbf{z}_d(t-\tau_{d2}),t)-\mathbf{B}(\mathbf{z}^l(t),\mathbf{z}^l(t-\tau_{d2}),t)$.

Substituting (11) into (14) gets

$$\begin{aligned}\mathbf{u}^{k+1}=&\sum_{m=1}^{N}(1-\mu)\mathbf{P}_m(t)\mathbf{u}^l+\mu\mathbf{u}^0\\&+\sum_{m=1}^{N}\left\{\mathbf{Q}_m\left(\mathbf{y}^l(t),t\right)\bar{\mathscr{H}}\mathbf{e}^l+\mathbf{R}_m\left(\mathbf{y}^l(t),t\right)\bar{\mathscr{H}}\dot{\mathbf{e}}^l\right.\\&\left.+\mathbf{S}_m\left(\mathbf{y}^l(t),t\right)\bar{\mathscr{H}}\int_0^t\mathbf{e}^l ds\right\}.\end{aligned}\tag{23}$$

Furthermore, substituting (18), (20), (21), and (22) into (23) yields

$$\begin{aligned}\delta\mathbf{u}^{k+1}&=\mathbf{u}_d-\mathbf{u}^{k+1}\\&=\sum_{m=1}^{N}(1-\mu)\mathbf{P}_m(t)\mathbf{u}_d-\sum_{m=1}^{N}(1-\mu)\mathbf{P}_m(t)\mathbf{u}^l+\mu\mathbf{u}_d-\mu\mathbf{u}^0\\&\quad-\sum_{m=1}^{N}\left\{\mathbf{Q}_m\left(\mathbf{y}^l(t),t\right)\bar{\mathscr{H}}\mathbf{e}^l+\mathbf{R}_m\left(\mathbf{y}^l(t),t\right)\bar{\mathscr{H}}\dot{\mathbf{e}}^l\right.\\&\quad\left.+\mathbf{S}_m\left(\mathbf{y}^l(t),t\right)\bar{\mathscr{H}}\int_0^t\mathbf{e}^l ds\right\}\\&=\sum_{m=1}^{N}(1-\mu)\mathbf{P}_m(t)\delta\mathbf{u}^l+\mu\delta\mathbf{u}^0-\sum_{m=1}^{N}\mathbf{Q}_m\left(\mathbf{y}^l(t),t\right)\bar{\mathscr{H}}\delta\mathbf{g}^l\\&\quad-\sum_{m=1}^{N}\mathbf{S}_m\left(\mathbf{y}^l(t),t\right)\bar{\mathscr{H}}\int_0^t\delta\mathbf{g}^l ds-\sum_{m=1}^{N}\mathbf{R}_m\left(\mathbf{y}^l(t),t\right)\bar{\mathscr{H}}\\&\quad\cdot\left\{\delta\mathbf{g}_{\mathbf{z}^l}\dot{\mathbf{z}}_d+\mathbf{g}_{\mathbf{z}^l}\left(\delta\mathbf{f}^l+\delta\mathbf{B}^l\mathbf{u}_d+\mathbf{B}^l\delta\mathbf{u}^l-\mathbf{w}^l\right)+\delta\mathbf{g}_{t^l}\right\}\\&=\sum_{m=1}^{N}\left\{(1-\mu)\mathbf{P}_m(t)-\mathbf{R}_m\left(\mathbf{y}^l(t),t\right)\bar{\mathscr{H}}\mathbf{g}_{\mathbf{z}^l}\mathbf{B}^l\right\}\delta\mathbf{u}^l\\&\quad+\mu\delta\mathbf{u}^0-\sum_{m=1}^{N}\mathbf{Q}_m\left(\mathbf{y}^l(t),t\right)\bar{\mathscr{H}}\delta\mathbf{g}^l\\&\quad-\sum_{m=1}^{N}\mathbf{S}_m\left(\mathbf{y}^l(t),t\right)\bar{\mathscr{H}}\int_0^t\delta\mathbf{g}^l ds-\sum_{m=1}^{N}\mathbf{R}_m\left(\mathbf{y}^l(t),t\right)\bar{\mathscr{H}}\\&\quad\cdot\left\{\delta\mathbf{g}_{\mathbf{z}^l}\dot{\mathbf{z}}_d+\mathbf{g}_{\mathbf{z}^l}\left(\delta\mathbf{f}^l+\delta\mathbf{B}^l\mathbf{u}_d-\mathbf{w}^l\right)+\delta\mathbf{g}_{t^l}\right\}.\end{aligned}\tag{24}$$

Considering the condition (19), taking norms on both sides of (24) has

$$\begin{aligned}\left\|\delta\mathbf{u}^{k+1}\right\|\le&\sum_{m=1}^{N}\beta_m\left\|\delta\mathbf{u}^l\right\|+\mu\left\|\delta\mathbf{u}^0\right\|+\sum_{m=1}^{N}b_Q\kappa_\mathbf{g}\left\|\delta\mathbf{z}^l\right\|\\&+\sum_{m=1}^{N}b_S\kappa_\mathbf{g}\int_0^t\left\|\delta\mathbf{z}^l\right\|ds+\sum_{m=1}^{N}b_R\\&\cdot\left\{\left(b_{\dot{z}_d}\kappa_{\mathbf{g}_z}+\kappa_{\mathbf{g}_t}+b_{g_z}\left(\kappa_{\mathbf{f}1}+b_{u_d}\kappa_{\mathbf{B}1}\right)\right)\left\|\delta\mathbf{z}^l\right\|+b_{g_z}b_w\right\}\\&+\sum_{m=1}^{N}b_R\left\{b_{g_z}\kappa_{\mathbf{f}2}\left\|\delta\mathbf{z}^l(t-\tau_{d1})\right\|+b_{g_z}b_{u_d}\kappa_{\mathbf{B}2}\left\|\delta\mathbf{z}^l(t-\tau_{d2})\right\|\right\}\\\le&\eta_0+\sum_{m=1}^{N}\beta_m\left\|\delta\mathbf{u}^l\right\|+\mu\left\|\delta\mathbf{u}^0\right\|+\eta_1\sum_{m=1}^{N}\left\|\delta\mathbf{z}^l\right\|\\&+\eta_2\sum_{m=1}^{N}\int_0^t\left\|\delta\mathbf{z}^l\right\|ds+\sum_{m=1}^{N}b_R\\&\cdot\left\{b_{g_z}\kappa_{\mathbf{f}2}\left\|\delta\mathbf{z}^l(t-\tau_{d1})\right\|+b_{g_z}b_{u_d}\kappa_{\mathbf{B}2}\left\|\delta\mathbf{z}^l(t-\tau_{d2})\right\|\right\},\end{aligned}\tag{25}$$

where

$$\begin{aligned}b_{u_d}&=\sup_{t\in[0,T]}\|\mathbf{u}_d(t)\|,\\\eta_0&=Nb_Rb_{g_z}b_w,\\\eta_1&=b_Q\kappa_\mathbf{g}+b_R\left(b_{\dot{z}_d}\kappa_{\mathbf{g}_z}+\kappa_{\mathbf{g}_t}+b_{g_z}\kappa_{\mathbf{f}1}+b_{g_z}b_{u_d}\kappa_{\mathbf{B}1}\right),\\\eta_2&=b_S\kappa_\mathbf{g}.\end{aligned}\tag{26}$$

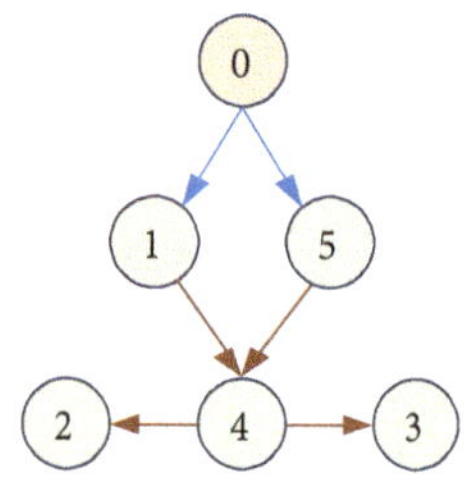

FIGURE 1: Communication topology graph.

Writing the integral expression for $\mathbf{z}^l$ and taking norms, we get

$$\left\|\delta\mathbf{z}^l\right\| = \left\|\delta\mathbf{z}^l(0) + \int_0^t \delta\dot{\mathbf{z}}^l ds\right\| \le \left\|\delta\mathbf{z}^l(0)\right\| + \int_0^t \Big\{ (\kappa_{f1} + b_{u_d}\kappa_{B1})\left\|\delta\mathbf{z}^l\right\| + \kappa_{f2}\left\|\delta\mathbf{z}^l(t-\tau_{d1})\right\| + b_{u_d}\kappa_{B2}\left\|\delta\mathbf{z}^l(t-\tau_{d2})\right\| + b_B\left\|\delta\mathbf{u}^l\right\| + b_w \Big\} ds. \quad (27)$$

In the light of Definition 1 and the fact that $\|\mathbf{y}(t-\tau_d)\|_\lambda \le \|\mathbf{y}(t)\|_\lambda$, we have

$$\sup_{t\in[0,T]} e^{-\lambda t}\int_0^t \|\mathbf{y}(s-\tau_d)\|_\lambda ds \le \|\mathbf{y}(t)\|_\lambda o(\lambda)e^{-\lambda\tau_d}. \quad (28)$$

Considering (28) and taking λ norm on both sides of (27) obtain

$$\left\|\delta\mathbf{z}^l\right\|_\lambda \le \left\|\delta\mathbf{z}^l(0)\right\| + b_w T + \eta_3 o(\lambda)\left\|\delta\mathbf{z}^l\right\|_\lambda + b_B o(\lambda)\left\|\delta\mathbf{u}^l\right\|_\lambda, \quad (29)$$

where

$$\eta_3 = \kappa_{f1} + \kappa_{B1}b_{u_d} + \kappa_{f2}e^{-\lambda\tau_{d1}} + \kappa_{B2}b_{u_d}e^{-\lambda\tau_{d2}}. \quad (30)$$

Obviously, there $\eta_3 o(\lambda) < 1$ for a sufficiently large λ and then have

$$\left\|\delta\mathbf{z}^l\right\|_\lambda \le \frac{\|\delta\mathbf{z}^l(0)\| + b_w T + b_B o(\lambda)\|\delta\mathbf{u}^l\|_\lambda}{1-\eta_3 o(\lambda)}. \quad (31)$$

Similarly, taking λ norm on both sides of (25), we have

$$\left\|\delta\mathbf{u}^{k+1}\right\|_\lambda \le \eta_0 + \sum_{m=1}^N \beta_m\left\|\delta\mathbf{u}^l\right\|_\lambda + \mu\|\delta\mathbf{u}^0\|_\lambda + \eta_4\sum_{m=1}^N\left\|\delta\mathbf{z}^l\right\|_\lambda, \quad (32)$$

where

$$\eta_4 = \eta_1 + \eta_2 o(\lambda) + b_R b_{g_z}\left(\kappa_{f2}e^{-\lambda\tau_{d1}} + \kappa_{B2}b_{u_d}e^{-\lambda\tau_{d2}}\right). \quad (33)$$

Substituting (31) into (32) yields

$$\left\|\delta\mathbf{u}^{k+1}\right\|_\lambda \le \sum_{m=1}^N \overline{\beta}_m\left\|\delta\mathbf{u}^l\right\|_\lambda + \sigma, \quad (34)$$

where

$$\overline{\beta}_m = \beta_m + \eta_4\frac{b_B o(\lambda)}{1-\eta_3 o(\lambda)}, \quad (35)$$

$$\sigma = \eta_0 + \mu\left\|\delta\mathbf{u}^0\right\|_\lambda + \eta_4\sum_{m=1}^N \frac{\|\delta\mathbf{z}^l(0)\| + b_w T}{1-\eta_3 o(\lambda)}. \quad (36)$$

Considering the condition (19) again, we have $\overline{\beta}_m < 1$ for a sufficiently large λ, then $\overline{\beta} = \sum_{m=1}^N \overline{\beta}_m < 1$. Hence, according to the Lemma 1, it is obtained that

$$\lim_{k\to\infty}\left\|\delta\mathbf{u}^k\right\|_\lambda \le \frac{\sigma}{1-\overline{\beta}}. \quad (37)$$

Furthermore, we also have

$$\left\|\delta\mathbf{y}^k\right\|_\lambda \le \kappa_g\left\|\delta\mathbf{z}^k\right\|_\lambda. \quad (38)$$

From (31), (36), and (38), one can be observed that the tracking errors $\|\delta\mathbf{z}^k\|$, $\|\delta\mathbf{u}^k\|$, and $\|\delta\mathbf{y}^k\|$ are all relevant to the initial state error $\|\delta\mathbf{z}^k(0)\|$ and disturbance bound b_w. Furthermore, if μ, $\|\delta\mathbf{z}^k(0)\|$, and b_w tend to zero, the bound of tracking errors also tends to zero through the iterative operations of the iterative learning control algorithm. The proof completed.

Remark 3. Under the convergence conditions (18) and (19), one finds that the convergence of iterative learning control is not influenced by the disturbances, the initial state difference, and even the choice of μ. However, the bound of the final tracking errors is directly influenced by those factors. In addition, the iterative learning control algorithm itself as using alone cannot eliminate disturbances or initial state difference, but it can realize a perfect tracking once those factors will not appear any more in the coming iterations.

5. Simulation Analysis

In this section, we will provide some cases to illustrate the effectiveness of the high-order iterative learning control scheme (14). Consider a time-varying delayed multiagent systems with five agents labeled as 1, 2, 3, 4, and 5. The virtual leader agent is labeled as 0. The communication graph with five agents and one virtual agent is shown in Figure 1.

The adjacency matrix is

$$\mathscr{A} = \begin{pmatrix} 0 & 0 & 0 & 0 & 0 \\ 0 & 0 & 0 & 1 & 0 \\ 0 & 0 & 0 & 1 & 0 \\ 1 & 0 & 0 & 0 & 1 \\ 0 & 0 & 0 & 0 & 0 \end{pmatrix}, \quad \mathscr{B} = \operatorname{diag}\{1,0,0,0,1\}. \quad (39)$$

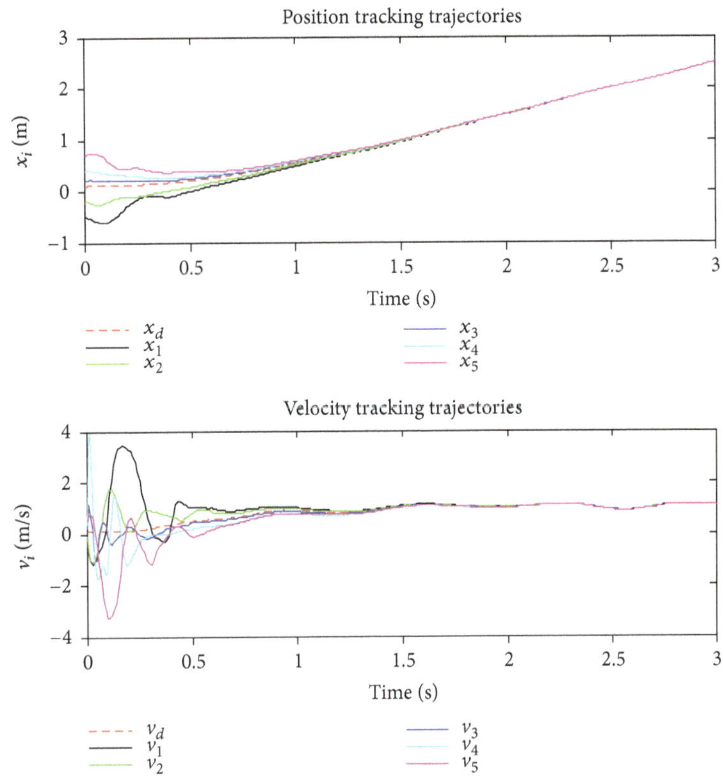

FIGURE 2: State tracking trajectories for Case 1.

The dynamics of agent i at the kth iteration are described as

$$\dot{z}_i^k(t) = \begin{bmatrix} v_i^k(t) \\ \sin\left(v_i^k(t)\right) + x_i^k(t-\tau_{d1}) \end{bmatrix} + \begin{bmatrix} 0 \\ 0.2\cos\left(v_i^k(t-\tau_{d2})\right) \end{bmatrix} u_i^k(t) + \begin{bmatrix} e^{-k}\sin(t) \\ e^{-k}\sin(t) \end{bmatrix},$$
$$y_i^k(t) = \begin{bmatrix} x_i^k(t) \\ v_i^k(t) \end{bmatrix}, \tag{40}$$

where $z_i^k(t) = [x_i^k(t), v_i^k(t)]^T$ and $i = 1, 2, 3, 4, 5$.

The dynamics of a virtual leader agent are given as

$$\dot{z}_d(t) = \begin{bmatrix} v_d(t) \\ \sin(x_d(t-\tau_{d1})) \end{bmatrix} + \begin{bmatrix} 0 \\ 0.2\cos(v_d(t-\tau_{d2})) \end{bmatrix} u_d(t),$$
$$y_d(t) = \begin{bmatrix} x_d(t) \\ v_d(t) \end{bmatrix}, \tag{41}$$

where $z_d(t) = [x_d(t), v_d(t)]^T$, and (41) indicates that the desired output trajectory is obtained by a desired input $u_d(t)$ and state $z_d(t)$.

The parameters are set as follows: the initial position and velocity of five agents are given as $x^0(0) = [-0.5, -0.2, 0.2, 0.4, 0.7]^T$(m) and $v^0(0) = [-0.3, -0.1, 0.2, 0.3, 0.5]^T$(m/s), respectively; the initial input $u^0(0) = [0.1, 0.1, 0.1, 0.1, 0.1]^T$; the simulation time $t \in [0, 3\text{ s}]$ and time-varying delays $\tau_{d1} = 0.2$(s) and $\tau_{d2} = 0.3$(s); the initial states of a virtual agent are $x_d(0) = 0.1$(m) and $v_d(0) = 0.1$(m/s); the desired input $u_d(t) = \sin(2\pi t)$; the term $\mu = 0.5$; and $k_{\max} = 50$.

According to (40), we have

$$\begin{aligned} \mathbf{g}_\mathbf{z}(\mathbf{z}(t), t) &= \frac{\partial \mathbf{g}(\mathbf{z}(t), t)}{\partial \mathbf{z}(t)} = \mathbf{I}_{10}, \\ \mathbf{B}(\mathbf{z}(t), \mathbf{z}(t-\tau_{d2}), t) &= \operatorname{diag}\left\{B(z_1(t), z_1(t-\tau_{d2}), t), \ldots, B(z_5(t), z_5(t-\tau_{d2}), t)\right\} \\ &\in \left[\operatorname{diag}\left\{[0, -0.2]^T, \ldots, [0, -0.2]^T\right\}, \operatorname{diag} \cdot \left\{[0, 0.2]^T, \ldots, [0, 0.2]^T\right\}\right]. \end{aligned} \tag{42}$$

In order to analyze the effectiveness of a high-order iterative learning control scheme (14), the following cases are discussed. It should be noted that the parameters $\mathbf{P}_m(\cdot)$, $\mathbf{Q}_m(\cdot,\cdot)$, $\mathbf{R}_m(\cdot,\cdot)$, and $\mathbf{S}_m(\cdot,\cdot)$ $(m \in \{1, 2, 3\})$ are obtained by the trial-and-error method.

Case 1. In this case, $N = 1$, $\mathbf{P}_1 = \mathbf{I}_5$, $\mathbf{Q}_1 = 1.5(\mathbf{I}_5 \otimes \mathbf{1}_{1\times 2})$, $\mathbf{R}_1 = 0.6(\mathbf{I}_5 \otimes \mathbf{1}_{1\times 2})$, and $\mathbf{S}_1 = 0.5(\mathbf{I}_5 \otimes \mathbf{1}_{1\times 2})$.

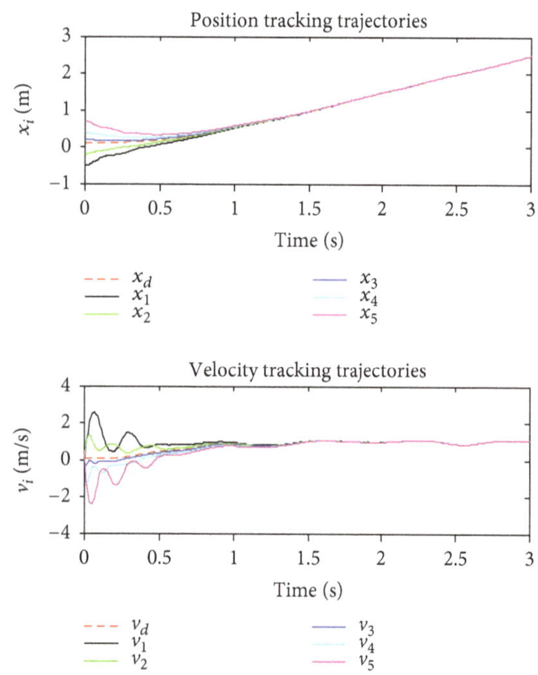

FIGURE 3: State tracking trajectories for Case 2.

Case 2. In this case, $N = 2$, $\mathbf{P}_1 = 0.5\mathbf{I}_5$ and $\mathbf{P}_2 = 0.5\mathbf{I}_5$, $\mathbf{Q}_1 = 1.5(\mathbf{I}_5 \otimes \mathbf{1}_{1\times 2})$ and $\mathbf{Q}_2 = (\mathbf{I}_5 \otimes \mathbf{1}_{1\times 2})$, $\mathbf{R}_1 = 0.6(\mathbf{I}_5 \otimes \mathbf{1}_{1\times 2})$ and $\mathbf{R}_2 = 0.4(\mathbf{I}_5 \otimes \mathbf{1}_{1\times 2})$, and $\mathbf{S}_1 = 0.5(\mathbf{I}_5 \otimes \mathbf{1}_{1\times 2})$ and $\mathbf{S}_2 = 0.6(\mathbf{I}_5 \otimes \mathbf{1}_{1\times 2})$.

Case 3. In this case, $N = 3$; $\mathbf{P}_1 = 0.5\mathbf{I}_5$, $\mathbf{P}_2 = 0.25\mathbf{I}_5$, and $\mathbf{P}_3 = 0.25\mathbf{I}_5$; $\mathbf{Q}_1 = 1.5(\mathbf{I}_5 \otimes \mathbf{1}_{1\times 2})$, $\mathbf{Q}_2 = (\mathbf{I}_5 \otimes \mathbf{1}_{1\times 2})$, and $\mathbf{Q}_3 = 0.5(\mathbf{I}_5 \otimes \mathbf{1}_{1\times 2})$; $\mathbf{R}_1 = 0.6(\mathbf{I}_5 \otimes \mathbf{1}_{1\times 2})$, $\mathbf{R}_2 = 0.4(\mathbf{I}_5 \otimes \mathbf{1}_{1\times 2})$, and $\mathbf{R}_3 = 0.2(\mathbf{I}_5 \otimes \mathbf{1}_{1\times 2})$; and $\mathbf{S}_1 = 0.5(\mathbf{I}_5 \otimes \mathbf{1}_{1\times 2})$, $\mathbf{S}_2 = 0.6(\mathbf{I}_5 \otimes \mathbf{1}_{1\times 2})$, and $\mathbf{S}_3 = 0.7(\mathbf{I}_5 \otimes \mathbf{1}_{1\times 2})$.

Checking the convergence conditions (18) and (19) in Theorem 1, we get the following.

Case 1. $\mathbf{P}_1(t) = \mathbf{I}_5$, $\beta_1 = \|(1-\mu)\mathbf{P}_1(t) - \mathbf{R}_1(\mathbf{y}(t), t)\bar{\mathscr{H}}\mathbf{g}_\mathbf{z}\mathbf{B}\|_{\max} = 0.3847 < 1$.

Case 2. $\mathbf{P}_1(t) + \mathbf{P}_2(t) = 0.5\mathbf{I}_5 + 0.5\mathbf{I}_5 = \mathbf{I}_5$, $\beta_2 = \|(1-\mu)\ \mathbf{P}_2(t) - \mathbf{R}_2(\mathbf{y}(t), t)\bar{\mathscr{H}}\mathbf{g}_\mathbf{z}\mathbf{B}\|_{\max} = 0.3214$, $\beta_1 + \beta_2 = 0.7061 < 1$.

Case 3. $\mathbf{P}_1(t) + \mathbf{P}_2(t) + \mathbf{P}_3(t) = 0.5\mathbf{I}_5 + 0.25\mathbf{I}_5 + 0.25\mathbf{I}_5 = \mathbf{I}_5$, $\beta_3 = \|(1-\mu)\mathbf{P}_3(t) - \mathbf{R}_3(\mathbf{y}(t), t)\bar{\mathscr{H}}\mathbf{g}_\mathbf{z}\mathbf{B}\|_{\max} = 0.2596$, $\beta_1 + \beta_2 + \beta_3 = 0.9657 < 1$.

Then, we ensure the satisfaction of the convergence conditions (18) and (19).

The simulation results for Case 1, Case 2, and Case 3 at 50th iteration are shown in Figures 2, 3, 4, and 5, respectively.

Figures 2, 3, and 4 show the tracking trajectories of time-varying delayed multiagent systems with a high-order iterative learning control scheme under different orders at the 50th iteration, which indicate that the consensus is achieved by using the proposed control protocol. However, compared with Figures 3 and 4, the fluctuation and control effect of Figure 2 are the worst. For Figure 4, it is not difficult to find that the control effect is better than Figures 2 and 3. The results of Figures 2, 3, and 4 imply that the higher the orders and the better the control effect. Figure 5 gives the control input curves of five agents for Case 3 and implies that the input of agents can converge asymptotically to the desired input value.

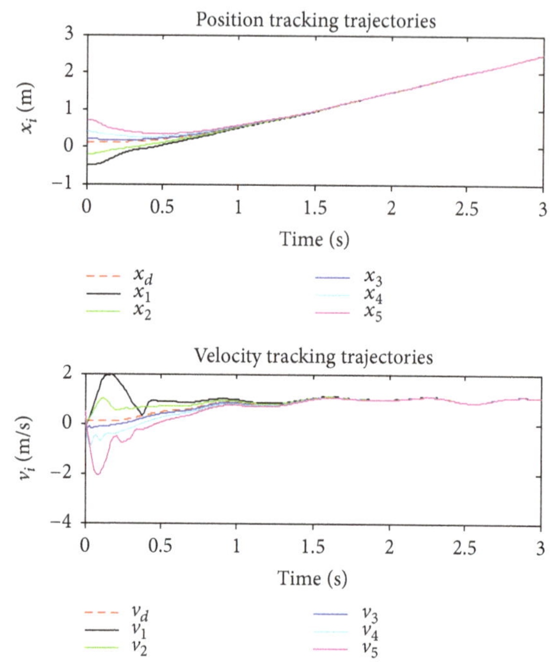

FIGURE 4: State tracking trajectories for Case 3.

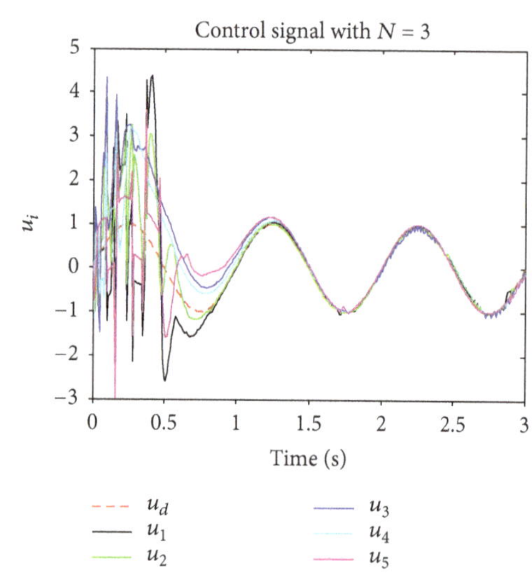

FIGURE 5: Control signal for Case 3.

Case 4. Comparison and analysis with other control protocols.

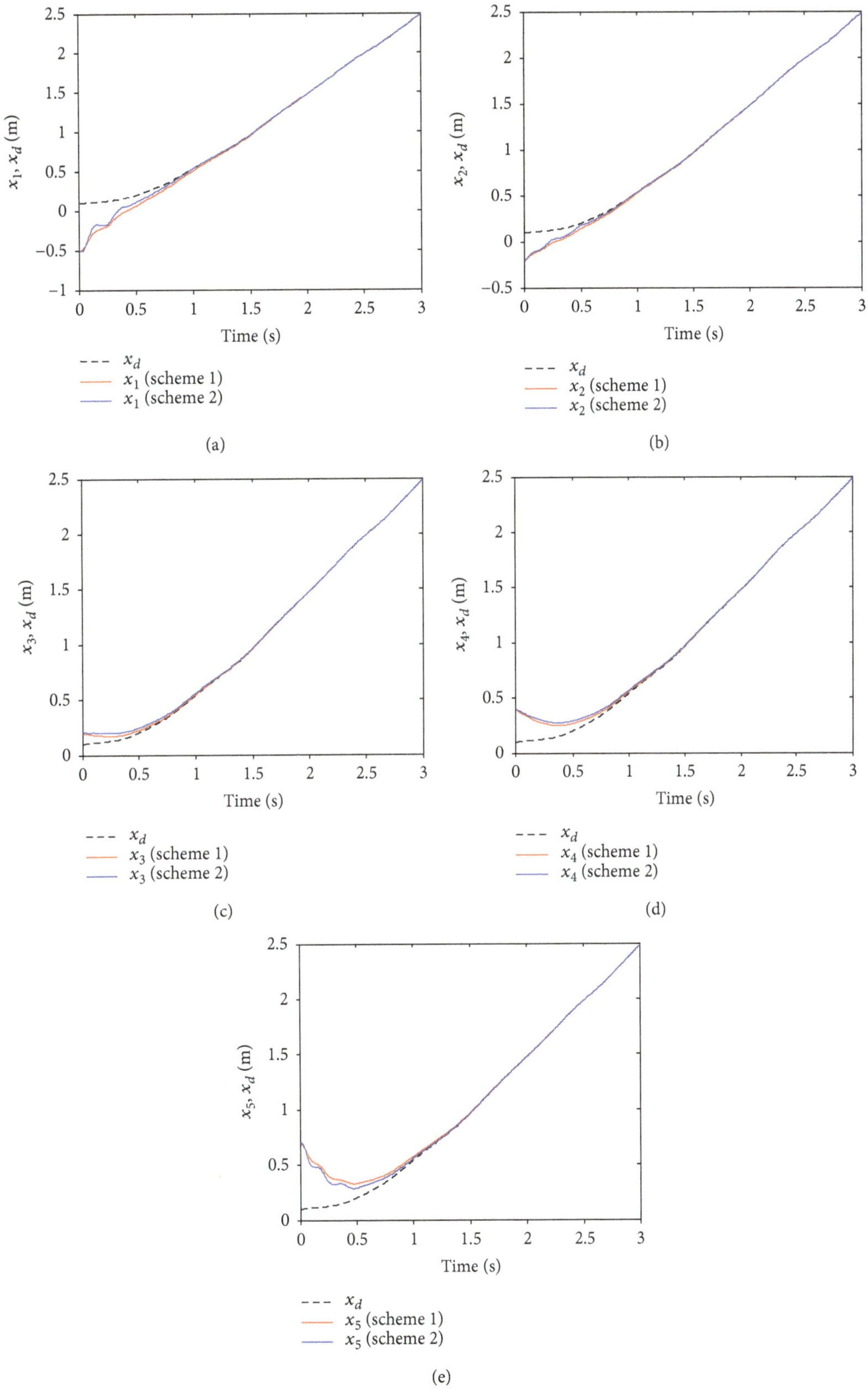

FIGURE 6: Position tracking trajectories.

In order to illustrate the control effect, another control law is selected to compare with the proposed control scheme (14) (scheme 1) of this paper. Consider the high-order iterative learning control updating law (7) (scheme 2) in [27], which did not include specific provision for the first iteration term. Let the order $N = 2$, the parameter setting is same as the

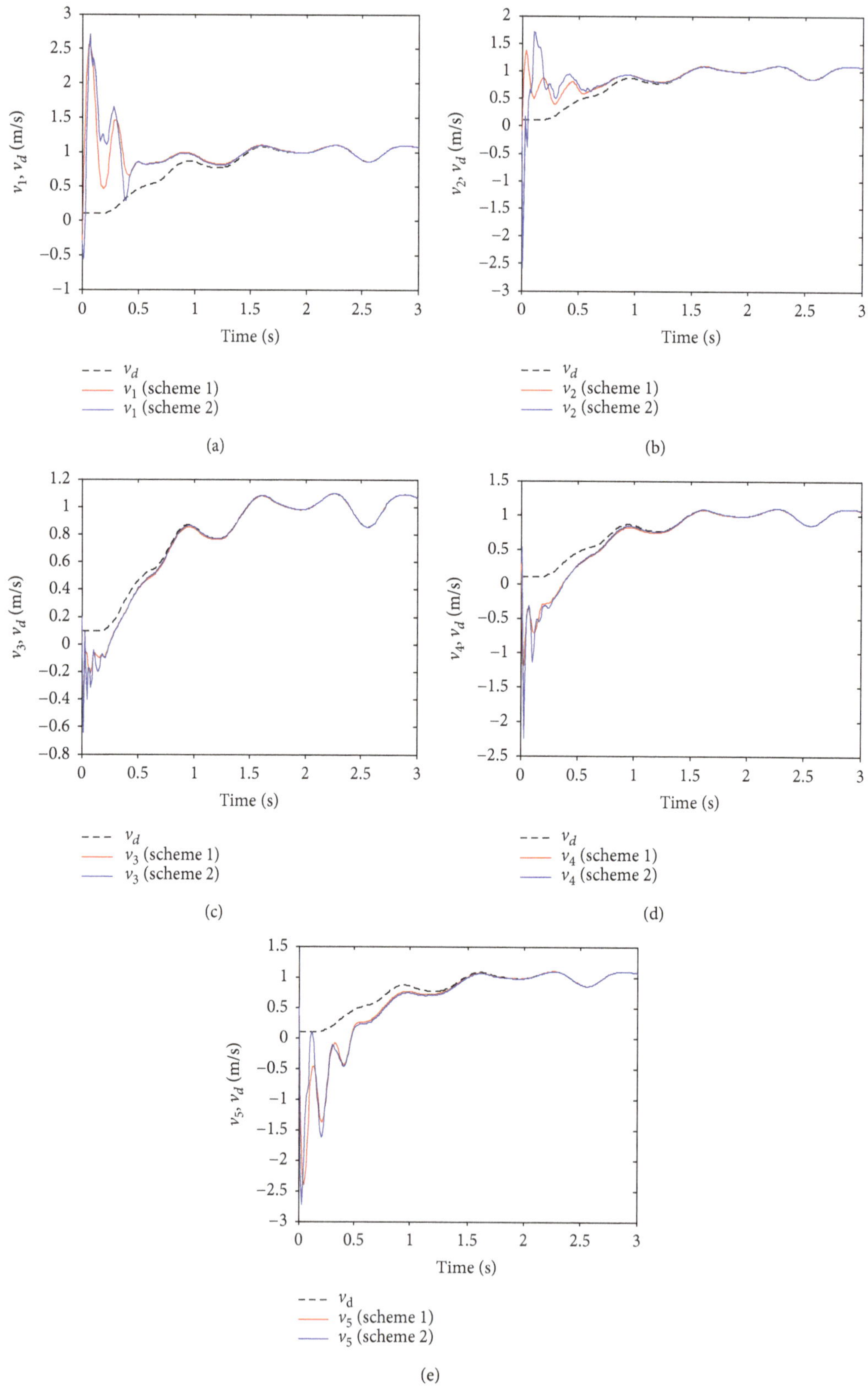

FIGURE 7: Velocity tracking trajectories.

Case 2 of this paper. Check the convergence conditions (27), (28) of Theorem 9 in [27], then have $\mathbf{P}_1(t) + \mathbf{P}_2(t) = 0.5\mathbf{I}_5 + 0.5\mathbf{I}_5 = \mathbf{I}_5$, $\rho_1 = \|\mathbf{P}_1(t) - \mathbf{R}_1(\mathbf{y}(t), t)\bar{\mathscr{H}}\mathbf{g}_\mathbf{z}\mathbf{B}\|_{\max} = 0.4257 < 1$, $\rho_2 = \|\mathbf{P}_2(t) - \mathbf{R}_2(\mathbf{y}(t), t)\ \bar{\mathscr{H}}\mathbf{g}_\mathbf{z}\mathbf{B}\|_{\max} = 0.2416 < 1$, and $\rho_1 + \rho_2 = 0.6673 < 1$, which mean the conditions are satisfied. The simulation results are shown in Figures 6 and 7.

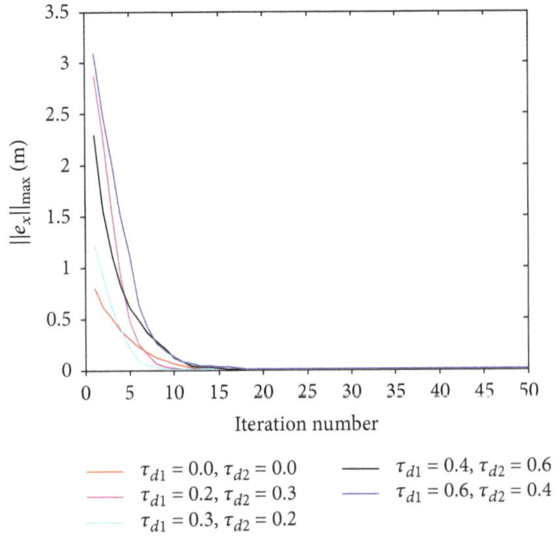

FIGURE 8: Maximum tracking errors of position with different time delays.

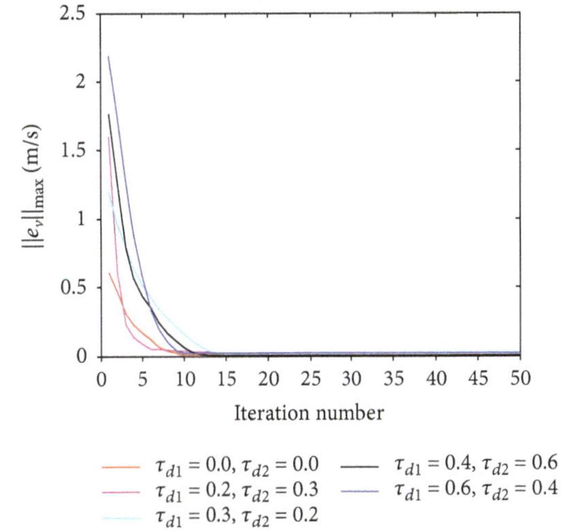

FIGURE 9: Maximum tracking errors of velocity with different time delays.

Figures 6 and 7 show the comparison results of position and velocity with two control schemes. It is implied that the consensus of multiagent systems with the two control laws can be achieved. However, the control effects with the consensus law of this paper are better than the control law of [27].

In addition, the maximum tracking errors with different time delays for $N = 3$ are also analyzed. The simulation results are shown in Figures 8 and 9.

Figures 8 and 9 give the maximum tracking errors of position and velocity with different time delays. Although the time delays are different, the consensus of multiagent systems with a high-order iterative learning control scheme is still achieved. That is to say, the convergence of the proposed control protocol is not affected by time delays.

6. Conclusions

In this paper, we have discussed the consensus control problem of time-varying delayed multiagent systems with directed communication topology. An improved high-order iterative learning control scheme is applied to solve the consensus tracking problem of multiagent systems. The desired tracking trajectory is produced by a virtual leader agent. Under the given convergence conditions, the consensus can be achieved by using the proposed control protocol. Based on simulation results, the control effect can be improved by utilizing a high-order iterative learning control scheme. Also, it is implied that the higher the orders, the better the control effect. In addition, compared with another control law, the better control effect with the proposed control scheme of this paper can be obtained. Furthermore, through analyzing the maximum state tracking errors of a multiagent with different time delays, the delays do not play a significant role in the convergence of high-order iterative learning control. In our future work, the consensus problem of multiagent systems with control input delays will be considered.

Conflicts of Interest

The authors declare that there is no conflict of interests regarding the publication of this paper.

Acknowledgments

This work is supported by the Aeronautical Science Foundation of China (Grand no. 20155896025).

References

[1] G. Yang, Q. Yang, V. Kapila, D. Palmer, and R. Vaidyanathan, "Fuel optimal manoeuvres for multiple spacecraft formation reconfiguration using multi-agent optimization," *International Journal of Robust and Nonlinear Control*, vol. 12, no. 2-3, pp. 243–283, 2002.

[2] R. Dong and Z. Geng, "Consensus for formation control of multi-agent systems," *International Journal of Robust and Nonlinear Control*, vol. 25, no. 14, pp. 2481–2501, 2015.

[3] Z. Qiu, S. Liu, and L. Xie, "Distributed constrained optimal consensus of multi-agent systems," *Automatica*, vol. 68, pp. 209–215, 2016.

[4] B. Wang, J. Wang, L. Zhang, and B. Zhang, "Robust adaptive consensus tracking for higher-order multi-agent uncertain systems with nonlinear dynamics via distributed intermittent communication protocol," *International Journal of Adaptive Control and Signal Processing*, vol. 30, no. 3, pp. 511–533, 2016.

[5] W. Li and W. Shen, "Swarm behavior control of mobile multi-robots with wireless sensor networks," *Journal of Network and Computer Applications*, vol. 34, no. 4, pp. 1398–1407, 2011.

[6] R. Olfati-Saber, "Flocking for multi-agent dynamic systems: algorithms and theory," *IEEE Transactions on Automatic Control*, vol. 51, no. 3, pp. 401–420, 2006.

[7] Q. Ma, F. L. Lewis, and S. Xu, "Cooperative containment of discrete-time linear multi-agent systems," *International Journal of Robust and Nonlinear Control*, vol. 25, no. 7, pp. 1007–1018, 2015.

[8] H. Haghshenas, M. A. Badamchizadeh, and M. Baradarannia, "Containment control of heterogeneous linear multi-agent systems," *Automatica*, vol. 54, pp. 210–216, 2015.

[9] L. Zeng and G. D. Hu, "Consensus of linear multi-agent systems with communication and input delays," *Acta Automatica Sinica*, vol. 39, no. 7, pp. 1133–1140, 2013.

[10] H. Du, Y. Cheng, Y. He, and R. Jia, "Second-order consensus for nonlinear leader-following multi-agent systems via dynamic output feedback control," *International Journal of Robust and Nonlinear Control*, vol. 26, no. 2, pp. 329–344, 2016.

[11] Q. Song, J. Cao, and W. Yu, "Second-order leader-following consensus of nonlinear multi-agent systems via pinning control," *Systems & Control Letters*, vol. 59, no. 9, pp. 553–562, 2010.

[12] C. Wang, J. Li, N. Jing, J. Wang, and H. Chen, "A distributed cooperative dynamic task planning algorithm for multiple satellites based on multi-agent hybrid learning," *Chinese Journal of Aeronautics*, vol. 24, no. 4, pp. 493–505, 2011.

[13] J. Feng and G. X. Wen, "Adaptive NN consensus tracking control of a class of nonlinear multi-agent systems," *Neurocomputing*, vol. 151, pp. 288–295, 2015.

[14] J. Liu, Z. Chen, X. Zhang, and Z. Liu, "Neural-networks-based distributed output regulation of multi-agent systems with nonlinear dynamics," *Neurocomputing*, vol. 125, pp. 81–87, 2014.

[15] S. Liu, L. Xie, and H. Zhang, "Distributed consensus for multi-agent systems with delays and noises in transmission channels," *Automatica*, vol. 47, no. 5, pp. 920–934, 2011.

[16] Y. Qian, X. Wu, J. Lü, and J.-a. Lu, "Consensus of second-order multi-agent systems with nonlinear dynamics and time delay," *Nonlinear Dynamics*, vol. 78, no. 1, pp. 495–503, 2014.

[17] J. Xu, H. Zhang, and L. Xie, "Input delay margin for consensusability of multi-agent systems," *Automatica*, vol. 49, no. 6, pp. 1816–1820, 2013.

[18] F. Xiao and L. Wang, "State consensus for multi-agent systems with switching topologies and time-varying delays," *International Journal of Control*, vol. 79, no. 10, pp. 1277–1284, 2006.

[19] Y. Chen, J. H. Lü, and X. H. Yu, "Robust consensus of multi-agent systems with time-varying delays in noisy environment," *Science China Technological Sciences*, vol. 54, no. 8, pp. 2014–2023, 2011.

[20] D. A. Bristow, M. Tharayil, and A. G. Alleyne, "A survey of iterative learning control," *IEEE Control Systems Magazine*, vol. 26, no. 3, pp. 96–114, 2006.

[21] M.-X. Sun and Q.-Z. Yan, "Error tracking of iterative learning control systems," *Acta Automatica Sinica*, vol. 39, no. 3, pp. 251–262, 2013.

[22] I. Rotariu, M. Steinbuch, and R. Ellenbroek, "Adaptive iterative learning control for high precision motion systems," *IEEE Transactions on Control Systems Technology*, vol. 16, no. 5, pp. 1075–1082, 2008.

[23] X. Bu, F. Yu, Z. Hou, and F. Wang, "Iterative learning control for a class of nonlinear systems with random packet losses," *Nonlinear Analysis: Real World Applications*, vol. 14, no. 1, pp. 567–580, 2013.

[24] B. Wu, D. Wang, and E. K. Poh, "High precision satellite attitude tracking control via iterative learning control," *Journal of Guidance, Control, and Dynamics*, vol. 38, no. 3, pp. 528–534, 2015.

[25] Y. Chen, C. Wen, J.-X. Xu, and M. Sun, "High-order iterative learning identification of projectile's aerodynamic drag coefficient curve from radar measured velocity data," *IEEE Transactions on Control Systems Technology*, vol. 6, no. 4, pp. 563–570, 1998.

[26] X. Bu, Z. Hou, and F. Yu, "Stability of first and high order iterative learning control with data dropouts," *International Journal of Control, Automation and Systems*, vol. 9, no. 5, pp. 843–849, 2011.

[27] X. Bu, F. Yu, Z. Fu, and F. Wang, "Stability analysis of high-order iterative learning control for a class of nonlinear switched systems," *Abstract and Applied Analysis*, vol. 2013, Article ID 684642, 13 pages, 2013.

[28] S. Yang, J.-X. Xu, D. Huang, and Y. Tan, "Optimal iterative learning control design for multi-agent systems consensus tracking," *Systems & Control Letters*, vol. 69, pp. 80–89, 2014.

[29] J. Li and J. Li, "Adaptive fuzzy iterative learning control with initial-state learning for coordination control of leader-following multi-agent systems," *Fuzzy Sets and Systems*, vol. 248, pp. 122–137, 2014.

[30] D. Meng and Y. Jia, "Iterative learning approaches to design finite-time consensus protocols for multi-agent systems," *Systems & Control Letters*, vol. 61, no. 1, pp. 187–194, 2012.

[31] Y. Liu and Y. Jia, "An iterative learning approach to formation control of multi-agent systems," *Systems & Control Letters*, vol. 61, no. 1, pp. 148–154, 2012.

[32] D. Meng, Y. Jia, J. Du, and J. Zhang, "On iterative learning algorithms for the formation control of nonlinear multi-agent systems," *Automatica*, vol. 50, no. 1, pp. 291–295, 2014.

[33] C. Liu, J. Xu, and J. Wu, "On iterative learning control with high-order internal models," *International Journal of Adaptive Control and Signal Processing*, vol. 24, no. 9, pp. 731–742, 2010.

[34] G. Heinzinger, D. Fenwick, B. Paden, and F. Miyazaki, "Robust learning control," in *Proceedings of the 28th IEEE Conference on Decision and Control*, pp. 436–440, Tampa, Florida, USA, December 1989.

14

Consideration of Three-Dimensional Attitude and Position Control for a Free-Floating Rigid Body using Three Thrusters

Takashi Matsuno[1] **Yasuhiro Yoshimura,**[2] **and Shinji Hokamoto**[2]

[1]*Department of Mechanical and Aerospace Engineering, Tottori, Tottori University, 4-101 Koyama-cho, Tottri 680-8552, Japan*
[2]*Department of Aeronautics and Astronautics, Fukuoka, Kyushu University, 744 Motooka, Nishi-ku, Fukuoka 819-0395, Japan*

Correspondence should be addressed to Takashi Matsuno; matsu@mech.tottori-u.ac.jp

Academic Editor: Paolo Gasbarri

The study examines the control algorithm of a three-dimensional attitude and position of a free-floating rigid body with three thruster forces in which the force directions are fixed with respect to the body. This study provides a theory to develop a control method of an underactuated satellite with the minimum thruster number. In the procedure, three switching controllers are used in conjunction with motion planning in the final angular-rate deceleration phase to individually control the six state variables to the target values. The switching controllers have a hierarchical structure by using invariant manifolds as switching surfaces. The state variables in higher class manifolds that include lower class ones are adjusted by repeatedly adding intentional disturbances while the lower class state variables are returned to the original values by using lower class invariant manifolds. This study describes methods to define the invariant manifolds and also the intentional disturbance for achieving the forementioned control strategy. Finally, the motion planning in the angular-rate deceleration phase from a remained single-axis rotation finalizes the six state values of the body to the target values. Numerical simulations verify the proposed method.

1. Introduction

What is the minimum number of thrusters required to control both the attitude and position of a rigid satellite? What strategies can be applied and what kind of motion is controlled for a satellite in which there are few remaining thrusters? These questions motivated us to pursue the current investigation because nonlinear underactuated control makes these satellites fault-tolerant with respect to thruster malfunctions and enables continuity of the respective missions.

Most satellites are equipped with a sufficient number of thrusters to control the attitude and position in three dimensions because of the following reasons. In typical thrusters, the force directions are fixed to a satellite body and the magnitudes are restricted as positive. This implies that another thruster is required to generate a force in the opposite direction. Consequently, more than twelve thrusters are required for a satellite to guarantee the six degrees of freedom in motion and to ensure the robustness of the system in simple control logic. However, the number of thrusters can be reduced by applying nonlinear control methods. Hence, combined attitude and position control based on the nonlinear motion of a satellite can achieve a significant reduction in the number of actuators.

A few underactuated systems in which the degrees of freedom of motion exceed the number of actuators can be potentially completely controlled by utilizing their nonholonomic constraints [1, 2]. A free-floating satellite is a type of an underactuated system, and its attitude and position dynamics are represented as nonintegrable second-order differential equations that act as nonholonomic constraints [3]. Several intensive studies focused on the attitude motion control of a satellite by utilizing nonholonomic constraints. Several extant studies indicate that two independent torques successfully control a three-dimensional satellite attitude. Tsiotras and Longuski [4] proposed new attitude parameters termed as "wz-parameters," and then Tsiotras and Doumtchenko [5] developed a discontinuous control law that utilized the parameters. In another study, Morin and

Samson [6] designed a smooth time-varying feedback control. Recent studies shifted to more practical ones including a controller for the underactuated CMG satellite system [7], implementation of the kinematic planning scheme [8], and an optimal process for state stabilization [9]. The strategies indicate that two bidirectional control torques (i.e., both positive and negative directional torques are generated) with independent magnitude can control three attitude parameters of a satellite. Therefore, the problem of satellite attitude control is transformed into the number of thrusters that are required to generate independent torques with respect to the two principal axes. Sidi [10] presented a pioneering study related to this problem that numerically demonstrated that four thrusters produce independent torques with respect to three principal axes. In [11], we analytically confirmed the forementioned results and indicated that three thrusters that satisfy a configuration produce bidirectional torques with respect to two independent axes. This implies that when the nonholonomic dynamics of a satellite are considered, three thrusters control the three-dimensional attitude of a satellite. Hence, Matsuno et al. [11] and Yoshimura et al. [12] demonstrated an attitude control strategy and developed feedback controllers although a zero or unidirectional (only positive or negative directional) torque was generated under the thruster configuration around the third axis owing to the coupling effect with the other two axes.

The present study focuses on the simultaneous attitude and position control of an underactuated satellite. In two-dimensional motion, at least two thrusters are required to accelerate and decelerate the rotation of a satellite. Essentially, the same type of controllers in [13–15] is applied to obtain a controller with two thrusters for the in-plane attitude and position motion of a free-floating satellite. The study numerically verified that a switching controller that utilizes an invariant manifold achieves a two-dimensional attitude and position control by using positive albeit variable magnitudes of thruster forces. The study was extended in [16] for two on/off types of thrusters in which the input magnitudes are fixed.

In contrast to the two-dimensional cases, only a few studies discuss three-dimensional attitude and position control for underactuated satellites. With respect to a satellite with four thrusters, the previous study [17] described a design procedure of motion planning for the simultaneous control of the attitude and position. However, in the case with three thrusters, the three-dimensional attitude and position control is quite challenging because of the following reason. Three thruster configurations that guarantee bidirectional control torques for two principal axes generate dependent torque with respect to the third principal axis of a satellite. Specifically, the attitude control proposed in the present requires a series of complex nonlinear processes. Furthermore, thruster forces are unidirectional, and thus, the controllability of the system is uncertain. It should be noted that although the theorem of Sussman [18] or Goodwine and Burdick [19] provides sufficient conditions for the controllability of nonholonomic systems, they cannot be applied in this study owing to the nonnegative restriction of the thruster forces.

Given the above background, the present study develops a switching controller by using three invariant manifolds and provides a motion planning scheme in the final phase. This paper deals with the problem to control a free-floating rigid body in three-dimensional space using three thruster forces (hereafter, we refer to this type of rigid body model as "FFRB"). In this simplified problem, the proposed method achieves both attitude and position control by using only three thrusters in which the directions are fixed with respect to the body. The effectiveness of the method is numerically certified in ideal conditions (constant mass and nonlimited positive thruster force). This study is the first to demonstrate that the three-dimensional attitude and position of a rigid body can be controlled with three fixed thrusters, although the following problems remain in the future. The proposed controller includes a few heuristic processes; rigorous certification of the control strategy is required, and it sometimes results in an extremely long controlling time to the target states. We believe that the concept of this controller explained in this study contributes to significantly decreasing number of thrusters in practical underactuated satellites.

The remainder of this paper is organized as follows. In Section 2, the formulations of the kinematic and dynamic equations of a FFRB are discussed. The procedure consists of a combination of the three hierarchical feedback steps and motion planning. Furthermore, the effects of the design parameters in each step are discussed. In Section 4, numerical simulations are conducted to verify the proposed controller, and a summary of the study is presented in Section 5.

2. Three-Dimensional Model of a FFRB (Free-Floating Rigid Body with Three Force Inputs)

2.1. Thruster Configuration. In order to simplify the discussion in Section 3, this subsection defines the three-thruster configuration for controlling a FFRB. It should be noted that the main purpose of this study is to demonstrate that a range of three-thruster configurations controls both the position and the attitude of a FFRB to the target values.

It is assumed that a rigid body has the three thrusters that satisfy the following conditions (i)–(iii) for its torques.

(i) Bidirectional control torques are generated about principal axes I and II of the FFRB.

This condition implies that the attitude motion in three dimensions is controlled by applying nonlinear control methods utilizing nonholonomic constraint of a FFRB unless the inertial moments of principal axes I and II are equal. (It is widely known that the attitude motion around the principal axis III is uncontrollable when the inertial moments are equal.)

(ii) With respect to the principal axis III, a unidirectional control torque is generated whenever control torques are applied around the principal axes I and II.

This condition simplifies the discussion in Section 3. Furthermore, with respect to the translational motion control, the following condition is assumed.

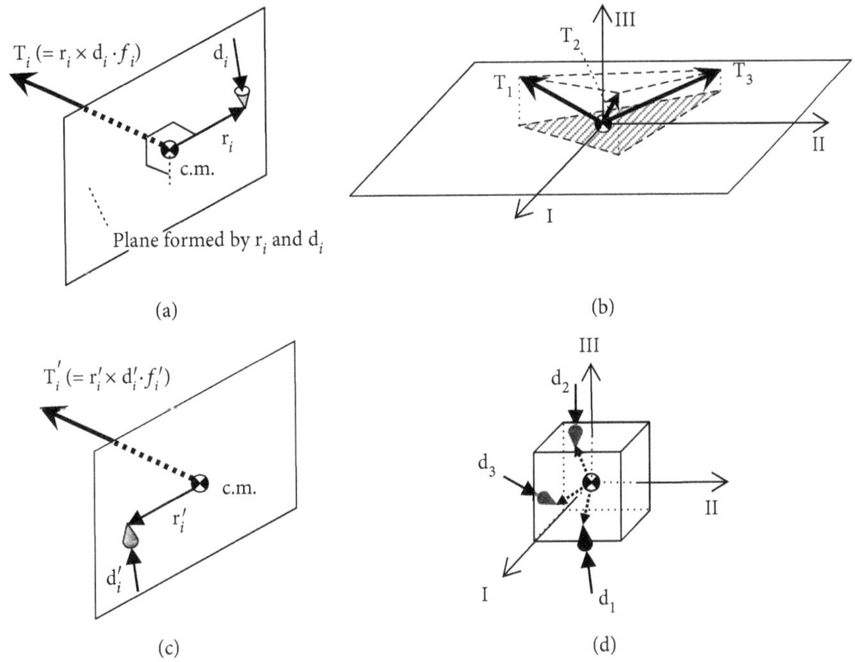

Figure 1: Three-thruster configuration for attitude and position control. (a) Direction of the torque vector. (b) Geometry formed with the three torque vector. (c) Another thruster placement. (d) Three-thruster configuration.

(iii) In the thruster configuration, the induced force along the principal axis III is bidirectional.

First, it is proven below that a range of three-thruster configurations satisfy the conditions (i) and (ii). We consider a thruster placed at $\mathbf{r}_i$ and oriented to $\mathbf{d}_i$ with respect to the body frame (see Figure 1(a)), and it generates a torque moment $\mathbf{T}_i = \mathbf{r}_i \times \mathbf{d}_i \cdot f_i$ with respect to the mass center. Three thrusters form a tetrahedron with their moments, and a range of three-thruster configurations have a projection of the tetrahedron on the I–II plane including the mass center inside the body as shown in Figure 1(b). Subsequently, the combination of the three forces in positive or zero values generates their resultant torque within the extension of the tetrahedron. Thus, this type of three-thruster configurations satisfies the conditions (i) and (ii). It is noted that a vertex of the tetrahedron is the mass center of the FFRB, and therefore, it is impossible to generate bidirectional control torques for all three principal axes with three thrusters as shown in [11, 12].

Next, it is shown that there is a three-thruster configuration that satisfies the condition (iii) under the conditions (i) and (ii) as follows. As described above, the moment vector $\mathbf{T}_i$ is defined as the outer product between the thruster's position vector $\mathbf{r}_i$ and orientation vector $\mathbf{d}_i$. However, as shown in Figure 1(c), there are numerous combinations for the two vectors in the plane perpendicular to the vector $\mathbf{T}_i$. This implies that the condition (iii) is satisfied as well as the conditions (i) and (ii) when one of $\mathbf{d}_i (i = 1 \sim 3)$ has a negative component along the axis III (as shown in Figure 1(a)) and one of the others has a positive component (as shown in Figure 1(c)). Figure 1(d) illustrates an example of this configuration. It should be noted the three-thruster configuration used in the simulations in this study satisfies the three conditions.

2.2. Attitude Equations. We consider a FFRB with constant moments of inertia and with the absence of external disturbance forces. Euler equations of the FFRB are described with its principal axes as follows:

$$\mathbf{J}\dot{\boldsymbol{\omega}} = \mathbf{T} + \tilde{\boldsymbol{\omega}}\mathbf{J}\boldsymbol{\omega}, \tag{1}$$

where

$$\mathbf{J} = \begin{bmatrix} J_{\mathrm{I}} & 0 & 0 \\ 0 & J_{\mathrm{II}} & 0 \\ 0 & 0 & J_{\mathrm{III}} \end{bmatrix}, \quad \tilde{\boldsymbol{\omega}} = \begin{bmatrix} 0 & \omega_{\mathrm{III}} & -\omega_{\mathrm{II}} \\ -\omega_{\mathrm{III}} & 0 & \omega_{\mathrm{I}} \\ \omega_{\mathrm{II}} & -\omega_{\mathrm{I}} & 0 \end{bmatrix}. \tag{2}$$

The attitude is defined by the Euler parameter $\mathbf{q}$ as follows:

$$\mathbf{q} = [q_{\mathrm{I}} \quad q_{\mathrm{II}} \quad q_{\mathrm{III}} \quad q_{\mathrm{IV}}]^T = \left[a_{\mathrm{I}} \sin\left(\frac{\psi}{2}\right) \quad a_{\mathrm{II}} \sin\left(\frac{\psi}{2}\right) \quad a_{\mathrm{III}} \sin\left(\frac{\psi}{2}\right) \quad \cos\left(\frac{\psi}{2}\right)\right]^T. \tag{3}$$

The differentiation of $\mathbf{q}$ is expressed with the angular velocity $\boldsymbol{\omega}$ as follows:

$$\dot{\mathbf{q}} = \mathbf{Q}(\mathbf{q})\boldsymbol{\omega}, \tag{4}$$

where

$$\mathbf{Q}(\mathbf{q}) = \frac{1}{2}\begin{bmatrix} q_{\mathrm{IV}} & -q_{\mathrm{III}} & q_{\mathrm{II}} \\ q_{\mathrm{III}} & q_{\mathrm{IV}} & -q_{\mathrm{I}} \\ -q_{\mathrm{II}} & q_{\mathrm{I}} & q_{\mathrm{IV}} \\ -q_{\mathrm{I}} & -q_{\mathrm{II}} & -q_{\mathrm{III}} \end{bmatrix}. \tag{5}$$

The total torque input $\mathbf{T}$ is given by the summation of $\mathbf{T}_i$ as follows:

$$\mathbf{T} = \sum_{i=1}^{3}\mathbf{T}_i = \sum_{i=1}^{3}\mathbf{r}_i \times \mathbf{d}_i \cdot f_i = \mathbf{B}\mathbf{f}, \tag{6}$$

where $\mathbf{f} = [f_1 \ \ f_2 \ \ f_3]^T$, $\mathbf{B} = [\mathbf{b}_1 \ \ \mathbf{b}_2 \ \ \mathbf{b}_3]$, and $\mathbf{b}_i = \mathbf{r}_i \times \mathbf{d}_i$.

2.3. Translational Equations. The translational equation of the FFRB is expressed as follows:

$$M\ddot{\mathbf{p}} = \mathbf{F}. \tag{7}$$

In (7), the translational force $\mathbf{F}$ induced by the thrusters is related to the control torque $\mathbf{T}$. For this purpose, we first consider the force vector $\mathbf{F}'$ that is expressed with respect to the body fixed frame. Given that $\mathbf{F}'_i = \mathbf{d}_i \cdot f_i$, the total translational force vector is expressed as follows:

$$\mathbf{F}' = \sum_{i=1}^{3}\mathbf{F}'_i = \sum_{i=1}^{3}\mathbf{d}_i \cdot f_i = \mathbf{D}\mathbf{f}, \tag{8}$$

where $\mathbf{D} = [\mathbf{d}_1 \ \ \mathbf{d}_2 \ \ \mathbf{d}_3]$. The matrix $\mathbf{B}$ in (6) is nonsingular when the three control torques are independent, and thus, the force vector $\mathbf{f}$ is formulated as $\mathbf{B}^{-1}\mathbf{T}$. Thus, the transformation matrix $\mathbf{A}(\mathbf{q})$ is used, and the translational force $\mathbf{F}$ in the inertial frame is expressed with the control torque $\mathbf{T}$ as follows:

$$\mathbf{F} = \mathbf{A}(\mathbf{q})\mathbf{D}\mathbf{B}^{-1}\mathbf{T}, \tag{9}$$

where

$$\mathbf{A}(\mathbf{q}) = \begin{bmatrix} 2(q_{\mathrm{IV}}^2 + q_{\mathrm{I}}^2) - 1 & 2(q_{\mathrm{I}}q_{\mathrm{II}} - q_{\mathrm{IV}}q_{\mathrm{III}}) & 2(q_{\mathrm{I}}q_{\mathrm{III}} + q_{\mathrm{IV}}q_{\mathrm{II}}) \\ 2(q_{\mathrm{I}}q_{\mathrm{II}} + q_{\mathrm{IV}}q_{\mathrm{III}}) & 2(q_{\mathrm{IV}}^2 + q_{\mathrm{II}}^2) - 1 & 2(q_{\mathrm{II}}q_{\mathrm{III}} - q_{\mathrm{IV}}q_{\mathrm{I}}) \\ 2(q_{\mathrm{I}}q_{\mathrm{III}} - q_{\mathrm{IV}}q_{\mathrm{II}}) & 2(q_{\mathrm{II}}q_{\mathrm{III}} + q_{\mathrm{IV}}q_{\mathrm{I}}) & 2(q_{\mathrm{IV}}^2 + q_{\mathrm{III}}^2) - 1 \end{bmatrix}. \tag{10}$$

Equation (9) indicates that the translational motion of the considering FFRB is induced dependently with the rotational motion.

3. Controller Design

3.1. Strategy. Without loss of generality, the target state can be set as $\mathbf{q} = [0\,0\,0 \pm 1]^T$ and $\mathbf{p} = [0\,0\,0]^T$. Furthermore, at the target attitude, the principal axis III of the FFRB coincides with the Z-axis of the inertial flame.

The switching controller proposed in this study uses the following three invariant manifolds: $S_1 = 0$, $S_2 = 0$, and $S_3 = 0$. The first one, namely $S_1 = 0$, corresponds to the manifold and indicates that the FFRB rotates about its principal axis III that coincides with the Z-axis. Hereafter, we refer to this state as "single spinning." The values of S_2 and S_3 indicate the remained velocity and position errors, respectively, along the Z-axis when the single spinning stops. Figure 2 shows the schematic image of the control process. It should be noted that maintaining a single spinning ($S_1 = 0$) under $T_{\mathrm{I}} = 0$ and $T_{\mathrm{II}} = 0$ after achieving $S_2 = 0$ and $S_3 = 0$ causes S_2 and S_3 to be invariant (this property is explained in the following section). Thus, various time histories of torque T_{III} are available to stop the single spinning at the target attitude while maintaining $S_2 = 0$ and $S_3 = 0$. Proper planning of the T_{III} history controls the position and velocity along X and Y to the target.

Figure 3 schematically illustrates the switching controller. The controller for achieving $S_1 = 0$, $S_2 = 0$, and $S_3 = 0$ is composed of the following three hierarchical control steps: Step A, Step B, and Step C. In Step A, $S_1 = 0$ (i.e., q_{I}, q_{II}, ω_{I}, and $\omega_{\mathrm{II}} = 0$) is achieved using feedback law. In Step B, intentional disturbances and, in Step A, controls are repeated such that $S_2 = 0$ is satisfied. Thus, $S_1 = 0$ and $S_2 = 0$ are satisfied at the end of Step B. Step C achieves $S_1 = 0$, $S_2 = 0$, and $S_3 = 0$ by disturbing S_1 and S_2 to nonzero values and by repeating Step B to satisfy $S_1 = 0$ and $S_2 = 0$ again. When $S_1 = 0$, $S_2 = 0$, and $S_3 = 0$ are satisfied, Step D finally controls the single spinning angle to the target angle by the time profile of T_{III}, which is planned beforehand such that the position (X and Y) in the inertial frame coincides with the target positions. These steps are explained individually below given the assumption that the resultant torque component along the principal axis III is always positive, i.e., the three-thruster configuration satisfies $T_{III} > 0$ as shown in Figure 1(b).

3.2. Step A: Control to the Single Spinning about the Axis III. Define S_1 as follows:

$$S_1 = q_I^2 + q_{II}^2 + \omega_I^2 + \omega_{II}^2. \tag{11}$$

This implies that at $S_1 = 0$, only q_{III}, q_{IV}, and ω_{III} correspond to nonzero values. Thus, achieving $S_1 = 0$ implies that the FFRB motion is led into a spinning motion around the axis III, and the axis coincides with the Z-axis of the inertial frame.

The following control law for T_I and T_{II} as proposed in [11] is adopted to achieve $S_1 = 0$ in the present study.

$$T_I = -K_I\omega_I - \frac{J_I g_I}{2} - J_I\bar{K}\omega_{II}, \tag{12}$$

$$T_{II} = -K_{II}\omega_{II} - \frac{J_{II} g_{II}}{2} + J_{II}\bar{K}\omega_I, \tag{13}$$

where

$$g_I = q_I q_{IV} + q_{II} q_{III} + 2\frac{J_{II} - J_{III}}{J_I}\omega_{II}\omega_{III}, \tag{14}$$

$$g_{II} = q_{II} q_{IV} - q_I q_{III} + 2\frac{J_{III} - J_I}{J_{II}}\omega_{III}\omega_I. \tag{15}$$

K_I and K_{II} are arbitrary positive gains. $\bar{K}$ is determined to provide a negative $\dot{\omega}_{III}$ anytime such that positive T_{III} reduces the single spinning rate. Reference [11] specifies the $\bar{K}$ as follows:

$$\bar{K} = \frac{\omega_I\omega_{II}((K_I/J_I) + (K_{II}/J_{II})) + (g_I/2)\omega_{II} + (g_{II}/2)\omega_I - \omega_{III}(\sigma_I\omega_{II}^2 + \sigma_{II}\omega_I^2)}{\omega_I^2 - \omega_{II}^2} - \kappa\,\mathrm{sgn}\,((\omega_I^2 - \omega_{II}^2)\sigma_{III}), \tag{16}$$

where

$$\sigma_I = \frac{J_{II} - J_{III}}{J_I}, \quad \sigma_{II} = \frac{J_{III} - J_I}{J_{II}}, \quad \sigma_{III} = \frac{J_I - J_{II}}{J_{III}}, \quad \kappa > 0. \tag{17}$$

It should be noted that when $\omega_I^2 - \omega_{II}^2$ approaches zero, the magnitude of $\bar{K}$ approaches infinity, and subsequently, the magnitudes of T_I and T_{II} become extremely high. Thus, in order to avoid this singularity, the control inputs in (12) and (13) should be designed to saturate when $\omega_I^2 - \omega_{II}^2$ becomes quite low. Furthermore, by setting $\bar{K} = 0$, the control torques T_I and T_{II} are used as disturbances. The reason is explained in Step B.

3.3. Step B: Control for the Translational Velocity in the Z-Direction. Step B begins after the FFRB is in single spinning, and thus, the principal axis III coincides with the Z-axis in the inertial frame and the control torques around the axes I and II are $T_I = T_{II} = 0$. Therefore, the third component of (7) is expressed from (9) as follows:

$$M\ddot{Z} = \gamma_3 T_{III}, \tag{18}$$

where $[\gamma_1 \ \ \gamma_2 \ \ \gamma_3]^T = \mathbf{D}\mathbf{B}^{-1}[0 \ \ 0 \ \ 1]^T$ denotes the effect of the control torque T_{III} on the translational motion in the single spinning. Furthermore, from (1) in the single spinning, (18) is reexpressed as follows:

$$M\ddot{Z} = \gamma_3 T_{III} = \gamma_3 J_{III}\dot{\omega}_{III}. \tag{19}$$

The above equation implies that the translational motion in the Z-direction depends on the time profile of the angular acceleration around the axis III. When the single spinning ceases, (19) is analytically integrated to derive the velocity variation $\Delta\dot{Z}$ as follows:

$$\Delta\dot{Z} = \frac{\gamma_3}{M}\left[J_{III}\omega_{III}'\right]_{\omega_{III}'=\omega_{III}}^{\omega_{III}'=0} = -\frac{\gamma_3}{M}J_{III}\omega_{III}. \tag{20}$$

Therefore, the translation velocity along the Z-axis in the inertial frame when the single spinning ceases is expressed as follows:

$$S_2 = \dot{Z} - \frac{\gamma_3}{M}J_{III}\omega_{III}. \tag{21}$$

Conversely, we consider the time differential of S_2 when a control torque vector $\mathbf{T} = [0, 0, T_{III}]^T$ is applied to the FFRB in the single spinning. It is expressed from the third components of (1), (7), and (9) as follows:

$$\begin{aligned}\dot{S}_2 = \ddot{Z} - \frac{\gamma_3}{M}J_{III}\dot{\omega}_{III} &= \frac{1}{M}(\gamma_1 T_I + \gamma_2 T_{II} + \gamma_3 T_{III}) \\ &\quad - \frac{\gamma_3}{M}(T_{III} + (J_I - J_{II})\omega_I\omega_{II}) \\ &= \frac{1}{M}\{(\gamma_1 T_I + \gamma_2 T_{II}) - \gamma_3(J_I - J_{II})\omega_I\omega_{II}\}.\end{aligned} \tag{22}$$

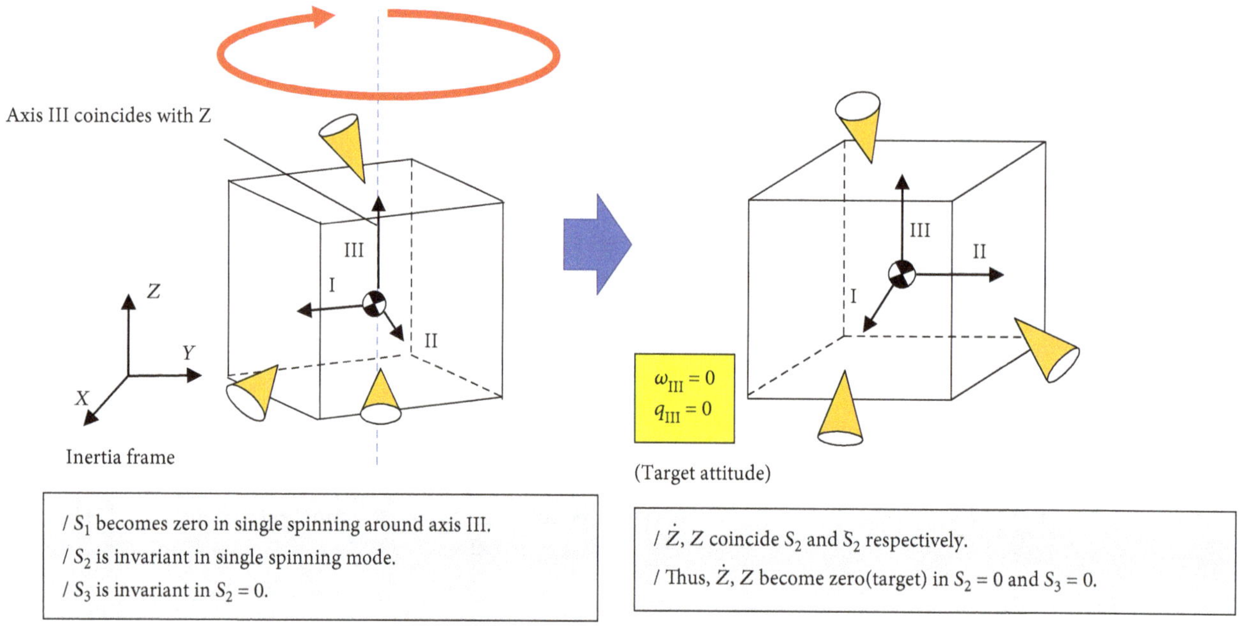

FIGURE 2: Schematic image of the S_1, S_2, and S_3.

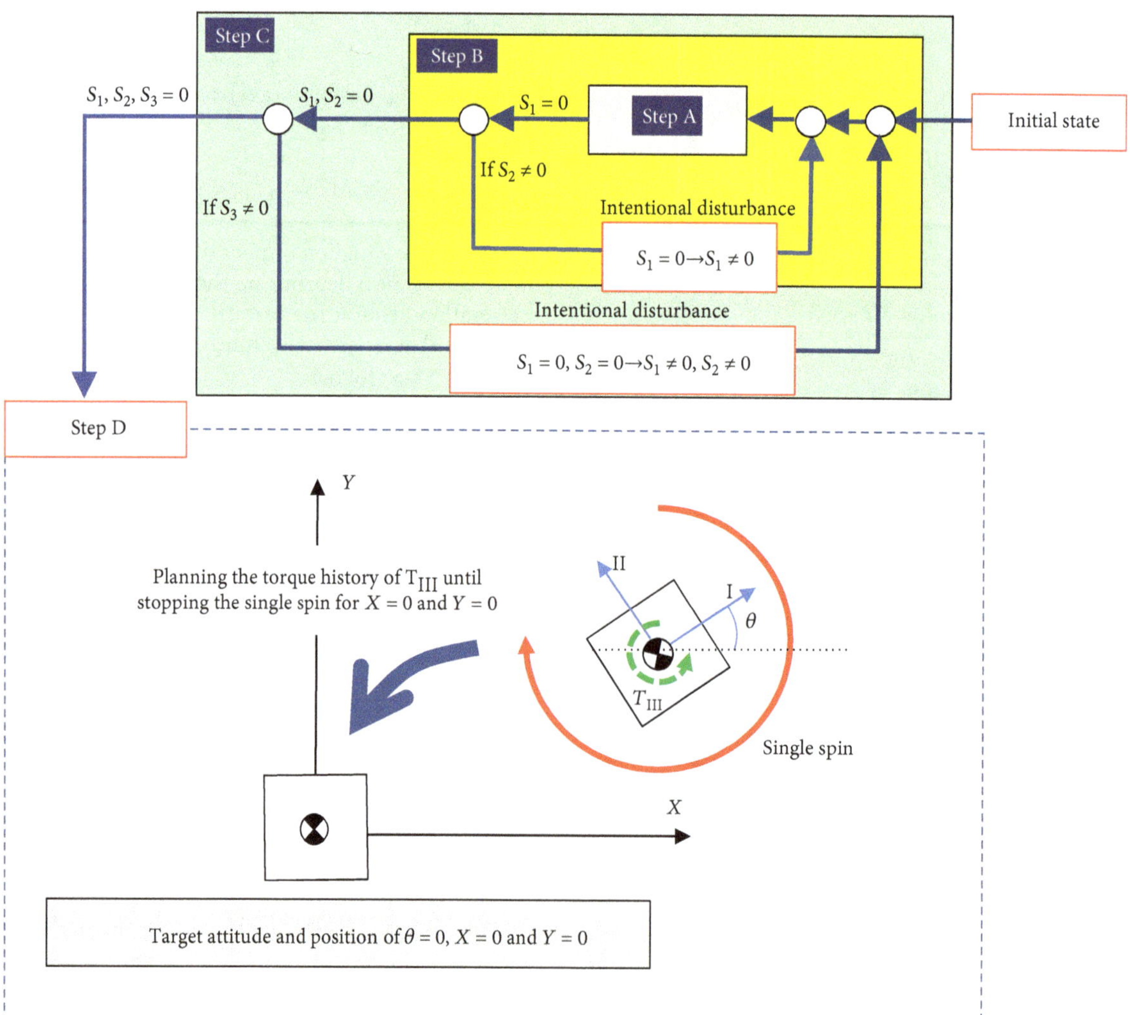

FIGURE 3: Schematic of the control sequence.

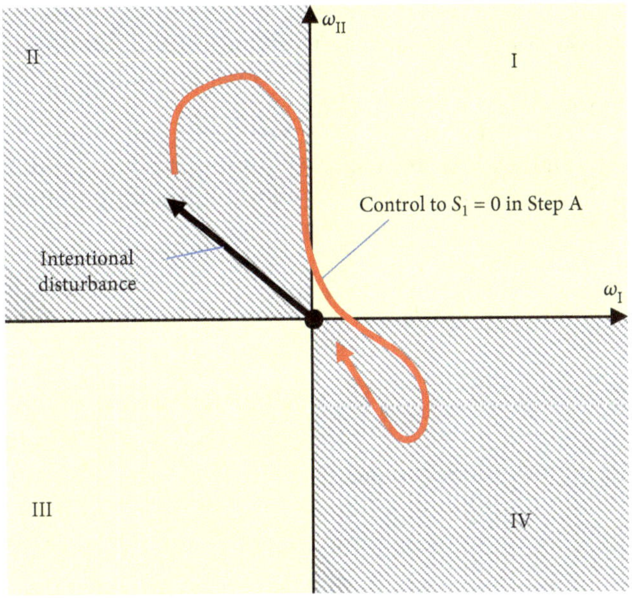

FIGURE 4: Trajectory of ω_I and ω_{II} in Step B in the case of $J_I > J_{II}$.

This implies that the variation of S_2 is generated by T_I, T_{II}, ω_I, and ω_{II}. Thus, the value of S_2 does not change when the FFRB's single spinning is maintained. Conversely, a disturbance that makes $S_1 \neq 0$ is necessary to change the value of S_2. Thus, Step B requires an intentional disturbance that results in a nonzero value of S_1 and the process of Step A for returning S_1 to zero. The value of S_2 should approach zero during a series of these processes.

Intentional disturbances that satisfy the above process are shown as follows. Given that $T_{III} > 0$ is assumed in this section, the angular rate ω_{III} around the axis III should approach zero and satisfy $\omega_{III} < 0$. This implies that time integration of $\dot{\omega}_{III}$ must be negative and from the third component of (1), the following expression is obtained:

$$\dot{\omega}_{III} = \frac{T_{III}}{J_{III}} + \sigma_{III}\omega_I\omega_{II}. \tag{23}$$

The second term ($\sigma_{III}\omega_I\omega_{II}$) in the right-hand side of the above equation should be negative for a high portion of the trajectory for $S_1 \neq 0$ and $S_1 = 0$. Therefore, when $J_I > J_{II}$ is assumed for the FFRB, the trajectory shown in Figure 4 on the ω_I–ω_{II} plane is a solution for the intentional disturbance and mainly in the second or fourth quadrant. However, σ_{III} $\omega_I\omega_{II}$ is not analytically integrable, and thus, it is not possible to determine whether this trajectory always provides a negative value in the integration of $\sigma_{III}\omega_I\omega_{II}$. (In the simulation of Section 4, the negative integration of $\sigma_{III}\omega_I\omega_{II}$ of the trajectory for the designed intentional disturbances is numerically confirmed.) It should be noted that the rotational direction (clockwise or counterclockwise) of the designed trajectory in the ω_I–ω_{II} plane dictates the sign of ΔS_2 where ΔS_2 denotes a variation of S_2 in a sequence $S_1 = 0 \rightarrow S_1 \neq 0 \rightarrow S_1 = 0$. Thus, according to the value of (21), the direction to pull S_2 towards zero is specified. Additionally, lower disturbance trajectories result in lower variations for S_2. Thus, by adjusting the direction and magnitude of intentional disturbance trajectories, S_2 converges to zero without overshooting.

To derive the above intentional disturbance, law as a general form requires rigorous mathematical analysis. Thus, as an alternative, we preliminarily determine the constants, D_{Ib}, D_{IIb}, and D_{S2}, in the following disturbance laws of (24), (25), and (26), respectively, by using numerical simulations as follows:

$$T_I = \frac{D_{Ib}\ \mathrm{sgn}\,(S_2)}{|\Omega_{III}|}, \tag{24}$$

$$T_{II} = \frac{D_{IIb}\ \mathrm{sgn}\,(S_2)}{|\Omega_{III}|}, \tag{25}$$

for the time duration corresponding to the following:

$$t_{S2} = D_{S2}|S_2|, \tag{26}$$

where Ω_{III} denotes the angular velocity around the axis III when intentional disturbances are applied. The disturbances are divided by the magnitude of Ω_{III} to decrease the effect on ΔS_2 (the variation in S_2 is proportional to the magnitude of Ω_{III} owing to the gyro effect). Equation (26) denotes that the intentional disturbance decreases with the magnitude of S_2 under positive D_{S2}. The calculation of ΔS_2 requires numerical integration in $S_1 \neq 0$, and thus, the setting process of D_{Ib}, D_{IIb}, and D_{S2} is heuristic. Therefore, the saturation of t_{S2} is desirable to avoid unexpected ΔS_2 behavior preliminary calculated by numerical simulations.

In order to cancel the effect of the Ω_{III} derivation, the following simple feedback law backs ω_{III} to the initial Ω_{III} as follows:

$$T_{III} = -K_{III}(\omega_{III} - \Omega_{III}), \tag{27}$$

where K_{III} denotes the positive gain. As shown in (22), (27) does not affect S_2.

3.4. Step C: Control of the Position in the Z-Direction. Step C controls the FFRB's position in the Z-direction in addition to Step B by achieving $S_1 = 0$, $S_2 = 0$, and $S_3 = 0$. In this step, S_3 is defined as the FFRB position when the single spinning ceases after the maneuver in Step D. As given below, $\theta = 2n\pi$ is assumed as an identified preferable attitude angle in Step D where n denotes an arbitrary integer that indicates a rotation number in the single spinning.

The rotational angle θ about axis III is expressed as follows:

$$\theta = 2\tan^{-1}\left(\frac{q_{III}}{q_{IV}}\right) + 2n\pi. \tag{28}$$

If the stopping maneuver for the spinning begins at $\theta = \theta_3 + 2n\pi$ and ends at $\theta = 2n\pi$ in the single spinning, the variation of the Z-directional position, ΔZ, is derived as follows:

$$\Delta Z = \dot{Z}\tau_d + \left[\frac{J_{\mathrm{III}}\gamma_3}{M}\theta'\right]_{\theta'=\theta_3+2n\pi}^{\theta=2n\pi} = \dot{Z}\tau_d - \frac{J_{\mathrm{III}}\gamma_3}{M}\theta_3, \quad (29)$$

where τ_d denotes the time of drift motion before the final stopping maneuver in Step D. From $S_2 = 0$ in Step C, the drift interval is calculated from (21) as follows:

$$\tau_d = \frac{\theta_3 + 2n\pi - \theta}{\omega_{\mathrm{III}}} = \frac{J_{\mathrm{III}}\gamma_3}{M}\frac{\theta_3 + 2n\pi - \theta}{\dot{Z}}. \quad (30)$$

Substituting (30) into (29) yields the following expression for ΔZ:

$$\Delta Z = \frac{J_{\mathrm{III}}\gamma_3}{M}(-\theta + 2n\pi). \quad (31)$$

Thus, S_3 is defined as follows:

$$S_3 = Z + \Delta Z = \xi_1 + \xi_2 n, \quad (32)$$

where

$$\xi_1 = Z - \frac{J_{\mathrm{III}}\gamma_3}{M}\theta, \qquad \xi_2 = \frac{2\pi J_{\mathrm{III}}\gamma_3}{M}. \quad (33)$$

It should be noted that S_3 does not include θ_3. Subsequently, the time derivative of S_3 is derived as follows:

$$\dot{S}_3 = \dot{\xi}_1 = \dot{Z} - \frac{\gamma_3}{M}J_{\mathrm{III}}\omega_{\mathrm{III}} = S_2. \quad (34)$$

This equation indicates that S_3 is invariant under $S_2 = 0$ ($S_1 = 0$ is also satisfied by the control in Step B).

In (32), the value of S_3 depends on n. However, specifying a number for n in advance is not possible, because the position variation is uncertain until $S_3 = 0$ is achieved. In order to solve this problem, $S_3 = 0$ are divided by ξ_2 to obtain $\xi_1/\xi_2 + n = 0$. Subsequently, in order to ensure that n is an integer, S_3 is modified without n as follows:

$$S_3 = \begin{cases} \mathrm{MOD}(\xi_1, \xi_2) & \text{for } \xi_1 < \dfrac{\xi_2}{2}, \\ \mathrm{MOD}(\xi_1, \xi_2) - \xi_1 & \text{for } \xi_1 \geq \dfrac{\xi_2}{2}, \end{cases} \quad (35)$$

where MOD indicates the remainder. It should be noted that the modified S_3 is also invariant under $S_2 = 0$, because ξ_2 is a constant value. Furthermore, (35) has two forms to avoid discontinuity for the feedback control because the term $\mathrm{MOD}(\xi_1, \xi_2)$ is not continuous at $\mathrm{MOD}(\xi_1, \xi_2) = 0$. This discontinuity significantly affects the convergence of $S_3 = 0$.

Given that S_3 is invariant under $S_2 = 0$, the repetition of the sequence $S_2 = 0 \rightarrow S_2 \neq 0 \rightarrow S_2 = 0$ is necessary to achieve $S_3 = 0$. The design procedure of the intentional disturbance for $S_2 \neq 0$ is almost identical to that for $S_1 \neq 0$ in Step B. The direction of the disturbance must be in the second or fourth quadrant on the ω_{I}–ω_{II} plane in the case of $J_{\mathrm{I}} > J_{\mathrm{II}}$, and the magnitude of the disturbance should be adjusted to avoid generating an excessive variation in ΔS_3 in a sequence.

Additionally, in Step B, we preliminary determine D_{Ic}, D_{IIc}, and D_{S3} in the following disturbance law of (36), (37), and (38), respectively, by using numerical simulations as follows:

$$T_{\mathrm{I}} = \frac{D_{\mathrm{Ic}}\,\mathrm{sgn}\,(S_3)}{|\Omega_{\mathrm{III}}|}, \quad (36)$$

$$T_{\mathrm{II}} = \frac{D_{\mathrm{IIc}}\,\mathrm{sgn}\,(S_3)}{|\Omega_{\mathrm{III}}|}, \quad (37)$$

for the following time duration:

$$t_{S3} = D_{S3}|S_3|. \quad (38)$$

3.5. Step D: Reaching the Target State. In Steps A, B, and C, $S_1 = 0$, $S_2 = 0$, and $S_3 = 0$ were achieved, respectively. This means that the attitude angles around the axes I/II and the velocity/position along the Z-direction were adjusted for a target state. Thus, the remaining states correspond to the attitude angle around the axis III and the FFRB positions along X- and Y-directions. The remaining attitude angle is easily controlled by T_{III}, and thus, Step D adjusts the FFRB velocities/positions along the X- and Y-directions to the target positions by using a properly designed T_{III} time profile until the single spinning ceases. Although there are various proper histories for T_{III}, this section proposes a solution for the motion planning problem.

The proposed motion planning procedure is composed of a series of control sequences. In each sequence, while $T_{\mathrm{I}} = T_{\mathrm{II}} = 0$ to maintain the single spinning state, the following T_{III} input is used:

$$T_{\mathrm{III}} = \eta\dot{\theta}, \quad (39)$$

where η denotes a constant and is selected to satisfy $T_{\mathrm{III}} > 0$. It is assumed that the control maneuver is applied from $t = t_0$ to $t = t_f$. In order to determine the proper η and the time duration, the variations of the velocity and position along the X- and Y-directions are calculated as follows. First, from (7) and (9) for the control torque in (39) with $T_{\mathrm{I}} = T_{\mathrm{II}} = 0$, the velocity variation $\Delta\dot{\hat{\mathbf{p}}}_j$ generated in the j-th control maneuver is calculated as follows:

$$\Delta\dot{\hat{\mathbf{p}}}_j = \frac{1}{M}\int_{t_{j0}}^{t_{jf}} \bar{\mathbf{A}}(\theta)[\gamma_1 \quad \gamma_2]^T T_{\mathrm{III}}dt = \frac{\eta_j}{M}\begin{bmatrix} \sin(\theta_{jf}) - \sin(\theta_{j0}) & \cos(\theta_{jf}) - \cos(\theta_{j0}) \\ -\cos(\theta_{jf}) + \cos(\theta_{j0}) & \sin(\theta_{jf}) - \sin(\theta_{j0}) \end{bmatrix}\begin{bmatrix}\gamma_1 \\ \gamma_2\end{bmatrix}, \quad (40)$$

where the subscripts 0 and f represent the values at the beginning and final instance of the j-th control maneuver, respectively. Subsequently, the position variation $\Delta\hat{\mathbf{p}}_j$ is obtained from the integration of (40) as follows:

$$\Delta\hat{\mathbf{p}}_j = \frac{J_{\mathrm{III}}}{M}\begin{bmatrix} \gamma_1 S_\theta\left(\theta_{j0}, \theta_{jf}, \dot{\theta}_{j0}, \eta_j\right) + \gamma_2 C_\theta\left(\theta_{j0}, \theta_{jf}, \dot{\theta}_{j0}, \eta_j\right) \\ -\gamma_1 C_\theta\left(\theta_{j0}, \theta_{jf}, \dot{\theta}_{j0}, \eta_j\right) + \gamma_2 S_\theta\left(\theta_{j0}, \theta_{jf}, \dot{\theta}_{j0}, \eta_j\right) \end{bmatrix} + \dot{\hat{\mathbf{p}}}_{j0}\Delta t_j, \tag{41}$$

where Δt_j denotes the time interval of this maneuver and $\dot{\hat{\mathbf{p}}}_{j0}$ denotes the velocity vector at the beginning of the maneuver. With respect to the single spinning control with (39), the interval Δt_j is described as follows:

$$\Delta t_j = \frac{J_{\mathrm{III}}}{\eta_j}\ln\left(1 + \frac{\theta_{jf} - \theta_{j0}}{\varphi_j}\right), \tag{42}$$

where

$$\varphi_j = \frac{J_{\mathrm{III}}\dot{\theta}_{j0}}{\eta_j}. \tag{43}$$

Furthermore, in (41), $S_\theta(\theta_{j0}, \theta_{jf}, \dot{\theta}_{j0}, \eta_j)$ and $C_\theta(\theta_{j0}, \theta_{jf}, \dot{\theta}_{j0}, \eta_j)$ are defined as the time integrations of $\sin(\theta)$ and $\cos(\theta)$ multiplied by η_j/J_{III}, respectively, and expressed as follows:

$$\begin{aligned} S_\theta&\left(\theta_{j0}, \theta_{jf}, \dot{\theta}_{j0}, \eta_j\right) \\ &= \frac{\eta_j}{J_{\mathrm{III}}}\int_{t_{j0}}^{t_{jf}} \sin(\theta)dt \\ &= \frac{\eta_j}{J_{\mathrm{III}}}\int_{t_{j0}}^{t_{jf}} \sin\left(\varphi_j \exp\left\{\frac{\eta_j}{J_{\mathrm{III}}}(t - t_{j0})\right\} + \theta_{j0} - \varphi_j\right)dt \\ &= \left\{\mathrm{Si}\left(\varphi_j + \theta_{jf} - \theta_{j0}\right) - \mathrm{Si}\left(\varphi_j\right)\right\}\cos\left(\theta_{j0} - \varphi_j\right) \\ &\quad + \left\{\mathrm{Ci}\left(\varphi_j + \theta_{jf} - \theta_{j0}\right) - \mathrm{Ci}\left(\varphi_j\right)\right\}\sin\left(\theta_{j0} - \varphi_j\right) \\ &\quad - \sin\left(\theta_{j0}\right)\ln\left(1 + \frac{\theta_{jf} - \theta_{j0}}{\varphi_j}\right), \end{aligned} \tag{44}$$

$$\begin{aligned} C_\theta&\left(\theta_{j0}, \theta_{jf}, \dot{\theta}_{j0}, \eta_j\right) \\ &= \frac{\eta_j}{J_{\mathrm{III}}}\int_{t_{j0}}^{t_{jf}} \cos(\theta)dt \\ &= \frac{\eta_j}{J_{\mathrm{III}}}\int_{t_{j0}}^{t_{jf}} \cos\left(\varphi_j \exp\left\{\frac{\eta_j}{J_{\mathrm{III}}}(t - t_{j0})\right\} + \theta_{j0} - \varphi_j\right)dt \\ &= \left\{\mathrm{Si}\left(\varphi_j + \theta_{jf} - \theta_{j0}\right) - \mathrm{Si}\left(\varphi_j\right)\right\}\sin\left(\theta_{j0} - \varphi_j\right) \\ &\quad + \left\{\mathrm{Ci}\left(\varphi_j + \theta_{jf} - \theta_{j0}\right) - \mathrm{Ci}\left(\varphi_j\right)\right\}\cos\left(\theta_{j0} - \varphi_j\right) \\ &\quad - \cos\left(\theta_{j0}\right)\ln\left(1 + \frac{\theta_{jf} - \theta_{j0}}{\varphi_j}\right), \end{aligned} \tag{45}$$

where $\mathrm{Si}(\cdot)$ and $\mathrm{Ci}(\cdot)$ denote the sine and cosine integral functions, respectively, which are defined as follows:

$$\begin{aligned} \mathrm{Si}(\theta) &= \int_0^\theta \frac{\sin(t)}{t}dt, \\ \mathrm{Ci}(\theta) &= -\int_\theta^\infty \frac{\cos(t)}{t}dt. \end{aligned} \tag{46}$$

The angular velocity $\dot{\theta}_{(j+1)0}$ at the beginning of the $(j+1)$-th maneuver in (43) is expressed in an algebraic form with η_j, θ_{j0}, and θ_{jf} as follows:

$$\dot{\theta}_{(j+1)0} = \dot{\theta}_{j0} + \int_{t_{j0}}^{t_{jf}} \frac{\eta_j}{J_{\mathrm{III}}}\frac{d\theta}{dt}dt = \dot{\theta}_{j0} + \frac{\eta_j}{J_{\mathrm{III}}}\left(\theta_{jf} - \theta_{j0}\right), \tag{47}$$

where $\Delta\hat{\mathbf{p}}_j$ and $\Delta\dot{\hat{\mathbf{p}}}_j$ are determined by η_j, θ_{j0}, and θ_{jf}. In this study, three control sequences ($j = 1 - 3$) are used to design the history of θ. The schematic image of the three sequences is shown in Figure 5. The details to specify the control parameters in each sequence are as follows:

The first sequence causes the velocities on the X–Y plane to correspond to zero. With respect to this control, η_1, θ_{10}, and θ_{1f} are determined from (32) such that the following equation is satisfied:

$$\frac{\eta_1}{M}\begin{bmatrix} \sin\left(\theta_{1f}\right) - \sin\left(\theta_{10}\right) & \cos\left(\theta_{1f}\right) - \cos\left(\theta_{10}\right) \\ -\cos\left(\theta_{1f}\right) + \cos\left(\theta_{10}\right) & \sin\left(\theta_{1f}\right) - \sin\left(\theta_{10}\right) \end{bmatrix}\begin{bmatrix} \gamma_1 \\ \gamma_2 \end{bmatrix} = -\dot{\hat{\mathbf{p}}}_{10}, \tag{48}$$

where θ_{10} and θ_{1f} are specified based on the direction of $\dot{\hat{\mathbf{p}}}_{10}$ and η_1 is determined from the norm of $\dot{\hat{\mathbf{p}}}_{10}$. Following the maneuver, the FFRB is in rotational drift motion with $T_{\mathrm{III}} = 0$ until the second sequence begins.

Next, the combination of the second and the third sequences achieves $\hat{\mathbf{p}} = 0$. In the second sequence, with respect to θ_{2f} satisfying $\theta_{2f} - \theta_{20} = -2\pi$, (32) indicates that $\Delta\dot{\hat{\mathbf{p}}}_2$ corresponds to zero (thus, $\dot{\hat{\mathbf{p}}}_{30}$ also corresponds to zero). Subsequently, given that $\dot{\hat{\mathbf{p}}}_{20} = \mathbf{0}$, $\Delta\hat{\mathbf{p}}_2$ is expressed as follows:

$$\Delta\hat{\mathbf{p}}_2 = \left|\Delta\hat{\mathbf{p}}_2\right|\bar{\mathbf{A}}\left(\theta_{20} + \phi\left(2\pi, 0, \dot{\theta}_{20}, \eta_2\right)\right)[1 \quad 0]^T, \tag{49}$$

where

$$\left|\Delta\hat{\mathbf{p}}_2\right| = \frac{J_{\mathrm{III}}}{M}\sqrt{\left(\gamma_1^2 + \gamma_2^2\right)\left\{S_\theta^2\left(2\pi, 0, \dot{\theta}_{20}, \eta_2\right) + C_\theta^2\left(2\pi, 0, \dot{\theta}_{20}, \eta_2\right)\right\}}, \tag{50}$$

$$\phi\left(\theta_{j0}, \theta_{jf}, \dot{\theta}_{j0}, \eta_j\right) = \tan^{-1}\left(\frac{S_\theta\left(\theta_{j0}, \theta_{jf}, \dot{\theta}_{j0}, \eta_j\right)}{C_\theta\left(\theta_{j0}, \theta_{jf}, \dot{\theta}_{j0}, \eta_j\right)}\right), \tag{51}$$

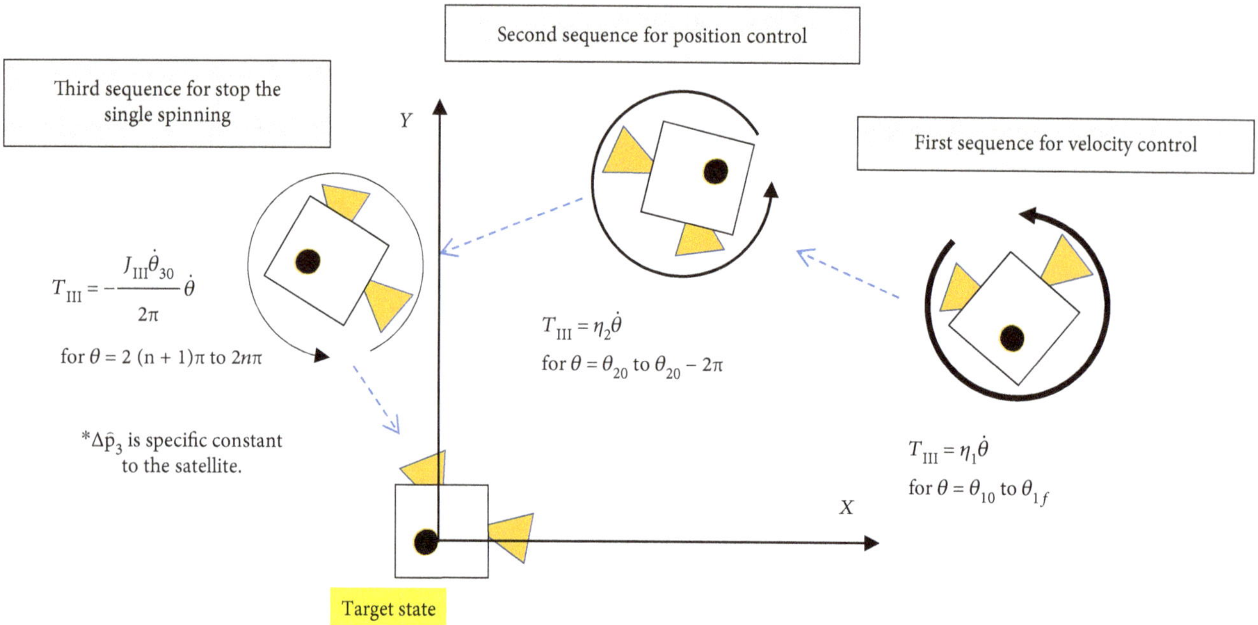

FIGURE 5: Schematic images of motion planning with the three sequences.

where $\phi(\theta_{j0}, \theta_{jf}, \dot{\theta}_{j0}, \eta_j)$ denotes the direction of the position variation. From (44), (45), (50), and (51), the value of η_2 affects the norm of $\Delta\widehat{\mathbf{p}}_2$ and the value of θ_{20} affects the direction of $\Delta\widehat{\mathbf{p}}_2$. Prior to detailing the procedure to determine η_2 and θ_{20}, we explain the third sequence because the value of $\Delta\widehat{\mathbf{p}}_3$ is required to determine η_2 and θ_{20}. The third sequence begins when the drift motion brings the FFRB towards $Z = 0$ within a full rotation of θ. In the third sequence, the interval $\theta_{3f} - \theta_{30}$ is set as -2π for $\Delta\dot{\widehat{\mathbf{p}}}_3 = 0$ and θ_{3f} must correspond to $2(n+1)\pi$ to achieve the target attitude when the single spinning is stopped. Thus, η_3 is specified from (47) as follows:

$$\eta_3 = \frac{J_{\mathrm{III}}\dot{\theta}_{30}}{2\pi}, \tag{52}$$

and $\Delta\widehat{\mathbf{p}}_3$ is expressed as follows:

$$\Delta\widehat{\mathbf{p}}_3 = \frac{J_{\mathrm{III}}}{M}\sqrt{\left(\gamma_1^2 + \gamma_2^2\right)\left\{S_\theta^2\left(2\pi, 0, \dot{\theta}_{30}, \eta_3\right) + C_\theta^2\left(2\pi, 0, \dot{\theta}_{30}, \eta_3\right)\right\}}\,\bar{\mathbf{A}} \cdot \left(\phi\left(2\pi, 0, \dot{\theta}_{30}, \eta_3\right)\right)\left[1 \quad 0\right]^T. \tag{53}$$

Thus, substituting (52) into (43) derives $\varphi_3 = -2\pi$. This indicates that $S_\theta(2\pi, 0, \dot{\theta}_{30}, \eta_3)$, $C_\theta(2\pi, 0, \dot{\theta}_{30}, \eta_3)$, and $\phi(2\pi, 0, \dot{\theta}_{30}, \eta_3)$ in (53) are not dependent on η_3 and $\dot{\theta}_{30}$, and thus, $\Delta\widehat{\mathbf{p}}_3$ is a constant vector specified by a FFRB model and not by control maneuvers. Therefore, η_2 and θ_{20} are determined as follows. In order to obtain $\widehat{\mathbf{p}} = \mathbf{0}$, $\Delta\widehat{\mathbf{p}}_2 + \Delta\widehat{\mathbf{p}}_3$ must correspond to $-\widehat{\mathbf{p}}_{20}$. Subsequently, $\Delta\widehat{\mathbf{p}}_3$ is a constant vector, and thus, $\Delta\widehat{\mathbf{p}}_2$ is adjusted to satisfy $\Delta\widehat{\mathbf{p}}_2 + \Delta\widehat{\mathbf{p}}_3 = -\widehat{\mathbf{p}}_{20}$. Consequently, η_2 is numerically calculated from (50), and (49) and (51) then yield the value of θ_{20} for the required $\Delta\widehat{\mathbf{p}}_2$.

3.6. Final Form of the Controller with a Supplemental Sequence. It should be noted that $|\Delta\widehat{\mathbf{p}}_2|$ has an upper limit that depends on $\dot{\theta}_{20}$. When a required $|\Delta\widehat{\mathbf{p}}_2|$ exceeds the upper limit, Step D does not achieve the target state. In order to avoid such a situation, minimizing $\widehat{\mathbf{p}} + \Delta\widehat{\mathbf{p}}_3$ is mandatory prior to Step D.

An idea for this minimization involves adjusting the timing when a sequence of the intentional disturbance and Step A ($S_1 = 0 \rightarrow S_1 \neq 0 \rightarrow S_1 = 0$) begins. As shown above, the attitude angle θ does not affect the controls of Steps A, B, and C. Thus, we switch on the control sequences while numerically predicting $\widehat{\mathbf{p}}$ and $\widehat{\mathbf{p}}$ to minimize the following V:

$$V = \begin{cases} \dot{\mathbf{P}}\left(\mathbf{P}^T + \Delta\widehat{\mathbf{p}}_3\right)^T & \text{for } V' = \left|\mathbf{P}^T + \Delta\widehat{\mathbf{p}}_3\right|^2 \geq L, \\ \left|\dot{\mathbf{P}}\right|^2 & \text{for } V' = \left|\mathbf{P}^T + \Delta\widehat{\mathbf{p}}_3\right|^2 < L, \end{cases} \tag{54}$$

where $\mathbf{P}$ and $\dot{\mathbf{P}}$ denote the positions and velocities, respectively, immediately after each sequence $S_1 = 0 \rightarrow S_1 \neq 0 \rightarrow S_1 = 0$ and V' denotes the squared distance between $\widehat{\mathbf{p}} = -\Delta\widehat{\mathbf{p}}_3$ and $\widehat{\mathbf{p}} = \mathbf{P}$. Numerical calculations are performed during the drift motions. The value L was heuristically determined given the following consideration. With respect to $V' \geq L$, V denotes the time derivative of $|\mathbf{P} + \Delta\widehat{\mathbf{p}}_3|^2$ after each sequence. Thus, it implies that the maneuver begins at the attitude angle that minimizes the velocity vector for $\widehat{\mathbf{p}} = -\Delta\widehat{\mathbf{p}}_3$. With respect to $V' < L$, the maneuver begins at

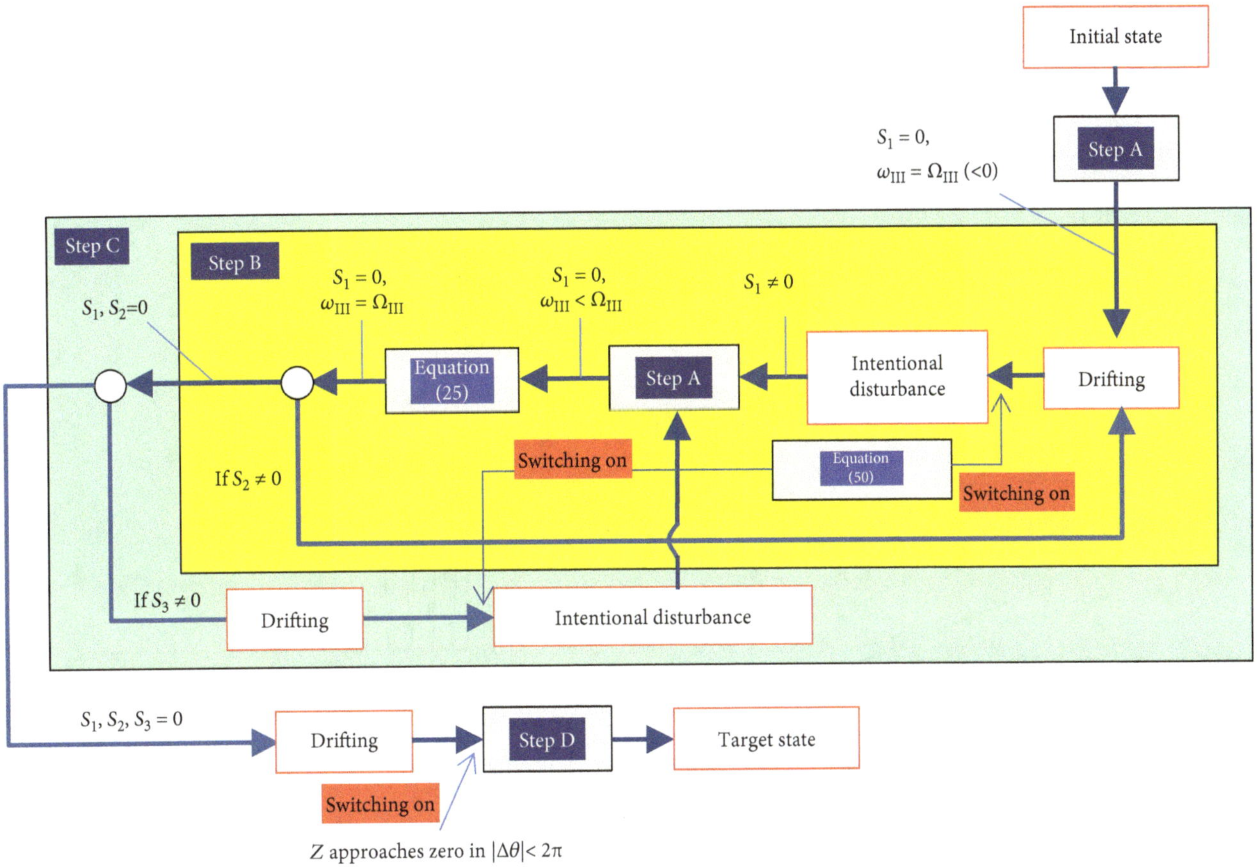

FIGURE 6: Diagram of the final control form.

TABLE 1: Initial state values for the numerical simulations.

	q_I	q_{II}	q_{III}	q_{IV}	X (m)	Y (m)	Z (m)
Case 1	0.4	−0.25	0.1	0.876	−200	700	−350
Case 2	0.8	−0.25	0.1	0.536	−200	700	−350
Case 3	0.4	−0.25	0.1	0.876	0.0	0.0	−500

the attitude angle minimizing the norm of $\dot{\widehat{\mathbf{p}}}$. For reference, we also attempted to minimize $|\mathbf{P} + \Delta\widehat{\mathbf{p}}_3|^2 + |\dot{\mathbf{P}}|^2$, which is a unified form of (54).

We emphasize that this supplement sequence is not always successful. Specifically, L requires severe tuning through trial and error in numerical simulations. This is the weak point in our proposed controller, and it is important to solve this problem in a future study.

The diagram of these sequences is shown as the "final control form" in Figure 6 as a summary of all the control sequences.

4. Example of the Proposed Controller with Numerical Simulation

This section describes the numerical validation of the control method shown in Section 3. The FFRB model and the controllers used in the simulations are summarized below, and the results of the numerical simulations are then shown.

4.1. FFRB Model and Initial State Value. According to the basic assumption of the study, the FFRB is assumed as with a constant mass, inertia, and no-upper-limitation thruster forces. In this example, the inertial moments of the FFRB are assumed as $J_I = 15$ kg·m^2, $J_{II} = 10$ kg·m^2, and $J_{III} = 20$ kg·m^2. Furthermore, the FFRB mass M is 20 kg. The placements and directions of the thrusters are set as follows:

$$\text{Thruster 1: } \mathbf{r}_1 = [0.0\ 1.0 - 1.0]^T, \mathbf{d}_1 = [0.0\ 0.0 - 1.0]^T$$

$$\text{Thruster 2: } \mathbf{r}_2 = [-1.0\ 0.0 - 1.0]^T, \mathbf{d}_2 = [0.0\ 0.0 - 1.0]^T$$

$$\text{Thruster 3: } \mathbf{r}_3 = [0.01 - 0.01 - 1.3]^T, \quad \mathbf{d}_3 = [0.43 - 0.25\ 0.87]^T$$

The forces of these thrusters are only positive, and their magnitudes are assumed as continuously changeable and without any upper limits. It should be noted again that the thruster configuration satisfies the conditions shown in Section 2.1. These model parameters result in $\Delta\widehat{\mathbf{p}}_3 = [668 \quad 383]^T$ for Step D.

Table 1 shows the initial state for the three cases used in simulations. Only the initial attitude is different between cases 1 and 2, and the initial position is different between

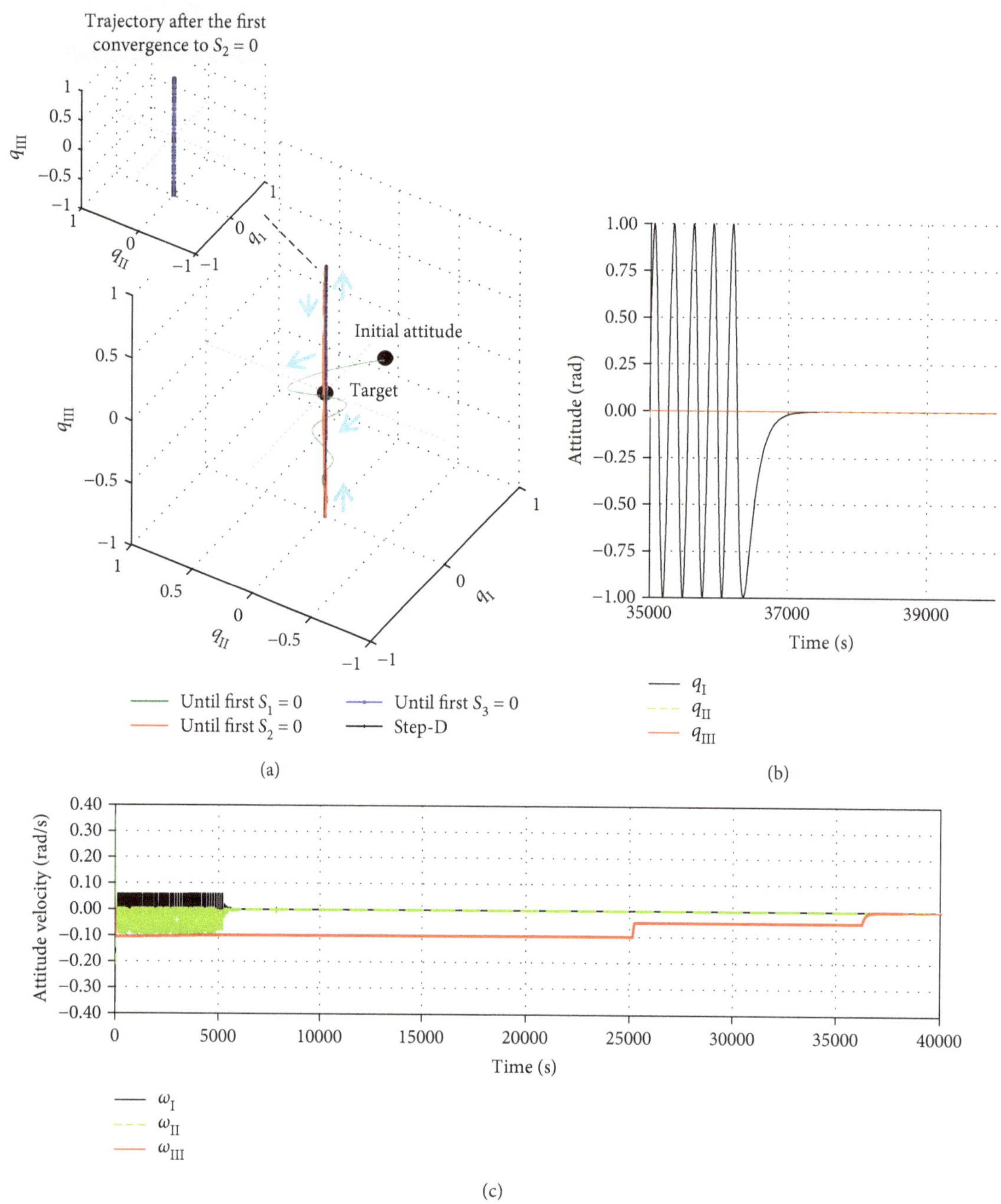

FIGURE 7: Attitude in case 1. (a) Attitude trajectory to the target. (b) Attitude history around the target. (c) Time history of attitude velocity.

cases 1 and 3. The angular and translational velocities are zero in all cases.

4.2. Controller

Step A. The parameters for the controller in (13) and (14) are set as $K_1 = 3.0$, $K_2 = 3.0$, and $\kappa = 1.0$, respectively.

Step B. The parameters for disturbance law in (24), (25), and (26) are set as $D_{Ib} = -0.3$, $D_{IIb} = 0.3$, and $D_{S2} = 0.5$, respectively. Saturation is applied as $t_{S2} = 0.1$ s for $|S_2| > 0.1$. Additionally, K_{III} in (27) is set to 30.

Step C. The parameters for the disturbance law in (36), (37), and (38) are set as $D_{Ic} = 0.1$, $D_{IIc} = -0.1$, and $D_{S3} = 20$, respectively. Saturation is applied as $t_{S3} = 0.1$ s for $|S_3| > 0.02$.

Step D. As shown in the next subsection for simulations, the maneuver minimizing V that results from switch timing adjustments by using (54) leads to $\dot{\hat{\mathbf{p}}} = 0$ after Step C in all the three cases. Thus, the first sequence to achieve zero velocities along the X- and Y-directions is omitted. The values for θ_{20} and η_2 specified are shown in the following subsection.

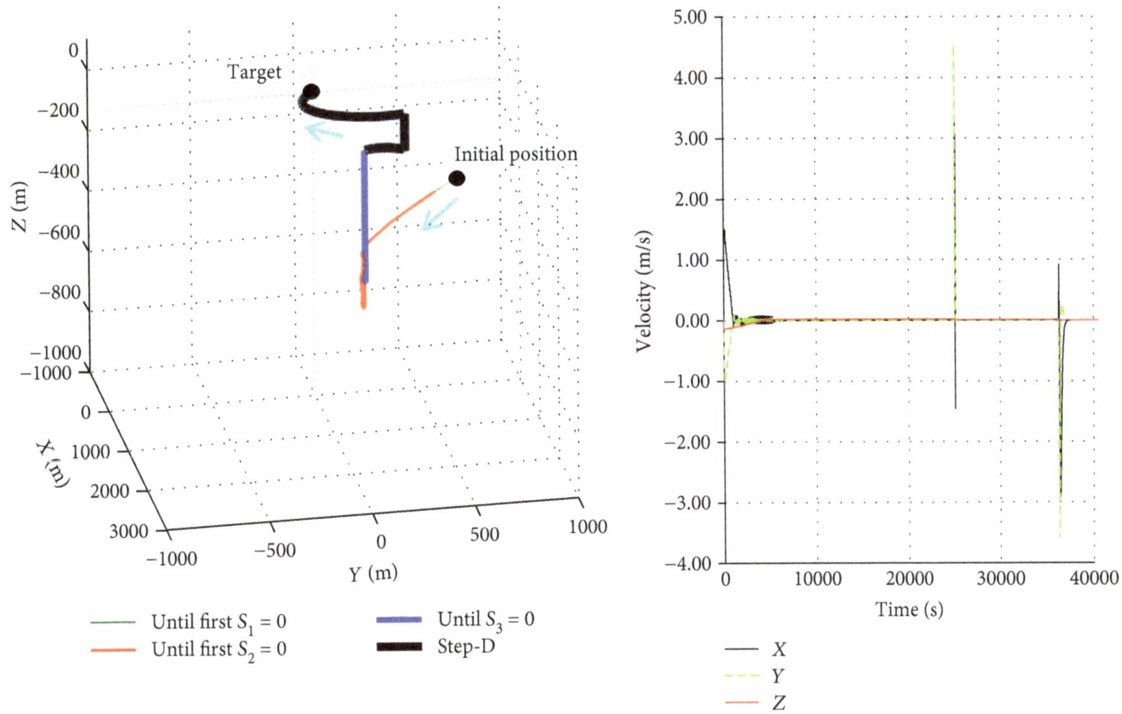

(a) (b)

FIGURE 8: Position in case 1. (a) Position trajectory to the target. (b) Velocity histories.

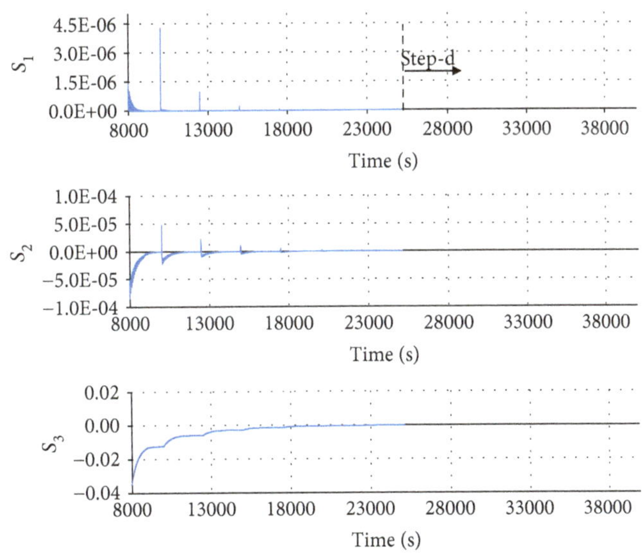

FIGURE 9: Time history of the three invariant manifolds after first achieving $S_2 = 0$ in case 1.

Supplemental sequence using (54): L in (54) is set to 200,000.

4.3. Simulation Results. Figures 7–10 show the results for case 1. As shown in Figures 7(a) and 7(b), the attitude successfully achieved the target state. Figure 7(a) indicates that the single spinning was dominant after initially achieving $S_1 = 0$. Figure 7(c) shows that the attitude velocities were also controlled to zero. Figure 8 verifies the success of the control of the positions. Figure 8 shows the time histories of S_1, S_2, and S_3 after initially achieving $S_2 = 0$. They confirm that the repetitions of the sequences, disturbances, and controls for $S_1 = 0$ and $S_2 = 0$ cause S_3 to quasi-asymptotically converge to zero. The thruster forces maintained positive values as shown in Figure 10, and the values of θ_{20} and η_2 were set as 3.27 rad and -0.174, respectively.

Cases 2 and 3 also show the success of the controls as illustrated in Figures 11–14. A significantly long period is involved in the process in case 2 owing to the low velocity in the Z-direction. This problem is caused by the small $\dot{Z}(\approx\Omega_{\mathrm{III}})$, which lengthens the drifting time for approaching

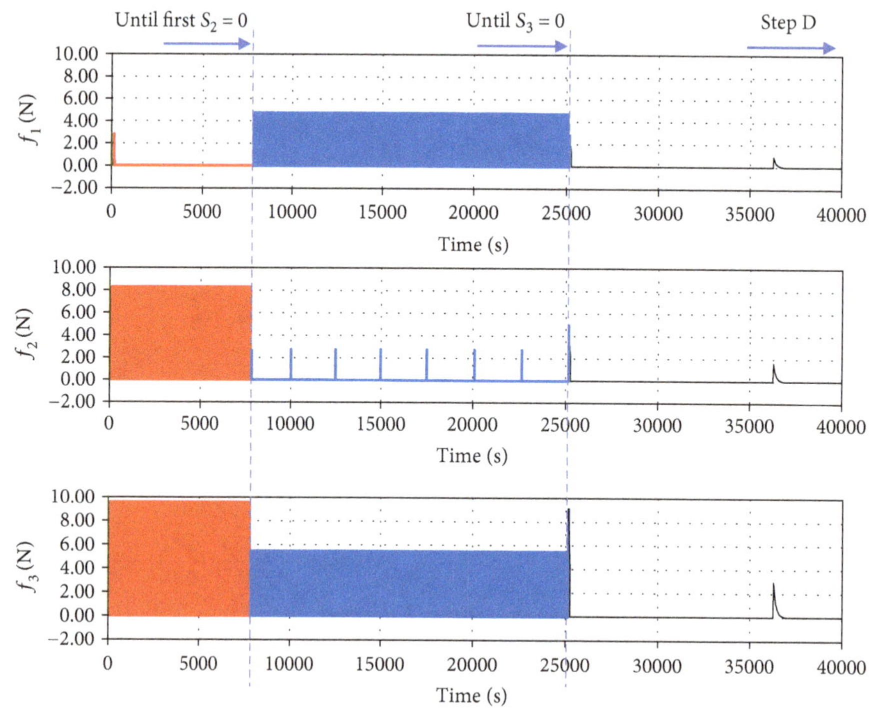

FIGURE 10: Time history of thruster forces in case 1. Thruster 1 generates f_1, thruster 2 generates f_2, and thruster 3 generates f_3.

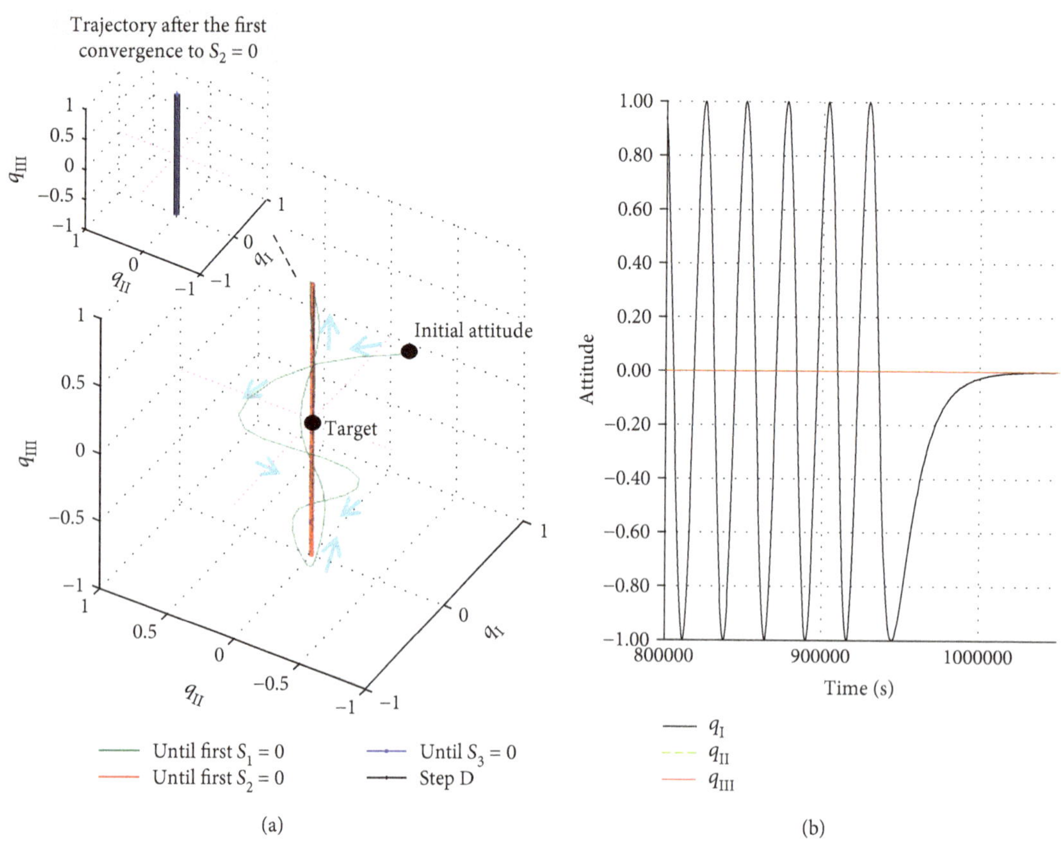

FIGURE 11: Attitude in case 2. (a) Attitude trajectory to the target. (b) Attitude history around the target.

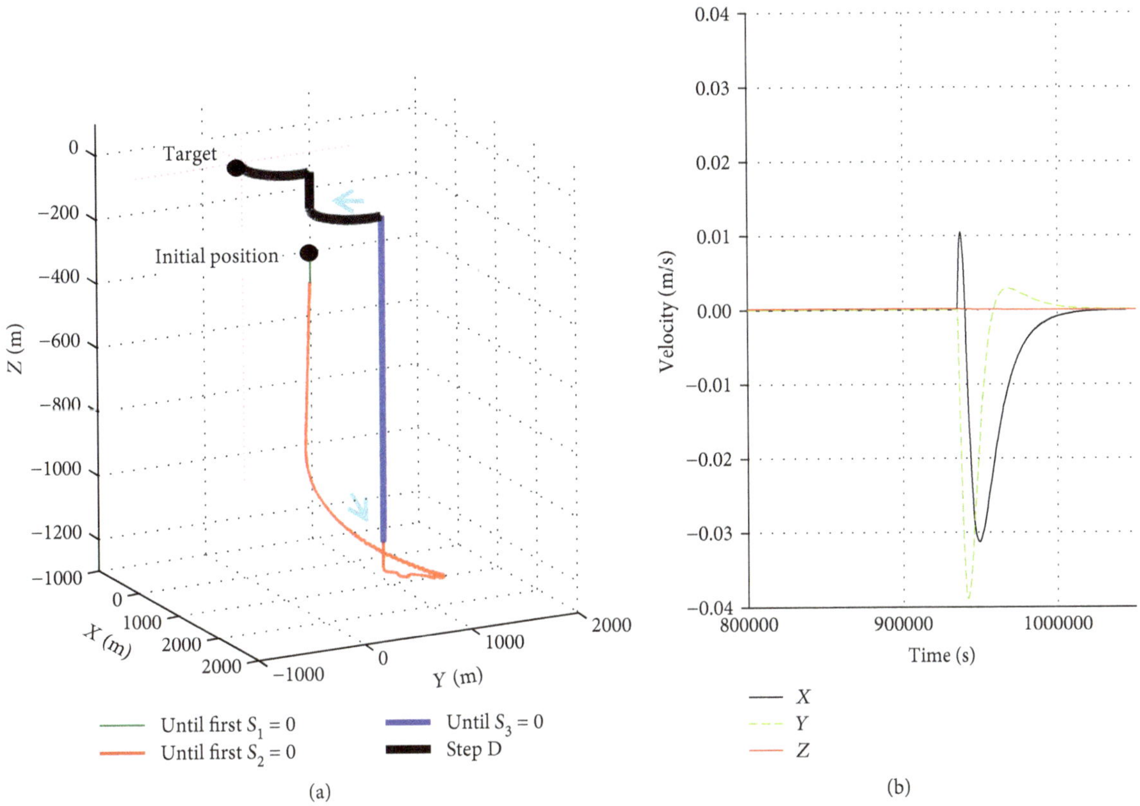

FIGURE 12: Position in case 2. (a) Position trajectory to the target. (b) Velocity histories around the target.

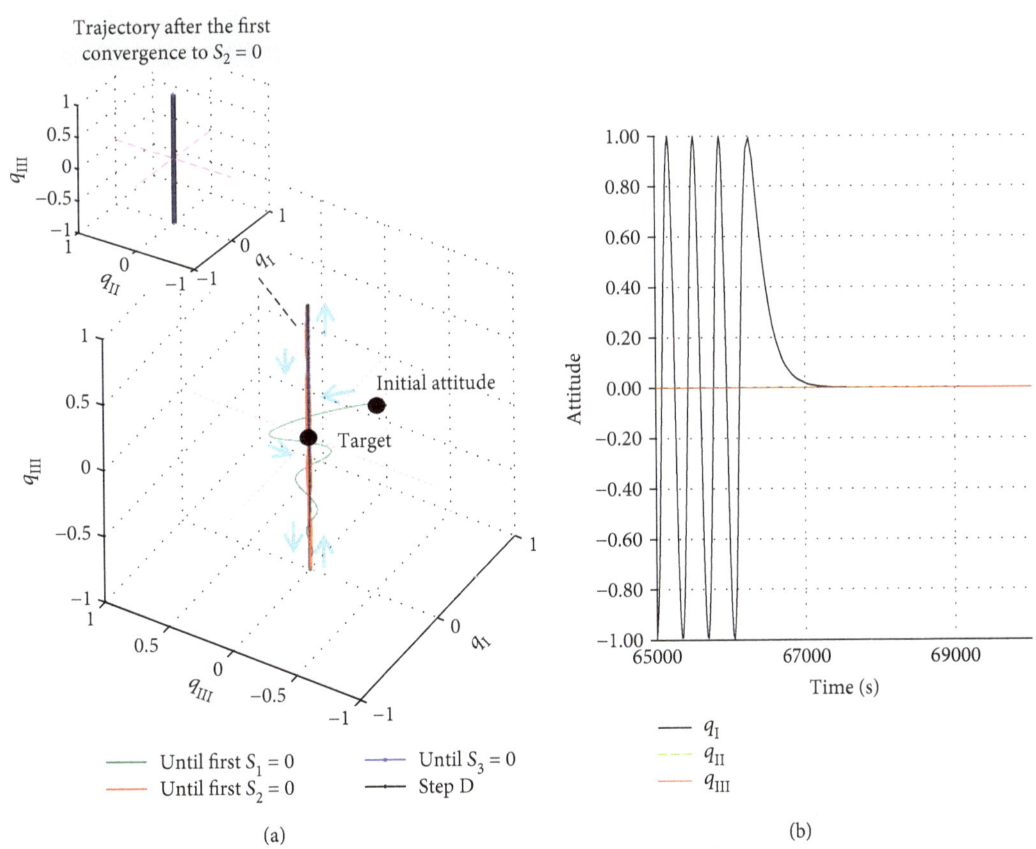

FIGURE 13: Attitude in case 3. (a) Attitude trajectory to the target. (b) Attitude history around the target.

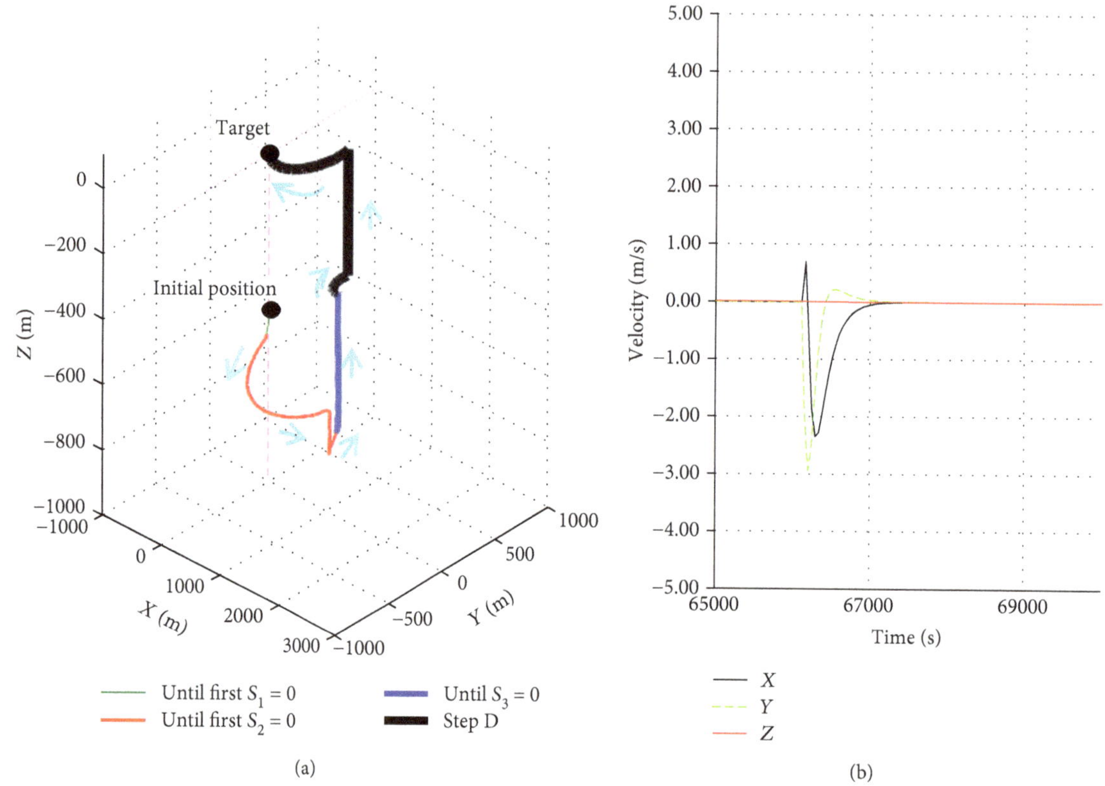

FIGURE 14: Position in case 3. (a) Position trajectory to the target. (b) Velocity histories around the target.

the target position, after the first Step A for $S_1 = 0$. The problem can be solved by adding acceleration step for Ω_{III} or tuning the parameters for Step A based on the initial states. The thruster forces maintained positive values in both cases 2 and 3 (the figures are omitted here). The values of θ_{20} and η_2 were set as 6.36 rad and −1.085, respectively, in case 2 and 4.55 rad and −0.179, respectively, in case 3.

5. Summary

This study presented a control method for three-dimensional attitude and position control of a free-floating rigid body using three thruster forces in which the directions are fixed to the body. First, the dynamics of the rigid body were described along with a thruster configuration to generate independent torques for two principal axes. The derived equations of motion for the rotation and translation confirmed that the equations were coupled. Next, a combination of switching controllers and motion planning was proposed. The switching controllers are composed of three hierarchical steps, and three invariant manifolds are used as switching surfaces. A process involving the application of an intentional disturbance and returning the state to on the lower class invariant manifolds is applied, and the states in upper class are individually controlled. In this control scheme, the magnitude of the intentional disturbance is tuned based on the distance to the invariant manifold. Motion planning in the final phase finalizes the state values to target values. The numerical simulation results verified the effectiveness of the proposed methods. In the simulations, the target attitudes and positions were successfully achieved in three cases with different initial states.

Although the limitations of this control method were not examined in detail, we first proposed a solution to this challenging problem. We believe that the methodology shown in this paper will contribute a practical satellite control in an underactuated situation due to thruster malfunction. A discussion of the generality of the method and overcoming the limitation will be addressed in the next phase of a future study.

Nomenclature

$\mathbf{a} = [a_{\text{I}}\ a_{\text{II}}\ a_{\text{III}}]^T$:	Rotational axis with respect to the inertial frame
$\mathbf{A}(\mathbf{q})$:	Coordinate transformation matrix from body coordinates to the inertial frame
$\bar{\mathbf{A}}(\theta)$:	Rotation matrix in the X-Y plane
$\mathbf{d}_i$:	Direction of the ith thruster ($i = 1 - 3$) with respect to the principal inertia axes of the free-floating rigid body (FFRB)
f_i:	Magnitude of force generated by the ith thruster ($i = 1 - 3$)
$\mathbf{F}$:	Total force vector generated by the three thrusters with respect to the inertial frame

$\mathbf{F}'$: Force vector generated by the three thrusters with respect to the body coordinates

$\mathbf{F}_i'$: Force vector generated by the i th thruster with respect to the body coordinates

$J_{\mathrm{I}}, J_{\mathrm{II}}, J_{\mathrm{III}}$: FFRB's principal moment of inertia

M: Mass of the FFRB

m, n: Arbitrary integer

$\mathbf{p} = [X \quad Y \quad Z]^T$: Position vector of the FFRB

$\widehat{\mathbf{p}} = [X \quad Y]^T$: Position vector of the FFRB on the X-Y plane

$\Delta\widehat{\mathbf{p}}_j, \Delta\dot{\widehat{\mathbf{p}}}_j$: Variations of $\widehat{\mathbf{p}}$ and $\dot{\widehat{\mathbf{p}}}$ caused by the jth sequence in Step D

$\mathbf{q} = [q_{\mathrm{I}}\, q_{\mathrm{II}}\, q_{\mathrm{III}}\, q_{\mathrm{IV}}]^T$: Euler parameter for attitude definition with respect to the inertial frame

$\mathbf{r}_i$: Placement of the ith thruster ($i = 1-3$) with respect to the principal inertia axes of the FFRB

Δt_j: Time duration for the maneuver in the jth sequence of Step D

$\mathbf{T} = [T_{\mathrm{I}}\, T_{\mathrm{II}}\, T_{\mathrm{III}}]^T$: Total torque vector generated by the thrusters around the principal inertia axes of the FFRB

$\mathbf{T}_i$: Torque vector generated by the i th thruster ($i = 1-3$) around the principal inertia axes of the FFRB

ψ: Rotational angle with respect to $[a_{\mathrm{I}}\, a_{\mathrm{II}}\, a_{\mathrm{III}}]^T$

θ: Attitude angle with respect to axis III during the single spin

τ_d: Drift time before a stopping maneuver for the single spin motion

$\boldsymbol{\omega} = [\omega_{\mathrm{I}} \quad \omega_{\mathrm{II}} \quad \omega_{\mathrm{III}}]^T$: Angular velocity vector with respect to the principal inertia axes of the FFRB

η_j, φ_j: Motion planning parameter in the jth sequence of Step D

sgn (): Signum function

MOD(a, b): Remainder of a/b

Subscription "$j0$" and "jf": Initial and final values of the jth sequence in Step D.

Conflicts of Interest

The authors declare that they have no conflicts of interest.

References

[1] R. W. Brockett, R. S. Milman, and H. J. Sussmann, Eds., "Asymptotic stability and feedback stabilization," in *Differential Geometric Control Theory*, pp. 181–191, Birkhauser, Boston, USA, 1983.

[2] A. Isidori, *Nonlinear Control Systems*, Springer, 3rd edition, 1995.

[3] S. Hokamoto and M. Iwase, "Position and attitude control of a planar satellite by two thrusters," in *AIAA Guidance, Navigation, and Control Conference*, pp. 4206–4211, Chicago, USA, August 2009.

[4] P. Tsiotras and J. M. Longuski, "A new parameterization of the attitude kinematics," *The Journal of the Astronautical Sciences*, vol. 43, no. 3, pp. 243–262, 1995.

[5] P. Tsiotras and V. Doumtchenko, "Control of spacecraft subject to actuator failures state-of-the-art and open problems," *Journal of Astronautical Sciences*, vol. 48, no. 2, pp. 337–358, 2000.

[6] P. Morin and C. Samson, "Time-varying exponential stabilization of a rigid spacecraft with two control torques," *IEEE Transactions on Automatic Control*, vol. 42, no. 4, pp. 528–534, 1997.

[7] S. Kasai, H. Kojima, and M. Satoh, "Spacecraft attitude maneuver using two single-gimbal control moment gyros," *Acta Astronautica*, vol. 84, pp. 88–98, 2013.

[8] A. Zavoli, G. De Matteis, F. Giulietti, and G. Avanzini, "Single-axis pointing of an underactuated spacecraft equipped with two reaction wheels," *Journal of Guidance, Control, and Dynamics*, vol. 40, no. 6, pp. 1465–1471, 2017.

[9] Y. Geng, D. Song, and R. Sun, "Inverse optimal stabilization of an underactuated spacecraft using two wheels," in *2016 8th International Conference on Intelligent Human-Machine Systems and Cybernetics (IHMSC)*, pp. 279–282, Hangzhou, China, August 2016.

[10] M. J. Sidi, *Spacecraft Dynamics & Control: A Practical Engineering Approach*, vol. 9, Cambridge University Press, 1997.

[11] T. Matsuno, Y. Yoshimura, and S. Hokamoto, "Geometric conditions of thrusters for 3D attitude control of a free-floating rigid spacecraft," *Mathematics in Engineering, Science & Aerospace (MESA)*, vol. 5, no. 1, pp. 83–95, 2014.

[12] Y. Yoshimura, T. Matsuno, and S. Hokamoto, "Three dimensional attitude control of an underactuated satellite with thrusters," *International Journal of Automation Technology*, vol. 5, no. 6, pp. 892–899, 2011.

[13] H. Khennouf and C. Canudas De Wit, "On the construction of stabilizing discontinuous controllers for nonholonomic systems," *IFAC Proceedings Volumes*, vol. 28, no. 14, pp. 667–672, 1995.

[14] R. Mukherjee and M. Kamon, "Almost smooth time-invariant control of planar space multibody systems," *IEEE Transactions on Robotics and Automation*, vol. 15, no. 2, pp. 268–280, 1999.

[15] T. Matsuno and S. Hokamoto, "Position and attitude feedback controller of a planar satellite with two thrusters," in *ICROS-SICE International Joint Conference*, pp. 4206–4211, Fukuoka, Japan, August 2009.

[16] T. Matsuno and S. Hokamoto, "Position and attitude control of a planar satellite by two constant force thrusters," in *AIAA/AAS Astrodynamics Specialist Conference*, Toronto, Ontario, Canada, August 2010.

[17] Y. Yoshimura, T. Matsuno, and S. Hokamoto, "Global trajectory design for position and attitude control of an underactuated satellite," *Transactions of the Japan Society for Aeronautical and Space Sciences*, vol. 59, no. 3, pp. 107–114, 2016.

[18] H. J. Sussmann, "A general theorem on local controllability," *SIAM Journal on Control and Optimization*, vol. 25, no. 1, pp. 158–194, 1987.

[19] B. Goodwine and J. Burdick, "Controllability with unilateral control inputs," in *Proceedings of 35th IEEE Conference on Decision and Control*, pp. 3394–3399, Kobe, Japan, December 1996.

15

Advanced Software Ground Station and UAV Development for NLoS Control Using Mobile Communications

Amr AbdElHamid,[1] Peng Zong,[2] and Bassant Abdelhamid[3]

[1] *College of Electronic and Information Engineering, Nanjing University of Aeronautics and Astronautics (NUAA), Nanjing 210016, Jiangsu, China*
[2] *Astronautics College, Nanjing University of Aeronautics and Astronautics (NUAA), Nanjing 210016, Jiangsu, China*
[3] *Electronics and Communications Department, Faculty of Engineering, Ain Shams University (ASU), Cairo 11517, Egypt*

Correspondence should be addressed to Amr AbdElHamid; amr@nuaa.edu.cn

Academic Editor: Francisco R. Villatoro

Over the last decades, Unmanned Aerial Systems (UASs) have gained much attention due to their various applications in different sections. However, their communication range is limited to utilized communication equipment. Therefore, utilization of GSM channels opens a new prospect towards long distance UAV missions and mobile command and control centers. This paper demonstrates new design and development of a small-scale UAV and a Ground Control Station (GCS) using GSM bidirectional communications for Non-Line of Sight (NLoS) long range control. GCSs are considered the front end node in UAV guidance process. Therefore, the proposed GCS employs a two-layer framework to consider all ground pilot requirements. Moreover, a new exploitation of global weather forecast data is added to the GCS. On the other hand, the proposed airborne system utilizes a new integration of different Commercial off-the-Shelf (COTS) components and excludes short range receivers. The ground and flight tests show that stable bidirectional GSM communication is established, reliable hardware integration is accomplished, real time performance is achieved, GCS functional fidelity is obtained, and low cost is maintained. Finally, some qualitative aspects of the proposed platform are presented to address the detailed features.

1. Introduction

Recently, Unmanned Aerial Vehicles (UAVs) have a great potential in Research and Development (R&D) sections, due to their vital role in academic, industrial, and military applications [1]. Their applications domains cover search and rescue, reconnaissance, intelligent traffic monitoring, border patrol, disaster monitoring, and fire detection (just to name a few) [2]. Small-scale UAV copters, such as quadcopters and helicopters, have attracted many researchers for their unique characteristics. Quadcopters have Vertical Takeoff and Landing (VTOL) capability and perform different flight missions ranging from hovering to cruising with quite good agility and maneuverability. It is considered an ideal platform for nonlinear control law research and a good test bench for trajectory planning algorithms for both remotely piloted and autonomous missions. Designing and implementing a small-scale UAV quadcopter is considered a challenging job [3], due to its role in reducing the gap between the theoretical studies and practical engineering methodologies. During the implementation, researchers might get stuck with many difficulties such as UAV platform specification, airborne systems, communication links, Ground Control Station (GCS) software capabilities [4], and the multiplexed data logging of different inputs and outputs for the Unmanned Aerial System (UAS). Owing to the advent of Microelectromechanical System (MEMS) technology, the airborne systems are shrinking in size and weight. Moreover, using the Commercial off-the-Shelf (COTS) components, it is possible to develop airborne avionics system for a small-scale UAV quadcopter with attributes of low cost, small size, light weight, and high performance. Moreover, Global System for Mobile (GSM) communications provides a remarkable connectivity between two mobile stations (MS) for voice or data, which represents a good infrastructure for different long range and Non-Line of Sight (NLoS) applications.

The main objectives of this paper are as follows:

(1) Develop an advanced GCS and a small-scale quadcopter using bidirectional terrestrial mobile networks and COTS components. The proposed scheme exploits GSM networks to provide a long NLoS control range, and COTS components maintain the low budget.

(2) Introduce a new vision of exploiting nonlimited range weather forecast data for different GCS operational procedures. This new feature is utilized as a replica for traditional weather equipment [5].

(3) Provide a large number of Pulse Width Modulation (PWM) control signals for payload control.

(4) Demonstrate a new centralized high accuracy online data logging algorithm.

Towards fulfilling these objectives, bidirectional cellular network communication using Point-to-Point Protocol (PPP) is introduced for exchanging the GCS flight commands and receiving the UAV feedback (surveillance data). The proposed scheme exploits VB.NET as a development language for the GCS software. The GCS is designed based on modular concept. The modular design is not only to simplify the future modifications [6] but also to minimize the budget [7]. The GCS software architecture employs two-layer framework [8]. The foreground layer contains an innovative Graphical User Interface (GUI) with geographic maps, mission planning, functional set of gauges, and state views. Additionally, a new weather forecast module is developed to support the mission designer decisions during the mission planning and the real flight. The background layer handles communication interfaces, web resources, and the data logging. The new data logging module is developed not only to record whole platform parameters, but also to measure the processing load of different software modules with a very high resolution up to 10^{-7} seconds. The GCS development utilizes many advanced programming techniques like Application Programming Interface (API), threads, Component Object Model (COM) technology, datasets, stopwatch, and many other components. On the other hand, COTS components are utilized to compose the new airborne system (avionics systems). The traditional short range RC receivers are replaced with COTS PWM controller [9, 10]. This integration facilitates controlling up to 32 PWM channels for different effective payloads. Moreover, an advanced Attitude Heading Reference System (AHRS) integrated with Global Positioning System (GPS) module is employed to deliver the UAV flight parameters to the GCS via the GSM channel.

The test procedures cover many characteristics such as (1) GSM communication latency using communication worst-case scenarios; (2) processing delays for different input/output (I/O) interface modules; (3) ground tests of the proposed avionics system using bidirectional GSM communication; and (4) flight test for whole platform for performance evaluation. Finally, some qualitative aspects of the proposed scheme and other similar systems are demonstrated. Results show that the developed GCS software achieves a NLoS bidirectional control for the UAV using PPP GSM communications at 9600 bps (as a communication worst-case scenario) with a delay less than 300 ms, the software processing delay is accurately measured and satisfies the system constraints, and the ground/flight tests validate the whole developed systems.

Many difficulties and challenges have been conquered toward fulfilling the mentioned objectives: (1) integration of different COTS components in a compatible scheme; (2) integration of real weather forecast data with the GCS platform; (3) instantaneous handling of multiple inputs and outputs interfaces by different software modules simultaneously; (4) high resolution online data logging module which is in order of 10^{-7} seconds; (5) preparation and evaluation procedures for the ground and flight tests; and (6) exploitation of GSM in PPP for exchanging bidirectional real time data.

Unlike [11–19], the proposed scheme (1) achieves a bidirectional GSM communication for both uplink and downlink with latency less than 300 ms, (2) opens new prospect towards controlling UAV from a moving GCS, (3) provides long range NLoS control for aerial vehicle, (4) provides a large number of controlled PWM channels (up to 32 channels) for more payload control, (5) obtains advanced capabilities via integrating real weather forecast data, (6) consider worst communication scenarios (low baud rates) for bidirectional communication channel, (7) provides online recording with high resolution, and (8) maintains high functional fidelity and low budget.

The rest of this paper is organized as follows: Section 2 presents the related work to the proposed system, Section 3 demonstrates the developed UAV description, Section 4 illustrates the control and communication scheme and ground station development, Section 5 comprehensively presents the conducted tests and results, Section 6 addresses some qualitative aspects of the proposed platform, and, finally, the conclusions are presented in Section 7.

2. Related Work

Software engineering plays an active role in the growth picture of UAV development due to its simple maintenance procedures and large Mean Time Between Failures (MTBF). In [11], a complete road surveillance system consisting of a UAV and over 30 Unattended Ground Sensors (UGSs) was designed and simulated. The system was also flight tested in a realistic operating environment. The system employs LoS Wi-Fi communications for data exchange and gathering. On the other hand, National University of Defense Technology had developed its own small-scale UAV based on Raptor 90 hobby helicopter, LoS RC transmitters and RF modems, and digital signal controller as an airborne CPU [12]. In [18], software and hardware architectures are used in the design and implementation of a small semiautonomous fixed-wing

UAV. The presented system utilizes RC transmitter/receiver (70 MHz LoS link) for guidance commands, video downlink (2.4 GHz LoS link) for onboard camera, and bidirectional data link (900 MHz LoS link) between the autopilot and the GCS. The data-logger commands the autopilot to store the state of the UAV for a specified period of time. When the log is completed, it is transmitted back to the ground station for viewing. In [15], the development of UAV system is introduced for remotely piloted missions using telemetry data link and Futaba RX Control Module (FCM) for up to two miles.

In [13], a web access to real time flight data for a modern GCS is achieved, in addition to some standard functions, such as receiving data in real time, storing data, and displaying data. Authors in [14] developed a systematic approach for Micro Aerial Vehicle (MAV) trajectory generation addressing the characteristic issues of MAV flights in winds. In [16], the guidance, navigation, and control system of a small Vertical Takeoff and Landing (VTOL) unmanned tripropeller air vehicle based on a six-degree-of-freedom nonlinear dynamic model is designed and developed. In [17], the development of a real time embedded onboard computer and GCS software system for a UAV helicopter is introduced. An embedded computer system is attached at the lower portion of HeLion helicopter to perform multiple tasks including data acquisition and measurement, servo driving, automatic flight control implementation, communications, and data logging, while the presented GCS gives a graphical user interface for monitoring the real time status of the UAV. The GUI consists of four views including command window view, curve view, state view, and 3D view. In [19], a balloon airship is controlled via Remote Control (RC) transmitter using two different communication links. The GSM and GPRS are utilized for uplink and downlink to establish the flight control and surveillance system, respectively.

Additionally, the National University of Singapore has presented UAV software designing [21] and hardware modules [22, 23]. At Georgia Institute of Technology, a small Navigation, Guidance, and Control (NGC) system was developed using Field Programmable Gate Array (FPGA) and Digital Signal Processor (DSP) technologies for advanced UAV behavior [24, 25]. Moreover, some other different institutes developed their own UAVs due to their impact on the growth picture of the scientific R&D roadmap [26–28].

According to the aforementioned literature, the surveyed platforms miss some key features like the following: (1) the long distance and NLoS operating range with low bidirectional latency; (2) the large number of PWM control signals at the onboard system for extended payload control; (3) the real weather forecast capabilities during mission planning and real operational flight; (4) the online ground based accurate data logging; (5) the functional fidelity of the developed GCS software; and (6) ease of development requirements.

Thus, this paper demonstrates a new advanced software-based GCS and UAV quadcopter implementation to facilitate advanced long range NLoS remotely piloted missions to fulfill the above-mentioned aspects.

FIGURE 1: Proposed UAV quadcopter.

TABLE 1: Quadcopter specification.

Specification	Value
Frame model	Tarot 650
Frame materials and propellers	Carbon fiber
Motors	T-motors MN-4012 480 KVA
ESC	60 A opto 2–6 LiPo/600 Hz
Battery	6-cell, 10000 mAH, 25C
Flight controller	Ardupilot MEGA 2.6
Communication system	Wavecom GSM modem
Developed COTS avionics system	GSM, AHRS plus GPS module, and PWM controller
Weight	2.75 Kg

3. Proposed UAV Description

This section focuses on demonstrating the UAV description with important onboard components integration such as (a) advanced AHRS with integrated GPS module, (b) multiple channel PWM signal controller, and (c) the proposed integrated avionics module with GSM communication device. Figure 1 illustrates the final form of the custom UAV quadcopter development, which consists of Tarot 650 carbon fiber frame. It is equipped with four Electronic Speed Controllers (ESC), four brushless motors, two pairs of carbon fiber propellers, and ardupilot mega 2650 as a flight controller. The power source is provided through Lithium-Polymer (Li-Po) battery pack with a capacity of 10000 mAh. The detailed specification of the utilized quadcopter is listed in Table 1. The following subsections illustrate the proposed avionics components and their final integration form to achieve the mentioned objectives.

3.1. Advanced Attitude and Heading Reference System Plus GPS Module. For small-scale quadcopter, it is necessary to

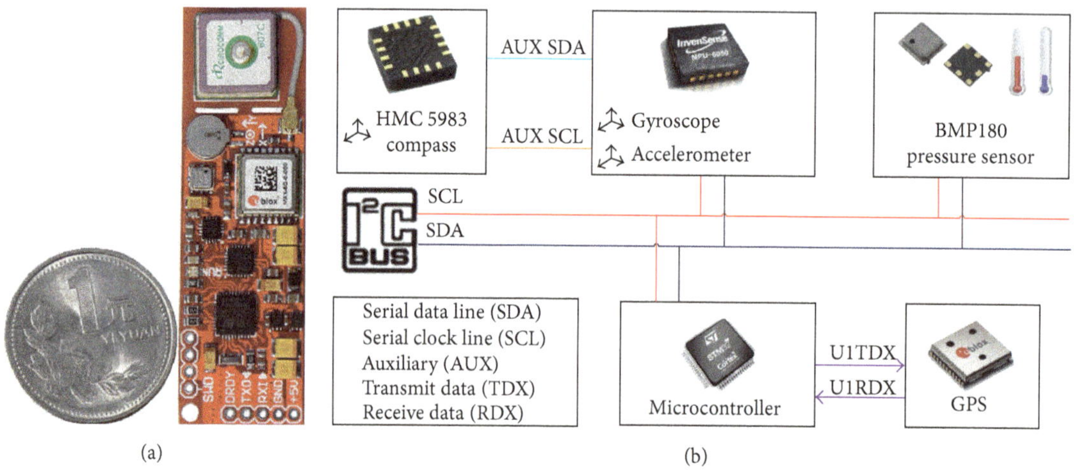

FIGURE 2: Attitude Heading Reference System. (a) Printed Board Layout (b) detailed components.

have the vital information about the vehicle during the mission, such as flight essential signals and the vehicle coordinates during the entire flight profile. Therefore, an advanced AHRS plus GPS module is provided. AHRS provides 3D orientation by integrating gyroscopes and fusing this data with accelerometer data and magnetometer data. With sensor fusion, drift from the gyroscopes integration is compensated by reference vectors, namely, gravity and the earth magnetic field. This makes AHRS a more cost effective solution than conventional high-grade Inertial Measurement Units (IMUs) that only integrate gyroscopes and rely on a very high bias-stability of the gyroscopes.

As shown in Figure 2(a), a light weight, small size compact enhanced COTS AHRS integrated with GPS on the same board is selected to deliver the UAV measurements. As illustrated in Figure 2(b), it consists of three-axis MEMS gyroscope, three-axis accelerometer, three-axis compass, and pressure/temperature sensors and integrated with a GPS receiver based on u-blox module. The manipulation of all components is performed using the ARM Cortex microcontroller (STM32F103). The specification of the COTS AHRS plus GPS module is shown in Table 2. This combination of MEMS sensors is manipulated by the ARM microcontroller to produce three types of data messages (labeled as A1, A2, and A3): (1) Message (A1) contains the attitude data, altitude, pressure, and temperature (18-byte data frame); (2) Message (A2) contains the raw sensors measurements gyroscope, accelerometer, and magnetic compass (22-byte data frame); (3) Message (A3) contains the GPS data longitude, latitude, altitude, number of locked satellites, heading, and speed (19-byte data frame). Furthermore, using TTL to RS232 converter, all these data are transmitted to the airborne GSM mode to be delivered for the GCS.

As the GPS signals in altitude channel miss some accuracy, the altitude signal from pressure sensor is important to correct any missed accuracy. The integrated barometer is able to provide altitude signal through detecting the change of air pressure. It can provide calibrated digital air pressure and temperature information. The formula for converting the air pressure to altitude is as follows [12]:

$$h_{\text{bar}} = h_0 + \frac{T_b}{L_b} \cdot \left[\left(\frac{P_{\text{bar}}}{P_b} \right)^{-(R \cdot L_b)/(g_0 \cdot M)} - 1 \right], \qquad (1)$$

where h_{bar} is the barometric altitude, h_0 is the height at the bottom of atmospheric layer (sea level height), T_b is the temperature at sea level in kelvin [K], L_b is the standard temperature lapse rate $= -0.0065$ [K/m], P_{bar} is the barometer pressure measurements, P_b is the pressure at sea level [Pa], R is universal gas constant $= 8.31432$ $[(\text{N} \cdot \text{m})/(\text{mol} \cdot \text{K})]$, g_0 is the gravitational acceleration constant $= 9.80665$ $[\text{m/s}^2]$, and M is the molar mass of Earth's air $= 0.0289644$ [kg/mol].

Hence, the air pressure is affected by various factors; then barometric attitude will be also affected. As a result, the measurement of barometric altimeter cannot be used directly. According to real measurements for a fixed location over different time, barometric altitude is found quite different. However, due to the short flight time, such difference can be modeled as a constant bias [12]. This calculation is performed by the GCS which has more processing capabilities. It is worthily noted that the onboard GPS antenna can be replaced with another external one upon the user demands (if this module will be mounted on a metal case).

3.2. PWM Servo Motor Controller. The majority of UAVs utilizes servo motors for controlling the control surfaces or ESCs for brushless motors. These pieces of equipment utilize PWM signals as a control input. As shown in Figure 3(a), an integrated COTS PWM servo motor controller is selected to provide the required PWM signals for the flight controller [9]. The selected PWM controller can provide up to 32 control signals simultaneously. Additionally, it is interfaced via RS-232 directly with custom user adjustment for the baud rate. Moreover, the PWM signals supply voltage can be powered from the controller power or can have external power supply. As providing RC guidance device with 32 channels costs

TABLE 2: COTS AHRS plus GPS specifications.

Specification	Value	Specification	Value
Yaw accuracy	0.1°	Three-axis accelerometer	Programmable full scale range of ±2 g, ±4 g, ±8 g, and ±16 g
Pitch accuracy	0.1°	Longitude accuracy	$10^{-6\circ}$
Roll accuracy	0.1°	Latitude accuracy	$10^{-6\circ}$
Altitude accuracy	0.1 m	Size	64 mm × 20 mm × 5 mm
Temperature accuracy	0.1°C	Operating temperature	−10°C–60°C
Pressure accuracy	10 Pa	Weight	20 grams

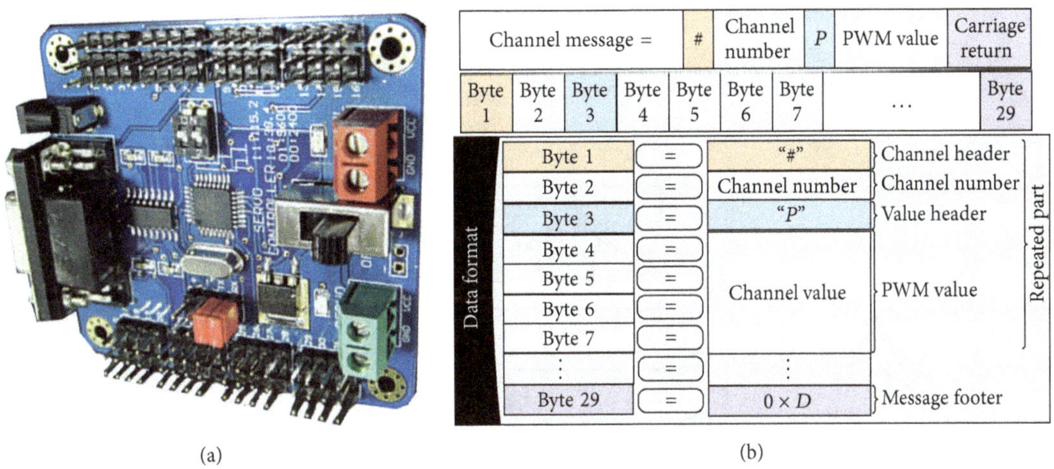

FIGURE 3: Multiple channel PWM controller. (a) Printed Board Layout [9], (b) four-channel data format sample.

a large budget, designing the proposed UAV quadcopter with this PWM controller provides a large number of channels with low cost. This is considered one of the conquered challenges in this paper. As shown in Figure 3(b), the data format for such controller consists of fixed headers and footer and dynamic part, which represent the channel number, and the PWM value. As a result, when the GCS captures the RC guidance device commands, it encapsulates the received data into the illustrated format to be delivered to the PWM controller through the GSM link.

3.3. Proposed Integrated Avionics Module. Finally, the airborne system could be integrated as shown in Figure 4. The developed avionics module consists of three layers' structure. The GSM modem is at the bottom layer, the PWM controller is in the middle layer, and the AHRS plus GPS module is at the top layer (for maximum signal strength of the GPS signals). All these modules are mounted on Tarot 650 carbon fiber base. Typically, the integration between GSM module and the PWM controller is considered a replica for traditional RC receiver to obtain long range of NLoS control with more control channels.

4. Proposed Control and Communication Scheme

Before describing the proposed scheme, it is mandatory to mention targeted platform requirements to know the complete scope of the proposed UAS. Taking into consideration the general UAS requirements in [29], the proposed platform has to fulfill the following requirements rather than other systems in the literature. These requirements could be summarized as follows:

(1) Achieve NLoS latency for both uplink and downlink less than 300 ms.

(2) Consider communication worst-case scenario by using 9600 bps baud rate.

(3) Support an operational altitude up to 1000 m.

(4) Software modules responsible of I/O operations (RC interface module and GSM interface module) should consume less than 1 ms to be negligible with respect to the NLoS latency (300 ms).

(5) Avoid airborne data logging to prevent data loss in case of any crash. It should be allocated at GSC with resolution up to 10^{-7} seconds.

(6) Provide bidirectional NLoS communications for the proposed COTS avionics (PWM controller, AHRS), while whole system is running to check that sensors measurements are consistent and noise free.

(7) Achieve good functional fidelity via advanced GUI capabilities (mapping, gauges, synthetic weather map, etc.) during real flight missions.

Finally, the proposed platform should be validated via a complete flight test to ensure the performance and the

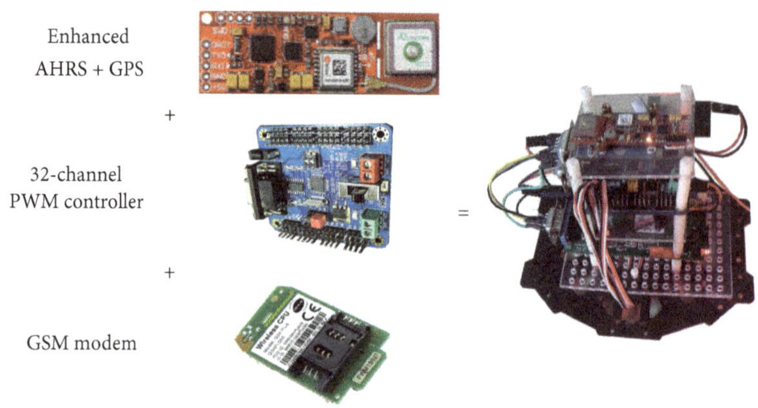

FIGURE 4: Integrated airborne system structured layers.

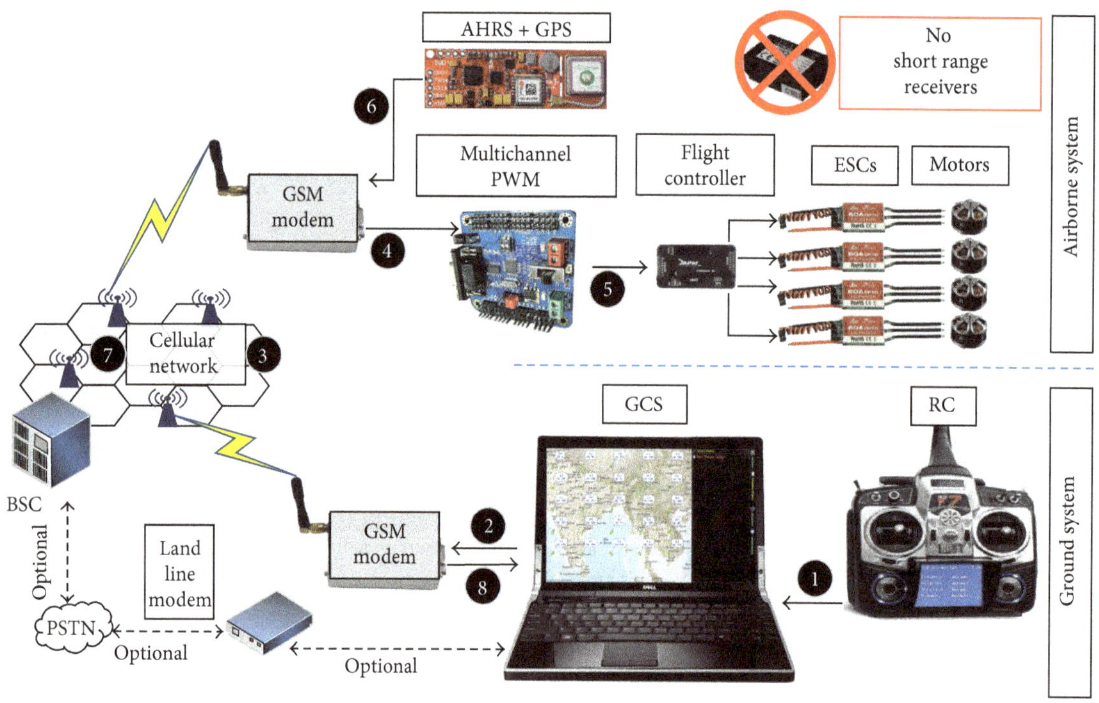

FIGURE 5: Proposed control and communication scheme.

consistency of different sensors measurements (such as GPS, barometer, and magnetometer data).

Towards fulfilling the mentioned requirements, Figure 5 demonstrates the proposed control and communication scheme, which consists of two main systems: (1) the ground system which contains standard RC transmitter, portable laptop equipped with the developed GCS software package, and the communication device (GSM modem) to communicate with the UAV; (2) airborne system which consists of UAV quadcopter (motors, ESCs, and flight controller), integrated AHRS plus GPS, multiple PWM channel controller, and an airborne GSM modem. In Figure 5 the labeled numbers demonstrated the data flow sequence of the proposed scheme which is as follows. (1) The GCS software captures the ground pilot guidance commands from the connected RC transmitter. (2) After reformatting the guidance command, it is transmitted to the GSM modem. (3) The GSM modem passes the commands to the cellular network to the airborne modem (uplink communication). (4) The airborne GSM modem sends the data to the PWM controller to generate the required signals to the flight controller in (5). (6) The AHRS module sends the measured data (feedback) to the airborne GSM modem. (7) The airborne modem passes the feedback to the cellular network to be delivered to the GCS (downlink communication). (8) The GCS manipulates the feedback data to provide the ground pilot with the required advanced visualization.

It can be deduced that, after the GSM modems establish an active bidirectional communication link, the ground pilot commands are directly captured by the GCS software and

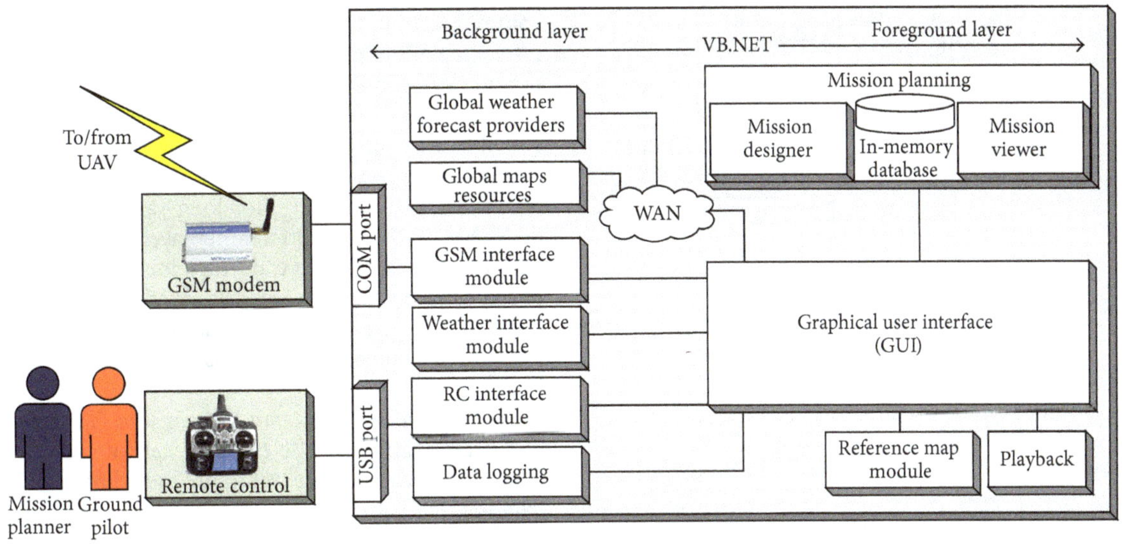

FIGURE 6: Ground Control Station block diagram.

transmitted to the GSM modem. As a result, the operation UAV range is not limited to the transmission range of the RC transmitter any more. The proposed platform can control the UAV at any point covered by the terrestrial cellular network using NLoS communications. Typically, for traditional RC transmitter/receiver, the more channels are obtained, the more budget is required. However, the proposed scheme replaces the traditional limited channel RC receiver with the 32-channel COTS PWM controller and maintains the low cost criteria. The proposed scheme facilitates dynamic movement of both the GCS and the UAV, which represents a feasible, low cost, and real time solution towards mobile command and control centers. Also, these features are considered to be some of the conquered challenges in this paper. This communication topology is validated for UAS for its tiny time constraints, but not limited to this, and the proposed GSM bidirectional communication can be extended to other ground or surface remote control vehicles (conditional to GSM coverage).

On the other hand, to build a GUI for the targeted GCS, several commercial and open source existing frameworks can be used such as visual studio.NET (including VB.NET and C#) and python. In this paper, VB.NET is selected as a high level programming language due to its various capabilities such as hardware interfacing, in-memory database, threading, APIs, stopwatches, data storage/manipulation, and COM technology.

As shown in Figure 6, the proposed GCS block diagram is illustrated, which represents a two-layer framework (foreground and background layers). The GCS software consists of interfacing modules, geographical modules, web resources manipulation, data logging, and playback. Since the proposed GCS extremely relies on software modules and components, therefore, the main development components are demonstrated in the following subsections.

4.1. GCS Graphical User Interface Console. The most important component in the GUI is the geographic reference map. It allocates all main important objects in the GCS (GCS coordinates, UAV coordinates, and waypoints). For globalization purpose, the geographic reference map should be in a standard format and style, for more compatibility with other commercial packages. A standard mapping module is developed to facilitate the use of maps, routing, geocoding from map providers. The maps module achieves a mouse move accuracy for the GUI console up to 10^{-6} degrees. Via the utilization of the map provider APIs, this module can achieve accessibility to the global map resources.

Typically, there is no successful mission training without a powerful and accurate mission planning module. Therefore, whatever the type of the operation (real/simulated or remotely piloted/autonomous), mission planning is considered an important feature of the GUI console. The mission trajectory and the way points design are overlaid over the geographic map, while a distance measurement module is developed to calculate the designed trajectory length. The mission designer stores the trajectories into a standard Keyhole Markup Language (KML) file [29] which provides a bidirectional importing/exporting of designed mission with standard packages like Google Earth. Moreover, during the real flight, the mission viewer module loads the targeted mission into in-memory database for rapid access.

4.2. Remote Control Guidance Device Interface. A complete software module is developed to capture the RC transmitter commands according to the delivered Interface Control Document (ICD). This RC interface module is implemented in a separate thread to be independent of the GCS activities. The complete flowchart of RC interface thread is illustrated in Figure 7. It is clear that the RC interface thread is an endless one, starts when the GUI application starts, detects the event which indicates that Human Interface Device (HID)

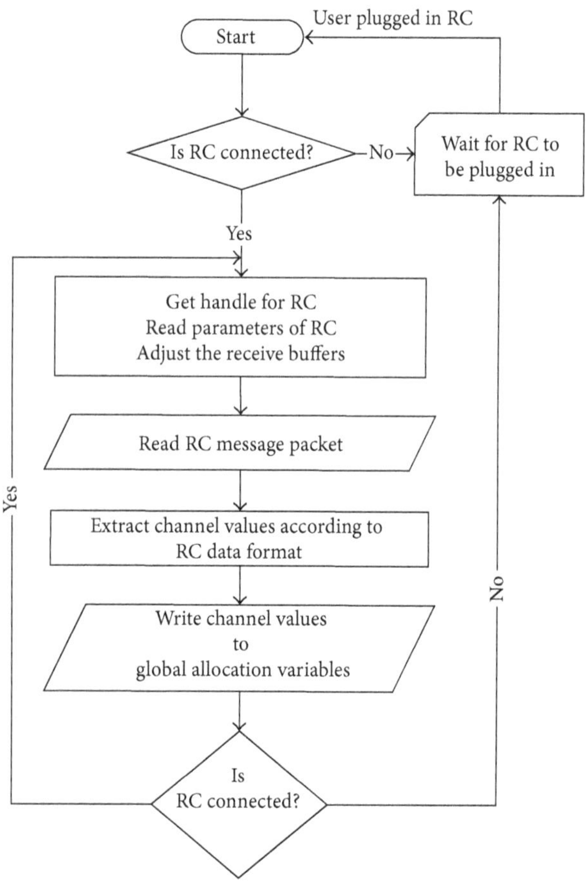

FIGURE 7: Remote control guidance device interface flowchart.

is connected, gets handle of the RC HID object, reads the HID device parameters to adjust the receive buffers, reads RC data message, extracts the guidance commands from the message, saves the extracted commands to global memory, and repeats the loop for continuous capture of ground pilot commands. If the RC device is disconnected at any time, the thread will wait until it is connected again and immediately trigger a red warning in the GUI status bar.

4.3. Global Weather Forecast Capability. As the proposed scheme targets long distance control via NLoS GSM communications, it is remarkable to achieve a weather data along the mission trajectory (not only at the GCS site). Therefore, by selecting an appropriate weather provider (local country or worldwide), the weather forecast data could be achieved. Obviously, this could be done after accomplishing some registration and interfacing procedures. Most of worldwide weather providers use APIs for supplying weather data to customers. The proposed GCS exploits "WorldWeatherOnline" as a worldwide instantaneous weather provider (as a sample to validate the concept). The weather provider online API explains the structure and formats of this data exchange service.

Figure 8(a) demonstrates the flowchart of the proposed weather interface module (which is implemented as a standalone thread). This module is responsible for the web communication with the weather providers through the implementation of the provider's APIs. Furthermore, this module parses the provider responses into useful information and stores them into the in-memory database. These activities are controlled from mission planning tab. However, Figure 8(b) represents the useful weather data and its effect on the mission planning (for long range UAVs). It can be deduced that the overlaid weather layer supports the mission designer to setup a safe trajectory for the mission. Because of the existence of such weather layer, the mission designer avoids a side wind of speed 12 m/sec coming from direction 216 (which is illustrated in red dashed circle).

4.4. Centralized Online Data Logging and Playback Features. The data logging module plays a vital role in recording all incoming and outgoing traffic for the whole platform. The data logging is developed using an accurate time stamping for all events. The resolution unit is measured in ticks (where 1 Tick = 10^{-7} seconds). All GCS interfaces, flight control commands, UAV feedback, and exchanged data among different modules are recorded. Threads are used to achieve the data logging goals to avoid any conflict with the GCS modules. However, multiple threads utilization may cause data violation and data loss. Therefore, an efficient technique is implemented to assure the consistency and integrity of all data and to maintain the high recording time resolution. This implementation relies on the thread safe containers that exist in VB.NET, which grantees the read and write operations without data violation. Algorithms 1, 2, and 3 represent 3 simultaneous threads to implement the proposed centralized data logging. It records different data sources using thread safe containers. This technique stores one multiplexed file for all received data.

As a result, the playback functionalities will become more accurate, because accessing one unified time stamped file is much efficient, convenient, and faster than accessing multiple files. This software module achieves advanced data logging capabilities without the need for extra hardware equipment. Typically, this is considered to be one of the conquered challenges in this paper. The resultant high resolution data logging capabilities clearly appear in all recording activities (as will be shown in Section 5).

4.5. Bidirectional GSM Communication (Ground and Airborne Modems). The GSM data communications could be established through many setup profiles. The proposed scheme utilizes the PPP mode for controlling the UAV. The PPP consists of two communication parties, one of them is the master (who initiates the communication request), while the other one is slave (who answers the call). Typically, through the standard AT commands [30], the GCS can play the master role and can initiate the communication request. Figure 9 illustrates the flowchart of GSM interface module. It can be inferred that handling the dial-up connections requires more programming effort and sensitive data handling (rather than the leased line or wireless modem), due to the switching between the data and commands modes and detecting any disconnect events. As a benefit for the proposed NLoS architecture, the GCS call can be initiated either from a GSM

(a) (b)

FIGURE 8: Integration of global weather forecasting reports with the mission planning process (a) weather interface module flowchart and (b) mission planning using synthetic graphical weather layer.

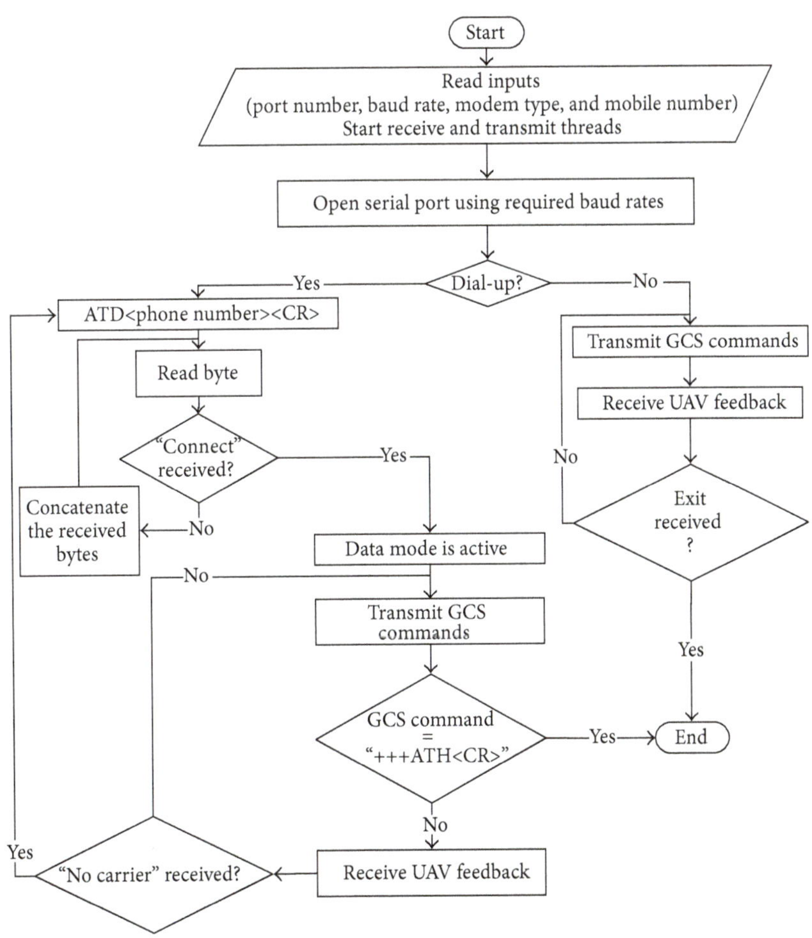

FIGURE 9: Detailed flowchart of GSM interface module.

```
(1) Lock Access (Global_Buffer Array List)
(2)     Add Time Stamp in Ticks
(3)     Add Recording Source Label
(4)     Add Recording Source Data
(5) Unlock Access (Global_Buffer Array List)
```

ALGORITHM 1: Proposed collecting data thread.

```
(1) Initialize Global_Buffer of the Type (Synchronized Array List)
(2) Initialize File_Buffer of the Type (Synchronized Array List)
(3) Start File Manipulation Thread
(4) Loop
(5)     If Global_Buffer Elements >0 then
(6)         Lock Access (Global_Buffer Array List)
(7)         Copy Global_Buffer → File Buffer
(8)         Clear All Global_Buffer
(9)         Unlock Access (Global_Buffer)
(10)    Else
(11)        GoTo (4)
(12)    End if
(13) End loop
```

ALGORITHM 2: Proposed data logging thread.

```
(1) Loop
(2)     If File_Buffer Elements >0 then
(3)         Lock Access (File Buffer)
(4)         Copy File_Buffer → Temp_Buffer
(5)         Clear All File_Buffer
(6)         Unlock Access (File_Buffer)
(7)     Else
(8)         GoTo (1)
(9)     End if
(10)    Loop (Temp_Buffer)
(11)        Seek to end of File
(12)        Write Temp_Buffer Data
(13)        Clear Temp_Buffer Data
(14)    End Loop
(15) End loop
```

ALGORITHM 3: Proposed file manipulation thread.

modem or from a traditional land line modem using the Public Switched Telephone Network (PSTN) (as shown in Figure 5). This feature is an option for the stationary GCS, while for mobile GCS the proposed GSM modem is much suitable.

5. Testing Results and Evaluation

The testing of UASs usually passes through different procedures to avoid any hazard during flight missions. This section describes the conducted tests and methods that are used to evaluate the proposed scheme, which is divided into two major sections: ground tests and flight test. The ground tests include different procedures to measure various parameters, validate the data logging module, and measure critical modules processing delay (I/O interface modules) for better evaluation [31–33]. On the other hand, the flight test is conducted to demonstrate the GUI capabilities and to verify the performance of the developed avionics, check the consistency of different sensors measurements, and test the whole proposed control and communication scheme.

5.1. Ground Tests

5.1.1. GSM Communication Latency and I/O Interface Modules Processing Delay. Communication delay is considered one of

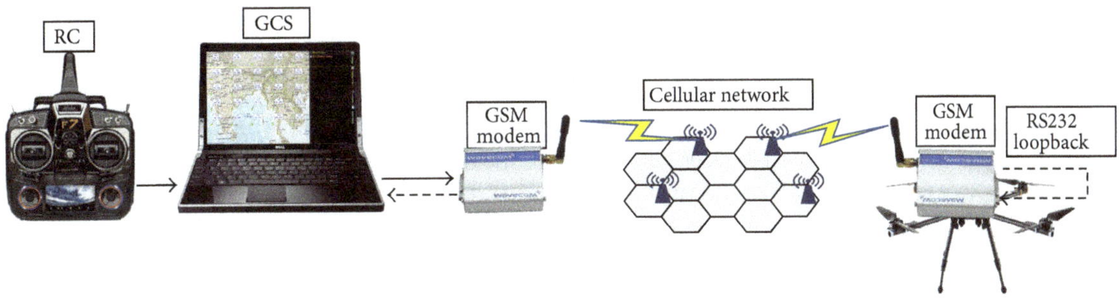

FIGURE 10: Experimental infrastructure for GSM delay measurements.

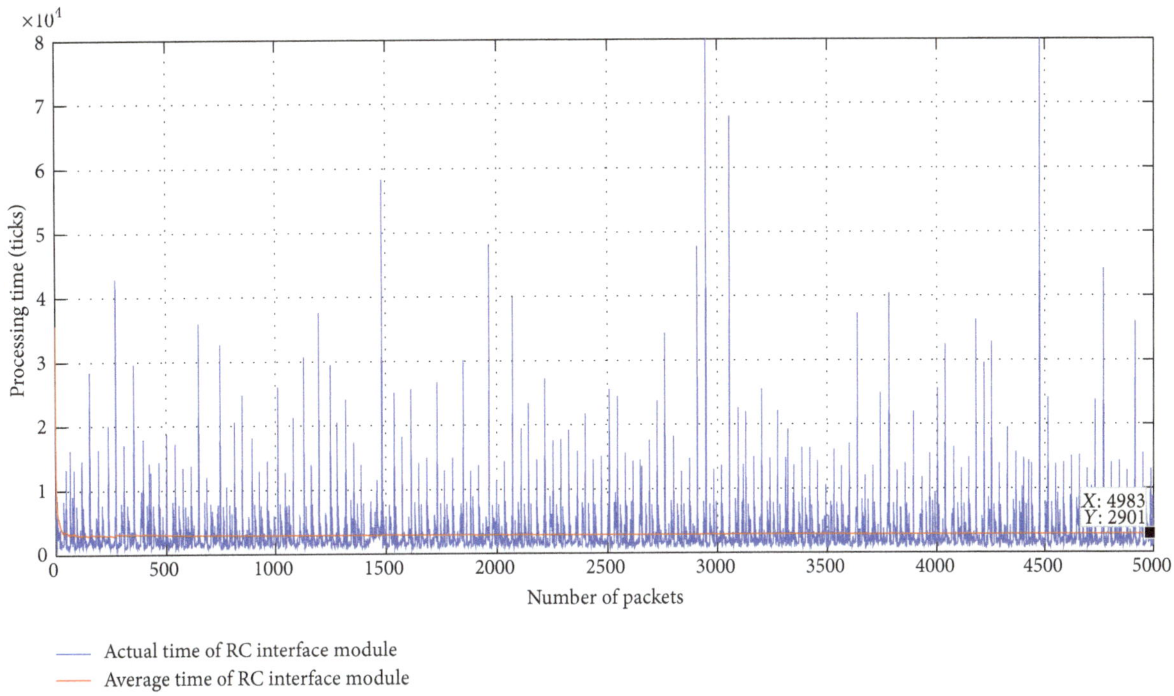

FIGURE 11: Processing delay of RC interface module.

the most challenges for UAV uplink flight controls and downlink surveillance data. Thus, as shown in Figure 10, a practical experiment is conducted to measure the round trip delay of GSM communication channel. The illustrated infrastructure shows that the airborne GSM modem is connected with RS-232 loopback, while the submitted data are the RC guidance commands as in real flight. The GSM in PPP is established between the GSM modems. The data logging module is adapted to compare the transmitted and received data and to measure the round trip delay in ticks. Table 3 lists a data logging sample for the loopback data transfer. It can be deduced that this experiment validates the proposed high resolution online data logging module, considers the communication worst-case scenario (9600 bps), and measures the average uplink/downlink latency to be in order of 286 ms. Therefore, according to these results, many requirements are achieved. Moreover, it is worthily noted that employing cellular data modems for UAS is experimentally approved for heights up to 1000 m [20].

As the proposed GCS relies deeply on software, thus, it is important to measure the processing delay of different data manipulation software modules. The most critical processing delays are the interface modules (I/O modules), which are responsible for the RC transmitter and GSM data manipulation. These two modules are responsible for capturing the serial data, parsing and extracting the required parameters, and, finally, submitting data to demanded destination. As shown in Figure 11, the average processing delay to manipulate the mentioned RC data format is 2901 ticks (0.29 ms). On the other hand, the processing delay of the GSM software interface module is divided into uplink delay (for sending UAV commands) and downlink delay (for receiving UAV flight parameters). As shown in Figure 12, the average processing delays of GSM uplink and downlink are 2545 ticks (0.255 ms) and 8150 ticks (0.815 ms), respectively. It can be concluded that I/O processing delays are considered negligible with respect to the GSM round trip delay, which satisfies the mentioned requirements in Section 4.

Table 3: GSM communication delay measurements.

Msg size (B)	Transmit time (ticks)	Receive time (ticks)	Round trip delay (ticks)	Average uplink/downlink delay (ticks)	Average uplink/downlink delay (ms)
29	635551560641752484	635551560647604261	5851777	2925888.5	292.5888
29	635551560647634952	635551560653605129	5970177	2985088.5	298.5088
29	635551560653640420	635551560659494641	5854221	2927110.5	292.7110
29	635551560659532802	635551560665385290	5852488	2926244	292.6244
29	635551560665416008	635551560670875085	5459077	2729538.5	272.9538
29	635551560670913652	635551560676948354	6034702	3017351	301.7351
29	635551560676984607	635551560682878280	5893673	2946836.5	294.6836
29	635551560682916203	635551560688272173	5355970	2677985	267.7985
29	635551560688310562	635551560693719847	5409285	2704642.5	270.4642
29	635551560693754578	635551560699582413	5827835	2913917.5	291.3917
29	635551560699615279	635551560705263850	5648571	2824285.5	282.4285
29	635551560705294142	635551560710838137	5543995	2771997.5	277.1997
29	635551560641752484	635551560647604261	5851777	2925888.5	292.5888
29	635551560647634952	635551560653605129	5970177	2985088.5	298.5088
29	635551560653640420	635551560659494641	5854221	2927110.5	292.7110

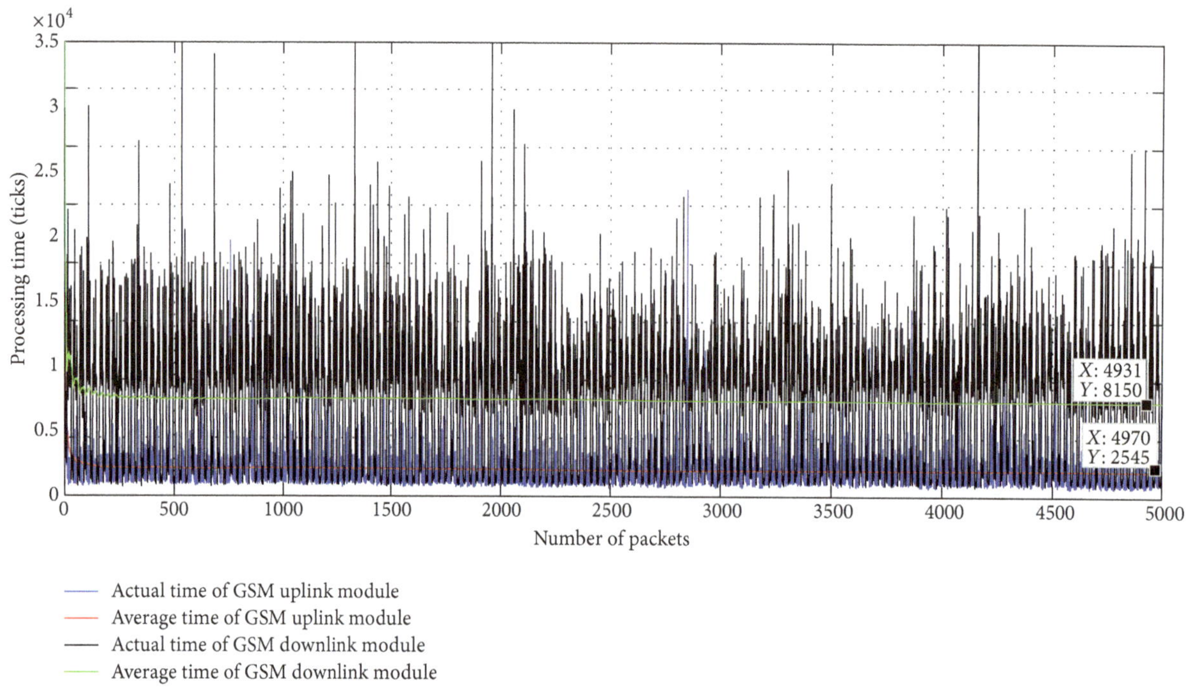

Figure 12: Software processing delay of GSM data manipulation module.

5.1.2. Avionics Module and Bidirectional Data Transmission. The purpose of this test is to ensure that there is no malfunctioning of the proposed avionics functionality, while whole pieces of equipment are powered up and motors are running. In other words, it is required to verify the PWM controller outputs and to collect AHRS data via GSM communications while motors are running.

During this test, the engine was powered on at 70% throttle, while the rotor blades were not installed to follow the safety procedures. The UAV quadcopter was allocated in an arbitrary orientation and position about 200 meters away from the GCS. The GCS established a call route to UAV SIM card as mentioned in Section 4. After establishing the connection at 9600 bps, the data logging was activated to record

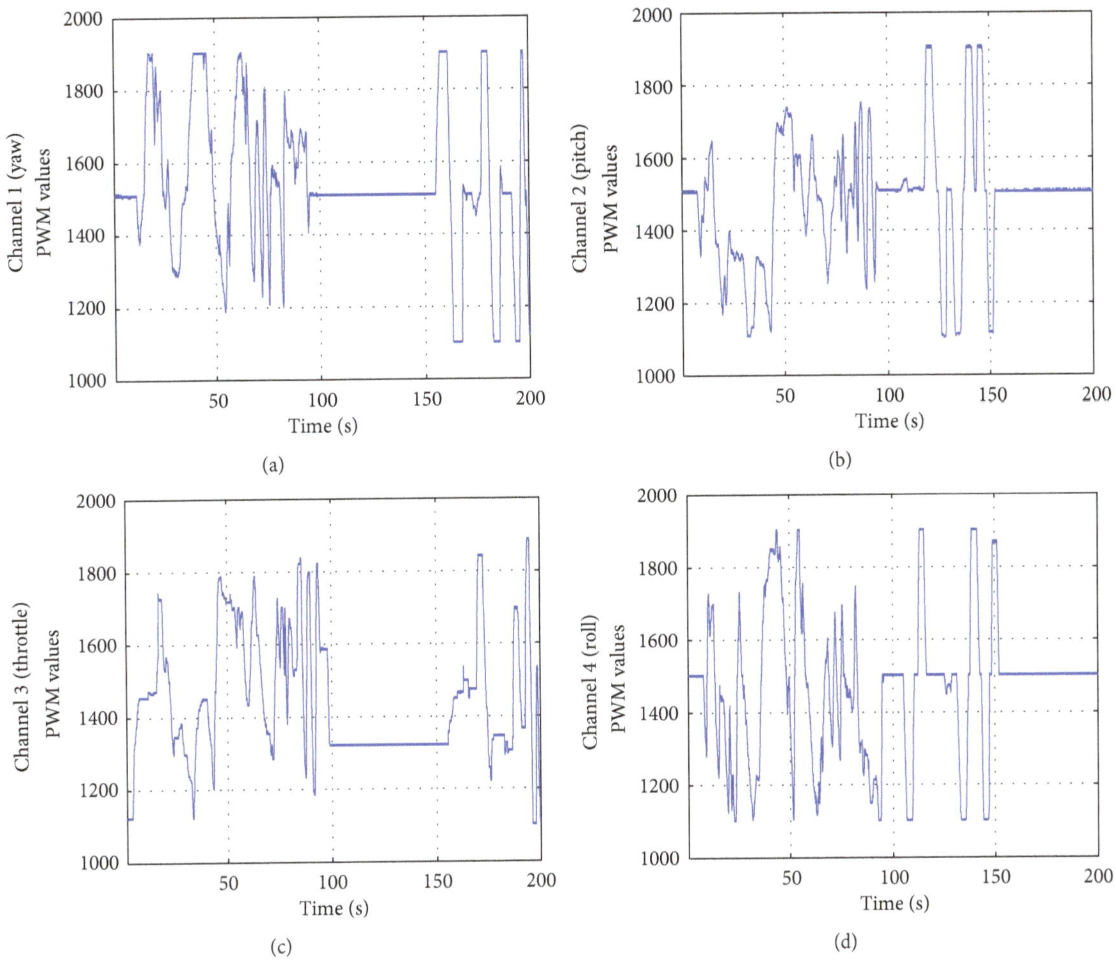

FIGURE 13: Flight control PWM signals in the ground test.

all RC guidance device commands (which is transmitted via the GSM uplink) and avionics feedback (which is received via the GSM downlink). This test lasts continuously for 15 minutes. Figure 13 demonstrates a recorded sample for the GCS transmitted commands (PWM signals), which shows the capability of the COTS controller to generate the correct PWM signals similar to the traditional RC receiver signals. On the other hand, Figure 14 presents consistent receiving of AHRS data with tiny vibration effect due to motor running with no interference on AHRS sensors measurements, which fulfills the requirements in Section 4.

5.2. Flight Tests. This section introduces the proposed GUI capabilities during flight test and also presents a complete set of surveillance data, which validates the overall performance and proves the consistency of different sensors measurements. Following the ground test, multiple remotely piloted flight test missions were conducted via the bidirectional GSM communication to verify the performance of the proposed software and hardware components.

In Figure 15(a), the GCS is located at the home symbol, while the proposed UAV is controlled to follow the planned trajectory, via bidirectional NLOS GSM channel over a distance more than 5 Km. The innovative GUI allocates all static and dynamic entities on the geographic reference map, while the detailed UAV feedback is presented in tabular form in the right panel. Additionally, a set of convenient gauges are presented to simplify the guidance process and to increase the functional fidelity of the proposed GCS [34]. Moreover, Figure 15(b) demonstrates a zoom in view for the weather forecast layer overlaid over the geographic map; it shows the wind speed and its direction along the mission trajectory. The illustrated weather layer provides more real atmospheric conditions for the guidance operator for better environmental situational awareness. All demonstrated software capabilities can be exploited with other UASs with minor modifications in the interface modules, which should be adapted to manipulate the new data protocols.

As shown in Figure 16, the raw inertial data of COTS AHRS module are captured and transmitted efficiently through the GSM modems to the GCS. Blue line is used to represent the raw data from accelerometer signals directly and red line is used to represent the average value. The raw data show a significant vibration; however, when integrating the data, it delivers a smooth and stable position, velocity, and attitude values. Figure 17 illustrates the quadcopter attitude

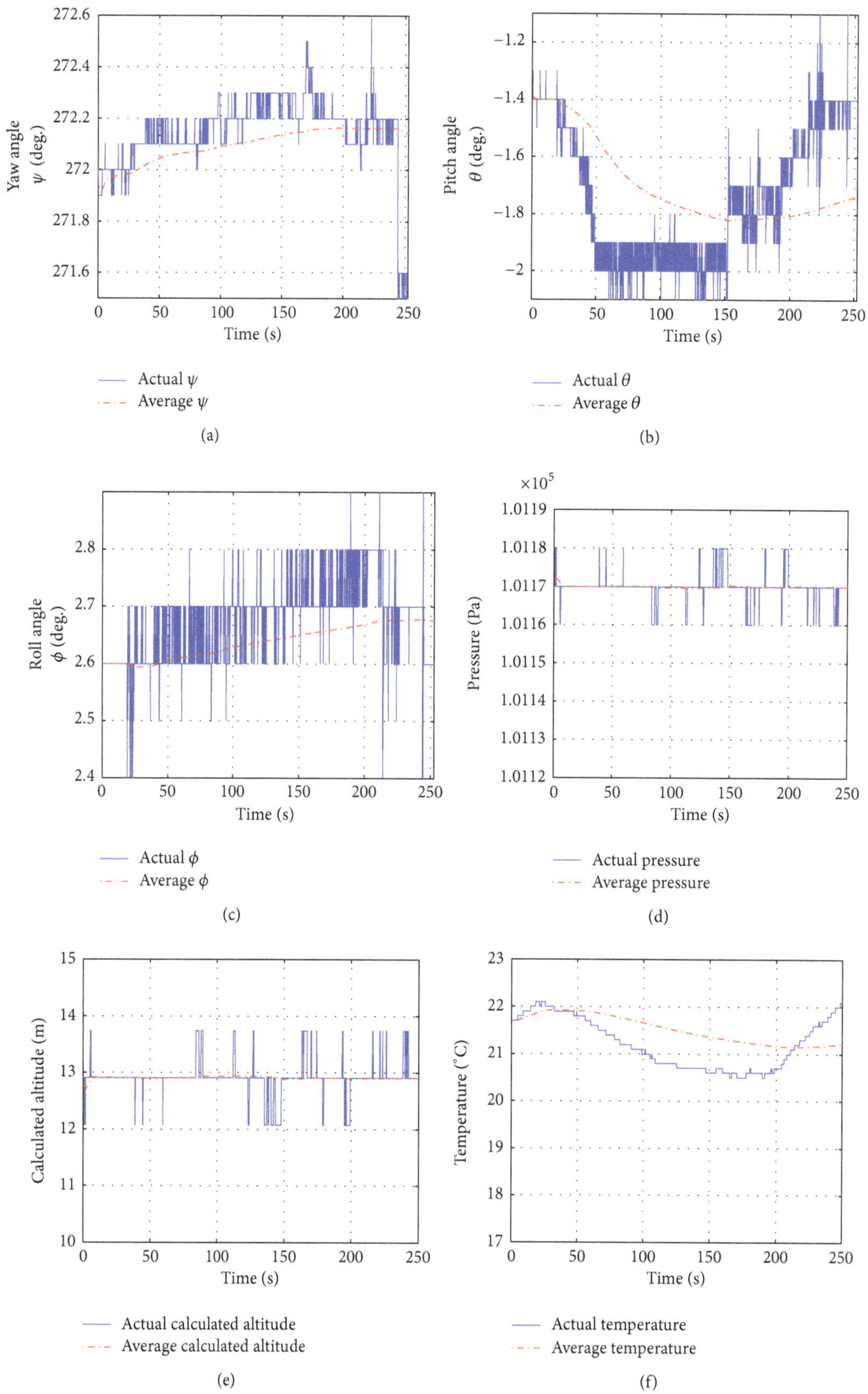

FIGURE 14: Ground test signals playback for the proposed UAV platform via GSM data communication channels.

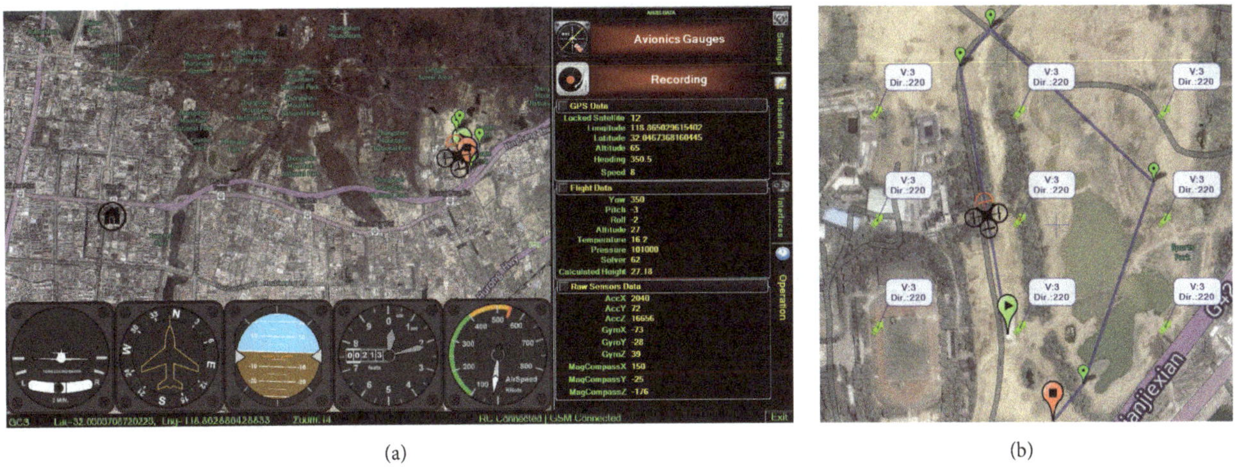

FIGURE 15: Proposed Ground Control Station GUI snapshot. (a) Operational panel and (b) mission trajectory with synthetic weather data layer.

data (Euler angles) over 200 seconds. It can be observed from Figure 17(a) that the GPS heading and obtained AHRS sensors yaw measurements are correlated, which ensure the proposed platform performance, and prove the consistency of different sensors measurements. In Figures 17(b) and 17(c), it can be observed that the UAV roll and pitch angles are successfully captured from the ARHS sensors for further visualization by the attitude indicator gauge.

Furthermore, the environmental data (pressure and temperature) are presented in Figure 18. The temperature accuracy (0.1°C) provides accurate measurements over different altitudes (as shown in Figure 18(a)), while Figure 18(b) presents (10 Pa) pressure resolution from barometer sensor during the flight. It is clear that the collected data are consistent, as the decrease in pressure is equivalent to incremental altitude, which leads to temperature decrease.

According to Section 3 (1), UAV altitude can be calculated from barometer readings. This feature could be used to correct the GPS altitude, or as an alternative altitude data source in case of GPS malfunctioning. Figure 19 shows the GPS altitude and the biased barometric altitude during the flight test. Finally, Figure 20(a) presents the GPS flight trajectory, while Figure 20(b) demonstrates the locked GPS satellites. GPS module is able to receive up to 12 satellite readings along the mission without the need for large GPS external antenna.

From all previous flight test results, it can be concluded that the proposed quadcopter, avionics, GCS, and the whole control and communication scheme parameters (illustrated in Section 4) are validated.

6. Qualitative Aspects of the Proposed GSM Based UAV Platform

As presented in Table 4, this subsection lists a brief comparison between the functions and features of the proposed UAV platform versus other different existing platforms. It can be observed from this table that the proposed platform achieves the following: (1) new bidirectional NLoS long communication range using challenging low bandwidth for a considerable height; (2) a large number of controllable PWM signals for extended payload control; (3) new environmental awareness during operational mission; (4) advanced GUI features; (5) online data logging; and (6) higher functional fidelity.

7. Conclusions

This paper illustrates the detailed development of a new bidirectional long range UAV flight control using mobile communications and COTS component. A new software package is developed for long range quadcopter UAV missions. The airborne system used employs a set of COTS components. Through the proposed control and communication scheme, traditional short range remote control receivers are no longer needed. On the other hand, the ground station software is proposed as a two-layer framework. The background layer is utilized for interfacing and data exchange, while the foreground layer is employed for the visualization and the operator activity. A new global real weather interface module is developed to support the human operator decisions during both mission planning and real operation. The software system has been thoroughly tested through different ground tests and real flight tests. The results show that the GCS established a stable data link with the UAV with latency average latency less than 300 ms. Moreover, developed software modules satisfied the real time processing requirements, GUI achieves a high functional fidelity, and high resolution data logging module facilitates an accurate and powerful evaluation. The flight tests validate the whole platform performance using GSM communications. Finally, as the development of universal software framework has a great potential in the near future, the illustrated software solution can also be used in other unmanned vehicles with minimal modifications.

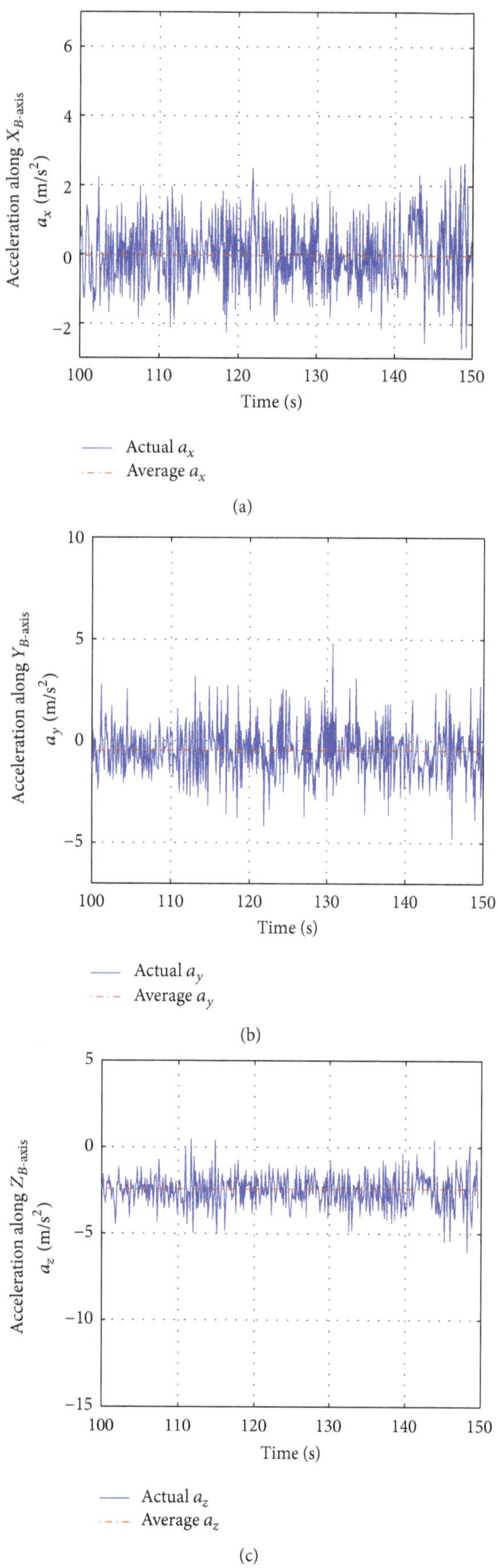

FIGURE 16: Flight test accelerometer outputs.

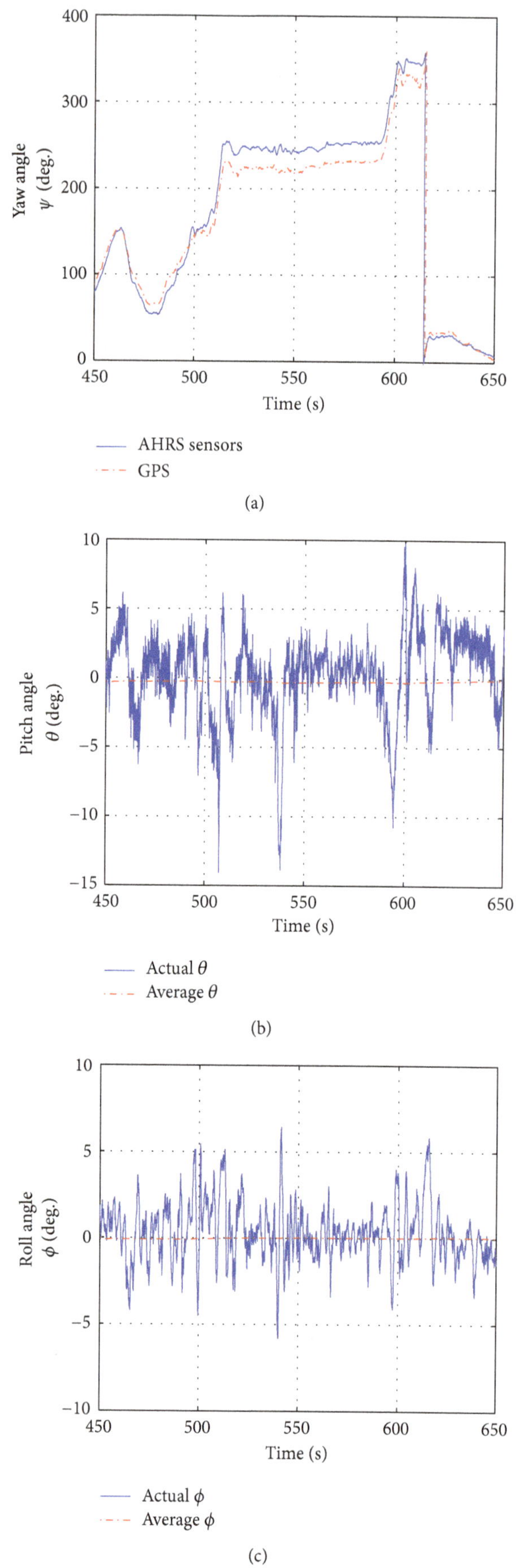

FIGURE 17: Euler angles during flight test.

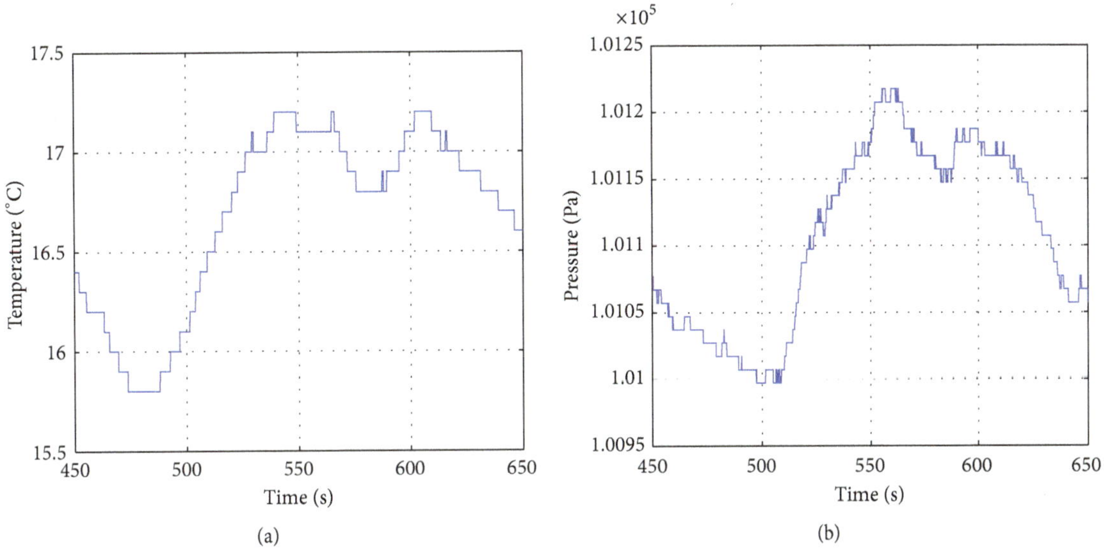

FIGURE 18: Environmental online data measurements via GSM channels.

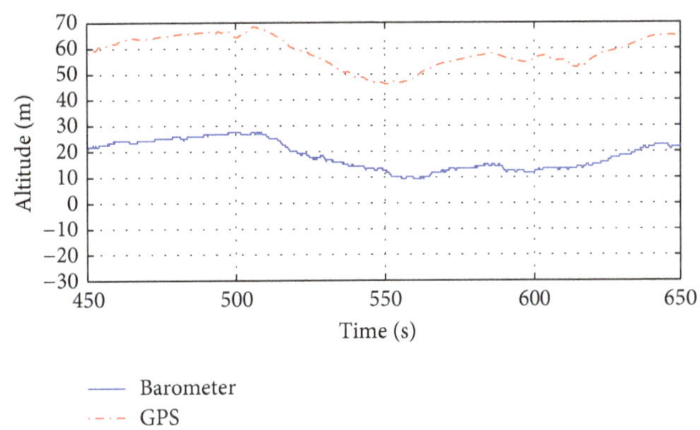

FIGURE 19: Flight test altitude profile.

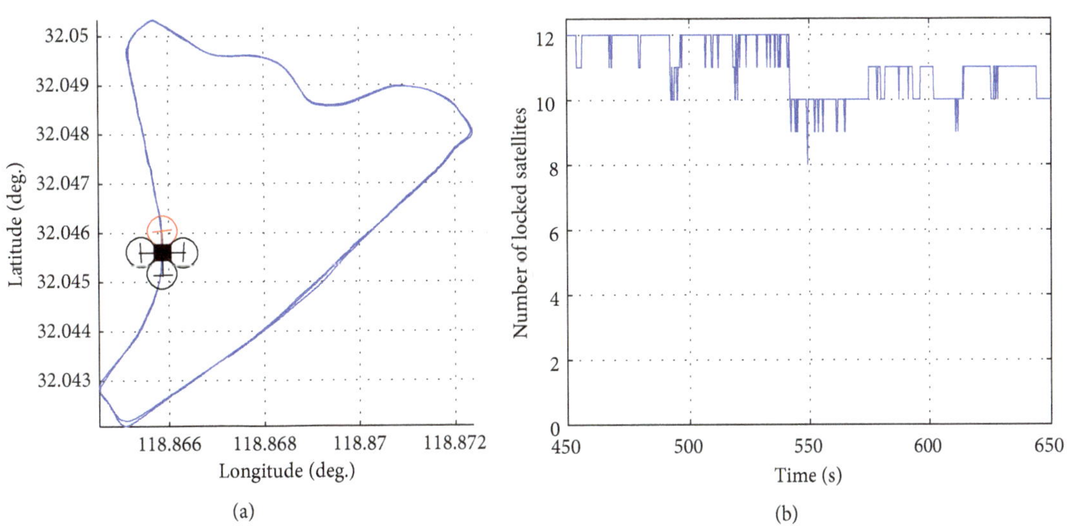

FIGURE 20: Flight test GPS measurements: (a) GPS trajectory, (b) GPS locked satellites.

TABLE 4: Proposed GSM based UAV platform versus other different existing platforms (see Figure 21).

Criteria	Proposed platform	Helicopter platform [12]	Balloon platform [19]
UAV type	Quadcopter	Raptor 90 Hoppy Helicopter	Balloon
GCS	GCS: 1 laptop, airborne system: COTS components	GCS: 1 laptop, avionics: digital signal controller TMS320F28335	GCS: 1 workstation, airborne system: ARM CPU + C8051
Communication device	2 GSM modems or 1 GSM + land line modem (optional)	RC guidance device + 2 Digi XTend-PKG RF modems	2 GSM modems + PSTN connection
Communication architecture	Uplink: GSM in PPP Downlink: GSM in PPP	Traditional 2.4 GHz receiver + wireless modems	Uplink: GSM in PPP Downlink: GPRS using TCP/IP
AT commands	Implemented	Not used	Implemented
Communication range	The cellular service provider coverage	Up to 64 Km in open field environment	The cellular service provider coverage
Link between GCS and UAV	NLoS	LoS	NLoS
Communication baud rates	Uplink: 9.6 Kbps Downlink: 9.6 Kbps	Up to 115.2 Kbps	Uplink: 9.6 Kbps Downlink: 9.6 Kbps
Average delay	Uplink: <0.3 sec Downlink: <0.3 sec	Not explicitly mentioned	Uplink: <0.3 sec Downlink: <0.85 sec
Ceiling altitude according to [20]	600–1000 m for consistent signal	Not explicitly mentioned	600–1000 m for consistent signal
Traditional short range receivers	Not existing	Exist	Not explicitly mentioned
Ease of development	AHRS plus GPS, PWM controller, and GSM modems are all COTS components	Need custom development using DSC (TMS320F28335)	Need custom development using ARM CPU + C8051
PWM channels	32	6	4
Real weather data for mission planning	Available	Not available	Not available
Real weather data during flight mission	Available	Not available	Not available
Real weather forecast data	Available up to 6 days	Not available	Not available
GCS capabilities	Geographical map Innovative set of gauges Mission planning Mission viewer Weather forecast features Hardware interfacing Precise recording (10^{-7} s) Playback	Compass gauges Text data for UAV position Curve style UAV motion trajectory	GIS display
Data logging location	GCS	Onboard system using SD card	Not explicitly mentioned
Development languages	VB.NET	Not explicitly mentioned	Not explicitly mentioned
Functional fidelity	Above medium	Medium	Medium

(a) (b) (c)

FIGURE 21: (a) Proposed platform. (b) Helicopter platform. (c) Balloon platform.

Conflict of Interests

The authors declare that there is no conflict of interests regarding the publication of this paper.

Acknowledgments

The authors would like to acknowledge the anonymous reviewers for their insightful comments and suggestions that led to significant enhancement in the paper quality. In addition, the authors would like to thank the Chinese government "National 863 Fund" (Reference 2014AA7010051) for their great support.

References

[1] A. AbdElHamid and P. Zong, "A novel software simulator model based on active hybrid architecture," *International Journal of Aerospace Engineering*, vol. 2015, Article ID 107301, 19 pages, 2015.

[2] A. Ollero and L. Merino, "Control and perception techniques for aerial robotics," *Annual Reviews in Control*, vol. 28, no. 2, pp. 167–178, 2004.

[3] A. Idries, N. Mohamed, I. Jawhar, F. Mohamed, and J. Al-Jaroodi, "Challenges of developing UAV applications: a project management view," in *Proceedings of the International Conference on Industrial Engineering and Operations Management (IEOM '15)*, pp. 1–10, IEEE, Dubai, United Arab Emirates, March 2015.

[4] M. Guerra, M. Rhudy, Y. Gu, B. Seanor, and M. R. Napolitano, "Mobile ground control station development for fault tolerant UAV research," in *Proceedings of the AIAA Guidance, Navigation, and Control Conference*, American Institute of Aeronautics and Astronautics, Minneapolis, Minn, USA, August 2012.

[5] M. Yeary, B. Cheong, J. Kurdzo, T.-Y. Yu, and R. Palmer, "A brief overview of weather radar technologies and instrumentation," *IEEE Instrumentation & Measurement Magazine*, vol. 17, no. 5, pp. 10–15, 2014.

[6] W. S. Davis and R. H. Fisher, *COBOL: An Introduction to Structured Logic and Modular Program Design*, Addison-Wesley, 1979.

[7] B. S. El-Haik and A. Shaout, *Software Design for Six Sigma: A Roadmap for Excellence*, John Wiley & Sons, Hoboken, NJ, USA, 2010.

[8] B. Fred, "UAV software architecture," in *Infotech@Aerospace 2012*, American Institute of Aeronautics and Astronautics, 2012.

[9] RobotBase, "32-Channel Servo Controller for Robotic Arm Biped Robot Kit," 2015, http://www.alsrobot.com/index.php?route=product/product&product%20id=62.

[10] Accomplish with GSM modems (Wavecom Instruments, GSM-GPRS Modems), 2015, http://www.wavecom.com.au/products.php?prod cat main=Modems.

[11] S. Rasmussen and D. Kingston, "Development and flight test of an area monitoring system using Unmanned Aerial Vehicles and unattended ground sensors," in *Proceedings of the International Conference on Unmanned Aircraft Systems (ICUAS '15)*, pp. 1215–1224, IEEE, Denver, Colo, USA, June 2015.

[12] S. Tang, X. Lu, and Z. Zheng, "Platform and state estimation design of a small-scale UAV helicopter system," *International Journal of Aerospace Engineering*, vol. 2013, Article ID 524856, 13 pages, 2013.

[13] Y. Liu, B. Xian, F. Wang, S. Liu, and X. Gu, "Development of the ground control station for a quadrotor unmanned aerial vehicle based on Java programming," in *Proceedings of the 31st Chinese Control Conference (CCC '12)*, pp. 5002–5007, IEEE, Hefei, China, July 2012.

[14] Y. Ketema and Y. J. Zhao, "Micro air vehicle trajectory planning in winds," *Journal of Aircraft*, vol. 47, no. 4, pp. 1460–1462, 2010.

[15] R. K. Rangel, K. H. Kienitz, and M. P. Brandão, "Development of a complete UAV system using COTS equipment," in *Proceedings of the IEEE Aerospace Conference*, pp. 1–11, IEEE, Big Sky, Mont, USA, March 2009.

[16] R. Huang, Y. Liu, and J. J. Zhu, "Guidance, navigation, and control system design for tripropeller vertical-takeoff-and-landing unmanned air vehicle," *Journal of Aircraft*, vol. 46, no. 6, pp. 1837–1856, 2009.

[17] M. Dong, B. M. Chen, G. Cai, and K. Peng, "Development of a real-time onboard and ground station software system for a UAV helicopter," *Journal of Aerospace Computing, Information and Communication*, vol. 4, no. 8, pp. 933–955, 2007.

[18] R. W. Beard, D. Kingston, M. Quigley et al., "Autonomous vehicle technologies for small fixed-wing UAVs," *Journal of Aerospace Computing, Information and Communication*, vol. 2, pp. 92–108, 2005.

[19] L. Chin, C. Hsu, and C. Li, "Verification of UAV flight control and surveillance using mobile communication," in *Proceedings of the 2nd AIAA 'Unmanned Unlimited' Conference and Workshop & Exhibit*, American Institute of Aeronautics and Astronautics, San Diego, Calif, USA, September 2003.

[20] B. K. Kevin, T. Hamilton, W. Christopher, M. Sumedha, and J. S. Larry, "A feasibility study of using cellular data modems for UAV tracking and control," in *Proceedings of the AIAA Infotech@Aerospace Conference (I@A '13)*, Boston, Mass, USA, August 2013.

[21] C. Guowei, B. M. Chen, T. H. Lee, and D. Miaobo, "Design and implementation of a hardware-in-the-loop simulation system for small-scale UAV helicopters," in *Proceedings of the IEEE International Conference on Automation and Logistics (ICAL '08)*, pp. 29–34, IEEE, Qingdao, China, September 2008.

[22] G. Cai, L. Feng, B. M. Chen, and T. H. Lee, "Systematic design methodology and construction of UAV helicopters," *Mechatronics*, vol. 18, no. 10, pp. 545–558, 2008.

[23] G. Cai, K. Peng, B. M. Chen, and T. H. Lee, "Design and assembling of a UAV helicopter system," in *Proceedings of the 5th International Conference on Control and Automation (ICCA '05)*, pp. 697–702, Budapest, Hungary, June 2005.

[24] C. Henrik, P. Wayne, K. Adrian, K. Suresh, and J. Eric, "Small adaptive flight control systems for UAVs using FPGA/DSP technology," in *Proceedings of the AIAA 3rd "Unmanned Unlimited" Technical Conference*, American Institute of Aeronautics and Astronautics, Chicago, Ill, USA, September 2004.

[25] J. S. Dittrich and E. N. Johnson, "Multi-sensor navigation system for an autonomous helicopter," in *Proceedings of the 21st Digital Avionics Systems Conference (DASC '02)*, vol. 2, pp. 8C1-1–8C1-19, Irvine, Calif, USA, January 2002.

[26] R. D. Garcia and K. P. Valavanis, "The implementation of an autonomous helicopter testbed," *Journal of Intelligent and Robotic Systems*, vol. 54, no. 1–3, pp. 423–454, 2009.

[27] Z. Taha, Y. R. Tang, and K. C. Yap, "Development of an onboard system for flight data collection of a small-scale UAV helicopter," *Mechatronics*, vol. 21, no. 1, pp. 132–144, 2011.

[28] B. Godbolt, N. I. Vitzilaios, and A. F. Lynch, "Experimental validation of a helicopter autopilot design using model-based

PID control," *Journal of Intelligent & Robotic Systems*, vol. 70, pp. 385–399, 2013.

[29] A. Chen, G. G. Leptoukh, and S. J. Kempler, "Using KML and virtual globes to access and visualize heterogeneous datasets and explore their relationships along the A-train tracks," *IEEE Journal of Selected Topics in Applied Earth Observations and Remote Sensing*, vol. 3, no. 3, pp. 352–358, 2010.

[30] F. Hargrave, *Appendixes A: AT Fax/Modem Commands*, Wiley-IEEE Press, 1st edition, 2001.

[31] A. M. Davis, E. H. Bersoff, and E. R. Comer, "A strategy for comparing alternative software development life cycle models," *IEEE Transactions on Software Engineering*, vol. 14, no. 10, pp. 1453–1461, 1988.

[32] H. Hashimi, A. Hafez, and M. Beraka, "A novel view of risk management in software development life cycle," in *Proceedings of the 12th International Symposium on Pervasive Systems, Algorithms, and Networks (I-SPAN '12)*, pp. 128–134, San Marcos, Tex, USA, December 2012.

[33] D. Horie, T. Kasahara, Y. Goto, and J. Cheng, "A new model of software life cycle processes for consistent design, development, management, and maintenance of secure information systems," in *Proceedings of the 8th IEEE/ACIS International Conference on Computer and Information Science (ICIS '09)*, pp. 897–902, Shanghai, China, June 2009.

[34] J. Craighead, R. Murphy, J. Burke, and B. Goldiez, "A survey of commercial & open source unmanned vehicle simulators," in *Proceedings of the IEEE International Conference on Robotics and Automation (ICRA'07)*, pp. 852–857, Roma, Italy, April 2007.

16

Optimal Configuration of Virtual Links for Avionics Network Systems

Dongha An, Kyong Hoon Kim, and Ki-Il Kim

Department of Informatics, Research Center for Aerospace Parts Technology, Gyeongsang National University, Jinju-daero 501, Jinju 660-701, Republic of Korea

Correspondence should be addressed to Kyong Hoon Kim; khkim@gnu.ac.kr

Academic Editor: Linda L. Vahala

As the bandwidth and scalability constraints become important design concerns in airborne networks, a new technology, called Avionics Full Duplex Switched Ethernet (AFDX), has been introduced and standardized as a part 7 in ARNIC 664. However, since previous research interests for AFDX are mainly bounded for analyzing the response time where flows information is given, configuration problem for both Maximum Transmission Unit (MTU) and Bandwidth Allocation Gap (BAG) over virtual links in AFDX networks has not been addressed yet even though it has great impact on required bandwidth. Thus, in this paper, we present two configuration approaches to set MTU and BAG values on virtual links efficiently while meeting the requirement of AFDX. The first is to search available feasible configuration (MTU, BAG) pairs to satisfy application requirements as well as AFDX switch constraints, and the second is to get an optimal pair to minimize required bandwidth through well-known branch-and-bound algorithm. We analyze the complexity of the proposed algorithm and then evaluate the proposed algorithm by simulation. Finally, we prove that the proposed schemes are superior to general approach in the aspects of speed and required bandwidth in AFDX networks.

1. Introduction

Due to increasing demands in Aircraft Data Networks (ADNs), such as high available bandwidth, minimum wiring to reduce the weight, and low development cost, typical communications technologies are expected to be replaced by new one. In other words, since ARNIC 429, MIL-STD-1553, and ARNIC 629 cannot accommodate the mentioned demands completely, a new technology, called Avionics Full Duplex Switched Ethernet (AFDX), has been implemented and then standardized for new ADN [1–3]. As a new technology for ADN, reliable and deterministic property as well as implementation cost should be considered. As a result, AFDX was introduced and recently deployed in data networks on the Airbus 380 aircraft.

In the point of technical issue, the AFDX follows original Ethernet for compatibility and scalability properties and extends it to ensure deterministic behavior and high reliability. In order to comply with the stringent requirements of ADNs as well as ensuring them, two new functions are introduced and implemented in AFDX. One is traffic control by guaranteeing the bandwidth of each application, and the other is duplicated transmission for reliability along dual redundant channels. While the former aims to bound the jitter and transmit latency within the deadline, the latter transmits the same data stream over disjoint networks concurrently. To achieve this goal, virtual links are introduced and created for each (source, destination) pair in AFDX. By controlling these virtual links, deterministic behaviors are completely guaranteed. Thus, determining flows for the same virtual link and its properties become the network designer's great task when it comes to configuring AFDX networks.

Two parameters of virtual links in AFDEX networks are important in terms of real-time requirements: BAG (Bandwidth Allocation Gap) and MTU (Maximum Transfer Unit). BAG is the time duration or period between two consecutive frames, where the value of BAG is 1 and 128 msec in a form of power of 2. MTU is the size of message in each frame. Thus, the AFDX configuration problem is to determine these two parameters of each virtual link without missing the deadline.

In [4], they provided several algorithms and integer-linear programming formulation in order to solve the transmission parameters of virtual links as well as data routing over links in the AFDEX networks. They provided optimal solutions to minimize reserved bandwidth as well as actually consumed bandwidth, respectively. In [5], modeling method for AFDX frame management was introduced to ascertain the reliability properties of design. Another important property of AFDX, reliability through redundant transmissions, was analyzed by formal method in [6].

Another related work is to analyze the system metric such as response time of AFDX networks. In [7], they applied network calculus, queuing networks simulation and model checking into evaluating bounding end-to-end delays on AFDX networks. In [8], they showed that Trajectory approach which analyzes the worst-case delays throughout message flows outperforms the network calculus method under industrial configuration. In [9], they obtain worst-case latency and output jitter for the network messages by defining a real-time model for a communications network based on AFDX. Another approach in [10] is performance evaluation system through simulation with popular NS-2 and analysis of impact of several system parameters such as scheduling algorithm.

Since most of existing research work has assumed the preconfigured networks in advance, configuration problem should be mentioned prior to deployment. To achieve this goal, in this paper, we focus on two main problems: (i) finding feasible BAG and MTU parameters of virtual links in an AFDX switch for given virtual links of messages and (ii) providing an optimal solution in feasible pairs to minimize the total bandwidth. We define two problems formally and then solve the problems using the branch-and-bound technique. While our earlier work [11] only provided an algorithm to solve the first problem, this paper improves and extends it to include optimal solution for the purpose of minimizing the total bandwidth. Finally, complexity of the proposed scheme and experimental results through diverse simulation scenarios are given to demonstrate the suitability of the proposed scheme.

The rest of this paper is organized as follows. The system model and the problem definition are provided in Section 2. The solution of a feasible configuration problem is dealt with in Sections 3 and 4, while the optimal algorithm is provided in Section 5. Performance evaluations are shown in Section 6. Finally, conclusion and further work followed in Section 7.

2. System Model and Problem Definition

2.1. System Model. In this paper, we assume that an avionics system is composed of many computing or sensing units and AFDX network switches to connect other computing units. Thus, an AFDX message is uniquely defined by UDP source and destination ports, as shown in Figure 1. Sine we focus on real-time AFDX messages, a message flow f_i is defined by (l_i, p_i), where l_i is the payload of the message in bytes and p_i is Message Transmit Cycle (MTC) of the message in msec. That is, a message of l_i bytes is generated every p_i time units and is delivered to the destination application.

The basic communication unit in AFDEX networks is defined as a *virtual link* (VL). For example, Figure 1 shows three virtual links among LRUs. These virtual links sharing physical links are scheduled in AFDX network switches. Furthermore, multiple applications transmit real-time messages throughout a common virtual link if their source and destination units are the same. In the example of Figure 1, two application messages are shared in the virtual link VL_3.

There are two important parameters in a virtual link. The first is Bandwidth Allocation Gap (BAG) to specify a periodic frame. In AFDX switches, a BAG is defined by a value of 2^k msec, where $k = 0, 1, \ldots, 7$. As all BAGs are 2^k msec, virtual links are multiplexed in AFDX switches. The second parameter is Maximum Transfer Unit (MTU) of the message in bytes at each frame. Payloads of applications in a virtual link are transmitted within maximum MTU bytes in a single frame. If the size of a payload is greater than the MTU, it is fragmented into multiple frames. Therefore, a virtual link VL_i is defined by (BAG_i, MTU_i, F_i) as follows:

(i) BAG_i: Bandwidth Allocation Gap or period of VL_i in a value of 2^k msec where $k = 0, 1, \ldots, 7$.

(ii) MTU_i: maximum transfer unit or message size of VL_i in bytes.

(iii) F_i: a set of message flows in VL_i, where the jth message flow is denoted as $f_{i,j} = (l_{i,j}, p_{i,j})$.

2.2. Problem Definition. For a given virtual link VL_i, MTU and BAG are configured so as to meet all the real-time requirements of message flows in the link. If the payload of a message is greater than the MTU size, it is transmitted in multiple fragmented packets. Since all BAGs of VLs are harmonic, the schedulability analysis is easily derived by utilization analysis. Thus, (1) tells the message constraint of VL_i with n_i messages to guarantee the real-time requirement of all message flows in the link [2]:

$$\sum_{j=1}^{n_i} \frac{\lceil l_{i,j}/\text{MTU}_i \rceil}{p_{i,j}} \le \frac{1}{\text{BAG}_i}. \tag{1}$$

Let us assume that the system has N VLs on an AFDX switch with B bandwidth in bps. Each VL_i is configured with (MTU_i, BAG_i), so that MTU_i bytes are transmitted every BAG_i msec. In addition, each VL message requires the overhead of 67 bytes as shown in Figure 2. Since the total bandwidth of VLs should not exceed the network bandwidth, the following bandwidth constraint should be met:

$$8\sum_{i=1}^{n} \frac{\text{MTU}_i + 67}{\text{BAG}_i} \times 10^3 \le B. \tag{2}$$

The last constraint of virtual link scheduling is about jitter. The maximum allowed jitter on each virtual link in

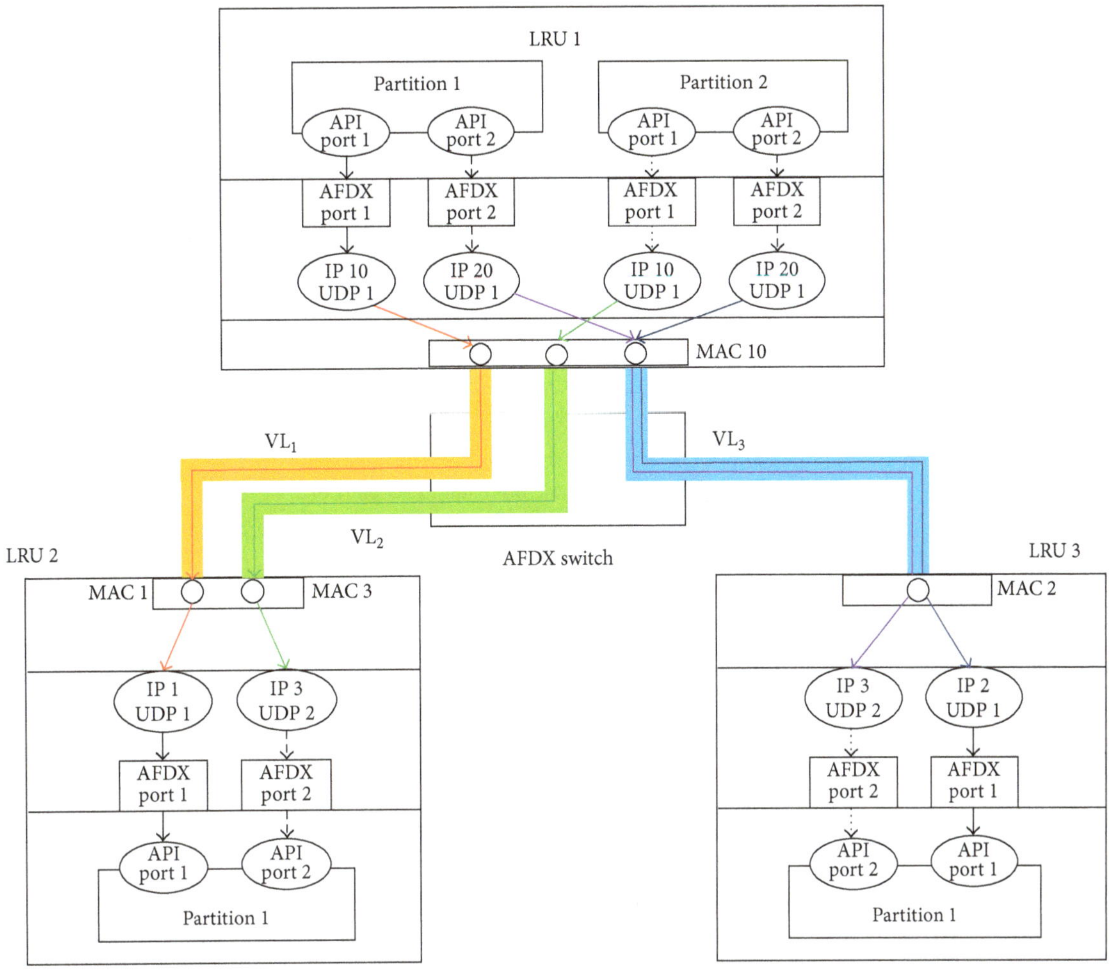

FIGURE 1: An example of virtual links in an AFDX switch.

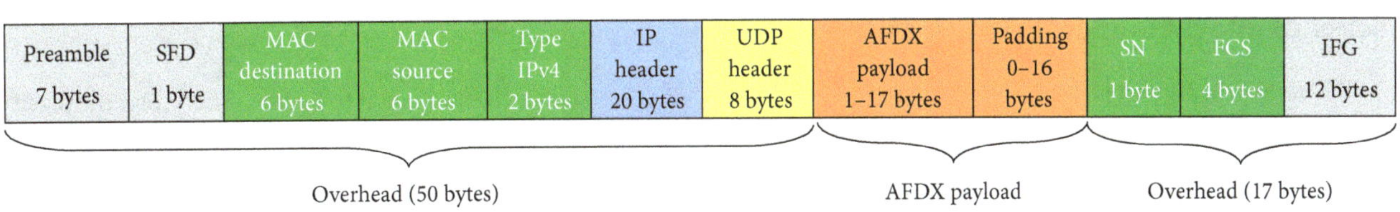

FIGURE 2: The AFDX frame structure and its overhead.

the ARINC 664 specification requires 500 μsec [2]. Thus, the following equation tells the jitter constraint, where 40 μsec is the typical technological jitter in hardware level to transmit an Ethernet frame:

$$40 + \frac{8\sum_{i=1}^{n}(67 + \mathrm{MTU}_i)}{B} \leq 500. \quad (3)$$

In this paper, we define two problems related to configuration of virtual links of an AFDX switch. The first problem is to find a feasible configuration of BAG and MTU pairs of virtual links, which satisfies three constraints of (1), (2), and (3). This is useful and important when an administrator finds a feasible configuration of a given set of real-time messages in the AFDX switch. The following tells the formal definition of the problem.

Definition 1. For a given set of virtual links $V = \{\mathrm{VL}_i \mid i = 1, \ldots, N\}$, the problem of *AFDX-CONF* is to determine $(\mathrm{BAG}_i, \mathrm{MTU}_i)$ of each VL_i so as to satisfy three constraints of (1), (2), and (3), where $\mathrm{BAG}_i \in \{1, 2, 4, 8, 16, 32, 64, 128\}$ and $\mathrm{MTU}_i \in \{1, 2, \ldots, 1471\}$.

The second one is to find an optimal configuration of virtual links for the purpose of minimizing the total bandwidth. The bandwidth remaining after reserving real-time flows is generally used for non-real-time network traffic. Thus, it is important to find a configuration with the least bandwidth as

Algorithm Find_Feasible_BAG_MTU (VL_i)
(1) $N_{step} \leftarrow \varnothing$
(2) **for each** message $f_{i,j}$ in VL_i **do**
(3) $\quad frag \leftarrow \lceil l_{i,j}/(\lceil l_{i,j}/p_{i,j}\rceil)\rceil$
(4) $\quad$ **while** $frag \geq 1$ **do**
(3) $\quad\quad \lceil m \leftarrow l_{i,j}/\text{frag}\rceil$
(5) $\quad\quad N_{step} \leftarrow N_{step} \cup \{m\}$
(6) $\quad\quad frag \leftarrow frag - 1$
(7) $\quad$ **endwhile**
(8) **endfor**
(9) **for** k **from** 0 **to** 7 **do**
(10) $\quad MTU_{i,k} \leftarrow 0$
(11) $\quad MTU_{i,k} \leftarrow$ the least $m \in N_{step}$ s.t. $\sum_{j=1}^{n_i}(\lceil l_{i,j}/m\rceil/p_{i,j}) \leq 1/2^k$
(12) **endfor**

ALGORITHM 1: Algorithm of feasible BAG and MTU pairs of a VL.

long as the configuration is feasible. The following tells the second problem definition.

Definition 2. For a given set of virtual links $V = \{VL_i \mid i = 1, \ldots, N\}$, the problem of *AFDX-BOPT* is to find (BAG_i, MTU_i) of each VL_i so as to minimize the total bandwidth and satisfy three constraints of (1), (2), and (3), where $BAG_i \in \{1, 2, 4, 8, 16, 32, 64, 128\}$ and $MTU_i \in \{1, 2, \ldots, 1471\}$.

We solve both problems of *AFDX-CONF* and *AFDX-BOPT* in two steps. The first step is to find the list of (BAG_i, MTU_i) which guarantees the schedulability of message flows in VL_i. Each (BAG_i, MTU_i) should be selected such that it satisfies the constraint of (1) (in Section 3). Then, we find the solution of given virtual links with consideration of two constraints of (2) and (3) (in Sections 4 and 5).

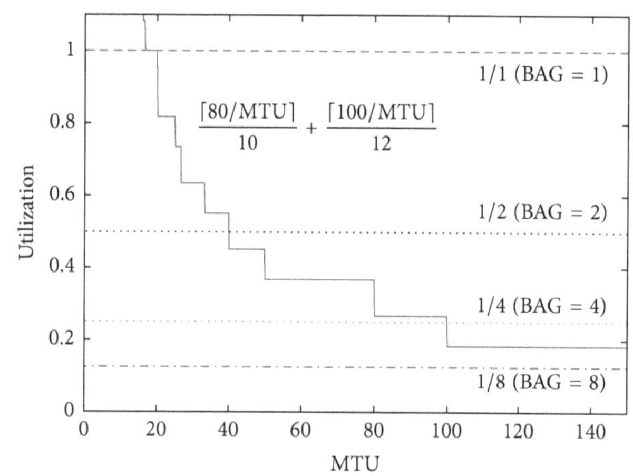

FIGURE 3: An example of feasible BAG and MTU of a virtual link.

3. Schedulable BAG and MTU Pairs of a VL

In AFDX networks, multiple applications should send their messages to a virtual link queue in the system, so that the scheduling issue for those messages is required. Let us consider a virtual link VL_1 with two message flows of $f_{1,1}(80, 10)$, $f_{1,2}(100, 12)$ as an example. The values of BAG and MTU of VL_1 are set to satisfy (1) in order to meet the real-time requirement of two messages. The left side of (1) is shown in Figure 3 as a step function, while 1/BAG is also drawn in the figure for different BAG values.

For a given BAG_i, there exist many MTUs which satisfy (1). For example, when $BAG_1 = 1$, all MTUs can be used if $MTU \geq 17$, as shown in Figure 3. Since a longer MTU size requires more bandwidth and jitter, the smallest value should be selected. Thus, MTU_1 of the example VL_1 is 17 bytes when BAG_1 is 1 msec. Similarly, MTUs of VL_1 for BAGs with 2 msec and 4 msec are given by 40 bytes and 100 bytes in each, as shown in Figure 3.

When the MTU size is greater than the maximum payload size of messages, the required utilization is not changed. For example, the lower bound of the utilization of VL_1 is given by about 0.1834 at MTU = 100. This implies that there is no MTU which guarantees the schedulability of two messages if BAG is greater than or equal to 8 msec. Therefore, the feasible solutions (BAG_1, MTU_1) of VL_1 are given by (1, 17), (2, 40), and (4, 100).

The pseudocode of Algorithm 1 describes how to obtain the set of feasible BAG and MTU pairs of a given virtual link VL_i. The first part of the algorithm gathers all step integers at which the utilization function begins a new piecewise constant due to the ceiling function. We denote the set of such step integers as N_{step}. For each message $f_{i,j}$, such step points are derived and added into N_{step} (lines 1–8).

Then, for each 2^k value, we find the minimum MTU which satisfies (1) (lines 9–12). We denote $MTU_{i,k}$ by the feasible MTU in case of $BAG_i = 2^k$ for a virtual link VL_i. If there is no feasible MTU, then $MTU_{i,k} = 0$. For a given n_i flows, the time complexity of Algorithm 1 is $O(n_i \cdot |N_{step}|)$ since we have to find and check the feasibility at each step point of messages.

4. Feasible BAG and MTU Pairs of VLs

4.1. Problem Definition. The problem of finding feasible BAG and MTU pairs of a given set of virtual links is not trivial. For

TABLE 1: An example of virtual links ($B = 1$ Mbps).

	Flows ($f_{i,j}$)	Payload ($l_{i,j}$)	MTC ($p_{i,j}$)	Feasible BAG and MTU pairs $s_{i,k}(\text{bag}_{i,k}, \text{MTU}_{i,k})$
VL_1	$f_{1,1}$	200	80	(1, 5), (2, 9), (4, 17), (8, 34), (16, 67), (32, 200)
	$f_{1,2}$	250	160	
VL_2	$f_{2,1}$	250	220	(1, 6), (2, 12), (4, 25), (8, 50), (16, 100), (32, 200)
	$f_{2,2}$	200	40	

example, let us consider the example of two virtual links of Table 1 where the network speed (B) is given by 1 Mbps. For each virtual link, the feasible BAG and MTU pairs are derived by Algorithm 1, as shown in the last column of Table 1. Now, a new problem arises about selecting appropriate BAG and MTU pairs of two virtual links so as to meet both constraints of (2) and (3).

There are some tradeoffs among feasible BAG and MTU pairs of a virtual link VL_i. Solutions with smaller BAG provide less jitter due to smaller MTU size, while they require more bandwidth due to overhead of fragmentation. For example, if we select (1, 5) and (1, 6) as (BAG, MTU) of two VLs of Table 1, it does not meet the bandwidth constraint of (2). On the contrary, if (2, 9) and (2, 12) are selected as (BAG, MTU) of two VLs, this configuration does not meet the jitter constraint of (3). The selection of (1, 5) and (2, 12) of VL_1 and VL_2 satisfies both constraints so that all messages in VLs meet their real-time requirements.

Let us assume that $\text{MTU}_{i,k}$ is derived from Algorithm 1 for each i and k. And, let us denote it by

$$X_{ik} = \begin{cases} 1 & \text{if BAG and MTU of } VL_i \text{ are set as } 2^k \text{ and } \text{MTU}_{i,k} \text{ in each,} \\ 0 & \text{otherwise.} \end{cases} \tag{4}$$

Then, the problem of *AFDX-CONF* can be stated as the following linear integer problem with $n \times 8$ binary variables and three constraints: to find X_{ik}

subject to

$$\sum_{k=0}^{7} X_{ik} = 1$$
$$\sum_{i=1}^{n} \left(\sum_{k=0}^{7} X_{ik} \cdot \frac{\text{MTU}_{i,k} + 67}{2^k} \right) \le \frac{B}{8000} \tag{5}$$
$$\sum_{i=1}^{n} \left(\sum_{k=0}^{7} X_{ik} \cdot (67 + \text{MTU}_{i,k}) \right) \le 460 \cdot \frac{B}{8}.$$

4.2. Algorithm. For given N virtual links, the exhaustive search of the problem *AFDX-CONF* takes $O(8^N)$ since each virtual link might have maximum eight solutions. In this paper, we provide an enhanced branch-and-bound algorithm proposed in [11] in order to find a feasible solution for given N virtual links with their feasible BAG and MTU pairs derived by Algorithm 1. In [11], we used the cumulative bandwidth and jitter for the pruning condition. In this work, however, we define the bound functions of bandwidth and jitter for the remaining levels in the search tree in order to find the pruning node as early as possible. The following is the pruning condition and branch-and-bound strategy of the proposed algorithm.

(i) Pruning Condition. The pruning condition is two constraints of (2) and (3). The algorithm examines whether the solutions in the subset satisfy both constraints by using the bound function of minimum values of remaining levels. The algorithm stops the search of the subset which already violates one of two constraints.

(ii) Branch-and-Bound Strategy. The algorithm searches a feasible solution in a leaf node in depth-first-search (DFS) manner. This algorithm finds a feasible solution when it reaches any leaf node in the search tree.

Algorithm 2 finds the set of feasible BAG and MTU pairs of each virtual link by calling the function Find_Feasible_BAG_MTU ( ) proposed in Algorithm 1 (lines 1, 2). Let us denote σ_i by the set of feasible BAG and MTU pairs ($s_{i,k}$) of virtual link i. Then, the bounds of bandwidth and jitter in virtual link i are found by selecting the minimum values (lines 4–7).

The function *EDFS_BandB* in Algorithm 2 is the recursive implementation at level i in the search tree. Two values of B_{prev} and J_{prev} are two bounds of bandwidth and jitter of sub-solutions from VL_i to VL_N. For each $s_{i,k} = (\text{bag}_{i,k}, \text{MTU}_{i,k})$, two constrains of (2) and (3) are checked including a new solution of VL_i (lines 14–16). If either of the two constraints is not satisfied, it is pruned. Otherwise, the depth-first-search is continued with two updated bound values (line 17).

When the search reaches a leaf node, the function returns *true* (line 12). The return value of calling *EDFS_BandB* is *true*; the final solution S is updated as to include $s_{i,k}$ (line 19) and the function returns *true*. Thus, the problem of *AFDX-CONF* is solved by Algorithm 2. If the return value of *DFS_BandB* (B_0, J_0, 1, S) is *true*, a feasible solution is stored

Algorithm Find_Feasible_Configurations (V)
/* $V = \{\mathrm{VL}_i \mid i = 1, \ldots, N\}$ */
for i **from** 1 **to** N **do**
  call **Find_Feasible_BAG_MTU** (VL_i)
$S \leftarrow \varnothing$
for i **from** 1 **to** N **do**
  $B_i \leftarrow \min_{s_{i,k} \in \sigma_i} \{(\mathrm{MTU}_{i,k} + 67)/\mathrm{bag}_{i,k}\}$
  $J_i \leftarrow \min_{s_{i,k} \in \sigma_i} \{\mathrm{MTU}_{i,k} + 67\}$
endfor
$B_0 \leftarrow \sum_{i=1}^{N} B_i$
$J_0 \leftarrow \sum_{i=1}^{N} J_i$
result $\leftarrow$ **EDFS_BandB** ($B_0, J_0, 1, S$)
return S
Function EDFS_BandB ($B_{\mathrm{prev}}, J_{\mathrm{prev}}, i, S$)
if $i = N + 1$ **then return** *true*
for each $s_{i,k}$ of VL_i **do**
  $B_{\mathrm{curr}} \leftarrow B_{\mathrm{prev}} - B_i + (\mathrm{MTU}_{i,k} + 67)/\mathrm{bag}_{i,k}$
  $J_{\mathrm{curr}} \leftarrow J_{\mathrm{prev}} - J_i + \mathrm{MTU}_{i,k} + 67$
  if $B_{\mathrm{curr}} \leq B/8000$ **and** $J_{\mathrm{curr}} \leq 460 \cdot B$ **then**
    result $\leftarrow$ **EDFS_BandB** ($B_{\mathrm{curr}}, J_{\mathrm{curr}}, i + 1, S$)
    if *result* = *true* **then**
      $S \leftarrow S \cup \{s_{i,k}\}$
      return *true*
    endif
  endif
endfor
return *false*

ALGORITHM 2: The proposed algorithm for *AFDX-CONF*.

in S. Otherwise, the empty set is returned, which implies no feasible configuration is found for a given set of virtual links.

5. Optimal Solution of BAG and MTU Pairs for Reducing Bandwidth

In this section, we focus on finding an optimal configuration of virtual links in terms of the used bandwidth of the configuration. It is important to minimize the total bandwidth of virtual links for real-time traffic because we can use the remaining bandwidth for non-real-time network traffic in avionics systems. The problem of *AFDX-CONF* finds a feasible solution which satisfies three constraints of AFDX switch as soon as possible, so that it does not minimize the bandwidth. However, *AFDX-BOPT* finds a feasible solution which minimizes the total bandwidth, so that the problem is rewritten as follows:

to minimize

$$8000 \sum_{i=1}^{n} \left(\sum_{k=0}^{7} X_{ik} \cdot \frac{\mathrm{MTU}_{i,k} + 67}{2^k} \right) \tag{6}$$

subject to

$$\sum_{k=0}^{7} X_{ik} = 1 \tag{7}$$

$$\sum_{i=1}^{n} \left(\sum_{k=0}^{7} X_{ik} \cdot \frac{\mathrm{MTU}_{i,k} + 67}{2^k} \right) \leq \frac{B}{8000} \tag{8}$$

$$\sum_{i=1}^{n} \left(\sum_{k=0}^{7} X_{ik} \cdot (67 + \mathrm{MTU}_{i,k}) \right) \leq 460 \cdot \frac{B}{8}. \tag{9}$$

For example, there are two feasible configurations in Table 1: {(1, 5), (2, 12)} and {(2, 9), (1, 6)} under the bandwidth constraint of 1 Mbps. The first configuration requires 892 kbps, while the second one uses 888 kbps. Thus, in this case, {(2, 9), (1, 6)} is the optimal configuration of two virtual links.

In order to solve the problem, we provide a branch-and-bound algorithm. The bound function is the same as one in Algorithm 2, so that we use the minimum bandwidth and jitter for bound values of each node. Thus, the pruning condition is two constraints of (8) and (9) using the bound functions. The branch strategy is different from DFS. Since the algorithm finds an optimal solution as soon as possible, we select a node with the minimum bandwidth bound function and branch its child nodes. The pseudocode of this is shown in Algorithm 3. Since the branch-and-bound technique searches all possible cases in the worst-case, the time complexity of Algorithm 3 is $O(2^N)$ for given N virtual links.

In Algorithm 3, a *node* has four data: *node.level* for the node level, *node.B* for the bandwidth bound, *node.J* for the jitter bound, and *node.S* for the selected configuration from the root to the node. We use a priority-queue for managing live nodes in the order of bandwidth bound. The function *insert_node_in_Q* (*i*, *B*, *J*, *S*) makes a node with four data and inserts it into the priority queue. Then, the function *get_first_in_Q* ( ) returns the minimum-bandwidth node in currently live nodes in the queue. When the algorithm reaches a leaf node (line 12), it finds an optimal solution and returns the configuration (line 13).

The branch strategy is implemented by the priority queue. The function call of *get_first_in_Q* ( ) selects such node to branch among live nodes. For all child nodes of the selected node, the bound functions of bandwidth and jitter are updated (lines 16-17). The pruning condition is checked in line 18. If a new child node is not pruned, it is added in the priority queue (line 19). The algorithm repeats this procedure until it finds an optimal solution or there is no live node in the queue. If there is no feasible configuration for virtual links, it returns the empty set for notifying no feasible solution (line 22).

6. Performance Evaluations

In this section, we show performance evaluation of the proposed algorithm. First, we evaluate the execution time of the proposed algorithms compared with the brute-force search and the previous work [11]. In the experiments, we generate five virtual links with two message flows in each virtual link. The payload of a message is randomly generated from 20 to 80 bytes. The MTC or period of a message is

```
Algorithm Find_Minimum_Bandwidth_Configuration (V)
for i from 1 to N do
    call Find_Feasible_BAG_MTU (VL_i)
for i from 1 to N do
    B_i ← log_{s_{i,k}∈σ_i} {(MTU_{i,k} + 67)/bag_{i,k}}
    J_i ← log_{s_{i,k}∈σ_i} {MTU_{i,k} + 67}
endfor
B_0 ← Σ_{i=1}^{N} B_i
J_0 ← Σ_{i=1}^{N} J_i
insert_node_in_Q (0, B_0, J_0, ∅);
while is_empty_Q ( ) = false do
    node ← get_first_in_Q ( );
    if node.level = N then
        return node.S;
    i ← node.level + 1
    for each s_{i,k} of VL_i do
        B_curr ← node.B − B_i + (MTU_{i,k} + 67)/bag_{i,k}
        J_curr ← node.J − J_i + MTU_{i,k} + 67
        if B_curr ≤ B/8000 and J_curr ≤ 460 · B then
            insert_node_in_Q (i, B_curr, J_curr, node.S ∪ {s_{i,k}});
    endfor
endwhile
return ∅
```

ALGORITHM 3: The *Bounded_BFS_BandB* algorithm.

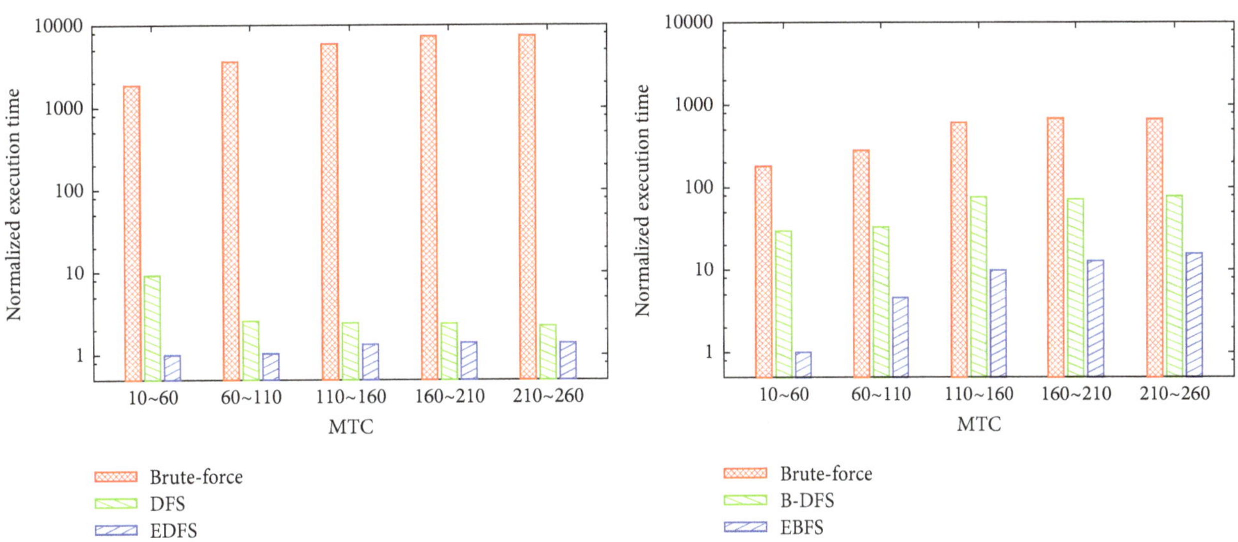

(a) AFDX-CONF

(b) AFDX-BOPT

FIGURE 4: Normalized execution times of algorithms.

randomly selected among five different intervals, as shown in Figure 4. The network bandwidth is set as 6 Mbps.

For each case of Figure 4, we generate 5000 random sets of five virtual links and measure the average execution time of algorithms. Figure 4(a) shows the normalized execution times of the brute-force search, *DFS* in [11], and *EDFS* in this paper for solving *AFDX-CONF*. Similarly, Figure 4(b) shows the performance of the proposed algorithm *EBFS* compared to the brute-force search and the modified version of DFS [11] for solving *AFDX-BOPT*.

As shown in Figure 4, the proposed algorithms run much faster than the exhaustive search algorithm due to the branch-and-bound technique. Let us note that the y-axis in Figure 4 is log-scale value. Table 2 shows the average number of BAG and MTU pairs per virtual link. As the MTCs become larger, the number of possible solutions of each virtual link is increased, which requires more execution time, as shown in Figure 4. Since we use bound functions for bandwidth and jitter, the proposed ones are faster than the previous algorithm in [11].

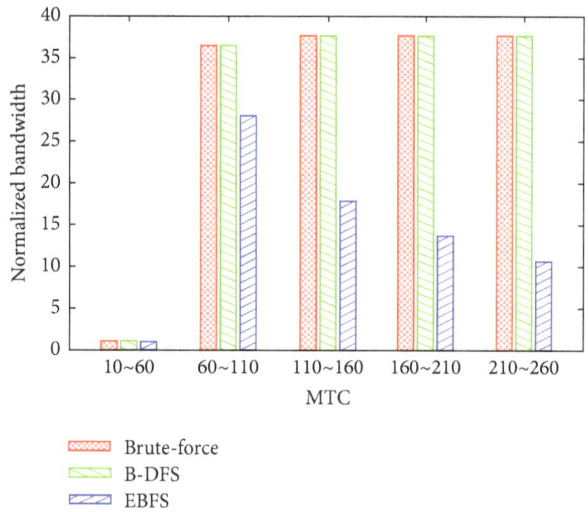

FIGURE 5: Bandwidth comparison.

TABLE 2: The average number of BAG and MTU pairs per VL.

MTC	10~60	60~110	110~160	160~210	210~260
BAG and MTU pairs per VL	4.7	6.0	6.8	7.0	7.0

Next, we compared the bandwidth reserved by the solutions of *AFDX-BOPT* with feasible solutions of *AFDX-CONF*. As shown in Figure 5, the proposed algorithm for *AFDX-BOPT* reduces much bandwidth, so that the remaining bandwidth can be used for other non-real-time network traffic.

We also analyzed the effect of algorithm execution times in terms of the number of virtual links. In Figure 6, we increased the number of virtual links from 1 to 6, where each virtual link includes two messages. As shown in Figure 6, the execution time of brute-force search increases exponentially since the search space is $O(2^N)$. Let us note that y-axis in Figure 6 is log-scale. Although the time complexity of the proposed algorithm is $O(2^N)$, both *EDFS* and *EBFS* are much faster than the previous algorithms in [11]. This is mainly because we use efficient bound functions in the algorithms. The execution time of *EBFS* in case of 6 virtual links is decreased since the algorithm decides that there is no solution in the upper-layer of solution trees.

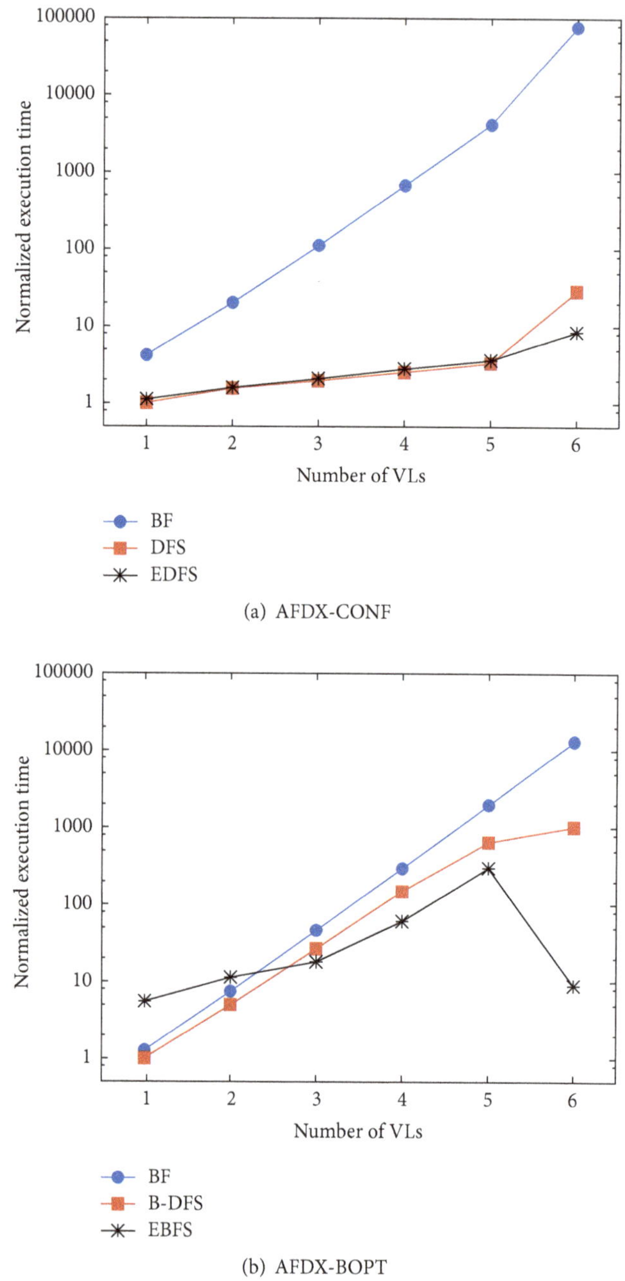

FIGURE 6: Normalized execution times with respect to the number of virtual links.

7. Conclusions

In this paper, we provided an efficient algorithm to find feasible configuration of an AFDX switch for the purpose of meeting the real-time requirements of all messages in avionics. Two important parameters of BAG and MTU of virtual links are considered in the configuration problem. The proposed algorithm is based on branch-and-bound technique so that the simulation results show that it is faster than the exhaustive search algorithm of the previous work.

We also provide an optimal solution of the problem for the purpose of reducing the total bandwidth of configuration. Throughout simulations, we showed that the algorithm runs fast and reduces much bandwidth compared with the algorithm to find a feasible set. The proposed algorithm is useful when the remaining bandwidth for real-time traffic is used for other network traffic.

Since the AFDX network configuration becomes an important issue in avionics systems, we will investigate many problems based on the results of this paper. For example, we will extend the problem into multiple AFDX switches or discuss the routing issues through the networks.

Conflict of Interests

The authors declare that there is no conflict of interests regarding the publication of this paper.

Acknowledgments

This work was supported by Basic Science Research Program through the National Research Foundation of Korea (NRF) funded by the Ministry of Science, ICT & Future Planning (no. NRF-2015R1A1A1A05001369), and the BK21 Plus Program (Research Team for Software Platform on Unmanned Aerial Vehicle, 21A20131600012) funded by the Ministry of Education (MOE, Korea) and National Rcsearch Foundation of Korea (NRF).

References

[1] R. L. Alena, J. P. Ossenfort, K. I. Laws, A. Goforth, and F. Figueroa, "Communications for integrated modular avionics," in *Proceedings of the IEEE Aerospace Conference*, pp. 1–18, IEEE, Big Sky, Mont, USA, March 2007.

[2] AFDX Tutorial, http://www.techsat.com/fileadmin/media/pdf/infokiosk/TechSAT_TUT-AFDX-EN.pdf.

[3] AFDX/ARNIC 664 Tutorial, http://www.cems.uwe.ac.uk/~ngunton/afdx_detailed.pdf.

[4] A. Al Sheikh, O. Brun, M. Chéramy, and P.-E. Hladik, "Optimal design of virtual links in AFDX networks," *Real-Time Systems*, vol. 49, no. 3, pp. 308–336, 2013.

[5] I. Saha and S. Roy, "A finite state modeling of AFDX frame management using spin," in *Formal Methods: Applications and Technology*, vol. 4346 of *Lecture Notes in Computer Science*, pp. 227–243, Springer, Berlin, Germany, 2007.

[6] J. Täubrich and R. von Hanxleden, "Formal specification and analysis of AFDX redundancy management algorithms," in *Computer Safety, Reliability, and Security*, vol. 4680 of *Lecture Notes in Computer Science*, pp. 436–450, Springer, Berlin, Germany, 2007.

[7] H. Charara, J.-L. Scharbarg, J. Ermont, and C. Fraboul, "Methods for bounding end-to-end delays on an AFDX network," in *Proceedings of the 18th Euromicro Conference on Real-Time Systems (ECRTS '06)*, pp. 193–202, Dresden, Germany, July 2006.

[8] M. Tawk, X. Liu, L. Jian, G. Zhu, Y. Savaria, and F. Hu, "Optimal scheduling and delay analysis for AFDX end systems," SAE Technical Paper 2011-01-2751, SAE International, 2011.

[9] J. J. Gutierrez, J. C. Palencia, and M. G. Harbour, "Response time analysis in AFDX networks with sub-virtual links and prioritized switches," in *XV Jornadas de Tiempo Real*, Santander, 2012.

[10] D. Song, X. Zeng, L. Ding, and Q. Hu, "The design and implementation of the AFDX network simulation system," in *Proceedings of the International Conference on Multimedia Technology (ICMT '10)*, pp. 1–4, Ningbo, China, October 2010.

[11] D. An, H. W. Jeon, K. H. Kim, and K. I. Kim, "A feasible configuration of AFDX networks for real-time flows in avionics systems," in *Proceedings of the International Workshop on Real-Time and Distributed Computing in Emerging Applications (REACTION '13)*, Vancouver, Canada, Decebmer 2013.

17

Evaluation of Matrix Square Root Operations for UKF within a UAV GPS/INS Sensor Fusion Application

Matthew Rhudy, Yu Gu, Jason Gross, and Marcello R. Napolitano

Department of Mechanical and Aerospace Engineering, West Virginia University, P.O. Box 6106, Morgantown, WV 26506, USA

Correspondence should be addressed to Matthew Rhudy, mbr5002@gmail.com

Academic Editor: Jinling Wang

Using an Unscented Kalman Filter (UKF) as the nonlinear estimator within a Global Positioning System/Inertial Navigation System (GPS/INS) sensor fusion algorithm for attitude estimation, various methods of calculating the matrix square root were discussed and compared. Specifically, the diagonalization method, Schur method, Cholesky method, and five different iterative methods were compared. Additionally, a different method of handling the matrix square root requirement, the square-root UKF (SR-UKF), was evaluated. The different matrix square root calculations were compared based on computational requirements and the sensor fusion attitude estimation performance, which was evaluated using flight data from an Unmanned Aerial Vehicle (UAV). The roll and pitch angle estimates were compared with independently measured values from a high quality mechanical vertical gyroscope. This manuscript represents the first comprehensive analysis of the matrix square root calculations in the context of UKF. From this analysis, it was determined that the best overall matrix square root calculation for UKF applications in terms of performance and execution time is the Cholesky method.

1. Introduction

The improvement of microprocessors and sensors has increased civilian use of Unmanned Aerial Vehicles (UAVs) for various applications, many of which requiring an accurate estimate of the aircraft attitude [1, 2]. For example, attitude estimation is an important requirement for UAV remote sensing applications such as 3D mapping with direct georeferencing [3] or constructing large mosaics [4]. Due to cost and weight restrictions [5, 6], high-quality military grade inertial navigation systems may not be practical [6, 7]. Therefore, attitude estimation algorithms that rely only on low-cost sensors have become essential for civilian applications. A popular approach to the attitude estimation problem involves fusing together information from a low-cost Inertial Navigation System (INS) with information from a Global Positioning System (GPS) receiver [8, 9]. Various formulations and analyses of GPS/INS sensor fusion exist in the literature [10–12], including a detailed comparison by the authors with respect to attitude estimation performance and computational cost [13].

The Unscented Kalman Filter (UKF) [14] is a well-known nonlinear estimator for use within GPS/INS sensor fusion [10, 11, 15]. Within this effort, the UKF is used to obtain an accurate estimate of the aircraft attitude, in particular the roll and pitch angles. For this type of problem, an alternative nonlinear estimator is the Extended Kalman Filter (EKF) [16], whose differences are discussed in detail in various sources [11, 12, 15, 17], including a sensitivity analysis by the authors with respect to different design parameters [18]. One primary difference between the EKF and UKF is that the UKF requires the calculation of a matrix square root, which is a computationally demanding operation [19]. In particular, computation is an important consideration for small UAV systems, due to resource limitations onboard the aircraft, for example, power, weight, and cost [5, 6]. The matrix square root computation is also an important consideration when designing a UKF algorithm because there are many different ways to compute the square root of a matrix, potentially with different accuracy and computational requirements [20].

The selection of matrix square root operation in the UKF differs among authors. Julier and Uhlmann state that "*the sigma points capture the same mean and covariance irrespective of the choice of matrix square root which is used. Numerically efficient and stable methods such as the Cholesky decomposition can be used* [14]." Crassidis [10] and Wendel et al. [21] also recommended the use of Cholesky decomposition. These claims, however, are provided without any theoretical or empirical justification. Stastny et al. found that using the Cholesky decomposition method caused divergence; therefore, they used the Schur method instead [22]. Some authors using the UKF do not explicitly state which matrix square root operation was used [23–26]. A different means of handling the square root operation was developed by van der Merwe and Wan called the "square-root UKF (SR-UKF)" method [19]. This method provides a prediction and update of the square root of the covariance matrix directly at each time step, which reduces computational requirements of the algorithm [19]. A simulated example was used to show that the SR-UKF and UKF performances were the same [19]. Since there are inconsistencies in the selected matrix square root method for UKF applications, a detailed comparison of available approaches is necessary. Although some matrix square root comparison studies exist, for example, by Higham [20] and Meini [27], these studies are mostly theoretical, with a few examples using known matrices, such as the Moler, Chebyshev-Vandermonde, and Frobenius matrices. These comparison studies, therefore, do not consider the potential error propagation effects that are introduced by application in a recursive filter such as the UKF.

This paper aims to expand upon the existing matrix square root comparison studies through an example application of the matrix square root within a UKF-based GPS/INS sensor fusion algorithm for attitude estimation that relies on experimentally collected flight data. By analyzing different matrix square root methods in the context of the UKF, a matrix square root is required at each discrete time step in the algorithm, allowing for a more general comparison since many different matrices are considered. In addition this recursive filtering application introduces the effects of the propagation of uncertainties in the matrix square root computation. Furthermore, the flight data used for this study were selected from a large library of data in order to obtain diversity with respect to different flight conditions, thus providing an additional level of generalization. The use of experimental data for this study provides a necessary empirical approach to a problem that has previously only been considered in a theoretical context. To evaluate the performance of the attitude estimation algorithm, the aircraft pitch and roll estimates are compared with measurements acquired independently from the IMU and GPS with a high-quality mechanical vertical gyroscope. This sensor fusion study builds upon previous work conducted by various West Virginia University (WVU) researchers [13, 18, 28, 29].

The work presented herein represents the first comprehensive analysis of matrix square root calculations used for UKF applications. This study utilizes a diverse set of experimental flight data to analyze the performance speed and accuracy of different matrix square root calculations in the context of a UKF-based GPS/INS sensor fusion algorithm. The rest of the paper is organized as follows. Section 2 describes the theoretical background required for this study. Section 3 discusses the UAV attitude estimation problem, including information about the UAV research platform as well as the problem formulation. Section 4 presents the results, and finally Section 5 provides a summary and discussion of the results.

2. Theoretical Background

2.1. Unscented Kalman Filter (UKF). Formulations of the UKF for GPS/INS sensor fusion are well known and can be found in a number of publications [10, 11, 14]. For this application of the UKF, the state vector is augmented to include navigation states, x, and process noise states, ν, leading to the augmented state vector, $x^a = [\,x^T \; \nu^T\,]^T$, where the superscript a denotes augmentation, and the process noise states are assumed to be zero mean Gaussian and mutually independent. The total dimension of the augmented state vector is denoted by l. The source of process noise modeled in this formulation is additive upon an input vector comprised of IMU measurements: $x_{k|k-1} = f(x_{k-1|k-1}, u_k + \nu_k)$, where the state prediction equations corresponding to the process noise states are simply $\nu_k = \nu_{k-1} + w_k$, where w is a vector of mutually independent zero mean Gaussian noise. These states are initialized to zero and remain constant with time. The effect of the process noise, however, is implemented implicitly by the use of the unscented transformation. Due to the state augmentation, the state covariance matrix is augmented accordingly to include the noise characteristics of the process noise states.

The UKF uses an unscented transformation [14] to obtain the *a posteriori* estimates of mean and covariance. The first step of the UKF implementation is to calculate $2l + 1$ sigma points based on the square-root of the augmented state covariance matrix:

$$\begin{aligned}\chi^a_{k-1|k-1} &= \begin{bmatrix}\chi^x_{k-1|k-1}\\ \chi^\nu_{k-1|k-1}\end{bmatrix}\\ &= \Big[x^a_{k-1|k-1} x^a_{k-1|k-1} + \eta\sqrt{P^a_{k-1|k-1}} x^a_{k-1|k-1}\\ &\quad -\eta\sqrt{P^a_{k-1|k-1}}\Big],\end{aligned} \tag{1}$$

where χ is a matrix of state sigma points, and η is the sigma point spread parameter, given by [11]

$$\eta = \sqrt{l+\lambda}, \tag{2}$$

$$\lambda = l(\alpha^2 - 1), \tag{3}$$

where λ is the compound sigma point parameter, and α is the primary sigma point scaling parameter, which is suggested to vary between 0.001 and 1 [11].

The prediction step for the UKF consists of passing each state sigma point through the state prediction equations:

$${}^i\chi^x_{k|k-1} = f\left({}^i\chi^x_{k-1|k-1}, u_k + {}^i\chi^\nu_{k-1|k-1}\right), \quad i = 0, 1, \ldots, 2l, \tag{4}$$

where the superscript i denotes the $(i + 1)$th column of the matrix. Then, the state mean and covariance are predicted using a weighted average of the transformed sigma points using

$$x_{k|k-1} = \sum_{i=0}^{2l} w_m^i \, {}^i\chi^x_{k|k-1}, \tag{5}$$

$$P_{k|k-1} = \sum_{i=0}^{2l} w_c^i \left({}^i\chi^x_{k|k-1} - x_{k|k-1}\right)\left({}^i\chi^x_{k|k-1} - x_{k|k-1}\right)^T, \tag{6}$$

where w_m and w_c are weight vectors [11]. For the measurement update, this paper considers a problem that has linear output equations; therefore a linear Kalman filter update can be used [30].

2.2. Matrix Square Root Algorithms. An important requirement of the UKF algorithm is the calculation of the square root of the state covariance matrix, P. A covariance matrix by definition is both symmetric and positive semidefinite. To calculate the square root of a positive semidefinite $n \times n$ matrix, A, various methods can be used. The matrix principal square root, $A^{1/2}$, exists only for positive semidefinite matrices and is defined by [20]

$$A = A^{1/2} \cdot A^{1/2}. \tag{7}$$

If A is symmetric, it can be diagonalized using a similarity transformation [31], and the principal square root can be calculated using the diagonalization method:

$$A^{1/2} = X\Lambda^{1/2}X^{-1}, \tag{8}$$

where $\Lambda^{1/2}$ is a diagonal matrix with the square roots of the eigenvalues of A along the main diagonal, and X is a matrix containing a corresponding set of eigenvectors of A.

Another common matrix square root method is the Schur method, which uses the Schur decomposition:

$$A = UTU^*, \tag{9}$$

where T is an upper triangular matrix and U is a unitary matrix whose columns form the Schur basis of A [32], and the $(*)$ denotes the complex conjugate transpose of a matrix. Once in this form, the matrix square root can be calculated from

$$A^{1/2} = UT^{1/2}U^*, \tag{10}$$

where $T^{1/2}$ can be calculated algebraically since T, and therefore also $T^{1/2}$, are upper triangular matrices. Let $S = T^{1/2}$. The diagonal elements of S are calculated directly from the diagonal elements of T such that [20]

$$s_{ii} = \sqrt{t_{ii}}, \quad i = 1, \ldots, n. \tag{11}$$

The strictly upper triangular elements are then calculated using [20]

$$s_{ij} = \frac{t_{ij} - \sum_{k=i+1}^{j-1} s_{ik}s_{kj}}{u_{ii} + u_{jj}}, \quad j = 2, \ldots, n, \; i < j. \tag{12}$$

In addition to analytical methods for calculating a matrix square root, various iterative methods have been derived. One of the most common iterative methods is Newton's method, which can be written as

$$X_{k+1} = \frac{1}{2}\left(X_k + X_k^{-1}A\right), \quad X_0 = A, \tag{13}$$

where the matrix X converges quadratically to $A^{1/2}$ under certain conditions [33]. One variant of Newton's method is known as Denman Beavers (DB) iteration, given by [20]

$$\begin{aligned} X_{k+1} &= \frac{1}{2}\left(X_k + Y_k^{-1}\right), \quad X_0 = A, \\ Y_{k+1} &= \frac{1}{2}\left(Y_k + X_k^{-1}\right), \quad Y_0 = I, \end{aligned} \tag{14}$$

where the matrix X converges to $A^{1/2}$, and the matrix Y converges to $A^{-1/2}$. A product form of the Denman-Beavers iteration was identified by Cheng et al., which is a more efficient version of the Denman-Beavers iteration [34]:

$$\begin{aligned} M_{k+1} &= \frac{1}{2}\left(I + \frac{M_k + M_k^{-1}}{2}\right), \quad M_0 = A, \\ X_{k+1} &= \frac{1}{2}X_k\left(I + M_k^{-1}\right), \quad X_0 = A, \\ Y_{k+1} &= \frac{1}{2}Y_k\left(I + M_k^{-1}\right), \quad Y_0 = I, \end{aligned} \tag{15}$$

where M converges to the identity matrix, I, X converges to $A^{1/2}$, while Y converges to $A^{-1/2}$. Another iterative method is the cyclic reduction (CR) iteration [27]:

$$\begin{aligned} Y_{k+1} &= -Y_k Z_k^{-1} Y_k, \quad Y_0 = I - A, \\ Z_{k+1} &= Z_k + 2Y_{k+1}, \quad Z_0 = 2(I + A), \end{aligned} \tag{16}$$

where Y converges to 0 and Z converges to $4A^{1/2}$. A variant of the cyclic reduction iteration is the IN iteration given by [20]

$$\begin{aligned} X_{k+1} &= X_k + E_k, \quad X_0 = A, \\ E_{k+1} &= -\frac{1}{2}E_k X_{k+1}^{-1} E_k, \quad E_0 = \frac{1}{2}(I - A), \end{aligned} \tag{17}$$

where X converges to $A^{1/2}$, and E converges to 0.

All of the matrix square root algorithms discussed up to this point are methods of calculating the principle square root of a matrix. Another form of the matrix square root is found using the Cholesky decomposition [35]:

$$A = LL^T, \tag{18}$$

where L is a lower triangular matrix which can be considered as a form of the matrix square root. Although most numerical methods of calculating the Cholesky decomposition of a matrix require positive definiteness, there are ways to calculate this decomposition for positive semidefinite matrices [36, 37], although in general this result is not unique. For this application, the state covariance matrix, P, was positive definite at each time step. In general, however, it is possible in a given application for the state covariance matrix to be positive semidefinite; in particular this can occur if some linear combinations of the states are known perfectly, that is, zero uncertainty in that combination of states.

TABLE 1: Matrix square root algorithm computational requirement summary.

Algorithm	FLOPs
Diagonalization	$49n^3/3 - n^2/2 + 43n/6$
Schur	$85n^3/3$
Newton's iteration	$8n^3/3$ (per iteration)
Denman-Beavers (DB) iteration	$4n^3$ (per iteration)
Product DB iteration	$4n^3$ (per iteration)
Cyclic Reduction (CR) iteration	$14n^3/3$ (per iteration)
IN iteration	$14n^3/3$ (per iteration)
Cholesky	$n^3/3$

The computational complexity of the matrix square root operation is important for use in the UKF because it is a significantly expensive part of the UKF algorithm [19]. To analyze the computational complexity of the different matrix square root algorithms, the number of floating point operations (FLOPs) as a function of the matrix dimension, n, was determined from various sources [20, 35, 38]. These results were derived theoretically based on the fundamental requirements of the algorithm and are summarized in Table 1. Note that the number of FLOPs listed for the iterative methods is the number of FLOPs required for one iteration of the algorithm.

It is shown in Table 1 that each of the matrix square root algorithms requires computations on the order of n^3, but scaled by different factors. A graphical comparison of these requirements is presented later in the paper in Figure 3(b).

3. UAV Attitude Estimation Problem

3.1. GPS/INS Sensor Fusion Formulation. The sensor fusion formulation considered for this study has states given by the local Cartesian position r_x, r_y, and r_z, local Cartesian velocity, V_x, V_y, and V_z, and the attitude of the aircraft represented by the Euler angles, ϕ, θ, and ψ, corresponding to the roll, pitch, and yaw angles, respectively. Therefore, the state vector is given by $x = [r_x \; r_y \; r_z \; V_x \; V_y \; V_z \; \phi \; \theta \; \psi]^T$. Since the position and velocity states are given in the same coordinate frame, their relationship defines the position state equations:

$$\begin{bmatrix} \dot{r}_x \\ \dot{r}_y \\ \dot{r}_z \end{bmatrix} = \begin{bmatrix} V_x \\ V_y \\ V_z \end{bmatrix}. \tag{19}$$

An Inertial Measurement Unit (IMU) provides measurements of the linear accelerations as specific force, $a_{\tilde{x}}$, $a_{\tilde{y}}$, and $a_{\tilde{z}}$, and angular rates, p, q, and r. The IMU data are measured in the aircraft body frame, which is distinguished from the local Cartesian frame using $(\sim)$. The rotation between the aircraft body frame and the local Cartesian frame is determined from the aircraft attitude angles. When this rotation is applied to the acceleration, a coupled relationship between the velocity and attitude states is defined. This coupling is critical because correction of the velocity states allows for implicit regulation of the attitude states. This relationship defines the velocity state equations:

$$\begin{bmatrix} \dot{V}_x \\ \dot{V}_y \\ \dot{V}_z - g \end{bmatrix} = \begin{bmatrix} \cos\theta\cos\psi & -\cos\phi\sin\psi + \sin\phi\sin\theta\cos\psi & \sin\phi\sin\psi + \cos\phi\sin\theta\cos\psi \\ \cos\theta\sin\psi & \cos\phi\cos\psi + \sin\phi\sin\theta\sin\psi & -\sin\phi\cos\psi + \cos\phi\sin\theta\sin\psi \\ -\sin\theta & \sin\phi\cos\theta & \cos\phi\cos\theta \end{bmatrix} \begin{bmatrix} a_{\tilde{x}} \\ a_{\tilde{y}} \\ a_{\tilde{z}} \end{bmatrix}. \tag{20}$$

The acceleration due to gravity, g, must be removed from the IMU accelerometer measurements, since those measurements are relative to free fall. Since the direction of g varies in the aircraft body frame, it is not easy to remove this effect directly from the IMU accelerometer measurements. However, since g acts in the negative local Cartesian z-direction, g is instead subtracted from the local Cartesian z-acceleration in order to compensate for this effect, as shown in +(20).

The IMU measurements of roll, pitch, and yaw rates, p, q, and r, are related to the corresponding Euler angles, ϕ, θ, and ψ using the attitude state equations, also known as the Inverted Kinematic equations [39]:

$$\begin{bmatrix} \dot{\phi} \\ \dot{\theta} \\ \dot{\psi} \end{bmatrix} = \begin{bmatrix} 1 & \sin\phi\tan\theta & \cos\phi\tan\theta \\ 0 & \cos\phi & -\sin\phi \\ 0 & \sin\phi\sec\theta & \cos\phi\sec\theta \end{bmatrix} \begin{bmatrix} p \\ q \\ r \end{bmatrix}. \tag{21}$$

The state equations (19)–(21) are initially provided in the continuous time domain. In order to implement the discrete time data from the IMU, the state equations are discretized using a first-order approximation [40]:

$$x_k = f(x_{k-1}, u_k) = x_{k-1} + T_s f_c(x_{k-1}, u_k), \tag{22}$$

where T_s is the sampling time of the IMU, k is the discrete time index, f is the discrete state function, f_c is the continuous state function defined using (19)–(21), and u is the input vector given by the IMU measurements: $u = [a_{\tilde{x}} \; a_{\tilde{y}} \; a_{\tilde{z}} \; p \; q \; r]^T$. The source of process noise for this formulation is this input vector consisting of IMU measurements. Since information is available about the noise characteristics of the IMU, the process noise covariance matrix is approximated based on precalculated values of the variance of each component of the IMU measured in a static setting:

$$Q = \operatorname{diag}\left[\sigma^2(a_{\tilde{x}}) \; \sigma^2\left(a_{\tilde{y}}\right) \; \sigma^2(a_{\tilde{z}}) \; \sigma^2(p) \; \sigma^2(q) \; \sigma^2(r)\right]. \tag{23}$$

FIGURE 1: WVU YF-22 research UAVs.

Measurements from GPS are used to calculate the position and velocity of the receiver in the Earth-Centered-Earth-Fixed (ECEF) coordinate frame. These calculations are done within the GPS receiver. For navigation purposes, the position and velocity are transformed into a local Coordinate frame with origin at the WVU flight testing facility. Therefore, the GPS is used to obtain calculations of local position, r_x, r_y, and r_z, and local velocity, V_x, V_y, and V_z. These values are represented as the vector: $z = [r_x^{\text{GPS}}\ r_y^{\text{GPS}}\ r_z^{\text{GPS}}\ V_x^{\text{GPS}}\ V_y^{\text{GPS}}\ V_z^{\text{GPS}}]^T$. Since the GPS does not provide any information directly related to the aircraft attitude, the outputs of the system are defined as the position and velocity states of the system, leading to the following output equations:

$$y = h(x,u) = Hx = \begin{bmatrix} r_x & r_y & r_z & V_x & V_y & V_z \end{bmatrix}^T, \qquad (24)$$

where h is the observation function, which for this formulation due to linearity can be written using the observation matrix, $H = [\ I_{6x6}\ 0_{6x3}\ ]$, where I is an identity matrix, and 0 is a matrix of zeros with given dimensions. Due to the simplicity of the observation equations, the measurement noise comes directly from the GPS position and velocity calculations. The GPS noise is assumed to be zero mean Gaussian and mutually independent, which leads to a diagonal measurement noise covariance matrix. Similar to the process noise covariance matrix, this matrix is assumed constant and approximated based on precalculated values of the variance of each component of the GPS:

$$R = \text{diag}\left[\sigma^2(r_x^{\text{GPS}}),\ \sigma^2\left(r_y^{\text{GPS}}\right),\ \sigma^2(r_z^{\text{GPS}}),\ \sigma^2(V_x^{\text{GPS}}),\ \sigma^2\left(V_y^{\text{GPS}}\right),\ \sigma^2(V_z^{\text{GPS}})\right]. \qquad (25)$$

Note that this assumption does not model any direct correlation between the uncertainty in GPS position and velocity, which is reasonable since the measurements are obtained independently with the studied GPS receiver.

3.2. Software Implementation. The results of this study were derived using an off-line postprocessing analysis of previously recorded flight data. The attitude estimation algorithm using UKF was coded manually in-house using MATLAB. Different versions of the attitude estimation code were created for each matrix square root calculation, keeping all other portions of the code constant. All postprocessing was done on the same HP Pavilion Elite HPE Series desktop PC with Intel CORE i5 processor, which has processor speed of 2.67 GHz and 6 GB RAM.

3.3. Research Platform. This sensor fusion study was conducted using flight data collected on the three WVU YF-22 research UAVs (Green, Red, and Blue), shown in Figure 1. The WVU YF-22 platforms have been utilized for various projects, including the validation of GPS-based formation flight control laws [41–43] and fault-tolerant flight control laws [44].

Flight data were collected with two different avionics system architectures. Avionics system number1 [41] consists of a Crossbow VG400 series IMU and a NovAtel OEM4 GPS receiver, which reports a 1.8-meter Circular Error Probable (CEP) for position measurements. One version of avionics system number 1 was used in each of the three UAVs. Avionics system number 2 [45] is a more recently developed system and was used in the retrofitted Blue YF-22. The GPS used in system 2 was also a NovAtel OEM4, while the IMU was an Analog Devices ADIS-16405. In total, data collected with four different IMUs, that is, ADIS-16405, Crossbow VG400CA-200, DMU(VG400)-100, and IMU400CC-200, were used for this effort. The specifications for the different IMUs are shown in Table 2.

Independent roll and pitch measurements for both systems were obtained with a Goodrich VG34 mechanical vertical gyroscope. The VG34 has a ±90° sensitivity on the roll axis and ±60° sensitivity on the pitch axis and is sampled with 16-bit resolution. The VG34 has a self-erection system and reported accuracy of within 0.25° of true vertical. The output of the mechanical vertical gyroscope is used as the "truth data" for the sensor fusion study.

3.4. Flight Data Selection. To evaluate the sensitivity of the UKF, 23 sets of flight data were selected from a large library of flight data collected at WVU. Data were selected from flights on the WVU YF-22 platform, which consists of three UAVs. Each data set consists of three-axis IMU measurements, GPS position, and velocity calculations in local geodetic coordinates, along with roll and pitch angle measurements from a mechanical vertical gyroscope. Flights were selected with the goal of obtaining a diverse set of data in order to fairly compare sensitivity effects. The distribution of the flight selection is summarized in Tables 3 and 4.

3.5. Performance Evaluation. The performance of the UKF sensor fusion algorithm is evaluated from the estimated roll

TABLE 2: Inertial measurement unit (IMU) specifications.

Inertial measurement unit (IMU)	Crossbow VG400CA-200/VG400CC-200	Crossbow DMU(VG400)-100	Analog devices ADIS-16405
Resolution	16-bit	16-bit	14-bit
Rate gyroscope sensitivity	±200°/s	±100°/s	±150°/s
Accelerometer sensitivity	±10 g	±8 g	±18 g
Rate gyroscope random walk	$1.7°/h^{1/2}$	$0.85°/h^{1/2}$	$2.0°/h^{1/2}$
Accelerometer random walk	$0.5\ m/s/h^{1/2}$	$0.1\ m/s/h^{1/2}$	$0.2\ m/s/h^{1/2}$

TABLE 3: Summary of the distribution of flight systems.

System type	Distribution of flights		
UAV	(8) Green YF-22	(5) Red YF-22	(10) Blue YF-22
Avionics system	(21) Avionics system number 1	(2) Avionics system number 2	
Control system	(12) Pilot only	(11) Pilot with sections of autonomous flight	

TABLE 4: Summary of the distribution of flight conditions.

Approximate flight condition	Minimum	Maximum	Mean	Standard deviation
Temperature (°F)	38	86	65.1	13.9
Wind speed (mph)	0	11	5.8	3.5
Flight duration (sec)	375	640	485	76.0

and pitch angles. These estimates are compared with independent measurements obtained from a high-quality mechanical gyroscope. The gyroscope data are considered to be a "truth" value, and therefore the difference between the sensor fusion estimate and the corresponding gyroscope measurement represents the error. To quantify this error, the mean of the absolute value of the error and the standard deviation of the error are considered for both roll and pitch. To simplify performance comparisons, a scalar composite performance cost, J, is defined as a weighted average of these values such that

$$J = 0.3[\sigma(\phi_{\text{error}}) + \sigma(\theta_{\text{error}})] + 0.2[\text{mean}(|\phi_{\text{error}}|) + \text{mean}(|\theta_{\text{error}}|)], \quad (26)$$

where ϕ_{error} and θ_{error} represent the error of the roll and pitch estimates with respect to the corresponding vertical gyroscope measurement. The weights for this performance metric were selected such that their sum is unity, equal importance is given to the roll and pitch errors, and less importance is placed on the mean errors because of potential alignment errors between the IMU and vertical gyroscope. Since the weights add up to one and the units for each value are in degrees, the performance cost, J, can be considered as a generalized error value in degrees.

4. Results

4.1. Sensitivity to UKF Matrix Square Root Calculation. To analyze the sensitivity of this formulation of GPS/INS sensor fusion to the matrix square root operation, the UKF algorithm was executed for each set of flight data using different methods of calculating the matrix square root. In particular, the diagonalization method, Schur method, Cholesky method, and five different iterative methods were implemented. For each of the iterative methods, the UKF was executed for each set of flight data using a set number of iterations throughout the entire flight. This process was repeated for the number of iterations ranging from 5 to 20 by unit increments. For each individual case, results were evaluated based on performance cost, J, total execution time of the UKF, and the accuracy of the matrix square root calculation. A mean was taken of each of these values over all of the 23 flights in order to establish a generalized result. All of the matrix square root operations executed without error, except for certain cases of Newton's iteration which incurred matrix square root divergence errors on some flights when the number of iterations exceeded 16. In order to fairly compare the results, these cases of Newton's iteration are omitted from the data set. The performance cost of the different algorithms is shown in Figure 2.

In Figure 2, the performance curves were combined for some of the algorithms for clarity since those methods yielded nearly identical performance results. The performance of the noniterative methods is also included in Figure 2 to compare with the iterative methods. The plot on the right in Figure 2 shows a zoomed-in section to show the convergence of the algorithms. In terms of performance, the Cyclic Reduction (CR) algorithm showed the fastest convergence with respect to number of iterations out of the five considered iterative methods. Another interesting observation from these performance curves is the number of iterations required for each of the iterative algorithms to converge to the same performance as the noniterative methods. For this application, the CR iteration requires 12 iterations and the other four iterative methods require 13 iterations to

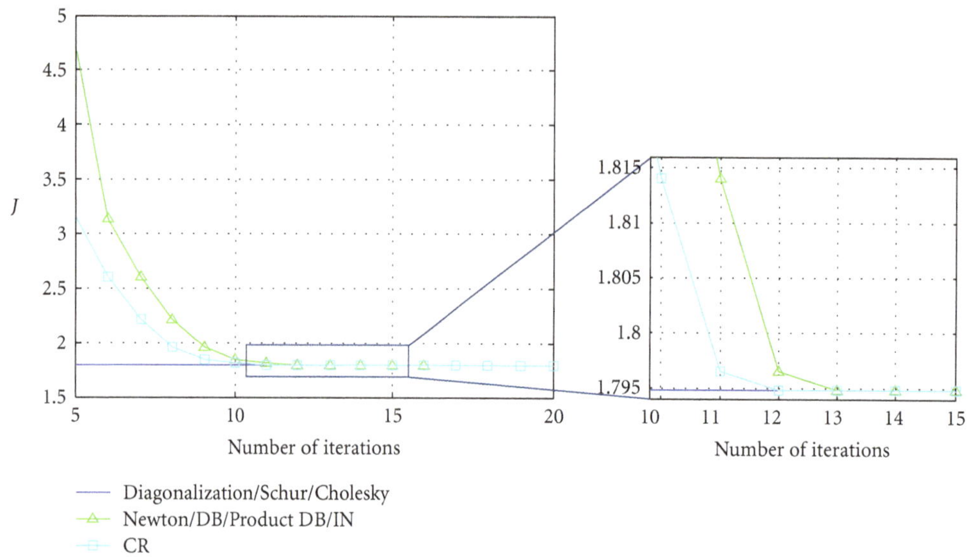

FIGURE 2: Performance cost of UKF for different matrix square root operations.

achieve similar performance as the noniterative methods to four significant figures. However, a smaller number of iterations can be used to reduce computations and still achieve reasonable performance. The CR method with 8 iterations could be used, for example, if performance cost of 2 degrees is acceptable for the application.

Since computational requirements are also important for a matrix square root algorithm, the actual execution time of the entire UKF algorithm was calculated for each of the different methods. These execution times are intended to provide an approximate empirical verification of the theoretical FLOP estimates in order to compare the computational requirements of the different algorithms. All execution time analyses were conducted using the same computer under approximately the same operating conditions. A mean of these execution times over each of the 23 flights is shown in Figure 3(a). Also in Figure 3(b) the corresponding estimated number of FLOPs using Table 1 is shown.

It is important to note that the number of FLOPs estimate is for the matrix square root operation only, while the execution times represent the run times of the entire sensor fusion algorithm. However, this is representative of the overall trend since the only difference between the curves is the matrix square root operation used. Because the diagonalization, Schur, and Cholesky methods do not require iterations, these algorithms are represented in Figure 3 by horizontal lines. Also, in Figure 3(b), the Denman-Beavers (DB) and Product DB are represented by a single line, since the estimated FLOPs for these algorithms are the same as listed in Table 1. The CR and IN iterations are similarly combined in Figure 3(b). Similar trends are observed between the estimated number of FLOPs and the UKF execution time with a few observations. First, the Denman-Beavers (DB) method demonstrates a longer execution time that grows at a steeper rate with the number of iterations than the product DB method, even though the FLOP estimations were the same. Another difference between the FLOP estimates and the UKF execution time is the location of the noniterative methods with respect to the iterative methods. With respect to the empirical execution time results, the iterative methods are all found to be more efficient in execution time than the Schur method for cases up to 15 iterations, with the exception of the DB iteration. This is an important result because previously, in Figure 2, it was shown that the iterative methods all achieve performance accuracy to four significant figures by using at most 13 iterations. This indicates the potential value of using iterative methods over the Schur method. It is also shown in Figure 3 that the Cholesky method has the fastest execution time with respect to any of the tested cases.

In order to compare accuracy of the actual matrix square root calculation itself, the L_1 norm [46] of the matrix $|P^{1/2} \cdot P^{1/2} - P|_1$ was calculated as a measure of the accuracy of the matrix square root operation. This norm, which is equal to the maximum of the absolute column sums of the matrix, was calculated at each time step of the UKF algorithm. To analyze the overall accuracy of the matrix square root operation over an entire flight, only the maximum of this norm over all discrete time was considered for each flight. This maximum represents the worst matrix square root estimate that occurred over the entire flight. A mean was taken of each of these maximum norms over all 23 flights, and the results are shown in Figure 4.

Figure 4 shows the convergence of the iterative methods in terms of the matrix square root accuracy. For smaller numbers of iterations, all of the iterative methods except for CR are very close in accuracy. These curves start to separate only at higher numbers of iterations, as shown in the right side plot of Figure 4. All of these algorithms converge to very high matrix square root accuracy after a sufficient number of iterations, with the exception of Newton's iteration. Figure 4 demonstrates the divergence of Newton's iteration. After 13 iterations, the matrix square root accuracy starts to degrade

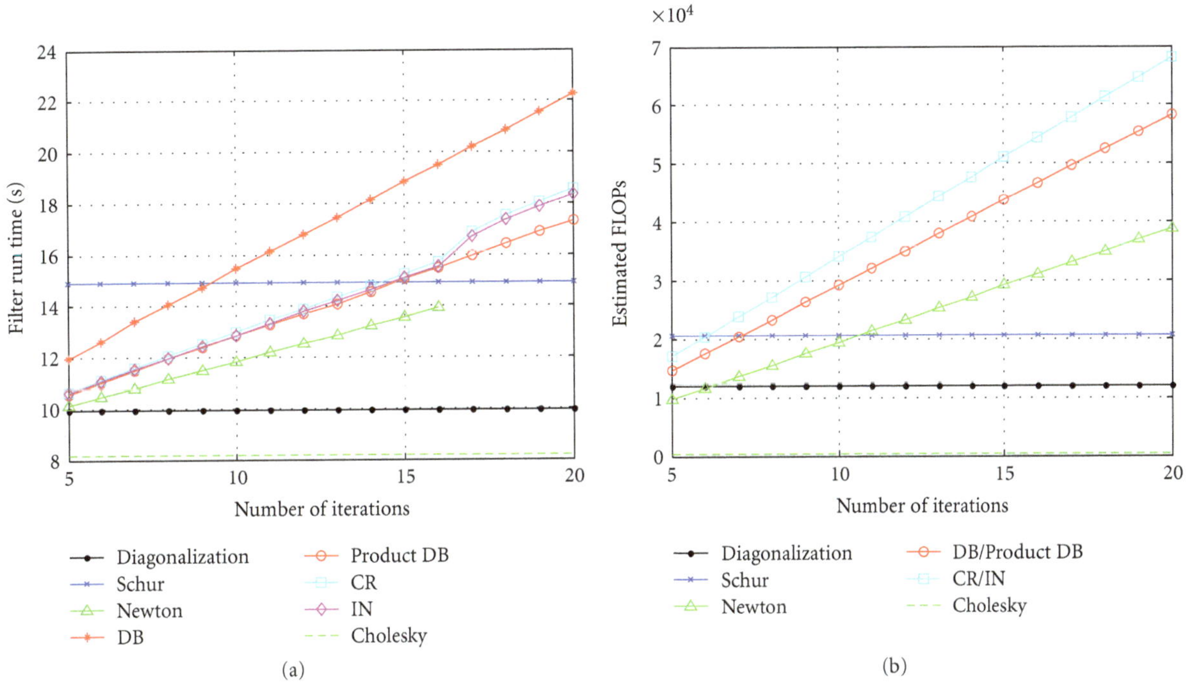

FIGURE 3: UKF computational requirements for different matrix square root operations.

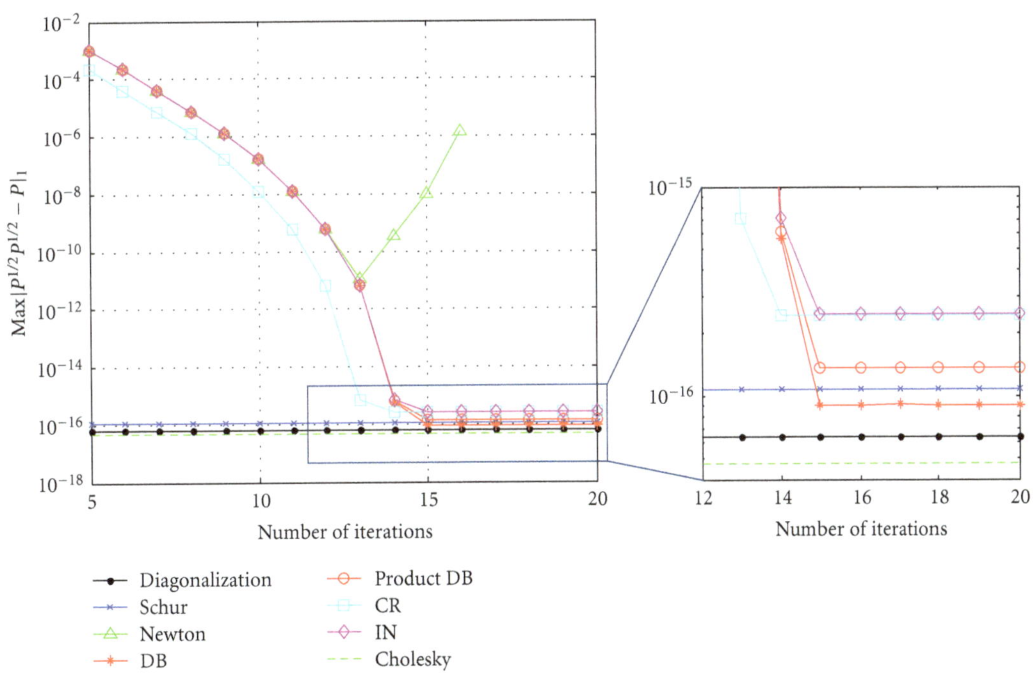

FIGURE 4: Matrix square root operation accuracy.

and eventually reaches a point where the matrix square root accuracy is too poor to use within the UKF algorithm. Because of the divergence issues associated with Newton's iteration, it is not recommended for UKF applications, even though it is the most computationally efficient iterative matrix square root method with respect to both FLOP estimate and execution time. Because the accuracy of the matrix square root operation has a direct effect on the accuracy of the prediction stage of the UKF, the relationship between the matrix square root accuracy and UKF performance accuracy for all considered matrix square root operations is shown in Figure 5.

As shown in Figure 5, there is a clear nonlinear relationship between the matrix square root accuracy and the UKF

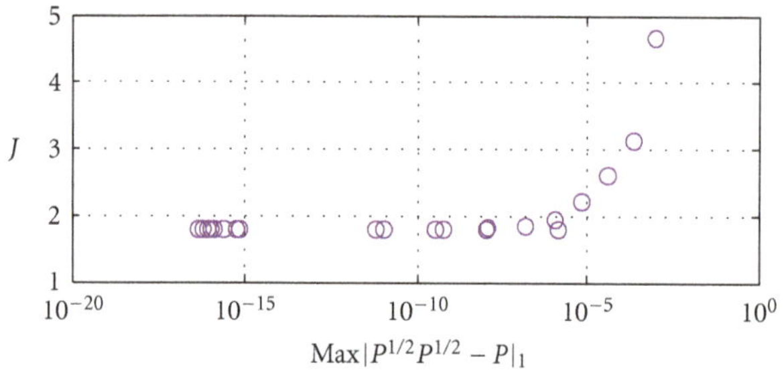

FIGURE 5: Relationship between matrix square root accuracy and UKF performance.

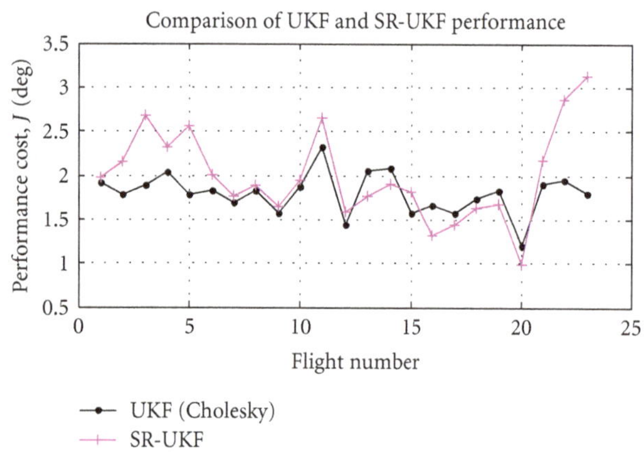

FIGURE 6: Comparison of UKF and SR-UKF performance.

performance for this application. This demonstrates the significant effect of the matrix square root accuracy on the performance of the UKF.

4.2. Comparison of Direct Matrix Square Root Methods to SR-UKF. A different method of handling the square root requirement of the UKF, named the "square-root UKF (SR-UKF)", was suggested by van der Merwe and Wan [19]. In this method, the square root of the state covariance matrix is estimated directly. This eliminates the need to refactorize the state covariance matrix at each time step, and instead it is updated using Cholesky updates. A significant advantage of this method is a decrease in computational complexity, which leads to a faster run time of the UKF. For comparison purposes, the Cholesky method was selected as a representative case of calculating the matrix square root at each time step from the state covariance estimate. Performance results were calculated using each of these two methods for each of the 23 flights, and the performance cost is plotted in Figure 6.

The mean performance costs of the two different methods are shown in Table 5. To determine if there is a statistically significant performance advantage of the UKF over the SR-UKF, a one-tailed paired samples hypothesis test [46] was done using the t-statistic to determine the probability that the SR-UKF has better performance than the UKF. Using this null hypothesis, the probability was calculated to be 1.49%, which is less than the commonly considered 5% null

TABLE 5: Comparison of UKF and SR-UKF.

Matrix square root method	Mean performance cost J (deg)	Mean execution time (sec)
UKF (Cholesky)	1.795	8.16
SR-UKF	1.996	5.73

hypothesis rejection criterion. Therefore, the UKF achieves statistically significant better performance than the SR-UKF for this application, at the cost of additional computational complexity. To compare the computational complexity of the two different algorithms, the mean execution time of the algorithms was calculated and is also shown in Table 5.

It is also interesting to note from Figure 6 that there are some cases where the SR-UKF method has better performance than the Cholesky method, for example, flight number 16. If this single flight alone was used to analyze results, the opposite conclusion could be drawn about the accuracy of this method. This demonstrates the value of using multiple data sets.

5. Conclusions

This paper presented a comparison of different matrix square root calculations within the UKF. The GPS/INS sensor fusion attitude estimation problem for UAV applications was used as an example to evaluate the performance with respect to matrix square root accuracy, computational cost, and attitude estimation performance. In terms of attitude estimation performance, the Cholesky, diagonalization, and Schur methods yielded the highest accuracy; however this same performance can be reached using a sufficient number of iterations in any of the iterative methods. Newton's iteration was found to diverge in certain instances and is therefore not recommended for UKF applications. The cyclic reduction (CR) iteration demonstrated the fastest performance convergence of the iterative methods. In terms of execution time, the SR-UKF is computationally efficient, but at the cost of performance. Overall, the Cholesky method was found to provide the best compromise in terms of both performance and execution time.

For real-time applications of the UKF, such as attitude estimation for small UAVs, computation is an important consideration factor. For most cases, the Cholesky method is the best suited matrix square root method due to its fast execution and high accuracy. If computational cost is even more important than the accuracy of the filter, the SR-UKF could be considered. The diagonalization and Schur methods are acceptable approaches for off-line applications, because the accuracy is similar to the Cholesky method, although they require more computation time. These methods also might be more desirable than the Cholesky method because they provide a more intuitive representation of the matrix square root, that is, the principle square root. Any of the iterative methods, except for Newton's iteration, could also be used with a sufficient number of iterations, though these methods are a bit less intuitive.

Acknowledgments

This research was partially supported by NASA Grant no. NNX07AT53A and Grant no. NNX10AI14G.

References

[1] L. Changchun, S. Li, W. Hai-bo, and L. Tianjie, "The research on unmanned aerial vehicle remote sensing and its applications," in *Proceedings of the IEEE International Conference on Advanced Computer Control (ICACC '10)*, pp. 644–647, Shenyang, China, March 2010.

[2] A. M. Jensen, M. Baumann, and Y. Chen, "Low-cost multispectral aerial imaging using autonomous runway-free small flying wing vehicles," in *Proceedings of the IEEE International Geoscience and Remote Sensing Symposium (IGARSS '08)*, pp. 506–509, Boston, Mass, USA, July 2008.

[3] M. Nagai, T. Chen, R. Shibasaki, H. Kumagai, and A. Ahmed, "UAV-borne 3-D mapping system by multisensor integration," *IEEE Transactions on Geoscience and Remote Sensing*, vol. 47, no. 3, Article ID 4783021, pp. 701–708, 2009.

[4] T. Suzuki, Y. Amano, and T. Hashizume, "Vision based localization of a small UAV for generating a large mosaic image," in *Proceedings of the SICE Annual Conference (SICE '10)*, pp. 2960–2964, Taipei, Taiwan, October 2010.

[5] S. Dascalu, "Remote sensing using autonomous UAVs suitable for less developed countries," *The International Archives of the Photogrammetry, Remote Sensing and Spatial Information Sciences, Enschede, Netherlands*, vol. 34, no. 30, 2006.

[6] G. Zhou and D. Zang, "Civil UAV system for earth observation," in *Proceedings of the IEEE International Geoscience and Remote Sensing Symposium (IGARSS '07)*, Barcelona, Spain, 2007.

[7] J. Everaerts, "The use of unmanned aerial vehicles (UAVs) for remote sensing and mapping," *The International Archives of the Photogrammetry, Remote Sensing and Spatial Information Sciences, Beijing*, vol. 36, no. B1, pp. 1187–1192, 2008.

[8] M. S. Grewal, L. R. Weill, and A. P. Andrew, *Global Positioning, Inertial Navigation & Integration*, John Wiley & Sons, New York, NY, USA, 2nd edition, 2007.

[9] E. Kaplan and C. Heagarty, *Understanding GPS Principles and Applications*, Arttech House, Norwood, Mass, USA, 2nd edition, 2006.

[10] J. L. Crassidis, "Sigma-point kalman filtering for integrated GPS and inertial navigation," in *Proceedings of the AIAA Guidance, Navigation and Control Conference and Exhibit*, San Francisco, Calif, USA, 2005.

[11] R. D. Van Merwe, E. A. Wan, and S. I. Julier, "Sigma-point kalman filters for nonlinear estimation and sensor-fusion—applications to integrated navigation," in *Proceedings of the AIAA Guidance, Navigation, and Control Conference*, pp. 1735–1764, Providence, RI, USA, August 2004.

[12] T. Fiorenzani et al., "Comparative Study of Unscented Kalman Filter and Extended Kalman Filter for Position/Attitude Estimation in Unmanned Aerial Vehicles," IASI-CNR, R. 08-08, 2008.

[13] J. Gross, Y. Gu, M. Rhudy, S. Gururajan, and M. Napolitano, "Flight test evaluation of GPS/INS sensorfusion algorithms for attitude estimation," *IEEE Transactions on Aerospace Electronic Systems*. In press.

[14] S. J. Julier and J. K. Uhlmann, "New extension of the Kalman filter to nonlinear systems," in *Signal Processing, Sensor Fusion, and Target Recognition VI*, vol. 3068 of *SPIE Proceedings Series*, pp. 182–193, July 1997.

[15] J. Gross et al., "A comparison of extended kalman filter, sigma-point kalman filter, and particle filter in GPS/INS sensor fusion," in *Proceedings of the AIAA Guidance, Navigation, and Control Conference*, Toronto, Canada, 2010.

[16] R. E. Kalman and R. S. Bucy, "New results in linear filtering and prediction theory," *Journal of Basic Engineering*, vol. 83, pp. 95–108, 1961.

[17] N. El-Sheimy, E. H. Shin, and X. Niu, "Kalman filter face-off: extended vs. Unscented kalman filters for integrated GPS and MEMS inertial," in *Proceedings of the International Symposium on Global Navigation Satellite Systems (GNSS '06)*, pp. 48–54, 2006.

[18] M. Rhudy, Y. Gu, J. Gross, and M. Napolitano, "Sensitivity analysis of EKF and UKF in GPS/INS sensor fusion," in *Proceedings of the AIAA Guidance, Navigation, and Control Conference*, Portland, Ore, USA, 2011.

[19] R. van der Merwe and E. A. Wan, "The square-root unscented Kalman filter for state and parameter-estimation," in *Proceedings of the International Conference on Acoustics, Speech, and Signal Processing*, pp. 3461–3464, Salt Lake City, Utah, USA, May 2001.

[20] N. J. Higham, *Functions of Matrices: Theory and Computation*, Society for Industrial and Applied Mathematics (SIAM), Philadelphia, Pa, USA, 2008.

[21] J. Wendel, J. Metzger, R. Moenikes, A. Maier, and G. F. Trommer, "A Performance comparison of tightly coupled GPS/INS navigation systems based on extended and sigma point kalman filters," *Journal of the Institute of Navigation*, vol. 53, no. 1, pp. 21–31, 2006.

[22] N. B. Stastny, R. A. Bettinger, and F. R. Chavez, "Comparison of the extended and unscented Kalman filters for angles based relative navigation," in *Proceedings of the AIAA/AAS Astrodynamics Specialist Conference and Exhibit*, Honolulu, Hawaii, USA, August 2008.

[23] Y. Wu, D. Hu, M. Wu, and X. Hu, "Unscented Kalman filtering for additive noise case: augmented vs. non-augmented," in *Proceedings of the American Control Conference (ACC '05)*, pp. 4051–4055, Portland, Ore, USA, June 2005.

[24] F. Sun, G. Li, and J. Wang, "Unscented kalman filter using augmented state in the presence of additive noise," in *Proceedings of the IITA International Conference on Control, Automation and Systems Engineering (CASE '09)*, pp. 379–382, Zhangjiajie, China, July 2009.

[25] M. C. VanDyke, J. L. Schwartz, and C. D. Hall, "Unscented kalman filtering for spacecraft attitude state and parameter estimation," in *Proceedings of the AIAA Space Flight Mechanics Meeting*, Maui, Hawaii, USA, 2004.

[26] S. A. Banani and M. A. Masnadi-Shirazi, "A new version of unscented kalman filter," *World Academy of Science, Engineering, and Technology*, vol. 20, pp. 192–197, 2007.

[27] B. Meini, "The matrix square root from a new functional perspective: theoretical results and computational issues," *SIAM Journal on Matrix Analysis and Applications*, vol. 26, no. 2, pp. 362–376, 2004.

[28] J. Gross, Y. Gu, and M. R. Napolitano, "A systematic approach for extended kalman filter tuning and low-cost inertial sensor calibration within a GPS/INS application," in *Proceedings of the AIAA Guidance, Navigation, and Control Conference*, Toronto, Canada, 2010.

[29] J. Jarrell, Y. Gu, B. Seanor, and M. Napolitano, "Aircraft attitude, position, and velocity determination using sensor fusion," in *Proceedings of the AIAA Guidance, Navigation and*

Control Conference and Exhibit, Honolulu, Hawaii, USA, August 2008.

[30] D. Simon, *Optimal State Estimation*, Wiley, New York, NY, USA, 2006.

[31] D. A. McQuarrie, *Mathematical Methods for Scientists and Engineers*, chapter 10, University Science Books, Sausalito, Calif, USA, 1st edition, 2003.

[32] M. M. Konstantinov, *Perturbation Theory for Matrix Equations*, Elsevier Science B.V., Amsterdam, The Netherlands, 2003.

[33] N. J. Higham, "Newton's method for the matrix square root," *Mathematics of Computation*, vol. 46, no. 174, pp. 537–549, 1986.

[34] S. H. Cheng, N. J. Higham, C. S. Kenney, and A. J. Laub, "Approximating the logarithm of a matrix to specified accuracy," *SIAM Journal on Matrix Analysis and Applications*, vol. 22, no. 4, pp. 1112–1125, 2001.

[35] L. N. Trefethen and D. Bau, *Numerical Linear Algebra*, SIAM, Philadelphia, Pa, USA, 1st edition, 1997.

[36] N. J. Higham, "Analysis of the cholesky decomposition of a semi-definite matrix," in *Reliable Numerical Computation*, pp. 161–185, Oxford University Press, Oxford, UK, 1990.

[37] N. J. Higham, *Accuracy and Stability of Numerical Algorithms*, Society for Industrial and Applied Mathematics (SIAM), Philadelphia, Pa, USA, 2nd edition, 2002.

[38] D. W. Tufts and C. D. Melissinos, "Simple, effective computation of principal eigenvectors and their eigenvalues and application to high-resolution estimation of frequencies," *IEEE Transactions on Acoustics, Speech, and Signal Processing*, vol. 34, no. 5, pp. 1046–1053, 1986.

[39] J. Roskam, *Airplane Flight Dynamics and Automatic Flight Controls*, chapter 1, DARcorporation, Lawrence, Kan, USA, 2003.

[40] F. L. Lewis and V. L. Syrmos, *Optimal Control*, chapter 6, Wiley & Sons, New York, NY, USA, 2nd edition, 1995.

[41] Y. Gu, B. Seanor, G. Campa, M. R. Napolitano, S. Gururajan, and L. Rowe, "Autonomous formation flight: Hardware development," in *Proceedings of the 14th Mediterranean Conference on Control and Automation (MED '06)*, Ancona, Italy, June 2006.

[42] B. Seanor, Y. Gu, M. R. Napolitano, G. Campa, S. Gururajan, and L. Rowe, "3-Aircraft formation flight experiment," in *Proceedings of the 14th Mediterranean Conference on Control and Automation (MED '06)*, Ancona, Italy, June 2006.

[43] Y. Gu, B. Seanor, G. Campa et al., "Design and flight testing evaluation of formation control laws," *IEEE Transactions on Control Systems Technology*, vol. 14, no. 6, pp. 1105–1112, 2006.

[44] M. G. Perhinschi, J. Burken, and G. Campa, "Comparison of different neural augmentations for the fault tolerant control laws of the WVU YF-22 model aircraft," in *Proceedings of the 14th Mediterranean Conference on Control and Automation (MED '06)*, Ancona, Italy, June 2006.

[45] J. Gross, Y. Gu, B. Seanor, S. Gururajan, and M. R. Napolitano, "Advanced Research Integrated Avionic (ARIA) system for fault-tolerant flight research," in *Proceedings of the AIAA Guidance, Navigation, and Control Conference and Exhibit*, Chicago, Ill, USA, August 2009.

[46] E. Kreyszig, *Advanced Engineering Mathematics*, chapter 20, Wiley & Sons, New York, NY, USA, 9th edition, 2006.

Permissions

All chapters in this book were first published in IJAE, by Hindawi Publishing Corporation; hereby published with permission under the Creative Commons Attribution License or equivalent. Every chapter published in this book has been scrutinized by our experts. Their significance has been extensively debated. The topics covered herein carry significant findings which will fuel the growth of the discipline. They may even be implemented as practical applications or may be referred to as a beginning point for another development.

The contributors of this book come from diverse backgrounds, making this book a truly international effort. This book will bring forth new frontiers with its revolutionizing research information and detailed analysis of the nascent developments around the world.

We would like to thank all the contributing authors for lending their expertise to make the book truly unique. They have played a crucial role in the development of this book. Without their invaluable contributions this book wouldn't have been possible. They have made vital efforts to compile up to date information on the varied aspects of this subject to make this book a valuable addition to the collection of many professionals and students.

This book was conceptualized with the vision of imparting up-to-date information and advanced data in this field. To ensure the same, a matchless editorial board was set up. Every individual on the board went through rigorous rounds of assessment to prove their worth. After which they invested a large part of their time researching and compiling the most relevant data for our readers.

The editorial board has been involved in producing this book since its inception. They have spent rigorous hours researching and exploring the diverse topics which have resulted in the successful publishing of this book. They have passed on their knowledge of decades through this book. To expedite this challenging task, the publisher supported the team at every step. A small team of assistant editors was also appointed to further simplify the editing procedure and attain best results for the readers.

Apart from the editorial board, the designing team has also invested a significant amount of their time in understanding the subject and creating the most relevant covers. They scrutinized every image to scout for the most suitable representation of the subject and create an appropriate cover for the book.

The publishing team has been an ardent support to the editorial, designing and production team. Their endless efforts to recruit the best for this project, has resulted in the accomplishment of this book. They are a veteran in the field of academics and their pool of knowledge is as vast as their experience in printing. Their expertise and guidance has proved useful at every step. Their uncompromising quality standards have made this book an exceptional effort. Their encouragement from time to time has been an inspiration for everyone.

The publisher and the editorial board hope that this book will prove to be a valuable piece of knowledge for researchers, students, practitioners and scholars across the globe.

List of Contributors

Fabio Balzano, Mario L. Fravolini and Michele Crispoltoni
Department of Engineering, University of Perugia, Perugia 06125, Italy

Marcello R. Napolitano and Stéphane d'Urso
Department of Mechanical and Aerospace Engineering, West Virginia University, Morgantown, WV 26506, USA

Giuseppe del Core
Department of Science and Technology, Parthenope University of Naples, Naples 80143, Italy

You Li and Yingxin Kou
Aeronautics and Astronautics Engineering College, Air Force Engineering University, Xi'an, Shaanxi 710038, China

Zhanwu Li
Aeronautics and Astronautics Engineering College, Air Force Engineering University, Xi'an, Shaanxi 710038, China
College of Electronic Communication, Northwestern Polytechnical University, Xi'an 710072, China

Xiaolin Gong
School of Instrumentation Science and Optoelectronics Engineering, Beihang University, Beijing 100191, China
Science and Technology on Inertial Laboratory, Beijing 100191, China

Xiaorui Zheng
School of Instrumentation Science and Optoelectronics Engineering, Beihang University, Beijing 100191, China

Xing-Gang Yan
School of Engineering and Digital Arts, University of Kent, Canterbury CT2 7NT, UK

Zhaoxing Lu
School of Instrumentation Science and Optoelectronics Engineering, Beihang University, Beijing 100191, China
Xi'an Institute of Hi-tech, Xi'an 710025, China

Nathan A. Tehrani and Jason N. Gross
Department of Mechanical and Aerospace Engineering, West Virginia University, Morgantown, WV, USA

Bo Yang, Jun Miao and Zichen Fan
School of Astronautics, Beihang University, Beijing, China

Jun Long and Xuhui Liu
Beijing Institute of Control Engineering, Beijing, China

Abdeldime Mohamed Salih Abdelgader
Southeast University, School of Information Science and Engineering, Nanjing 210096, China
Karary University, College of Engineering, Electrical and Computer Department, 12304 Khartoum, Sudan

Lenan Wu and Mohammed Mohsen Mohammed Nasr
Southeast University, School of Information Science and Engineering, Nanjing 210096, China

Xiaolin Gong and Haojie Liu
School of Instrumentation Science and Optoelectronics Engineering, Beihang University, Beijing 100191, China
Science and Technology on Inertial Laboratory, Beijing 100191, China

Xing-Gang Yan
School of Engineering and Digital Arts, University of Kent, Canterbury CT2 7NT, UK

Jikun Ye, Humin Lei and Jiong Li
Air and Missile Defense College, Air Force Engineering University, Shaanxi, Xi'an 710051, China

Peng Wang
Civil Aircraft Airworthiness and Repair Key Laboratory of Tianjin, Civil Aviation University of China, Tianjin 300300, China
Key Laboratory of Civil Aircraft Airworthiness Technology, Civil Aviation Administration of China, Tianjin 300300, China

Changxiao Zhao
School of Airworthiness, Civil Aviation University of China, Tianjin 300300, China

Fang Yan
Civil Aircraft Airworthiness and Repair Key Laboratory of Tianjin, Civil Aviation University of China, Tianjin 300300, China

Michael Felux and Mihaela-Simona Circiu
German Aerospace Center (DLR), Münchener Straße 20, 82234 Weßling, Germany

Jiyun Lee
Korea Advanced Institute of Science and Technology (KAIST), 291 Daehak-ro, Yuseong-gu, Daejeon 34141, Republic of Korea

Florian Holzapfel
Institute of Flight System Dynamics, Technische Universität München (TUM), Boltzmannstraße 15, 85748 Garching, Germany

You Li, Yingxin Kou and Yizhe Chang
Aeronautics and Astronautics Engineering College, Air Force Engineering University, Xi'an, Shaanxi 710038, China

Zhanwu Li and An Xu
Aeronautics and Astronautics Engineering College, Air Force Engineering University, Xi'an, Shaanxi 710038, China
College of Electronic Communication, Northwestern Polytechnical University, Xi'an 710072, China

Yuhang Yun, Jie Guo and Shengjing Tang
School of Aerospace Engineering, Beijing Institute of Technology, Beijing 100081, China

Xiongfeng Deng, Xiuxia Sun and Shuguang Liu
Equipment Management and Unmanned Aerial Vehicle Engineering College, Air Force Engineering University, Xi'an 710051, China

Ri Liu
Theory Training Department, Air Force Harbin Flight Academy, Harbin 150001, China

Takashi Matsuno
Department of Mechanical and Aerospace Engineering, Tottori, Tottori University, 4-101 Koyama-cho, Tottri 680-8552, Japan

Yasuhiro Yoshimura and Shinji Hokamoto
Department of Aeronautics and Astronautics, Fukuoka, Kyushu University, 744 Motooka, Nishi-ku, Fukuoka 819-0395, Japan

Index

www.ingramcontent.com/pod-product-compliance
Lightning Source LLC
LaVergne TN
LVHW080735210926
848485LV00003B/562